AF334748

Reviews and Protocols in DT40 Research

Subcellular Biochemistry
Volume 40

SUBCELLULAR BIOCHEMISTRY

SERIES EDITOR

J. ROBIN HARRIS, University of Mainz, Mainz, Germany

ASSISTANT EDITORS

B.B. BISWAS, University of Calcutta, Calcutta, India

P. QUINN, King's College London, London, U.K

Reviews and Protocols in DT40 Research

Subcellular Biochemistry
Volume 40

Edited by

Jean-Marie Buerstedde

GSF National Research Centre,
Munich,
Germany

and

Shunichi Takeda

Kyoto University,
Japan

Springer

This series is a continuation of the journal Sub-Cellular Biochemistry
Volume 1 to 4 of which were published quarterly from 1972 to 1975

ISBN-10 1-4020-4895-5 (HB)
ISBN-13 978-1-4020-4895-1 (HB)
ISBN-10 1-4020-4896-3 (e-book)
ISBN-13 978-1-4020-4896-8 (e-book)

CONTENTS

FOREWORD

Jean-Marie Buerstedde

The DT40 B cell line was established in the laboratory of Eric H. Humphries from an ALV induced B cell lymphoma. However, it was only after the demonstration of high ratios of targeted to random integration of transfected gene constructs that DT40 gained wide-spread popularity. Shunichi Takeda and I have been working with DT40 for the last 15 years and I remember Tatsuko Honjo once told us that we had a love affair with this cell line. If it is love, it was not love without doubts and crisis. It would be unfair to blame this on DT40 as it proved to be a reliable and robust companion with fast doubling time, easy clonability and a relatively stable karyotype. The problem was rather that some of the early knock-outs were technically demanding due to the lack of good chicken cDNA and genome resources. In addition, it is difficult to predict gene disruption phenotypes as seen by genes which are needed for DT40 proliferation, but whose homologues in yeast are not essential. All this would have been much harder to bear without the nice spirit in the DT40 research community. This is still a small, friendly world and many reagents in form of vectors and assays are freely shared among the laboratories even before publications.

It is with this in mind that the DT40 handbook has been perceived by Mike van den Bosch from the Springer publishing house. The intention is to give an up to date overview about the different facets of research, but also to help newcomers get started and avoid looming pitfalls. The collection of protocols which have been kindly provided by a number of laboratories will be particularly useful in this regard.

Research is fast paced and advances in RNA interference have recently opened up new opportunities for genetic experiments in human cell lines. However the possibility to easily modify the genome still remains a powerful tool to investigate the function of coding and regulatory sequences in the vertebrate genome. DT40 has never been a quick and easy road to fame. If this model system is going to flourish over the next 15 years, it will be thanks to ingenious and original researchers. They may feel as if they work outside the mainstream, but they can take heart by the fact that only the clever exploitation of diversity and conservation makes biological research both elegant and rewarding.

Chapter 1

DT40 GENE DISRUPTIONS: A HOW-TO FOR THE DESIGN AND THE CONSTRUCTION OF TARGETING VECTORS

Hiroshi Arakawa and Jean-Marie Buerstedde
GSF, Institute for Molecular Radiobiology, Ingolstaedter Landstr. 1, D-85764 Neuherberg-Munich, Germany

Abstract: Genome projects have provided comprehensive gene catalogs and locus maps for many model organisms. Although sequence comparison and protein domain searches may suggest evolutionary conserved gene functions, genetic systems are still needed to determine the role of genes within living cells. Due to high ratios of targeted to random integration of transfected DNA constructs, the chicken B cell line DT40 has been widely used as a model for gene function analysis by gene knockout. Targeting vectors need to be carefully designed to introduce defined mutations and to ensure high targeting rates. In this review we summarize general guidelines for the design of targeting vectors which can be used for single, multiple or conditional gene knockouts, as well as site-directed genome mutagenesis in DT40.

Key words: DT40, knockout, targeting vector, mutant loxP.

1.   INTRODUCTION

Targeted integration of DNA constructs by homologous recombination enables the inactivation of genes by disruption or deletion as well as the introduction of more subtle gene mutations. This approach, first pioneered for the yeast *S. cerevisiae* has been used extensively for gene modification in murine embryonic stem (ES) cells, which are subsequently used to produce mutant mouse strains. However, genes essential for cell proliferation and tissue development are often difficult to study, as a homozygous deletion

1

J.-M. Buerstedde and S. Takeda (eds.), Reviews and Protocols in DT40 Research, 1–9.
© 2006 *Springer.*

causes early embryonic lethality. Gene disruption in a cell line is an alternative to knockouts in murine ES cells, if the mutant phenotype can be studied in cell culture. The chicken B cell line DT40 is popular for these studies due to unusual high ratios of targeted to random integration (Buerstedde and Takeda, 1991).

The design of targeting vectors for DT40 studies requires information about the chicken target gene locus in form of restriction and exon-intron maps. The ideal situation, in which the sequence of the entire locus is available, is now often encountered, because more than 90% of chicken genomic sequence has been released to the public databases (International Chicken Genome Sequencing Consortium, 2004). In addition, large EST and full length cDNA sequence database from bursal cells (Abdrakhmanov et al., 2000; Caldwell et al., 2005) and other chicken tissue (Boardman et al., 2002) help to reveal the precise exon-intron structure of many loci (Caldwell et al., Chapter 3, this issue).

A series of versatile plasmid vectors have been developed to assist genetic engineering of DT40. Mutant loxP vectors enable the excision of the drug-resistance gene by Cre/loxP recombination for drug-resistance marker recycling (Arakawa et al., 2001). In this way vectors including the same drug resistance gene can be repeatedly used for the selection of stable transfectants. Other vectors allowing the cloning of cDNA's into a loxP flanked expression cassette can be useful for the complementation of knockout phenotypes and conditional gene expression (Arakawa et al., 2001). These tools are freely distributed through DT40 web site (http://pheasant.gsf.de/DEPARTMENT/ dt40.html), and have been commonly used by the DT40 community.

1.1   How to determine the exon-intron structure of the target locus

Information about the genomic locus of the gene of interest is critical for the design of the gene targeting construct. To determine the exon-intron structure of a gene, you can simply use the chicken cDNA sequence as input for a genome BLAT search (http://genome.cse.ucsc.edu/cgi-bin/hgBlat). If the sequence of the genomic locus is available, the cDNA sequence will be aligned along the genomic sequence, showing the location of the exons. If the chicken cDNA sequence is not available, it is still worth trying to use either the cDNA or the protein sequence of the gene ortholog from other species like human or mouse as input. Depending on the degree of sequence conservation this may indicate the locations of homologous exons. More detailed information on how to use the available genome resources is provided in another review in this issue (Randy Caldwell et al., Chapter 3, this issue).

1.2   How to design a gene knockout construct

Gene knockout constructs need to be carefully designed to ensure the introduction of the desired mutation at high ratios of targeted to random integration. The inactivation of a gene (null mutation) can either be achieved by gene deletion or gene disruption. If feasible, the deletion of the complete coding sequence is preferred, since this precludes interference of left over gene sequences with the mutant phenotype. Since the size of the deletion is determined by the positions of the 5' and 3' target arms within the genome sequence, it is not difficult to design constructs for large deletions, if the arm sequences are available. However, a large distance between the 5' and 3' target arms in the genome sequence most likely decreases the ratios of targeted to random integration of the construct. Although more than 20 kb of the immunoglobulin light chain locus could be deleted using a conventionally designed targeting construct (Arakawa et al., 2004), we have the feeling that the ratios of targeted to random integration are difficult to predict a priori for deletions of this size. We usually try complete gene deletions only, if the size of the target locus does not exceed 5 kb.

If the target locus covers large genomic distances, one may attempt to create a null mutation either by the deletion of exons encoding a critical domain of the protein or by deleting as much of the gene coding region as possible. We usually combine a partial gene deletion with the introduction of an in-frame stop codon near the 5' end of the gene coding sequence. This has the advantage that the translation of the remaining transcript will terminate at a defined position and produce a truncated peptide which is unlikely to have a function. However, the effects of the partial gene deletions must be carefully considered on a case by case basis, because it is difficult to predict the phenotypes with certainty due to possible variation in mRNA translation and splicing. Especially if a mutation does not produce a measurable effect, it becomes difficult to determine whether this is due to the incomplete inactivation of the gene or due to redundant gene function. In certain situations, the partial inactivation of a gene will be more informative than a null mutation and is the intended outcome of the gene targeting as shown below for the example of a PCNA mutation.

The size of target arms may influence targeting efficiency, and in general, longer target arms are believed to increase targeting efficiency. However, there is a trade-off as plasmids of larger size are more difficult to handle and possess a lower number of unique restriction sites. We usually design our constructs in such a way that the sizes of the individual 5' and 3' arm of are more than 1 kb, the combined size of the 5' and the 3' arms is more than 3 kb and a total plasmid size is less than 12 kb. Targeting vectors made according to these rules will usually give targeting efficiencies of 20-80% among stable transfectants.

We recommend amplification of the target arms by long range PCR using primers with attached restriction sites and genomic DNA from DT40 as template. This will produce arm sequences isogenic to at least one allele of DT40 and should enhance the targeting efficiency. A systematic design of knockout constructs enables the easy exchange of drug resistance marker cassettes. All of our drug resistance marker cassettes are flanked by *Bam*HI sites and can be conveniently cloned into *Bam*HI or *Bgl*II sites located between the 5' and the 3' arms. If feasible, we try to clone the 5' arm into the *Xho*I-*Bam*HI sites and the 3' arm into the *Bam*HI-*Spe*I sites of pBluescript plasmid vector. If one of those restriction sites is present internally in the arm sequences, other compatible sites for example *Sal*I instead of *Xho*I; *Xba*I, *Nhe*I and *Avr*II instead of *Spe*I; *Bgl*II and *Bcl*I instead of *Bam*HI might be chosen. The resulting targeting vector can usually be linearized in the plasmid polylinker by the rare cutter *Not*I, which is unlikely to cleave the target arms.

Efficient targeting of the first allele, but no recovery of homozygous mutant clones after transfection of the heterozygous clone may indicate an critical role of the target gene for cell growth. However, difficulties to recover clones targeted for both alleles in the second transfection cannot be taken as proof that the target gene is essential, because we have experienced situations in which the second allele could eventually be inactivated by a differently designed targeting construct. The most likely explanation for this phenomenon is that the original knock-out construct targeted preferentially the already targeted allele due to a stronger degree of sequence homology and/or due to the difference of deletion size. We therefore recommend to redesign the position of one target arm, if homozygous mutant are not recovered after transfection of heterozygotes and there remains doubt about whether the target gene is essential. The best way to prove that a gene is required for DT40 cell survival or proliferation is by conditional gene knockouts as described below.

1.3   Drug resistance marker recycle by Cre/loxP recombination

The generation of a homozygous mutant usually requires two different drug resistance markers for the stepwise targeting of both alleles and another drug resistance marker is needed if the mutant phenotype is complemented by the re-introduction of the gene. Modifications of the genome by additional transfections are limited by the number of available drug resistance markers. This problem can be solved by the excision of the drug resistance marker using a site-specific recombination system like Cre/loxP. We previously described floxed drug-resistance marker cassette vectors which can be recycled after Cre expression (Figure 1) (Arakawa et al., 2001).

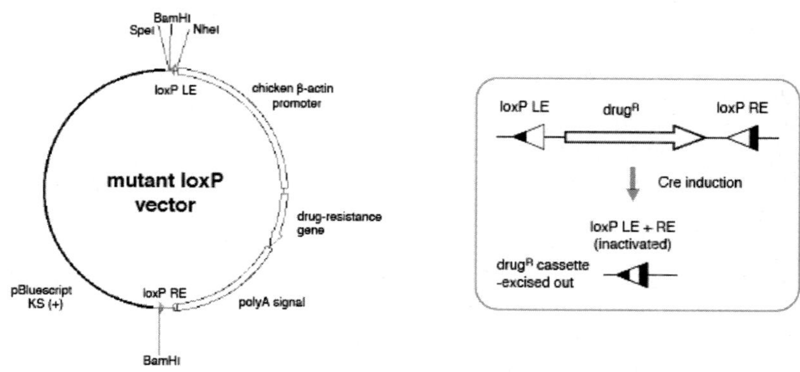

Figure 1-1. Drug resistant marker cassette flanked by mutant loxP sites. Left, Five vectors are available including different drug-resistance genes: pLoxNeo, pLoxPuro, pLoxBsr, pLoxGpt and pLoxHygro. Right, The principle of the mutant loxP system Rearrangement of mutant loxP sites creates a new site which is poorly recognized by Cre.

To prevent genetic instability due to the cutting of the loxP sites left-over after marker excision, the drug resistant markers are flanked by mutant loxP sites (loxP_RE and loxP_LE) (Albert et al., 1995). The Cre recombinase efficiently recombines the loxP_RE and loxP_LE sites leading to the deletion of the drug resistance marker gene. However, the new variant loxP_RE+LE site generated after the excision is only poorly recognized and cleaved by Cre. At this moment five of mutant loxP vectors are available: three vectors, pLoxNeo (neomycin resistance), pLoxPuro (puromycin resistance) and pLoxBsr (blasticidin S resistance) which have been reported previously (Arakawa et al., 2001), and two new vectors, pLoxGpt (mycophenolic acid resistance) and pLoxHygro (hygromycin resistance) which have been recently completed (unpublished results). All these vectors contain drug resistance genes driven by the chicken β-actin promoter. The selectable marker cassettes can be easily cloned into the *Bam*HI or *Bgl*II sites of targeting constructs and are freely distributed as part of the DT40 web site.

1.4   Conditional gene knockout

Mutants homozygous for genes which are essential for cell survival or proliferation cannot be produced by standard targeted gene disruption. One way to study the function of these genes is by stable transfection of conditional cDNA expression constructs followed by the knockout of both alleles. When the expression of the cDNA is terminated, such cell line will be converted to the homozygous mutant stage.

A vector, pExpress, has been designed for the conditional expression of cDNA inserts (Figure 2) (Arakawa et al., 2001). The expression of the cloned cDNA is controlled by the chicken β-actin promoter and a SV40 poly A signal. The cDNA of interest can be inserted into multiple cloning sites (*Hind*III, *Nhe*I, *Eco*RV, *Bgl*II, *Nco*I, *Sma*I sites). The pExpress vector can be combined with the mutant loxP vectors by cloning the cDNA expression cassette (*Spe*I cassette) either outside (*Spe*I site) or inside (*Nhe*I site: *Spe*I-compatible) of the marker cassettes. pExpress offers additional flexibility, because any one of five different drug-resistance marker genes (*neoR*, *puroR*, *bsr*, *gpt* and *hygroR*) can be chosen and expression constructs can be designed either for random or targeted integration.

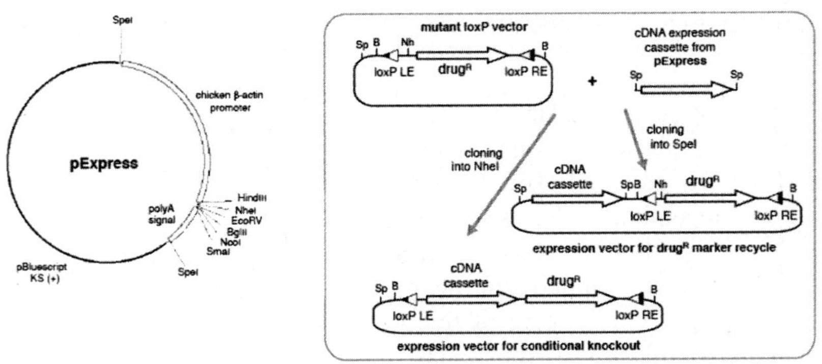

Figure 1-2. The cDNA expression vector, pExpress. After inserting a cDNA coding sequence into pExpress, the expression cassette can be excised as a *Spe*I cassette. Cloning of the expression cassette (*Spe*I cassette) into the *Spe*I site of mutant loxP vectors produces an expression vector whose drug-resistance marker can be recycled. Cloning of the expression cassette (*Spe*I cassette) into the *Nhe*I site of mutant loxP vectors produces a vector in which the drug-resistance marker and the expression cassette are located between mutant loxP sites and can be excised together. B: *Bam*HI, Nh: *Nhe*I, Sp: *Spe*I.

In addition, a new generation of conditional expression vectors is being developed which allows cDNA expression to be turned on (pTurnOn) or turned off (pIresGfp and pIresDrugR) by mutant-loxP based inversion or deletion. Multiple cloning sites are followed either by an internal ribosome entry site (IRES)-GFP sequence or an IRES-drug resistance gene enabling an estimation of the cDNA expression levels by measuring green fluorescence or drug-resistance. These vectors are currently being tested in pilot experiments. They will be released to the DT40 community as soon as their function is confirmed and documented.

1.5   Site-directed mutagenesis

Site directed mutagenesis of genes is possible in DT40 by the expression of mutant cDNA's after the disruption of the endogenous gene copies (Morrison et al., 1999). However, difficulties to fine tune gene expression using artificial cDNA expression vectors can be a problem, if a certain level of gene expression or cell cycle controlled gene expression is required. In these cases, it is preferable to introduce the desired mutation into the endogeneous loci of DT40 thereby preserving the transcriptional and post-transcriptional regulation of the mutated gene. Cre/loxP marker cassette recycle can be adapted to introduce subtle mutations into a locus accompanied by minimal additional sequence modifications. We tested this strategy in DT40 by studying the role of post-translational modification of the proliferating cell nuclear antigen (PCNA) whose expression is synchronized with DNA replication. Since this cell cycle dependent regulation of PCNA expression is likely to be required for normal proliferation rates of DT40 cells, a point mutation of codon 164 was introduced into the *PCNA* locus using a construct which carried the mutation in its 5' target arm (Figure 3).

Targeted integration of the PCNA mutagenesis construct accompanied by a crossing-over upstream of the mutation in the 5' target arm introduced the point mutation into the locus. Since the bsr marker gene cassette may perturb the splicing and transcription of *PCNA* gene, it was excised out by Cre induction leaving only a remnant loxP site in an intronic position. This procedure was repeated to modify the second allele. The presence of the desired mutations was confirmed by sequencing of *PCNA* locus.

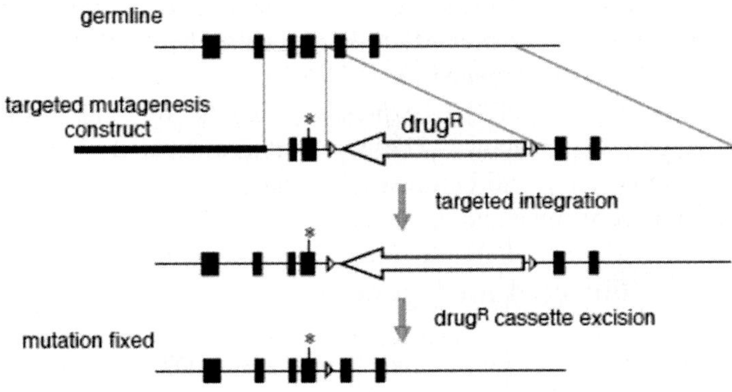

Figure 1-3. Site-directed mutagenesis of a target gene locus using PCNA as an example. Physical maps of the PCNA locus and the PCNA mutagenesis construct are shown together with the targeting strategy.

1.6   Future perspective

Work with murine embryonic stem cells has been a driving force to develop a wide range of technologies for targeted genome modifications. Most of these techniques could be successfully adapted to DT40. We are currently working on a new series of cloning vectors which should further facilitate studies in DT40. Among these are i) a new generation of user-friendly, conditional expression vectors, ii) a loxP system using mutant loxP sites for conditional gene knockout without cDNA complementation, iii) expression vector carrying tags for biochemical purification of the protein complexes, iv) a new version of mutant loxP vectors based on Gateway technology (Invitrogen) for restriction-site independent cloning of targeting vectors. Some of these futuristic vectors already exist and they will be released for general use as soon as quality checks and proper documentation are completed.

REFERENCES

Abdrakhmanov I, Lodygin D, Geroth P, Arakawa H, Law A, Plachy J, Korn B, Buerstedde JM. A large database of chicken bursal ESTs as a resource for the analysis of vertebrate gene function. Genome Res. 10, 2062-2069 (2000).

Albert H, Dale EC, Lee E, Ow DW. Site-specific integration of DNA into wild-type and mutant lox sites placed in the plant genome. Plant J. 7, 649-659 (1995).

Arakawa H, Lodygin D, Buerstedde J-M. Mutant loxP vectors for selectable marker recycle and conditional knock-outs. BMC Biotechnol. 1, 7 (2001).

Arakawa H, Saribasak H, Buerstedde J-M. Activation-induced cytidine deaminase initiates immunoglobulin gene conversion and hypermutation by a common intermediate. PLoS Biol. 2, E179 (2004).

Boardman PE, Sanz-Ezquerro J, Overton IM, Burt DW, Bosch E, Fong WT, Tickle C, Brown WR, Wilson SA, Hubbard SJ. A comprehensive collection of chicken cDNAs. Curr. Biol. 12, 1965-1969 (2002).

Buerstedde J-M, Takeda S. Increased ratio of targeted to random integration after transfection of chicken B cell lines. Cell 67, 179-188 (1991).

Caldwell RB, Kierzek AM, Arakawa H, Bezzubov Y, Zaim J, Fiedler P, Kutter S, Blagodatski A, Kostovska D, Koter M, Plachy J, Carninci P, Hayashizaki Y, Buerstedde JM. Full-length cDNA's from chicken bursal lymphocytes to facilitate gene function analysis. Genome Biol. 6, R6 (2005).

International Chicken Genome Sequencing Consortium. Sequence and comparative analysis of the chicken genome provide unique perspectives on vertebrate evolution. Nature 432, 695-716 (2004).

Morrison C, Shinohara A, Sonoda E, Yamaguchi-Iwai Y, Takata M, Weichselbaum RR, Takeda S. The essential functions of human Rad51 are independent of ATP hydrolysis. Mol. Cell. Biol. 1, 6891-6897 (1999).

Chapter 2

IMMUNOGLOBULIN GENE CONVERSION OR HYPERMUTATION: THAT'S THE QUESTION

Jean-Marie Buerstedde and Hiroshi Arakawa
GSF, Institute for Molecular Radiobiology, Ingolstaedter Landstr. 1, D-85764 Neuherberg-Munich, Germany

Abstract: Chicken B cells develop their primary immunoglobulin (Ig) gene repertoire by pseudogene templated gene conversion within the bursa of Fabricius. The DT40 cell line is derived from bursal B cells and continues to diversify its rearranged Ig light chain in cell culture. Ig gene conversion of DT40 requires expression of the AID gene which was earlier shown to be needed for Ig hypermutation and switch recombination in mammalian B cells. Interestingly, Ig hypermutation can be induced in DT40, if Ig gene conversion is blocked by the disruption of RAD51 paralog genes, the deletion of the nearby pseudogene locus or the disruption of the UNG gene. The ease of gene targeting and the compactness of the chicken Ig light chain locus makes DT40 an ideal model to study the molecular mechanism of AID induced gene conversion and hypermutation.

Key words: Immunoglobulin, gene conversion, somatic hypermutation, DT40, B cell, AID, UNG, DNA repair.

1.   INTRODUCTION

The immune system of vertebrates is unique, as B lymphocytes activate and diversify their Ig genes by recombination and hypermutation. Somatic modification of the Ig loci accomplishes the following tasks at different stages of B cell development: i) allelic exclusion assuring the expression of only one Ig light and heavy chain in each B cell, ii) development of a diverse primary antibody repertoire, iii) affinity maturation of antibodies leading to

11

J.-M. Buerstedde and S. Takeda (eds.), Reviews and Protocols in DT40 Research, 11–24.
© 2006 *Springer.*

tighter antigen binding and iv) isotype switching to produce antibodies with different heavy chain isotypes. These biological phenomena can be explained by four types of sequence alterations in the Ig loci: V(D)J recombination, gene conversion, hypermutation and switch-recombination.

Ig genes are not encoded in a functional form in the germline, but are assembled from gene segments by site-specific recombination. In humans and mice, functional V, D and J gene segments exist in large clusters and the random assortment of individual V, D and J segments generates considerable combinatorial diversity for the primary repertoire (Tonegawa, 1983).

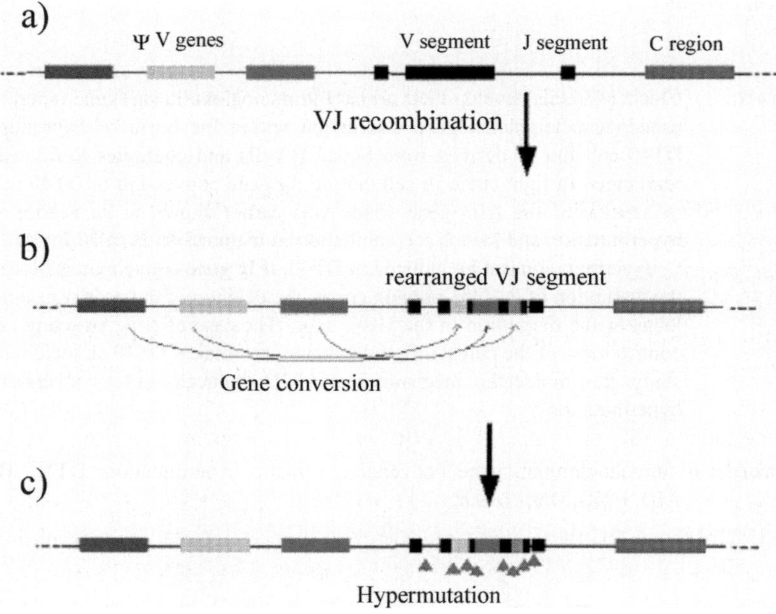

Figure 2-1. Development of a diverse Ig light chain gene repertoire by gene conversion and hypermutation (see plate 1).

In other vertebrate species, V(D)J recombination only ensures allelic exclusion. This was first demonstrated for the chicken where only a single functional V and J segment are rearranged in the light chain gene locus (Fig.2-1a). The rearranged chicken light chain V segment is diversified by gene conversion using nearby pseudo V (ψV) genes as donor sequences (Reynaud et al., 1987) (Fig.2-1b). A gene conversion scheme, in which the boundaries and the lengths of the conversion tracts vary and multiple conversion events superimpose on each other, is ideally suited for the development of a diverse primary Ig repertoire from a germline sequence pool. Not only avian species, but also rabbits, cattle, swine and horses create their primary Ig gene repertoire by this elegant mechanism (Butler, 1998).

Following antigen stimulation the rearranged V(D)J segments are further modified by the introduction of single untemplated nucleotide substitutions (Fig.2-1c). Some of the mutations increase the affinity of the encoded antibodies leading to the proliferation of the respective B cells (Jacob et al., 1991). All vertebrate species including chicken (Arakawa et al., 1996) employ Ig hypermutation for affinity maturation of antibodies. In parallel to hypermutation, class switch recombination introduces deletions between so-called switch regions in the Ig heavy chain locus resulting in the expression of a new isotype (Dudley et al., 2005).

1.1 Ig hypermutation and switch recombination in mice and men

A number of approaches have been tried to identify genes involved in Ig gene repertoire formation in mammals. B cell lines undergoing Ig gene diversification and other cell lines in which this phenomenon is induced by gene transfections are valuable for cell culture experiments. Mice and human patients suffering from hereditary immunodeficiencies can be searched for mutations in genes involved in Ig recombination or hypermutation. The function of candidate genes can be tested by disruption in murine embryonic stem cells and the analysis of lymphoid developments in animals derived from the embryonic stem cells.

The outlined studies were successful to define the lymphoid specific and general DNA repair genes and the mechanism of V(D)J recombination in molecular detail (Jones and Gellert, 2004, Dudley et al., 2005). In comparison, the genetics of Ig hypermutation and switch recombination remained poorly understood for a long time (Jakobs et al., 1998). However, the cloning of the AID gene from switch recombination active cells (Muramatsu et al., 1999) and the subsequent demonstration that Ig hypermutation and switch recombination is abolished in AID knockout mice (Muramatsu et al., 2000) and AID deficient patients (Revy et al., 2000) was a breakthrough. The hypothesis that the AID protein, which contains an evolutionary conserved cytosine deamination motif, directly deaminates cytosines to uracils inspired further genetic studies. Most important were the findings that inactivation of the UNG gene inhibited switch recombination and changed the spectrum of Ig hypermutations at G/C bases in mice (Rada et al., 2002) and human patients (Imai et al., 2004). Although hypermutations at A/T bases were not affected by UNG deficiency, these mutations were reduced in MSH2 and MSH6 deficient mice (Wiesendanger et al., 2000). UNG/MSH2 double knock-out mice showed no hypermutations at A/T bases and no switch recombination (Rada et al., 2004) consistent with the hypothesis that MSH2 is needed for A/T hypermutations and for the low level of switch recombinations still present in UNG deficient mice.

It remains still unresolved how hypermutations are targeted to the Ig loci. The analysis of hypermutations in B cells indicated that mutations are limited to a 2-kb region from the Ig transcription start site and that apart from rare exceptions, other transcribed genes are not mutated (Kotani et al., 2005). It was furthermore demonstrated that artificial Ig locus constructs in transgenic mouse lines required the presence of Ig enhancer sequences for hypermutation activity (Neuberger et al., 1998). Nevertheless, new results from AID expressing non B cell and even B cell lines suggest that hypermutations may not be strictly limited to Ig loci and that AID can act as a global genome mutator (Yoshikawa et al., 2002; Wang et al., 2004).

1.2   Ig gene conversion in DT40

Ig gene conversion in bursal B cells and the DT40 cell line has recently been reviewed (Arakawa and Buerstedde, 2004) and emphasis is placed on new insights and certain technical issues. Ig gene conversion is difficult to investigate in primary B cells from chicken or other farm animals, because these cells can be maintained only for short time in cell culture. In contrast, the ALV induced lymphoma cell line DT40 which seems to be arrested at the stage of bursal B cells continues gene conversion during in-vitro cell culture (Buerstedde et al., 1990; Kim et al., 1990). A high ratio of targeted to random gene integration (Buerstedde and Takeda, 1991) makes DT40 an ideal system to study the mechanism of Ig gene conversion by candidate gene disruptions and the modifications of cis-acting regulatory sequences.

Ongoing gene conversion can be detected in DT40 by sequencing rearranged light chain VJ segments from the progeny of a single cell after subcloning. In most cases, the VJ sequence of the progenitor cell is still dominant among the sequences and can be established as a consensus sequence to which all divergent sequences are compared. This approach has the advantages that conversion events can be detected within the whole VJ segments and the length of the tracts as well the likely pseudogene donors can be determined. However, an exact quantification of the Ig conversion activity in different genetic backgrounds is difficult by direct sequencing due to fluctuation effects within individual subclones. This problem can be addressed by using the surface Ig (sIg) reversion assay which relies on the repair of an Ig light chain frameshift by gene conversion (Buerstedde et al.,

1990) (Fig. 2-2). Wild-type DT40 cells are dominantly sIg(+), but spontaneously arising subclones have been isolated which have lost sIg expression due to frameshifts in the rearranged light chain V segment. Overlapping gene conversion events can repair these frameshifts leading to re-expression of sIg. The reversion from sIg(-) to sIg(+) status can be easily measured by fluorescence activated cell sorter (FACS) analysis in a large number of subclones. Furthermore, sIg(+) revertants from individual subclones can be isolated by preparative FACS sorts and their rearranged light chain V segments can be sequenced to confirm that the frameshift was indeed repaired by gene conversion.

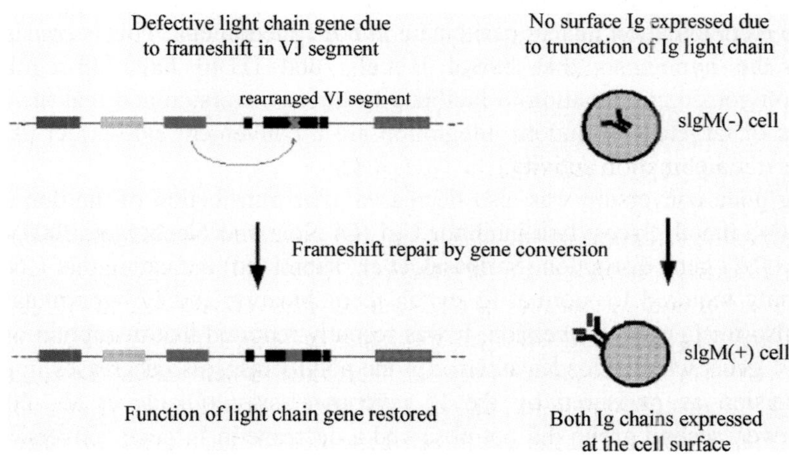

Figure 2-2. Surface Ig reversion assay to quantify Ig gene conversion activity (see plate 2).

The level of Ig gene conversion has been tested in many DT40 mutants in which general DNA repair and recombination factors had been inactivated. Reminiscent of Ig hypermutation, the deletion of the AID gene had most striking effect abolishing all detectable Ig gene conversion activity of DT40 (Arakawa et al., 2002; Harris et al., 2002). This result demonstrated that all three B cell specific mechanisms of Ig diversification - hypermutation, switch recombination and gene conversion - are tightly regulated by AID. Surprisingly, no evidence for Ig gene conversion events were detected in AID[-/-] cells by the Ig reversion assay even after prolonged culture given that most cells seem to perform intrachromosomal gene conversion at low frequency (Liskay and Stachelek, 1983). This suggests that Ig gene conversion is completely suppressed in the absence of AID expression despite the unusual density of potential conversion donor sequences in the neighborhood.

A reduction, but not a complete block in Ig gene conversion frequencies was demonstrated for a number of other DT40 mutants by the Ig reversion assay or by direct sequence analysis. Most of these mutants are either members of the RAD52 recombination repair pathway which is known to mediate DNA repair by homologous recombination (Bezzubova et al., 1997; Sale et al., 2001; Hatanaka et al., 2005) or are associated with this pathway (Hatanaka et al., 2005). Inactivation of two members of the Fanconi anemia pathway, FANCC and FANCD2, also reduced Ig gene conversion activity a few fold (Niedzwiedz et al., 2004; Yamamoto et al., 2005). Interestingly, all mutants mentioned above showed not only reduced Ig gene conversion, but also lower ratios of targeted to random gene integration suggesting that the same recombination factors participate in both phenomena. This is consistent with the hypothesis that bursal B cells and DT40 have up-regulated homologous recombination to facilitate Ig gene conversion and that the high ratios of targeted to random integration are a convenient side-effect of this hyper-recombination activity.

Ig gene conversion was also decreased after transfection of the dominant negative uracil glycosylase inhibitor Ugi (Di Noia and Neuberger, 2004) and after UNG gene disruption (Saribasak et al., submitted) indicating that UNG is not only required for normal Ig switch recombination and Ig hypermutation, but also for Ig gene conversion. It was recently reported that disruption of the REV1 gene, which encodes an error prone polymerase, also decreases Ig gene conversion as measured by the Ig reversion assay (Okada et al., 2005). However, second group did not observed a decrease in Ig gene conversion in REV1$^{-/-}$ cells by direct light chain gene sequencing (Simpson and Sale, 2003). Another finding which needs to be confirmed is a report that over-expression of the NBS1 gene stimulates Ig gene conversion in DT40 (Yabuki et al., 2005). No correlation between the amount of NBS1 over-expression and the stimulatory effect on Ig gene conversion were observed and it would be interesting to test how NBS1 gene disruption affects Ig gene conversion (Tauchi et al., 2002). Ig gene conversion is significantly enhanced after treating DT40 with the histone deacetylase inhibitor, trichostatin A, and this phenomenon could be exploited for the production of antigen specific antibodies in DT40 (Seo et al., 2005) .

1.3   Ig hypermutation in DT40

Wild-type DT40 diversifies its Ig gene predominantly by ψV templated gene conversions similar to bursal B cells. Nevertheless, single nucleotide substitutions, which cannot be accounted for by the known ψV genes of the chicken CB strain, are occasionally found in light chain VJ sequences of the same subclone. These apparently untemplated nucleotide substitutions have been considered Ig hypermutations (Simpson and Sale, 2003; Niedzwiedz et al., 2004), although it cannot be ruled out that they are templated by polymorphic ψV genes of DT40 or that they are a by-product of gene conversion events (Reynaud et al., 1997). Good evidence for a new Ig hypermutation activity is however found in certain DT40 mutants in which AID is expressed and Ig gene conversion is compromised at an early stage. A shift from Ig gene conversion to hypermutation were first reported for disruptions of the RAD51 paralog genes, XRCC2, XRCC3 and RAD51B, which are members of the RAD52 recombination repair pathway (Sale et al., 2001). Direct sequencing of the light chain VJ segments of mutant subclones revealed a decreased frequency of gene conversion tracts and high frequencies of single untemplated nucleotide substitutions at G/C bases.

This hypermutation activity is reflected by the fast appearance of sIg negative cells during the culture of subclones (Sale et al., 2001). In this Ig loss assay (Fig. 2-3), the precursor cell is sIg(+), and deleterious Ig hypermutations lead to the loss of sIg expression in some of its progeny which can be easily measured by FACS. Wild-type DT40 and non-hypermutating mutants generate only low numbers of sIg(-) negative cells, because Ig gene conversion events are usually accurate and the resulting sequence modify-cations rarely interfere with sIg expression. A correlation of sIg status and light chain VJ sequences in a situation, where only hypermutations of the light chain gene are expected, suggests that most sIg expression defects are not due to premature stop codons, but to mis-sense mutations which impair Ig light and heavy chain pairing (Arakawa et al., 2004). The sIg loss assay can be used to quantify Ig hypermutation activity easily in a large number of subclones and thereby minimize fluctuation effects. However, this assay is of questionable value for the quantification of Ig gene conversion activity (Yabuki et al., 2005) and the sIg reversion assay should be used instead for this purpose.

Truncation of the BRCA2 gene also decreased Ig gene conversion and activated Ig hypermutation indicating that BRCA2 acts in a similar way to the RAD51 paralogues. However, the RAD54 (Bezzubova et al., 1987), the FANCC (Niedzwiedz et al., 2004) and the FANCD2 (Yamamoto et al., 2005) disruption mutants showed no evidence for an increase of untemplated Ig mutations despite a reduced Ig gene conversion activity.

A complete block in Ig light chain gene conversion and a high frequency of hypermutations at G/C bases within and surrounding the light chain VJ segment were seen after the deletion of all ψV genes in the rearranged light chain locus (Arakawa et al., 2004). This demonstrated that only the ψV genes on the same chromosome are used as donors sequences for conversion

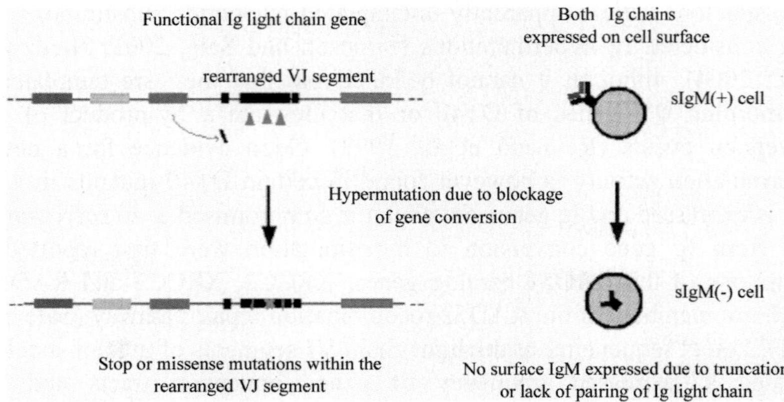

Figure 2-3. Surface Ig loss assay to quantify Ig hypermutation activity (see plate 3).

events. The hypermutation activity induced by the pseudogene deletion was shown to be AID dependent, Ig locus specific and focused to known Ig hypermutation hotspot sequences within the light chain V sequence. In contrast to the other hypermutating DT40 mutants, the ψV deleted variant of DT40 is DNA repair proficient and devoid of any background Ig gene conversion activity and therefore represents an ideal system to study the mechanism of Ig hypermutation in more detail.

1.4   The relationship of Ig gene conversion and hypermutation

The results cited above support a simple model explaining how Ig hypermutation and recombination may be initiated and regulated (Arakawa et al., 2004) (Fig.2-4). At the top of the events is a modification of the rearranged V(D)J segment induced by AID. The default processing of this alteration in the absence of nearby donors or high homologous recombination activity leads to Ig hypermutation in form of a single nucleotide substitution (Fig.2-4, right side). However, if donor sequences and homologous recombination factors are available, processing of the AID induced lesion can be divided into a stage before strand exchange, when a shift to Ig hypermutation is still possible and a stage after strand exchange when a commitment toward Ig gene conversion has been made (Fig.2-4, left side). Whereas completion of the first stage requires the participation of the RAD51 paralogues and BRCA2, the second stage involves other recombination factors like the Rad54 protein and the Fanconi anemia pathway.

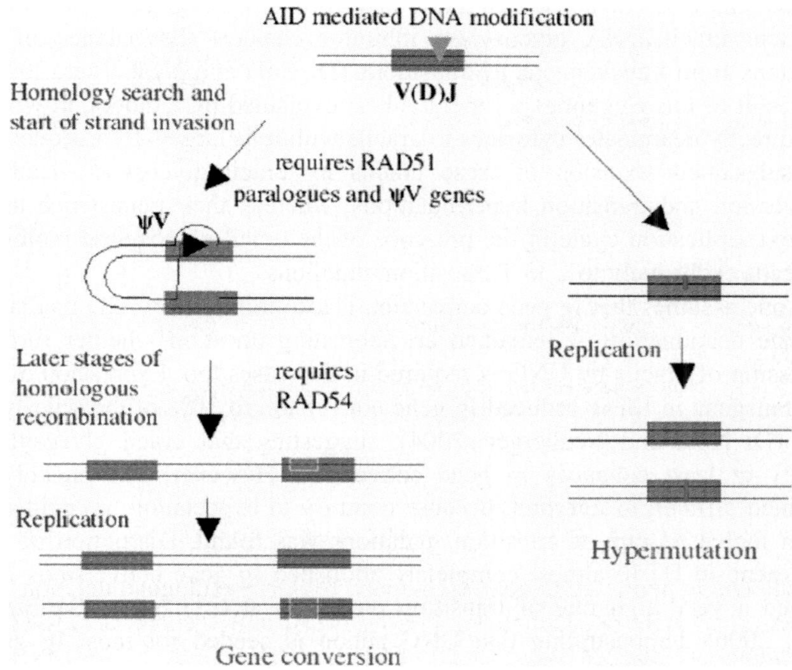

Figure 2-4. Model for Ig gene conversion and hypermutation (see plate 4).

The difference in commitment explains why disruptions of the RAD51 paralog and BRACA2 genes not only decrease Ig gene conversion, but also induce Ig hypermutation (Sale et al., 2001; Yabuki et al., 2005) whereas disruption of the RAD54 gene only decreases Ig gene conversion (Bezzubova et al., 1997). The model also predicts that a low homologous recombination activity precludes Ig gene conversion even in the presence of conversion donors. Such a lack of recombination proficiency might be the reason why human and murine B cells never use Ig gene conversion despite the presence of nearby candidate donors in form of unrearranged V segments and why chicken germinal center B cells have shifted the balance from Ig gene conversion to Ig hypermutation (Arakawa et al., 1996).

1.5 Insight into the mechanism of Ig hypermutation

The exciting discovery of Ig hypermutation in DT40 mutants of the RAD51 paralogues triggered further studies about the hypermutation mechanism. It was demonstrated that transfection of the Ugi gene encoding a dominant uracil DNA glycosylase inhibitor changes the balance of Ig mutations from transversions to transitions (Di Noia and Neuberger, 2002). The result of this elegant experiment is best explained by a model, in which AID directly deaminates cytosines to uracils within its target DNA sequence. The subsequent excision of these uracils by uracil glycosylase lead to transversion and transition hypermutations, whereas their persistence until the next replication cycle in the presence of the uracil glycosylase inhibitor Ugi leads exclusively to C to T transition mutations.

If one assumes that Ig gene conversion is also initiated by AID mediated cytosine deamination, it remained an interesting question whether further processing of uracils by UNG is required in this cases too. Expression of an *Ugi* transgene in DT40 reduced Ig gene conversion to 30% of the wild-type level (Di Noia and Neuberger, 2004), suggesting that uracil glycosylase activity at least enhances Ig gene conversion. However, this phenotype remained difficult to interpret, because contrary to expectation, no evidence for an increased rate of transition mutations was found. Disruption of the *UNG* gene in DT40 almost completely abolished Ig gene conversions and induced a very high rate of transition mutations at C/G bases (Saribasak et al., 2006) demonstrating that UNG action is needed for most Ig gene conversion activity. Nevertheless, a minor alternative pathway seems to be responsible for the few Ig gene conversion events still seen in UNG deficient cells.

These results in DT40 support a model in which AID first deaminates cytosine within the rearranged V(D)J segments and the resulting uracils are processed by UNG to give rise to either transition and transversion mutations or to gene conversions (Fig. 2-5). Inactivation of the *UNG* gene blocks the processing of uracils and leads exclusively to transition mutations when uracil pairs with adenine in the next replication cycle. UNG$^{-/-}$ cells accumulate light chain mutation seven times faster than ψV deleted cells, indicating that about one of seven AID-induced uracils is processed into a mutation. This frequent conversion of uracils into mutations suggests the presences of an unusual error-prone repair pathway which specifically recognises AID-induced uracils. Inactivation of the translesion polymerase REV1 reduces hypermutation in DT40 implying that REV1 participate in the repair of the abasic sites after UNG mediated uracil excision (Simpson and Sale, 2003).

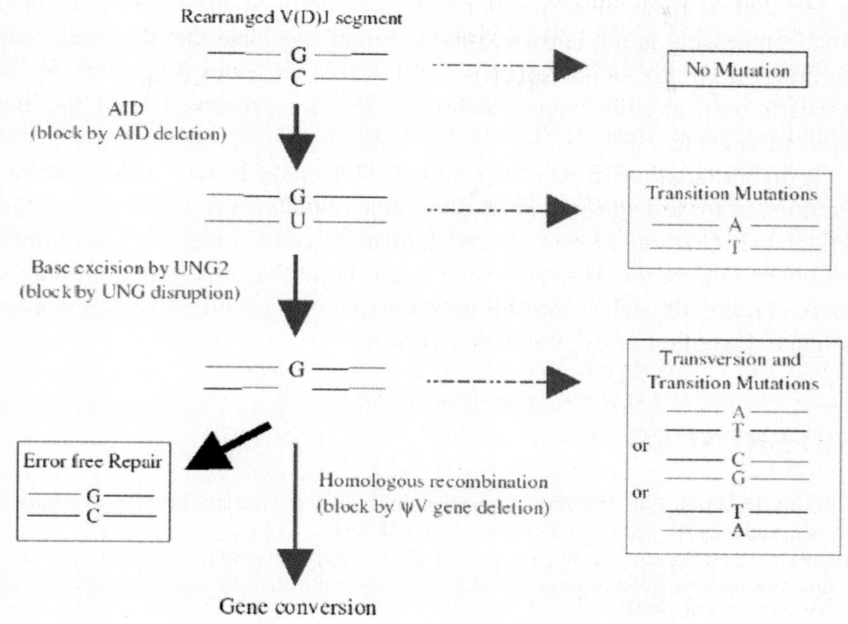

Figure 2-5. Phenotypes of DT40 mutants affecting Ig gene conversion and hypermutation (see plate 5).

1.6   Future perspectives

DT40 offers unique opportunities to investigate the molecular mechanisms of Ig gene conversion and hypermutations. The most intriguing questions are how hypermutations are targeted to the Ig loci and how events downstream of UNG mediated uracil excision lead alternatively to hypermutation or gene conversion.

An attractive working hypothesis postulates that AID is recruited to the Ig target genes by interactions with transcription factors and the RNA polymerase complex. The compact size of the rearranged Ig light chain locus in the hypermutating, ψV deleted DT40 mutant should facilitating the identification of cis-acting regulatory sequences by targeted deletions and substitutions. In parallel, the role of trans-acting DNA binding factors like for example the E2A transcription factor (Conlon and Meyer, 2004) can be investigated by gene disruption.

In the absence of gene conversion, AID induced uracils are most likely channeled into a pathway involving error prone DNA polymerases like REV1 (Simpson and Sale, 2003) and pol η (Martomo et al., 2005; Delbos et al., 2005). How these polymerases are recruited after UNG mediated uracil processing is not known, but one might speculate that the abasic sites resulting from the action of AID and UNG are not recognized by the standard base excision repair pathway, but are preserved until the next replication cycle.

Ig hypermutations in DT40 is a kind of sleeping beauty which has been overlooked for a long time. Even now, hypermutations in DT40 do not fully reflect the situation in murine and human B cells, since they accumulate mainly at G/C bases. However, one might hope that this limitation may be overcome one day when the still missing A/T mutator will be awoken either by gene disruption or by gene transfection.

REFERENCES

Arakawa, H, Buerstedde, JM. 2004. Immunoglobulin gene conversion: Insights from bursal B cells and the DT40 cell line. *Dev. Dyn.* 229: 458-464.

Arakawa, H, Saribasak, H, Buerstedde JM. 2004. Activation-induced cytidine deaminase initiates immunoglobulin gene conversion and hypermutation by a common intermediate. *PLoS Biol.* 2:967-974.

Arakawa H, Furusawa S, Ekino S, Yamagishi H. 1996. Immunoglobulin gene hyper-conversion ongoing in chicken splenic germinal centers. EMBO J. 15:2540-2546.

Arakawa H, Hauschild J, Buerstedde JM. 2002. Requirement of the Activation-Induced Deaminase (AID) Gene for Immunoglobulin Gene Conversion. Science. 295:1301-1306.

Bezzubova O, Silbergleit A, Yamaguchi-Iwai Y, Takeda S, Buerstedde JM. 1997. Reduced X-ray resistance and homologous recombination frequencies in a RAD54-/- mutant of the chicken DT40 cell line. Cell. 89:185-193.

Buerstedde JM, Reynaud CA, Humphries EH, Olson W, Ewert DL, Weill JC. 1990. Light chain gene conversion continues at high rate in an ALV-induced cell line. EMBO J. 9:921-927.

Buerstedde JM, Takeda S. 1991. Increased ratio of targeted to random integration after transfection of chicken B cell lines. Cell. 67:179-188.

Butler JE. 1998. Immunoglobulin diversity, B-cell and antibody repertoire development in large farm animals. Rev Sci Tech. 17:43-70.

Conlon TM, Meyer KB. 2004. Cloning and functional characterisation of avian transcription factor E2A. BMC Immunol. 5:11.

Delbos F, De Smet A, Faili A, Aoufouchi S, Weill JC, Reynaud CA. 2005. Contribution of DNA polymerase eta to immunoglobulin gene hypermutation in the mouse. J Exp Med. 201:1191-6.

Di Noia J, Neuberger MS. 2002. Altering the pathway of immunoglobulin hypermutation by inhibiting uracil-DNA glycosylase. Nature. 419:43-48.

Di Noia, JM, Neuberger MS. 2004. Immunoglobulin gene conversion in chicken DT40 cells largely proceeds through an abasic site intermediate generated by excision of the uracil produced by AID-mediated deoxycytidine deamination. *Eur J Immunol.* 34: 504-508.

Dudley DD, Chaudhuri J, Bassing CH, Alt FW. 2005. Mechanism and Control of V(D)J Recombination versus Class Switch Recombination: Similarities and Differences. Adv Immunol. 86:43-112.

Liskay RM, Stachelek JL. 1983. Evidence for intrachromosomal gene conversion in cultured mouse cells. Cell. 35:157-65.

Harris RS, Sale JE, Petersen-Mahrt SK, Neuberger MS. 2002. AID is essential for immunoglobulin V gene conversion in a cultured B cell line. Curr Biol. 12:435-8.

Hatanaka A, Yamazoe M, Sale JE, Takata M, Yamamoto K, Kitao H, Sonoda E, Kikuchi K, Yonetani Y, Takeda S. 2005. Similar effects of Brca2 truncation and Rad51 paralog deficiency on immunoglobulin V gene diversification in DT40 cells support an early role for Rad51 paralogs in homologous recombination. Mol Cell Biol. 25:1124-34.

Imai, K, Slupphaug, G, Lee, WI, Revy, P, Nonoyama, S, Catalan, N, Yel, L, Forveille, M, Kavli, B, Krokan, HE, Ochs, HD, Fischer, A, Durandy A. (2003). Human uracil-DNA glycosylase deficiency associated with profoundly impaired immunoglobulin class-switch recombination. *Nat. Immunol.* 4:1023-1028.

Jacobs H, Fukita Y, van der Horst GT, de Boer J, Weeda G, Essers J, de Wind N, Engelward BP, Samson L, Verbeek S, de Murcia JM, de Murcia G, te Riele H, Rajewsky K. Hypermutation of immunoglobulin genes in memory B cells of DNA repair-deficient mice. J Exp Med. 1998 187:1735-43.

Jacob J, Kelsoe G, Rajewsky K, Weiss U. 1991. Intraclonal generation of antibody mutants in germinal centres. Nature. 354:389-392.

Jones JM, Gellert M. The taming of a transposon: V(D)J recombination and the immune system. Immunol Rev. 2004 200:233-48.

Kim S, Humphries EH, Tjoelker L, Carlson L, Thompson CB. 1990. Ongoing diversification of the rearranged immunoglobulin light-chain gene in a bursal lymphoma cell line. Mol Cell Biol. 10:3224-31.

Kotani A, Okazaki IM, Muramatsu M, Kinoshita K, Begum NA, Nakajima T, Saito H, Honjo T. 2005. A target selection of somatic hypermutations is regulated similarly between T and B cells upon activation-induced cytidine deaminase expression. Proc Natl Acad Sci U S A. 102:4506-11.

Muramatsu M, Sankaranand VS, Anant S, Sugai M, Kinoshita K, Davidson NO, Honjo T. 1999. Specific expression of activation-induced cytidine deaminase (AID), a novel member of the RNA-editing deaminase family in germinal center B cells. J Biol Chem. 274:18470-18476.

Muramatsu M, Kinoshita K, Fagarasan S, Yamada S, Shinkai Y, Honjo T. 2000. Class switch recombination and hypermutation require activation-induced cytidine deaminase (AID), a potential RNA editing enzyme. Cell. 102:553-563.

Neuberger MS, Ehrenstein MR, Klix N, Jolly CJ, Yelamos J, Rada C, Milstein C. 1998. Monitoring and interpreting the intrinsic features of somatic hypermutation. Immunol Rev. 162:107-16.

Niedzwiedz W, Mosedale G, Johnson M, Ong CY, Pace P, Patel KJ. 2004. The Fanconi anaemia gene FANCC promotes homologous recombination and error-prone DNA repair. Mol Cell. 15:607-20.

Okada T, Sonoda E, Yoshimura M, Kawano Y, Saya H, Kohzaki M, Takeda S. 2005. Multiple Roles of Vertebrate REV Genes in DNA Repair and Recombination. Mol Cell Biol. 25:6103-11.

Rada C, Di Noia JM, Neuberger MS. (2004). Mismatch recognition and uracil excision provide complementary paths to both Ig switching and the A/T-focused phase of somatic mutation. Mol Cell. 16:163-171.

Rada, C, Williams, GT, Nilsen, H, Barnes, DE, Lindahl, T, Neuberger, MS. (2002). Immunoglobulin isotype switching is inhibited and somatic hypermutation perturbed in UNG-deficient mice. *Curr Biol.* 12:1748-1755.

Revy P, Muto T, Levy Y, Geissmann F, Plebani A, Sanal O, Catalan N, Forveille M, Dufourcq-Labelouse R, Gennery A, Tezcan I, Ersoy F, Kayserili H, Ugazio AG, Brousse N, Muramatsu M, Notarangelo LD, Kinoshita K, Honjo T, Fischer A, Durandy A. 2000. Activation-induced cytidine deaminase (AID) deficiency causes the autosomal recessive form of the Hyper-IgM syndrome. Cell. 102:565-575.

Reynaud CA, Anquez V, Grimal H, Weill JC. 1987. A hyperconversion mechanism generates the chicken light chain preimmune repertoire. Cell. 48:379-388.

Sale JE, Calandrini DM, Takata M, Takeda S, Neuberger MS. 2001. Ablation of XRCC2/3 transforms immunoglobulin V gene conversion into somatic hypermutation. Nature. 412:921-926.

Saribasak H, Saribasak NN, Ipek FM, Ellwart JW, Arakawa H, Buerstedde JM. 2006. UNG disruption blocks immunoglobulin gene conversion and induces transition mutations. J Immunol. 176: 365-371.

Seo H, Masuoka M, Murofushi H, Takeda S, Shibata T, Ohta K. 2005. Rapid generation of specific antibodies by enhanced homologous recombination. Nat Biotechnol. 23:731-735.

Simpson LJ, Sale JE. 2003. Rev1 is essential for DNA damage tolerance and non-templated immunoglobulin gene mutation in a vertebrate cell line. EMBO J. 22:1654-1664.

Tauchi H, Kobayashi J, Morishima K, Van Gent DC, Shiraishi T, Verkaik NS, VanHeems D, Ito E, Nakamura A, Sonoda E, Takata M, Takeda S, Matsuura S, Komatsu K. 2002. Nbs1 is essential for DNA repair by homologous recombination in higher vertebrate cells. Nature. 420:93-98.

Tonegawa S. 1983. Somatic generation of antibody diversity. Nature. 302:575-81.

Wang CL, Harper RA, Wabl M. 2004. Genome-wide somatic hypermutation. Proc Natl Acad Sci U S A. 101:7352-6.

Wiesendanger M, Kneitz B, Edelmann W, Scharff MD. 2000. Somatic hypermutation in MutS homologue (MSH)3-, MSH6-, and MSH3/MSH6-deficient mice reveals a role for the MSH2-MSH6 heterodimer in modulating the base substitution pattern. J Exp Med. 191:579-84.

Yabuki M, Fujii MM, Maizels N. 2005. The MRE11-RAD50-NBS1 complex accelerates somatic hypermutation and gene conversion of immunoglobulin variable regions. Nat Immunol. 6:730-6.

Yamamoto K, Hirano S, Ishiai M, Morishima K, Kitao H, Namikoshi K, Kimura M, Matsushita N, Arakawa H, Buerstedde JM, Komatsu K, Thompson LH, Takata M. 2005. Fanconi anemia protein FANCD2 promotes immunoglobulin gene conversion and DNA repair through a mechanism related to homologous recombination. Mol Cell Biol. 25: 34-43.

Yoshikawa K, Okazaki IM, Eto T, Kinoshita K, Muramatsu M, Nagaoka H, Honjo T. 2002. AID enzyme-induced hypermutation in an actively transcribed gene in fibroblasts. Science. 296:2033-6.

Chapter 3

GENOME RESOURCES FOR THE DT40 COMMUNITY

Randolph B. Caldwell and Andrzej M. Kierzek
GSF, Institute for Molecular Radiobiology, Ingolstaedter Landstr. 1, D-85764 Neuherberg-Munich, Germany; School of Biomedical and Molecular Sciences, University of Surrey, Guildford, GU2 7XH, UK, Institute of Biochemistry and Biophysics, Polish Academy of Sciences, Pawinskiego 5a 02-106 Warszawa, Poland

Abstract: The chicken B cell line DT40 is a tool that is uniquely situated to study the function of genes due to its high rate of homologous recombination. As any tool is only as good as the resources behind it, the recent boon and continuing output of available genomic information has only widened the possibilities for the DT40 community. Besides the release of the chicken genome, the public databases are expanding rapidly with a wealth of experimentally produced data including various chicken cell specific EST's, full-length cDNA's, non-coding RNA's and expressed SAGE tags. In addition, many laboratories have also taken it upon themselves to set up a web presence freely sharing and distributing information, program applications and software such as the DT40 website. Of course, the standard bearers in this are the institutes that have led the way in development of bioinformatics tools for sequence information, comparative genomics and proteomics such as The National Center for Biotechnology Information (NCBI), The European Molecular Biology Laboratory (EMBL), The DNA Database of Japan (DDBJ) and The Molecular Biology Server-Swiss Institute of Bioinformatics (ExPASy). These resources for virtual science can not replace the hands in the lab, but they can help at every turn along the path to discovery.

Key words: DT40 website, genome, alignment, annotation, internet, computer, programs, software.

1.   INTRODUCTION

To take advantage of a high rate of homologous recombination in the DT40 cell line (Buerstedde and Takeda, 1991) in a targeted gene inactivation

J.-M. Buerstedde and S. Takeda (eds.), Reviews and Protocols in DT40 Research, 25–37.
© 2006 *Springer.*

experiment, one needs to know the DNA sequence of the gene of interest. Thus the availability of the chicken genome sequence (International Chicken Genome Sequencing Consortium, 2004) complemented by a large collection of full-length cDNA sequences (Abdrakhmanov, et al., 2000, Boardman, et al., 2002, Caldwell, et al., 2004, Hubbard, et al., 2004) allows full exploitation of the DT40 model as a tool for functional genomics studies and biotechnology. In this chapter we will describe computational resources which may be used to mine vast amounts of genomic and transcriptomic sequences available for the DT40 cell line.

In the first part of this chapter we will describe the DT40 Web site (Abdrakhmanov, et al., 2000), a comprehensive database of bursal EST and full-length cDNA sequences. Next we will present other WWW based resources useful for research based around the DT40 model.

1.1   DT40 Web site

The recent release of the chicken genome sequence greatly benefits the DT40 research community. For the first time, the entire genome can be searched for sequences that are conserved during vertebrate evolution and whose function might be clarified after genetic modification in DT40. Moreover, non-coding sequences may also be subjected to experimental modification which is of paramount importance for investigation of gene regulatory mechanisms and other phenomena in which non-coding sequences play a role (e.g., segmental gene conversion using pseudo V genes as donors) (Pesole, et al., 1997; Arakawa, et al., 2004; Arakawa and Buerstedde, 2004). Modifications of both coding and non-coding regions require information of the intron-exon structure of genes. Unfortunately, *in silico* gene structure prediction methods have a high error rate and only full-length cDNA's unambiguously define the boundaries of the transcription units within whole genome assemblies. Cloned full-length cDNA's are also of immense practical value to complement mutant phenotypes and artificially express the encoded protein. For these reasons, whole genome sequencing projects have been complemented by large-scale efforts to obtain a maximum number of full-length cDNA's. The DT40 Web site presents a collection of full length cDNA's and EST's expressed in chicken bursal lymphocytes. Moreover, this resource contains a collection of SAGE (Serial Analysis of Gene Expression) tags from both the chicken bursal lymphocytes and DT40 cells (Wahl and Caldwell, et al., 2004). The DT40 website is backed by the FOUNTAIN software package and the following text demonstrates this software's invaluable use.

1.1.1　cDNA sequencing and annotation protocol

A cDNA library called 'riken1' was synthesized from bursal lymphocyte mRNA of a two weak old CB strain of chicken using the biotinylated cap trapper method which is optimized to produce a large percentage of full-length cDNA inserts (Carninci and Hayashizaki, 1999; Carninci et al., 2000). To assess the quality of the library and guide the selection of clones for full-length sequencing, the 5' ends of over 14,000 clone inserts were sequenced. BLAST (Altschul et al., 1997) searches against the public protein databases indicated that about 80% of the 11,116 high-quality EST's obtained showed significant homology to existing entries. More than 80% of these extended further upstream than the methionine start codon of their homolog in the databases. This indicated that the riken1 library indeed contains an extraordinary high percentage of full-length cDNA inserts. Only clones whose EST's showed significant BLAST matches against the public protein databases and covered the methionine start codon of their homolog were considered for full-length sequencing, as evolutionarily conserved genes are of the highest interest for the DT40 research community. The remaining EST's were clustered to remove duplicates corresponding to the same gene. In addition, EST's corresponding to already known chicken genes in the public databases were removed. Finally, the plasmids corresponding to 2,796 EST's were chosen for full insert sequencing by bidirectional primer walks. Once the end of the walks had been reached, the sequences of the full-length cDNA inserts were assembled. From the BLAST search results, the most likely methionine start codon was assigned to each sequence. About 15% of the cDNA sequences showed evidence for premature frame shifts in the form of short ORF's and stretches of conserved sequence in a different frame further 3'. If overlapping EST's were found in the public databases, the cDNA sequences were edited to correct the likely reverse transcription error, otherwise these sequences were discarded.

The riken1 cDNA's were compared with the collection of predicted chicken transcripts downloaded from the Ensembl ftp (Birney et al., 2004) site using the BLASTN program. Results of these searches have been used to link bursal cDNA's to the chicken genome sequence annotated by the ENSEMBL system. BLASTP software was used to compare translated ORF's with the protein sequences stored in the UniProt database (Apweiler et al., 2004). Functional domains were assigned by comparing riken1 cDNA's with sequence profiles representing Pfam domains (Bateman et al., 2004). This comparison was performed with RPSBLAST software (e-value cut-off of 10^{-6}) run on the binary database files downloaded from the National Center for Biotechnology Information (NCBI). Functional classes were assigned according to Pfam to GO (The Gene Ontology Consortium,

2004) mapping provided by the InterPro database (Mulder et al., 2003). The XML information exchange standard was used to interface the BLAST program outputs with the FOUNTAIN package.

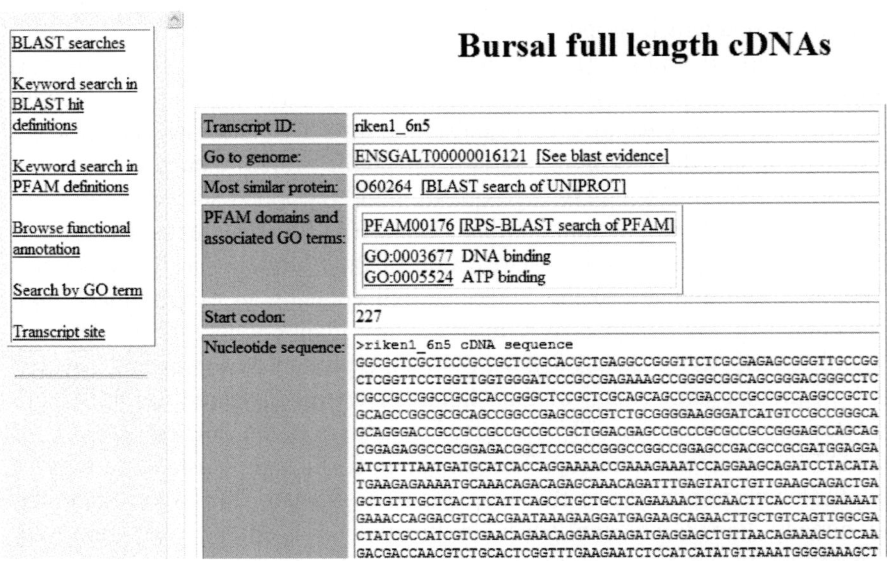

Figure 3-1. Example page of the DT40 website. The left frame lists basic functionalities of the system. The right frame presents information about single full-length cDNA.

1.1.2   cDNA page

An example page of the DT40 website, representing single full-length cDNA is shown in figure 1. Every cDNA in the database is linked to the appropriate region of the chicken genome sequence in the ENSEMBL system. Results of the BLASTN searches that have been used to create these links are also available. By following the link to ENSEMBL one can access information about predicted intron-exon structure of the gene and its genomic context. Additionally, ENSEMBL calculates gene orthology relationships between protein-coding gene sets of different genomes and following the link will display orthologous mapping of the chicken gene of interest.

The cDNA clones are also annotated by two other sequence similarity searches. First, the translated ORF (start codon listed on the cDNA page) is used in a BLASTP query of the UniProt database, a most comprehensive resource for protein sequences and functional information. The cDNA page provides links to the most similar UniProt entry and BLASTP output. The ORF sequence is also used as a query in an RPS-BLAST search of the library of PFAM protein domains. The cDNA page provides links to the alignment of the ORF with the PFAM domains and appropriate PFAM entries. The page also lists Gene Ontology terms associated with the domains detected within the cDNA sequence.

The cDNA page lists the full-length cDNA sequence, the 5' and 3'UTR sequences and the translated ORF. The sequences are displayed in FASTA format so they can be easily used as inputs for web-based and stand alone sequence analysis programs.

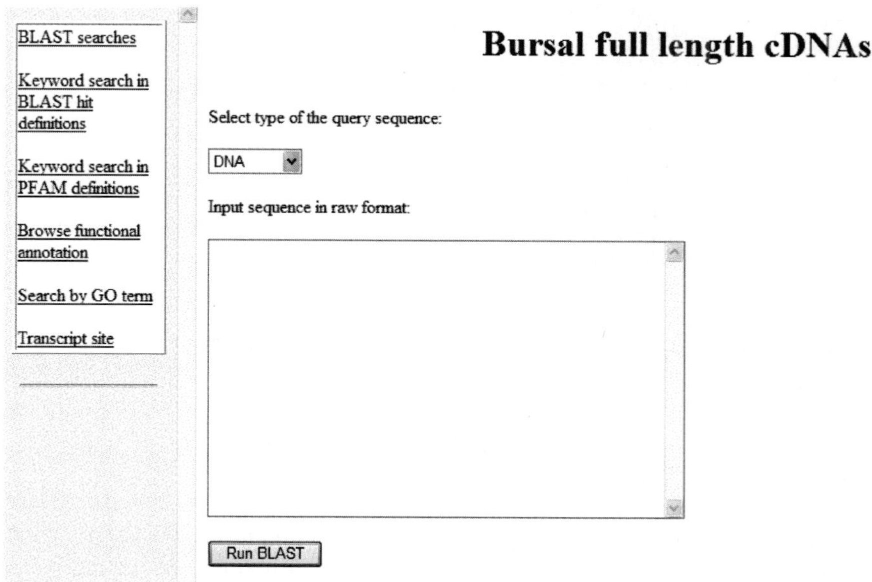

Figure 3-2. BLAST interface of DT40 web site.

1.1.3   Blast and keyword searches

The DT40 website implements both sequence similarity and keyword searches of the collection of annotated full-length cDNA's and EST's. A database can be searched by BLAST program (Figure 2) using either nucleic acid or amino acid sequences as the query. There are two ways in which sequence annotations can be searched by keywords. The keyword search can search the annotation lines of the BLAST output files, which are stored in the database for every full-length cDNA, or the annotation of the PFAM domains assigned to the full-length cDNA's. The database can also be searched by Gene Ontology terms.

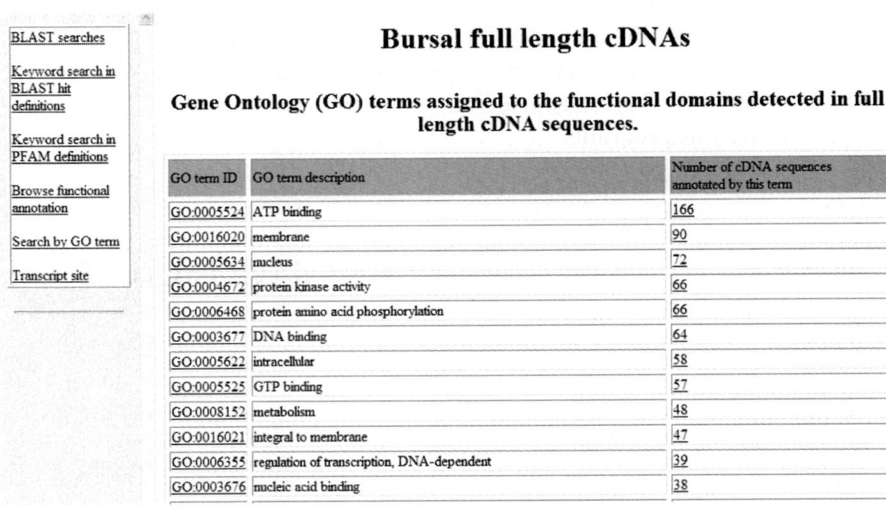

Bursal full length cDNAs

Gene Ontology (GO) terms assigned to the functional domains detected in full length cDNA sequences.

GO term ID	GO term description	Number of cDNA sequences annotated by this term
GO:0005524	ATP binding	166
GO:0016020	membrane	90
GO:0005634	nucleus	72
GO:0004672	protein kinase activity	66
GO:0006468	protein amino acid phosphorylation	66
GO:0003677	DNA binding	64
GO:0005622	intracellular	58
GO:0005525	GTP binding	57
GO:0008152	metabolism	48
GO:0016021	integral to membrane	47
GO:0006355	regulation of transcription, DNA-dependent	39
GO:0003676	nucleic acid binding	38

Figure 3-3. Functional annotation summary. For every GO term the number of sequences in which it occurs is listed. The numbers are linked to the appropriate sequence lists. GO terms are linked to Gene Ontology system.

1.1.4 Functional annotation summary

All full-length cDNA's were compared to the Pfam database, which stores sequence profiles representing functional domains. Subsequently, we have used the Gene Ontology (GO) annotation of Pfam domains provided by the InterPro database to assign functional descriptors to the domains detected in our sequences. It is important to note that the assignment of a GO term to a given cDNA sequence indicates only the presence of a functional domain rather than an orthologous relationship to other genes annotated by the term. Determination of orthologous relationships is best done at the level of whole-genome comparisons and could be accessed by following the links to the ENSEMBL system. The summary of Gene Ontology terms is displayed on a dedicated page of the DT40 web site (Figure 3). For every GO term, the number of occurrences of this term within the collection is listed thus giving an overview of the functional coverage of the database. The functional annotation summary may be used to browse the database in a search for interesting candidates for gene inactivation experiments.

1.1.5   SAGE sequences

In addition to full-length cDNA's and EST's, the DT40 web site contains a collection of SAGE sequences (Velculescu, et al., 1995; Madden, et al., 2000; Saha et al., 2002; Wahl and Caldwell, et al., 2004). Two SAGE libraries, one named busage and the other dt40sage, were made from the bursa of Fabricius and from DT40 cells using the Long SAGE technique which generates tags of 21 nucleotides in length and therefore decreases the likelihood of ambiguous matches. Of the 129,568 tags collected, about equal numbers were derived from the busage and the dt40sage libraries respectively. In total 38,212 unitags were derived from the SAGE tags of both libraries. SAGE tags having matches to full-length cDNA's, EST's or Ensembl predicted genes of the chicken genome have been identified and the collections have been hyper-linked. Subsequently, the number of occurrences of each SAGE tag in the two libraries has been compared. To evaluate the significance of the SAGE tag count differences between the libraries for each unitag, we used Fishers exact test (Benjamini and Hochberg, 1995) since it is most easy to use, has exact size and does not require specifying hyper-parameters. As usual, no method to account for multiple testing was used so p-values were simply used as a convenient tool to rank the unitags. Figure 4 shows a page of DT40 web site that lists all the tags ordered according to statistical significance of the count difference between the two libraries. All SAGE tags are also available for download.

General Information					
			SAGE unitags ordered by significance		
Download	1) Hits	busage = 1179	dt40sage = 22	0.0	sageUnitag_37270
Order SAGE tags by significance	2) Hits	busage = 1	dt40sage = 195	0.0	sageUnitag_37156
	3) Hits	busage = 127	dt40sage = 2	1.6128968E-34	sageUnitag_40568
	4) Hits	busage = 118	dt40sage = 1	1.1394287E-33	sageUnitag_40858
Transcript site	5) Hits	busage = 8	dt40sage = 110	8.697209E-26	sageUnitag_37846
	6) Hits	busage = 96	dt40sage = 287	1.1610697E-24	sageUnitag_37137
	7) Hits	busage = 3	dt40sage = 90	5.1857464E-24	sageUnitag_37996
	8) Hits	busage = 333	dt40sage = 624	1.8972513E-23	sageUnitag_37141
	9) Hits	busage = 15	dt40sage = 116	8.1047766E-22	sageUnitag_37286
	10) Hits	busage = 394	dt40sage = 162	8.696001E-22	sageUnitag_37332

Figure 3-4 SAGE tags ordered by significance of the count difference between busage and DT40 sage library.

1.2   Web based genome resources

As personal computing and the internet expanded, so did the nearly unlimited sources of information and applications for the molecular biology community. One of the first and best known sources for collecting and categorizing this wealth of information and services was Pedro's Bio-Molecular Research Tools (Harper, 1995). The website was run by Pedro Maldonado Coutinho, a student at Iowa State University of Science and Technology. The fate of Pedro's site, it has not been updated since 1995, demonstrates the extremely dynamic nature of the World Wide Web. It is for that reason that this will not be a laundry list of websites that are ever changing, but more a focus on what is available for the DT40 researcher. There are however several publications worth noting for current lists of resources as they may continue to be updated as part of a special issue series by the journal Nucleic Acid Research: *The Bioinformatics Links Directory: a compilation of molecular biology web servers* (Fox et al., 2005: Web Server issue) and *Database resources of the National Center for Biotechnology Information* (Wheeler et al., 2005: Database issue).

1.2.1   Comparative and analytical sequence tools

Ever since the first nucleotide sequence database was established in 1980 by EMBL (Hamm and Cameron, 1986), the need to accurately scan the stored information was immediately recognized. An algorithm published in 1983 by Wilbur and Lipman was the basis for the later described FastP and subsequent FastA programs for sequence similarity searches. To increase the sensitivity of the searches, a time consuming optimization methods described by Needleman-Wunsch (Needleman and Wunsch 1970) and Smith-Waterman (Smith-Waterman 1981) were also used. As the information within all the available sequence databases grew, the need for speed became apparent. Blast, **B**asic **L**ocal **A**lignment **S**earch **T**ool, has become the first tool of choice when investigating newly cloned and sequenced genes or targeting specific genes of interest. The Blast program debuted in 1990, five years after FASTA, but was much faster (Altschul et al., 1990; Pearson, 1991). The software behind Blast is an algorithm employing a heuristic search method. This allows for speed, but is less sensitive. Both the FastA and the BLAST programs have user controllable variables that can help define its application, but without a complete understanding behind the algorithms and settings, it is assumed that the majority of users simply stick with the default settings. As the number of genomes sequenced has increased, a new comparative analysis tool that was developed by Jim Kent (Kent, 2002), Genome BLAT (The BLAST-Like Alignment Tool), has proven to be quite useful. This program has been made freely available for internet use on The University of California Santa Cruz's Genome Bioinformatics Department's Genome Browser website and it is also available for download to run locally. This new program is a powerful tool for aligning a known sequence, complete or partial, to the source organism's genome in order to obtain its predicted or known intron-exon structure and chromosomal location or to identify possible homologs across species. The results returned link the information to experimental and in-silico references held within other databases previously mentioned and to Single Nucleotide Polymorphism (SNP) information supplied by the Beijing Genomics Institute (International Chicken Polymorphism Map Consortium, 2004).

1.2.2   Alignment and annotation tools

Medium to large scale sequencing projects are no longer the sole domain of large institutes or well funded labs as the costs are no longer as restrictive as they once were. Smaller ambitious labs are now taking on a major role in sequencing. This is especially true of sequencing EST's and more importantly full length cDNA's. Although the costs have been reduced, the reams of data produced can be daunting to process. Even a small database of EST's can hold many millions of letters. There are numerous online applications for the individual researcher, but this would be too cumbersome to process thousands of sequences manually. As with the evolution of all things computer, so it goes with the available software to semi or fully automatically accomplish the task of aligning bulk sequence data, building contiguous sequences and producing full length cDNA's ready for annotation. One of the best known packages of software to accommodate these tasks is the Staden Package (Staden, et al., 2001). However funding for the original group of developers of the Staden Package was stopped and at the time of this writing, only one developer is listed as still actively working on this software. Still, the Staden Package beta-test release v1.6.0 came out on the 15th of July, 2005 more than two years after discontinuation of funding by the MRC. As this software package is so popular and flexible, many add on modules have been individually developed that can enhance the overall performance of the Staden Package. These include the CAP2/3 and Phrap sequence assembly engines and an ABI base calling program called Phred that assigns quality scores (Huang, 1992; Huang, 1996; Huang and Madan, 1999; Ewing et al., 1998). Adding to their programs Phred and Phrap, the labs of Phil Green and Dave Gordon and others have continued developing programs such as cross match, swat, consed and autofinisher to enable enhanced flexibility, better automation and higher throughput (Ewing and Green 1998, Ewing et al., 1998). These program packages can be run together in various configurations to the specifications and needs of the individual laboratory.

For large scale projects, annotation of cDNA's needs to be automated and performed thoroughly as possible. The FOUNTAIN software package (Buerstedde and Prill, 2001) that backs the DT40 website includes just such an annotation package as previously described. Prior to that implementation, annotation in smaller labs meant manually performing BLAST searches of individual databases and cross comparing the results, a tedious and time consuming job to say the least. As more sequence data has come online, multi BLAST based automated programs have started to be developed that cross references various datasets for a more thorough output and in various formats such as that produced by FOUNTAIN. Many of the most well known labs, TIGR, Ensembl, ExPASy and NCBI to name a few, have started to make various programs available on line for the smaller labs as well.

Many of these web-based and downloadable programs are free or are free to academics. Obviously, not all possible software has been listed here and what is listed may not be there tomorrow. Whichever software or resource one uses, it is best to become completely familiar with all the functions in order to unlock the real potential in its use and to increase ones understanding of the meaning of any results obtained.

REFERENCES

Abdrakhmanov I, Lodygin D, Geroth P, Arakawa H, Law A, Plachy J, Korn B, Buerstedde JM: A large database of chicken bursal ESTs as a resource for the analysis of vertebrate gene function. Genome Res., 2000, 10:2062-2069.

Altschul SF, Gish W, Miller W, Myers EW, Lipman DJ. Basic local alignment search tool. J Mol Biol. 1990 Oct 5;215(3):403-10.

Altschul SF, Madden TL, Schaffer AA, Zhang J, Zhang Z, Miller W, Lipman DJ: Gapped BLAST and PSI-BLAST: a new generation of protein database search programs. Nucleic Acids Res 1997, 25:3389-3402.

Apweiler R, Bairoch A, Wu CH, Barker WC, Boeckmann B, Ferro S, Gasteiger E, Huang H, Lopez R, Magrane M, et al.: UniProt: the Universal Protein knowledgebase. Nucleic Acids Res 2004, 32 (Database issue):D115-D119.

Arakawa H, Buerstedde JM: Immunoglobulin gene conversion: insights from bursal B cells and the DT40 cell line. Dev Dyn 2004, 229:458-464.

Arakawa H, Saribasak H, Buerstedde JM. Activation-induced cytidine deaminase initiates immunoglobulin gene conversion and hypermutation by a common intermediate. PLoS Biol. 2004 Jul;2(7):E179. Epub 2004 Jul 13.

Bateman A, Coin L, Durbin R, Finn RD, Hollich V, Griffiths-Jones S, Khanna A, Marshall M, Moxon S, Sonnhammer EL, et al.: The Pfam protein families database. Nucleic Acids Res 2004, 32 (Database issue):D138-D141.

Benjamini Y, Hochberg Y: Controlling the False Discovery Rate – A Practical and Powerful Approach to Multiple Testing. J Roy Stat Soc B Met 1995, 57(1):289-300.

Birney E, Andrews TD, Bevan P, Caccamo M, Chen Y, Clarke L, Coates G, Cuff J, Curwen V, Cutts T, et al.: An overview of Ensembl. Genet Res 2004, 14:925-928.

Boardman PE, Sanz-Ezquerro J, Overton IM, Burt DW, Bosch E, Fong, WT, Tickle C, Brown WR, Wilson SA, Hubbard SJ: A comprehensive collection of chicken cDNAs. Curr. Biol., 2002, 12:1965-1969.

Buerstedde JM, Takeda S: Increased ratio of targeted to random integration after transfection of chicken B cell lines. Cell 1991, 67:179-188.

Buerstedde JM, Prill F: FOUNTAIN: a JAVA open-source package to assist large sequencing projects. BMC Bioinformatics 2001, 2:6.

Carninci P, Hayashizaki Y: High-efficiency full-length cDNA cloning. Methods Enzymol 1999, 303:19-44.

Carninci P, Shibata Y, Hayatsu N, Sugahara Y, Shibata K, Itoh M, Konno H, Okazaki Y, Hayashizaki Y: Normalization and subtraction of cap-trapper-selected cDNAs to prepare full-length cDNA libraries for rapid discovery of new genes. Genome Res., 2000, 10:1617-1630.

Ewing B, Hillier L, Wendl M, Green P: Basecalling of automated sequencer traces using phred. I. Accuracy assessment. Genome Res. 1998 Mar;8(3):175-85.

Ewing B, Green P: Basecalling of automated sequencer traces using phred. II. Error probabilities. Genome Res. 1998 Mar;8(3):186-94.

Fox JA, Butland SL, McMillan S, Campbell G, Ouellette BF. The Bioinformatics Links Directory: a compilation of molecular biology web servers. Nucleic Acids Res. 2005 Jul 1;33(Web Server issue):W3-24.

Gordon D, Abajian C, Green P. Consed: a graphical tool for sequence finishing. Genome Res. 1998 Mar;8(3):195-202.

Gordon D, Desmarais C, Green P. Automated finishing with autofinish. Genome Res. 2001 Apr;11(4):614-25.

Hamm GH, Cameron GN. The EMBL data library. Nucleic Acids Res. 1986 Jan 10;14(1): 5-9.

Harper R. World Wide Web resources for the biologist. Trends Genet. 1995 Jun;11(6): 223-8.

Hillier LW, Miller W, Birney E, Warren W, Hardison RC, Ponting CP, Bork P, Burt DW, Groenen MA, Delany ME, Dodgson JB, Chinwalla AT, Cliften PF, Clifton SW, Delehaunty KD, Fronick C, Fulton RS, Graves TA, Kremitzki C, Layman D, Magrini V, McPherson JD, Miner TL, Minx P, Nash WE, Nhan MN, Nelson JO, Oddy LG, Pohl CS, Randall-Maher J, Smith SM, Wallis JW, Yang SP, Romanov MN, Rondelli CM, Paton B, Smith J, Morrice D, Daniels L, Tempest HG, Robertson L, Masabanda JS, Griffin DK, Vignal A, Fillon V, Jacobbson L, Kerje S, Andersson L, Crooijmans RP, Aerts J, van der Poel JJ, Ellegren H, Caldwell RB, Hubbard SJ, Grafham DV, Kierzek AM, McLaren SR, Overton IM, Arakawa H, Beattie KJ, Bezzubov Y, Boardman PE, Bonfield JK, Croning MD, Davies RM, Francis MD, Humphray SJ, Scott CE, Taylor RG, Tickle C, Brown WR, Rogers J, Buerstedde JM, Wilson SA, Stubbs L, Ovcharenko I, Gordon L, Lucas S, Miller MM, Inoko H, Shiina T, Kaufman J, Salomonsen J, Skjoedt K, Wong GK, Wang J, Liu B, Wang J, Yu J, Yang H, Nefedov M, Koriabine M, Dejong PJ, Goodstadt L, Webber C, Dickens NJ, Letunic I, Suyama M, Torrents D, von Mering C, Zdobnov EM, Makova K, Nekrutenko A, Elnitski L, Eswara P, King DC, Yang S, Tyekucheva S, Radakrishnan A, Harris RS, Chiaromonte F, Taylor J, He J, Rijnkels M, Griffiths-Jones S, Ureta-Vidal A, Hoffman MM, Severin J, Searle SM, Law AS, Speed D, Waddington D, Cheng Z, Tuzun E, Eichler E, Bao Z, Flicek P, Shteynberg DD, Brent MR, Bye JM, Huckle EJ, Chatterji S, Dewey C, Pachter L, Kouranov A, Mourelatos Z, Hatzigeorgiou AG, Paterson AH, Ivarie R, Brandstrom M, Axelsson E, Backstrom N, Berlin S, Webster MT, Pourquie O,

Reymond A, Ucla C, Antonarakis SE, Long M, Emerson JJ, Betran E, Dupanloup I, Kaessmann H, Hinrichs AS, Bejerano G, Furey TS, Harte RA, Raney B, Siepel A, Kent WJ, Haussler D, Eyras E, Castelo R, Abril JF, Castellano S, Camara F, Parra G, Guigo R, Bourque G, Tesler G, Pevzner PA, Smit A, Fulton LA, Mardis ER, Wilson RK; International Chicken Genome Sequencing Consortium. Sequence and comparative analysis of the chicken genome provide unique perspectives on Vertebrate evolution. Nature. 2004 Dec 9;432(7018):695-716. Erratum in: Nature. 2005 Feb 17;433(7027):777.

Huang X. A contig assembly program based on sensitive detection of fragment overlaps. Genomics. 1992 Sep;14(1):18-25.

Huang X. An improved sequence assembly program. Genomics. 1996 Apr 1;33(1):21-31.

Huang X, Madan A. CAP3: A DNA sequence assembly program. Genome Res. 1999 Sep;9(9):868-77.

Hubbard SJ, Grafham DV, Beattie KJ, Overton IM, McLaren SR, Croning MD, Boardman PE, Bonfield JK, Burnside J, Davies RM, Farrell ER, Francis MD, Griffiths-Jones S, Humphray SJ, Hyland C, Scott CE, Tang H, Taylor RG, Tickle C, Brown WR, Birney E, Rogers J, Wilson SA. Transcriptome analysis for the chicken based on 19,626 finished cDNA sequences and 485,337 expressed sequence tags. Genome Res. 2005 Jan;15(1): 174-83.

Kent WJ. BLAT–the BLAST-like alignmei t tool. Genome Res. 2002 Apr;12(4):656-64.

Lee WH, Vega VB. Heterogeneity detector: finding heterogeneous positions in Phred/Phrap assemblies. Bioinformatics. 2004 Nov 1;20(16):2863-4. Epub 2004 May 6.

Lipman DJ, Wilbur WJ, Smith TF, Waterman MS. On the statistical significance of nucleic acid similarities. Nucleic Acids Res. 1984 Jan 11;12(1 Pt 1):215-26.

Madden SL, Wang CJ, Landes G: Serial analysis of gene expression: from gene discovery to target identification. Drug Discov Today 2000, 5:415-425.

Majoros WH, Pertea M, Antonescu C, Salzberg SL. Glimmer M, Exonomy and Unveil: three ab initio eukaryotic genefinders. Nucleic Acids Res. 2003 Jul 1;31(13):3601-4.

Mulder NJ, Apweiler R, Attwood TK, Bairoch A, Barrell D, Bateman A, Binns D, Biswas M, Bradley P, Bork P, et al.: The InterPro Database, 2003 brings increased coverage and new features. Nucleic Acids Res 2003, 31:315-318.

Needleman SB, Wunsch CD. A general method applicable to the search for similarities in the amino acid sequence of two proteins. J Mol Biol. 1970 Mar;48(3):443-53.

Pearson WR. Searching protein sequence libraries: comparison of the sensitivity and selectivity of the Smith-Waterman and FASTA algorithms. Genomics. 1991 Nov;11(3):635-50.

Pesole G, Liuni S, Grillo G, Saccone C: Structural and compositional features of untranslated regions of eukaryotic mRNAs. Gene 1997, 205:95-102.

The Gene Ontology Consortium.: The Gene Ontology (GO) database and informatics resource. Nucleic Acids Res 2004, 32 (Database issue):D258-D261.

Saha S, Sparks AB, Rago C, Akmaev V, Wang CJ, Vogelstein B, Kinzler KW, Velculescu VE: Using the transcriptome to annotate the genome. Nat Biotechnol 2002, 20:508-512.

Smith TF, Waterman MS. Identification of common molecular subsequences. J Mol Biol. 1981 Mar 25;147(1):195-7.

Staden R, Judge DP, Bonfield JK: Sequence assembly and finishing methods. Methods Biochem Anal 2001, 43:303-322.

Stabenau A, McVicker G, Melsopp C, Proctor G, Clamp M, Birney E. The Ensembl core software libraries. Genome Res. 2004 May;14(5):929-33. Review.

Velculescu VE, Zhang L, Vogelstein B, Kinzler KW: Serial analysis of gene expression. Science 1995, 270:484-487.

Wahl MB, Caldwell RB, Kierzek AM, Arakawa H, Eyras E, Hubner N, Jung C, Soeldenwagner M, Cervelli M, Wang YD, Liebscher V, Buerstedde JM. Evaluation of the chicken transcriptome by SAGE of B cells and the DT40 cell line. BMC Genomics. 2004 Dec 21;5(1):98.

Wheeler DL, Barrett T, Benson DA, Bryant SH, Canese K, Church DM, DiCuccio M, Edgar R, Federhen S, Helmberg W, Kenton DL, Khovayko O, Lipman DJ, Madden TL, Maglott DR, Ostell J, Pontius JU, Pruitt KD, Schuler GD, Schriml LM, Sequeira E, Sherry ST, Sirotkin K, Starchenko G, Suzek TO, Tatusov R, Tatusova TA, Wagner L, Yaschenko E. Database resources of the National Center for Biotechnology Information. Nucleic Acids Res. 2005 Jan 1;33(Database issue):D39-45.

Wilbur WJ, Lipman DJ. Rapid similarity searches of nucleic acid and protein data banks. Proc Natl Acad Sci U S A. 1983 Feb;80(3):726-30.

Wong GK, Liu B, Wang J, Zhang Y, Yang X, Zhang Z, Meng Q, Zhou J, Li D, Zhang J, Ni P, Li S, Ran L, Li H, Zhang J, Li R, Li S, Zheng H, Lin W, Li G, Wang X, Zhao W, Li J, Ye C, Dai M, Ruan J, Zhou Y, Li Y, He X, Zhang Y, Wang J, Huang X, Tong W, Chen J, Ye J, Chen C, Wei N, Li G, Dong L, Lan F, Sun Y, Zhang Z, Yang Z, Yu Y, Huang Y, He D, Xi Y, Wei D, Qi Q, Li W, Shi J, Wang M, Xie F, Wang J, Zhang X, Wang P, Zhao Y, Li N, Yang N, Dong W, Hu S, Zeng C, Zheng W, Hao B, Hillier LW, Yang SP, Warren WC, Wilson RK, Brandstrom M, Ellegren H, Crooijmans RP, van der Poel JJ, Bovenhuis H, Groenen MA, Ovcharenko I, Gordon L, Stubbs L, Lucas S, Glavina T, Aerts A, Kaiser P, Rothwell L, Young JR, Rogers S, Walker BA, van Hateren A, Kaufman J, Bumstead N, Lamont SJ, Zhou H, Hocking PM, Morrice D, de Koning DJ, Law A, Bartley N, Burt DW, Hunt H, Cheng HH, Gunnarsson U, Wahlberg P, Andersson L, Kindlund E, Tammi MT, Andersson B, Webber C, Ponting CP, Overton IM, Boardman PE, Tang H, Hubbard SJ, Wilson SA, Yu J, Wang J, Yang H; International Chicken Polymorphism Map Consortium. A genetic variation map for chicken with 2.8 million single-nucleotide polymorphisms. Nature. 2004 Dec 9;432(7018):717-22.

CHROMOSOME ENGINEERING IN DT40 CELLS AND MAMMALIAN CENTROMERE FUNCTION

William R.A. Brown[1], Margaret C.M. Smith[2], Felix Dafhnis-Calas, Sunir Malla and Zhengyao Xu
Institute of Genetics, Queens Medical Centre, University of Nottingham, Nottingham, NG7 2UH, UK.

Abstract: Chromosome engineering is the term given to procedures which modify the long range structure of a chromosome by homologous and site specific recombination or by telomere directed chromosome breakage. DT40 cells are uniquely powerful for chromosome engineering because mammalian chromosomes may be moved into them, efficiently modified and then moved back into a mammalian cell lines (Dieken et al., 1996). The high rate of sequence targeting seen in DT40 cells carrying human chromosomes is necessary but not sufficient for chromosome engineering. The ability to either delete or introduce long tracts of DNA subsequent to a sequence targeting reaction depends upon the use of site specific recombinases. We have made important progress in the development of this technology in the past few years and much of this review will be used to describe this work.

Key words: centromere, site-specific recombination, mini-chromosome vector

1.   INTRODUCTION

The use of DT40 cells as hosts for chromosome engineering studies has focussed on three areas. The first of these has been the development of artificial chromosome vectors. About fifteen years ago it was supposed that mammalian chromosomes would make useful vectors for introducing genes into mammalian cells and whole animals (Brown, 1992). Viral vectors have limited sequence capacity and, in the case of retroviral vectors, are insertional mutagens. Chromosome based vectors would avoid these problems. Mini-

J.-M. Buerstedde and S. Takeda (eds.), Reviews and Protocols in DT40 Research, 39–48.

chromosome vectors have been developed. In some cases these were simply generated by radiation induced fragmentation of human chromosomes (Tomizuka et al., 1997) (Tomizuka et al., 2000) and in other cases by telomere directed fragmentation of human chromosomes in cultured vertebrate cells (Shen et al., 1997) (Shen et al., 2000). Chromosome engineering in DT40 cells has been used for the manipulation of both of these types of vector (Kuroiwa et al., 2000). Mini-chromosomes produced by radiation induced fragmentation of human chromosomes have been used to produce human monoclonal antibodies in mice (Tomizuka et al., 2000) and other vertebrates. There are only a few proteins that need be produced in whole animals and require large tracts of DNA in order to do so and consequently this technique has not been used for other protein production problems in biotechnology. Despite this limitation the industrial production of human monoclonal antibodies is an important achievement and alone justifies some of the early hopes for mammalian artificial chromosomes. However moving engineered chromosomes between cells continues to rely upon the somatic cell genetic technique of micro-cell fusion which is very inefficient and so artificial chromosomes have not found any use as vectors in fundamental biology.

2.   DT40 VERSUS MAMMALIAN SYSTEMS

It has been argued that artificial chromosomes may be useful for engineering human stem cells (Kakeda et al., 2005; Katoh et al., 2004; Ren et al., 2005). However it is hard to see what would justify the effort required to use an artificial chromosome vector when most genes are small enough to be moved into stem cells as cDNA constructs using retroviral vectors. These considerations would suggest that mammalian artificial chromosome vectors are likely to have only limited application in the immediate future.

2.1   DT40 and chromosomal engineering challenges

The outstanding question posed by chromosome engineering in DT40 cells when applied to basic problems is; why bother? Mammalian chromosomes are likely to be no different from the chromosomes of single celled eukaryotes in their fundamental properties. Single celled eukaryotes offer the advantages of a short generation time and a powerful forward genetics. The only sort of fundamental problem that justifies the effort required to undertake chromosome engineering work in DT40 cells is therefore one that cannot be addressed in a single celled eukaryote. The range of fundamental problem that are worthwhile to address using chromosome engineering in DT40 cells is therefore extremely limited. In fact there are really only two problems that have used chromosome engineering in DT40 as a methodology. The first has been to use DT40 cells to investigate the far upstream sequence requirements for accurate gene expression. This set of experiments was carried out using the human β-globin (Dieken et al., 1996) and serpin (Marsden and Fournier, 2003) loci as model systems. The second has been to understand the sequence requirements for centromere function. This is the goal of the two laboratories that have used DT40 cells most extensively for chromosome engineering (Mills et al., 1999; Spence et al., 2002; Yang et al., 2000) and will be the focus of the rest of this review.

2.2   The problem of mammalian centromeric DNA

In almost all eukaryotes the centromeric DNA is a tandemly repeated sequence. In humans and other primates the centromeric repeats belong to a thoroughly studied family of repeated sequences called alphoid DNA with a unit repeat length of 170bp. In all metazoons in which the centromeric DNA is organized in this way the centromeric DNA poses a paradox: although this sequence mediates an evolutionarily conserved function it is one of the most rapidly evolving in the genome. Three hypotheses have been advanced to resolve this paradox:

1. The rapid evolution of centromeric DNA is apparent rather than real because the sequences found at the centromeres of different eukaryotes form similar tertiary or higher order structures despite their differences and it is this higher order structure that is recognized by the kinetochore proteins.

2. Possibly the most widely accepted is that the rapid evolution of centromeric DNA reflects evolutionary drift arising as a consequence of the tandemly repeated nature of the sequence compensating for the emergence of novel repeats with lower functionality. The rapid evolution of the centromeric DNA may also arise as a result of accelerated drift resulting from fluctuations in the selective forces acting on the sequence as a result of variation in the number of cells in the organism or individuals in the population.

3. Centromeric DNA evolves rapidly because it is subject to meiotic drive in the female germ line as a result of an asymmetric first meiotic division selecting for sequences that attach to the side of the spindle that ends up in the gamete (Malik et al., 2002) (Malik et al., 2002). This model is unsatisfactory because it is not clear what drives the continuing evolution of new sequences. In particular why should suppressors arise which favour the continuous emergence of new sequences? Thus once a particular sequence has emerged that

binds most favourably to the side of the female MI spindle that ends up in the egg there is no obvious reason why it should subsequently be replaced. One possibility is that a sequence that binds to the appropriate side of the MI female spindle is somehow antagonistic to successful male meiosis or to somatic mitosis. Why this should be so is unclear but a set of testable hypotheses would be useful if a drive theory is to be considered realistically.

Distinguishing between these three hypotheses requires experiment. It would be helpful to know whether there are differences in the ability of different centromeric DNA sequences to seed centromere formation and whether these differences are under selection. Such information might be useful in refining the various models and in establishing the extent to which each of the mechanisms is responsible for the rapid evolution of centromeric DNA. It might be wondered why this sort of experiment is necessary: mammalian chromosomes can be moved between cells of different species by somatic cell hybridization techniques and segregate quite accurately in heterospecific mono-chromosomal hybrids. Such experiments are however impossible to interpret properly because the chromosomes moved into the somatic hybrids have centromeres that are likely to be have been maintained by epigenetic processes. Thus their functionality in the hybrids reflects not their ability to function in the cells in question but their history. If we are to assay the functionality of a sequence it is necessary to establish a system in which the ability of any particular sequence to seed centromere formation de novo is measured.

Developing an assay for the ability of DNA to seed centromere formation has been an important goal for many laboratories for many years. The first real advance in this area was the demonstration, initially by Willard and colleagues (Harrington et al., 1997) and subsequently by Masumoto and colleagues (Ikeno et al., 1998) that transfection of human alphoid DNA into human HT1080 fibrosarcoma cells seeds the formation of mini-chromosomes in which the introduced alphoid DNA functions as a centromere. These experiments were justifiably regarded as a breakthrough but it has proved impossible to extend this approach. Centromere formation following transfection of centromeric DNA only occurs in human HT1080 cells and importantly the resulting mini-chromosomes have structures that are impossible to determine. Moreover the process occurs at efficiencies of 10-6/transfected cell making comparisons between different experiments statistically challenging. In retrospect the real importance of these experiments was to establish beyond argument that alphoid DNA was the functional centromeric DNA on human chromosomes. This may seem to be a small gain but for many years serious study of metazoan centromeres was inhibited by the prejudice held by yeast geneticists and cell biologists that the functional sequences at vertebrate centromeres were not composed of the repeats themselves but were embedded within the repeats. Such views ran counter to all that was well established about the evolution of tandemly repeated sequences and significantly impeded progress in the field. In summary the experiments of Willard and Masumoto represent the start of experimental analysis of metazoan centromeric DNA.

2.3   The use of DT40 cells to investigate mammalian centromere function

The ideal experimental approach for investigating the ability of a sequence to seed centromere function would allow one to investigate the ability of a defined sequence to seed centromere formation in a wide range of different vertebrate cell lines. In principle this could most easily be done by removing the pre-existing centromeric DNA to generate an acentric chromosome fragment and then introducing a defined test sequence. Unfortunately this strategy is impractical because it is not possible to maintain acentric chromosomes in most cell types. Thus in early telomere directed chromosome breakage experiments we, and others, were unable to recover chromosomes with less than about 90kb of centromeric DNA. Centromeres were found on all chromosome fragments and all fragments bigger than about 1Mb segregated accurately. In our experiments we were studying the alphoid DNA at the centromere of the human Y chromosome (Yang et al., 2000) while in the experiments of Farr and colleagues the results were very similar but the substrate was the X chromosome alphoid DNA (Spence et al., 2002). Interestingly fragments with 90kb tracts of alphoid DNA failed to form a centromere in mouse LA-9 cells (Shen et al., 2001) and consequently segregated very inaccurately in such cells. These fragments would thus seem to offer the possibility of assaying centromere function in mouse LA-9 cells although the interpretation of such experiments would be compromised by the presence of the limited amount of alphoid DNA on these chromosomes.

Given the demonstration that it is not possible to generate acentric chromosome fragments in DT40 cells we need to propose an alternative experiment. Such an alternative would be to substitute the pre-existing centromeric DNA with a range of test sequences of different sizes and sequence (Fig. 1A). How might this be done is explained in figure 1B. According to this proposal a chromosome is modified so that a candidate sequence is introduced adjacent to the pre-existing centromere on a chromosome in DT40 cells. The modified chromosome is then moved into cells of a variety of different species by somatic cell fusion techniques

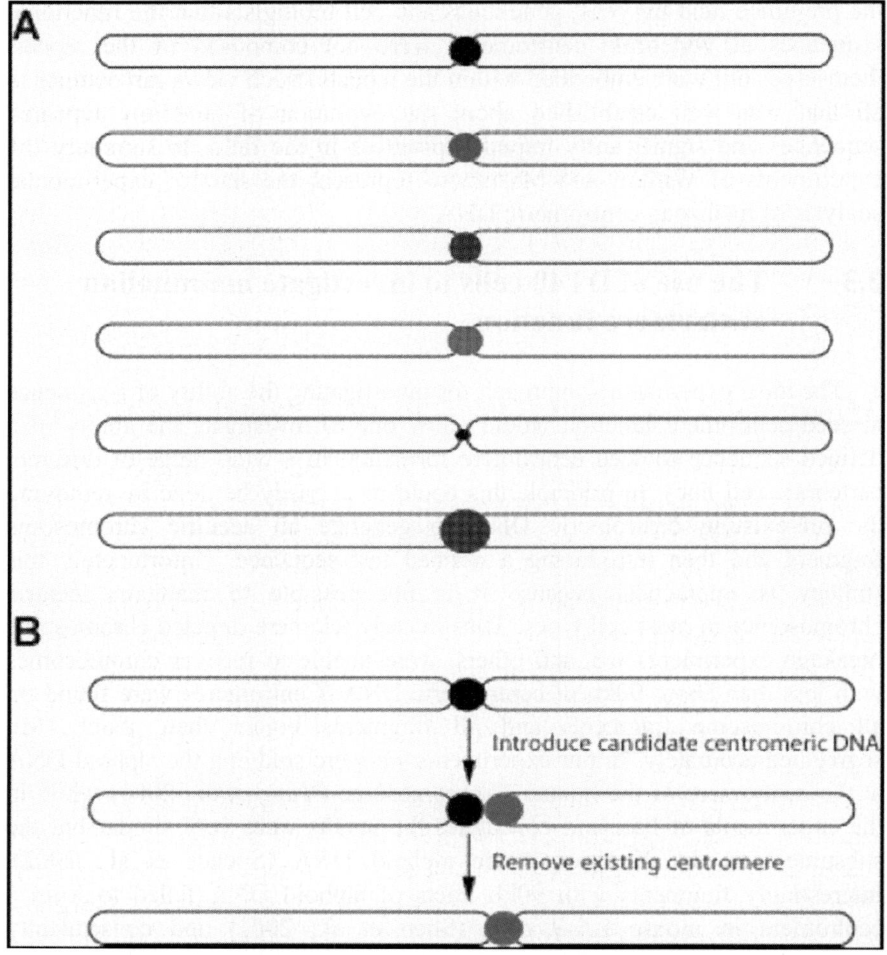

Figure 4-1. Exchanging sequences at the vertebrate centromere.

A. The ideal type of manipulation needed for the study of vertebrate centromeres is one
where the pre-existing centromeric DNA, represented by a solid disc is exchanged for a
variety of different sequences of varying sequence identity and size.

B. The methodological approach that we are attempting to achieve the goals set out in A.
A candidate centromeric sequence (red disc) is introduced near the pre-existing centromere
(black disc) and then the pre-existing centromere is removed. Both the construction of the
candidate centromeric DNA and removal of the pre-existing centromere depend upon the use
of unidirectional serine recombinases (see plate 6).

and then the candidate sequence assayed for function by excision of
the pre-existing centromere. This approach is practical because di-centric

chromosomes are stable in both human (Sullivan and Willard, 1998) and DT40 cells (Yang et al., 2000) if the two centromeres are within a few megabase pairs of one another. Implementation of this experimental approach however requires the development of a new set of chromosome engineering techniques that would allow firstly; the construction of a realistically sized centromeric sequence for assay and secondly; the ability to irreversibly excise the pre-existing centromere. We have spent the last five years developing such techniques. These may be of value to chromosome engineering projects in other organisms and we describe one of them that has been published below.

2.4   Using ϕC31 integrase to delete a human centromere

We also would like to use a unidirectional integrase to excise a human centromere and to demonstrate that of the two fragments produced the circular fragment containing a centromere is stable and the acentric linear fragment is unstable. This last point is important because it is not clear as to how frequently neo-centromeres form on vertebrate chromosomes but were this to be frequent event then it would make it impossible to assay centromeric DNA using the swap strategy outlined in Figure 1.

In order to address these questions we have targeted ϕC31 integrase attP and attB sites around the centromere of a human Y chromosome fragment (Fig. 2A) (Malla et al., 2005) contained in DT40 cells and then expressed the ϕC31 integrase in the resulting cells. The results were consistent with the outcome predicted in figure 2B; we were able to recover cells containing exclusively a small centric fragment and failed to isolate the complementary linear fragment lacking alphoid DNA. Thus these experiments established that the ϕC31 integrase can be used to remove a centromere from a human chromosome and that neo-centromere formation occurs infrequently. However these experiments revealed an unexpected complication: the deletion reaction did not go to completion because in about half of the cells the participating sites were damaged during the course of the reaction by deletions of sequences extending from the core of the attachment sites. This observation was interpreted to suggest that the site specific recombination intermediates had been damaged during an attempt at recombination and as a result had been unable to complete the reaction. Although DT40 cells maybe particularly efficient at identifying recombination intermediates as substrates for DNA repair establishing the degree to which this unproductive side reaction will prevent the use of the ϕC31 integrase for promoting long range re-arrangements in any cell type will require further work in different cell lines.

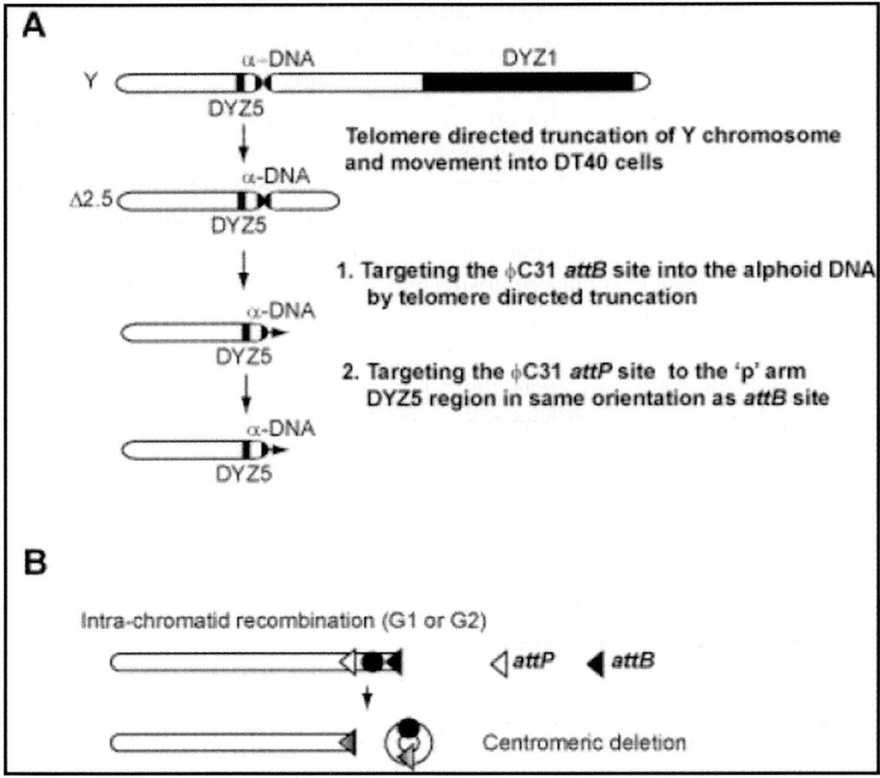

Figure 4-2. Deleting the centromere of the human Y chromosome using φC31 integrase (Malla et al., 2005).

A. Introducing attB and attP sites around the centromere of the human Y chromosome. The human Y chromosome isolated in a Chinese hamster ovary cell line was truncated by telomere directed chromosome breakage to yield a mini-chromosome termed Δ2.5. This mini-chromosome was moved into DT40 cells by micro-cell fusion. A further round of telomere directed chromosome breakage was used to target a φC31attB site to the centromeric array of alphoid DNA on the long arm side of the functional kinetochore. A subsequent targeting reaction placed an attP site within the DYZ5 array approximately 1Mb from the centromere in the same orientation as the attB site.

B. The cell line containing the Y chromosome derived mini-chromosome with the attachment sites flanking the centromere; generated as described in A, was stably transfected with the φC31 integrase expression construct which, as predicted, excised the centromere on a small circular fragment of approximately 1Mb in size. The residual linear fragment was not recovered. More complex products consistent with a site specific recombination reaction between sister chromatids were also recovered (Malla et al., 2005).

3.   THE FUTURE OF CHROMOSOME ENGINEERING IN DT40 CELLS

The study of vertebrate centromeric DNA has justified most chromosome engineering experiments that have used DT40 cells. The development of an approach that should enable experimental definition of the sequence requirements for centromere function in vertebrate cells should allow all of the major mechanistic questions posed by the sequence variation in centromeric DNA to be addressed at the DNA level and thus may allow this type of experiment to be brought to an end. If this hope is realized then what further utility will DT40 cells have as tools for chromosome engineering? It is possible that the testing of ideas about how centromeric DNA is recognized may benefit from a system where precise modification of both the centromeric DNA and of the genes involved in centromere assembly is easily achieved. If this type of experiment becomes attractive then it seems reasonable to suppose that DT40 cells and chromosome engineering may continue to play a role in the study of the centromere.

ACKNOWLEDGEMENTS

We thank the BBSRC (ExGen Initiative: EGH16108), EU FW5 (Genetics in a Cell line), Leverhulme Trust (F/00114M) and Nottingham University (Studentship to SKM) for funding.

REFERENCES

Brown, W. R. (1992). Mammalian artificial chromosomes. Curr Opin Genet Dev 2, 479-486.

Dieken, E. S., Epner, E. M., Fiering, S., Fournier, R. E. K., and Groudine, M. (1996). Efficient modification of human chromosomal alleles using recombination-proficient chicken human microcell hybrids. Nature Genetics 12, 174-182.

Harrington, J. J., VanBokkelen, G., Mays, R. W., Gustashaw, K., and Willard, H. F. (1997). Formation of de novo centromeres and construction of first-generation human artificial microchromosomes. Nature Genetics 15, 345-355.

Ikeno, M., Grimes, B., Okazaki, T., Nakano, M., Saitoh, K., Hoshino, H., McGill, N. I., Cooke, H., and Masumoto, H. (1998). Construction of YAC-based mammalian artificial chromosomes. Nature Biotechnology 16, 431-439.

Kakeda, M., Hiratsuka, M., Nagata, K., Kuroiwa, Y., Kakitani, M., Katoh, M., Oshimura, M., and Tomizuka, K. (2005). Human artificial chromosome (HAC) vector provides long-term therapeutic transgene expression in normal human primary fibroblasts. Gene Ther 12, 852-856.

Katoh, M., Ayabe, F., Norikane, S., Okada, T., Masumoto, H., Horike, S., Shirayoshi, Y., and Oshimura, M. (2004). Construction of a novel human artificial chromosome vector for gene delivery. Biochem Biophys Res Commun 321, 280-290.

Kuroiwa, Y., Tomizuka, K., Shinohara, T., Kazuki, Y., Yoshida, H., Ohguma, A., Yamamoto, T., Tanaka, S., Oshimura, M., and Ishida, I. (2000). Manipulation of human minichromosomes to carry greater than megabase-sized chromosome inserts. Nat Biotechnol 18, 1086-1090.

Malik, H. S., Vermaak, D., and Henikoff, S. (2002). Recurrent evolution of DNA-binding motifs in the Drosophila centromeric histone. Proc Natl Acad Sci USA 99, 1449-1454.

Malla, S., Dafhnis-Calas, F., Brookfield, J. F., Smith, M. C., and Brown, W. R. (2005). Rearranging the centromere of the human Y chromosome with phiC31 integrase. Nucleic Acids Res 33, 6101-6113.

Marsden, M. D., and Fournier, R. E. (2003). Chromosomal elements regulate gene activity and chromatin structure of the human serpin gene cluster at 14q32.1. Mol Cell Biol 23, 3516-3526.

Mills, W., Critcher, R., Lee, C., and Farr, C. J. (1999). Generation of an approximately 2.4 Mb human X centromere-based minichromosome by targeted telomere-associated chromosome fragmentation in DT40. Hum Mol Genet 8, 751-761.

Ren, X., Katoh, M., Hoshiya, H., Kurimasa, A., Inoue, T., Ayabe, F., Shibata, K., Toguchida, J., and Oshimura, M. (2005). A Novel Human Artificial Chromosome Vector Provides Effective Cell Lineage-Specific Transgene Expression in Human Mesenchymal Stem Cells. Stem Cells.

Shen, M. H., Mee, P. J., Nichols, J., Yang, J., Brook, F., Gardner, R. L., Smith, A. G., and Brown, W. R. (2000). A structurally defined mini-chromosome vector for the mouse germ line. Curr Biol 10, 31-34.

Shen, M. H., Yang, J., Loupart, M. L., Smith, A., and Brown, W. (1997). Human mini-chromosomes in mouse embryonal stem cells. Human Molecular Genetics 6, 1375-1382.

Shen, M. H., Yang, J. W., Yang, J., Pendon, C., and Brown, W. R. (2001). The accuracy of segregation of human mini-chromosomes varies in different vertebrate cell lines, correlates with the extent of centromere formation and provides evidence for a trans-acting centromere maintenance activity. Chromosoma 109, 524-535.

Spence, J. M., Critcher, R., Ebersole, T. A., Valdivia, M. M., Earnshaw, W. C., Fukagawa, T., and Farr, C. J. (2002). Co-localization of centromere activity, proteins and topoisomerase II within a subdomain of the major human X alpha-satellite array. Embo J 21, 5269-5280.

Sullivan, B. A., and Willard, H. F. (1998). Stable dicentric X chromosomes with two functional centromeres. Nature Genetics 20, 227-228.

Tomizuka, K., Shinohara, T., Yoshida, H., Uejima, H., Ohguma, A., Tanaka, S., Sato, K., Oshimura, M., and Ishida, I. (2000). Double trans-chromosomic mice: maintenance of two individual human chromosome fragments containing Ig heavy and kappa loci and expression of fully human antibodies. Proc Natl Acad Sci U S A 97, 722-727.

Tomizuka, K., Yoshida, H., Uejima, H., Kugoh, H., Sato, K., Ohguma, A., Hayasaka, M., Hanaoka, K., Oshimura, M., and Ishida, I. (1997). Functional expression and germline transmission of a human chromosome fragment in chimaeric mice. Nature Genetics 16, 133-143.

Yang, J. W., Pendon, C., Yang, J., Haywood, N., Chand, A., and Brown, W. R. (2000). Human mini-chromosomes with minimal centromeres. Hum Mol Genet 9, 1891-1902.

Chapter 5

FUNCTION OF RECQ FAMILY HELICASE IN GENOME STABILITY

Masayuki Seki, Shusuke Tada and Takemi Enomoto
Molecular Cell Biology Laboratory, Graduate School of Pharmaceutical Sciences, Tohoku University, Sendai 980-8578, Japan

Abstract: The *recQ* gene of *Escherichia coli* is the founding member of the RecQ family of helicases. Like *E. coli,* lower eukaryotic species also possess single RecQ proteins, such as Sgs1 and Rqh1 in budding and fission yeast, respectively. However, there are five RecQ helicases in human as well as in chicken cells. Three of the human RecQ helicases are encoded by *BLM, WRN* and *RECQL4* genes, defects of which give rise to the cancer predisposition disorders known as Bloom syndrome (BS), Werner syndrome (WS) and Rothmund–Thomson syndrome (RTS), respectively. The other two, *RECQL1* and *RECQL5*, have not been associated with human diseases. Characterization of RecQ family proteins in unicellular organisms has revealed that their defects confer genomic instability and impairment of homologous recombination. Although systematic genetic analysis of human BS, WS, and RTS cells must be useful to understand their functions, such approach is hampered by the difficulty of making cell lines with double gene disruptions. In this context, the chicken DT40 cell line is an ideal experimental tool for sophisticated approaches that illuminate the functions of vertebrate RecQ helicases. Here, we briefly review general features of RecQ helicases and describe their functions as revealed by analysis of DT40 cells.

Key words: RecQ; Bloom syndrome; Werner syndrome; Rothmund-Thomson syndrome; homologous recombination.

J.-M. Buerstedde and S. Takeda (eds.), Reviews and Protocols in DT40 Research, 49–74.
© 2006 *Springer.*

1.   INTRODUCTION

The *Escherichia coli recQ* gene was originally identified in 1984 in a search for mutants resistant to thymine starvation (Nakayama et al., 1984), and its product was subsequently revealed to be required both for the initiation of RecF-dependent recombination and for the suppression of illegitimate recombination. Ten years later, the first RecQ homologue in higher eukaryotic cells was independently identified by our group (Seki et al., 1994a) and by Puranam and Blackshear (1994). We initially purified a protein that we designated as DNA-dependent ATPase Q1 and later as DNA helicase Q1 because of its DNA helicase activity (Seki et al., 1994b), and working in parallel, Puranam and Blackshear identified a gene they named *RECQL*. Because of these circumstances and because of the existence of multiple RecQ homologues in higher eukaryotes, we proposed to designate this gene *RECQL1* (Enomoto, 2001).

A second *RECQ* homologue, *BLM*, and a third, *WRN*, were identified by positional cloning in 1995 and 1996 as the genes responsible for the rare genetic disorders Bloom syndrome (BS) and Werner syndrome (WS), respectively (Ellis et al., 1995; Yu et al., 1996). The fourth and fifth *RECQ* homologues, originally called *RecQ4* and *RecQ5* and recently renamed *RECQL4* and *RECQL5*, were cloned in 1998 after a search of the EST database for sequences similar to RecQ helicase motifs (Kitao et al., 1998). Mutations in the *RECQL4* gene have been found in some but not all patients with Rothmund-Thomson syndrome (RTS) (Kitao et al., 1999). At present, no genetic disorder has been attributed to mutations in the *RECQL1* and *RECQL5* genes.

Despite great advances in understanding the function of RecQ orthologues in *S. pombe* and *S. cerevisiae*, it has been difficult to define the functions of individual *RECQL* genes in human cells because unicellular eukaryotes possess only a single RecQ helicase. On the other hand, like human cells, the domestic chicken possesses five *RECQL* genes, and therefore, chicken DT40 cells provide a good experimental tool to study vertebrate RECQL paralogues.

2.   GENERAL FEATURES OF RECQL PROTEINS

2.1   Pathogenic and cellular phenotypes of human disorders caused by defects in *RECQL* genes

BS is characterized by growth retardation, sunlight sensitivity, immunodeficiency, male infertility, and predisposition to a wide variety of malignant tumors (German, 1993). The most characteristic feature of BS cells is genomic instability, which is manifested as elevated frequencies of chromosome breaks, interchanges between homologous chromosomes, and sister chromatid exchanges (SCEs) (German, 1993). Early studies of BS cells showed a slow replication fork progression and the accumulation of abnormal replication intermediates (Hand and German, 1975; Lonn et al., 1990).

WS patients suffer accelerated aging and an early onset of age-related diseases such as arteriosclerosis, melituria, and cataract (Epstein et al., 1966). WS is also associated with a predisposition to malignant tumors, although of a limited range compared with BS. Cells derived from WS patients show genomic instability and a shorter *in vitro* life span (Martin, 1977). The genomic instability of WS is manifested as a high frequency of chromosomal rearrangements, translocations, inversions, and deletions (Fukuchi et al., 1989). S phase of WS cells is prolonged (Poot et al., 1992), with a reduced frequency of replicon initiation (Takeuchi et al., 1982).

RTS, also rare, is associated with growth retardation, diverse skeletal abnormalities, skin disorders, sunlight sensitivity, and premature aging (Lindor et al., 2000). RTS patients are highly susceptible to osteosarcoma and squamous cell carcinoma. RTS cells show a high frequency of chromosomal rearrangements, trisomies, translocations and deletions, indicative of genomic instability (Lindor et al., 2000). Unlike what is seen for BS cells, the frequency of SCE is normal in RTS cells as well as in WS cells. Recently, it has been shown that the RTS protein is essential for the initiation of DNA replication in *Xenopus* egg extracts (Sangrithi et al., 2005). Thus, the functions of the RECQL helicases implicated in these diseases appear to be related to aspects of DNA replication.

Many reports have described the sensitivity of BS, WS, and RTS cells to DNA-damaging agents. Contradictory results have been obtained with BS and WS cells. For example, BS cells were reported to show either increased or decreased sensitivity to γ-irradiation (Aurias et al., 1985; Kuhn, 1980; Wang et al., 2001; Beamish et al., 2002). WS cells have increased sensitivity to 4-nitroquinoline 1-oxide (4NQO) (Prince et al., 1999; Poot et al., 2002) and DNA cross-linking agents such as mitomycin C (MMC) (Poot et al., 2001).

However, embryonic stem (ES) cells isolated from *WRN* knockout mice do not exhibit increased sensitivity to MMC, UV-irradiation, γ-irradiation, or methyl methanesulfonate (MMS), but have increased sensitivity to camptothecin (CPT), a DNA topoisomerase I inhibitor (Lebel and Leder, 1998). WS cells are reportedly slightly more sensitive to γ-irradiation than are wild-type cells (Yannone et al., 2001), although other studies indicate that WS cells are insensitive to γ-irradiation and H_2O_2 (Prince et al., 1999). Thus, despite much effort, it has been difficult to characterize the effects of DNA-damaging agents on human BS and WS cells.

RTS cells are also reported to have enhanced sensitivity to ionizing radiation and UV-irradiation (Smith and Paterson, 1982; Shinya et al., 1993). Taken together, BS, WS, and RTS cells appear to have defects in DNA repair or damage tolerance, but it has been difficult to identify the relevant pathways in which BLM, WRN, and RTS are involved by analysis of cells from BS, WS, and RTS patients.

2.2   Biochemical properties of human RecQ helicases

The RecQ family is defined by seven highly conserved motifs in the helicase domain, which is centrally located in general. The helicase domains of all human and chicken RECQLs, as well as, that of budding yeast Sgs1 and fission yeast Rqh1 share 40-50% amino acid identity with *E. coli* recQ. In contrast, the helicase domains of chicken RECQL1, BLM, WRN, RECQL4, and RECQL5 share 86%, 84%, 73%, 60%, and 78% identity with their respective human counterparts, underscoring the conservation of the five RECQLs in vertebrate cells. Motifs similar to those in the nuclease domain of certain proof-reading DNA polymerases are also present in the N-terminal regions of chicken and human WRN, and human WRN exhibits exonuclease activity *in vitro* (Huang et al., 1998). The biochemical properties of human RECQLs are summarized in Figure 5-1; more detailed information is provided in two reviews (Enomoto, 2001; Bachrati and Hickson, 2003), and we cite here results from several recent papers. All human RecQ helicases tested possess single-stranded DNA-stimulated or -dependent ATPase activity and DNA helicase activity that unwinds DNA in the $3' \rightarrow 5'$ direction with respect to the DNA strand to which the enzyme is bound. Human RecQ helicases cannot unwind intact duplex DNA, but in contrast, they efficiently unwind replication fork-like structures. Interestingly, both BLM and WRN unwind G-quartet DNA (G4 DNA), which is formed and stabilized by the non-Watson–Crick bonding of four guanines. Two types of homologous recombination intermediates are also good substrates for human RecQ helicases. RECQL1, BLM, and WRN can unwind D-loop structures, which are generated in an early step of recombination, and all RecQ helicases examined can unwind Holliday junctions, which appear later in recombination. BLM, but not RECQL1, WRN, or RECQL5, acts in concert with DNA topoisomerase IIIα (Top3α) to resolve recombination intermediates containing double Holliday junctions (Wu and Hickson, 2003; Wu et al., 2005). Finally, RECQL1, BLM, WRN, and RECQL5β appear to possess strand pairing activity that forms duplex DNA, a property that contrasts with their DNA helicase activity (Cheok et al., 2005; Sharma et al., 2005; Garcia et al., 2004).

2.3 The roles of yeast RecQ orthologues

In *Saccharomyces cerevisiae*, *SGS1* (*slow growth suppressor 1*) was identified as a mutation that suppresses the slow growth phenotype of DNA topoisomerase III (Top3) mutants (Gangloff et al., 1994). Like human RECQLs, Sgs1 has DNA helicase activity with a 3'→5' specificity. Sgs1 also unwinds G4 DNA and Holliday junctions, indicating a role in suppressing recombination (Enomoto, 2001). The N-terminal region of Sgs1 is responsible for direct interaction with Top3, and it has been reported that Sgs1 acts together with Top3 to suppress MMS sensitivity and hyper-recombination (Ui et al., 2001), suggesting the involvement of the topoi-somerase in conjunction with the helicase in recombination. Indeed, Sgs1 and Top3 are both required to remove double Holliday junctions that arise during HO endonuclease-induced recombination, in order to produce non-crossover products (Ira et al., 2003). The *top3* null mutation confers lethality in *S. pombe*, a phenotype that can be suppressed by deletion of the RecQ orthologue *rqh1+* (Goodwin et al., 1999). Thus, similar genetic interactions are observed in *S. pombe* and *S. cerevisiae*, suggesting that the RecQ/Top3 role is conserved in all eukayotic cells.

We found that *SGS1* is involved in the *RAD52* recombinational repair pathway by an epistasis analysis of the MMS sensitivity of *sgs1* disruptants (Onoda et al., 2001). Moreover, an epistatic relationship between *SGS1* and *RAD51* has been demonstrated (Wu et al., 2001). These findings, coupled with evidence for a direct physical interaction between Sgs1 and Rad51 (Wu et al., 2001), imply that Sgs1 plays a role in homologous recombination. We have found that interchromosomal heteroallelic recombination, which is induced in wild-type cells exposed to MMS, is not induced in *sgs1* disruptants, also implicating Sgs1 in homologous recombination (Onoda et al., 2001). In addition, spontaneous interchromosomal heteroallelic recombination is increased in *sgs1* disruptants (Miyajima et al., 2000). These results indicate dual functions for Sgs1, one in the suppression of spontaneous recombination

Activity / RECQLs	RECQL1	BLM	WRN	RECQL5β
DNA-dependent ATPase activity	O	O	O	O
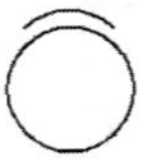 (circular ssDNA)	O	O	O	O
3' ss tailed	O	O	O	O
5' ss tailed	×	×	×	×
Blunt duplex	×	×	×	N.D.
Fork	O	O	O	O
G4 tetraplex	N.D.	O	O	N.D.
D-loop	O	O	O	N.D.
Holliday junction	O	O	O	O
Double Holliday junction	×	O	×	×
Strand pairing activity	O	O	O	O

Figure 5-1. Enzymatic activities of human RECQLs *in vitro.* DNA-stimulated ATPase activity, DNA unwinding activities with of various substrates, and single strand annealing activity are described. Open circles, activity demonstrated; X, no activity; ND, not determined.

and another in promoting recombination following DNA damage induction (Onoda et al., 2001, Gangloff et al., 2000). The *S. pombe rqh1* mutant exhibits greatly increased sensitivity to hydroxyurea, MMS, and UV- and γ-irradiation (Davey et al., 1998). Genetic data indicate that mutations in *rqh1+* are epistatic to mutations affecting *S. pombe RAD51* orthologues with respect to

UV sensitivity, suggesting that Rqh1 participates in recombinational repair (Murray et al., 1997).

Finally, analyses of yeast cells subjected to DNA damage have also indicated that Sgs1 belongs to an epistasis group that includes Rad53 and that it acts upstream of Rad53 to produce a signal that arrests the cell cycle when DNA replication is perturbed (Frei and Gasser, 2000). Since *S. pombe* checkpoint genes also interact genetically with *rqh1* (Murray et al., 1997), it is likely that the Sgs1 and Rqh1 functions are closely related to check-point control as well as to homologous recombination.

2.4   Proteins that interact with human RECQLs

Numerous reports have described proteins that interact with BLM or WRN; these proteins have been extensively reviewed elsewhere (Enomoto, 2001; Bachrati and Hickson, 2003). Here, we summarize those that interact with human RECQLs and describe DT40 *recql* mutants that were generated by our and other groups (Figure 5-2).

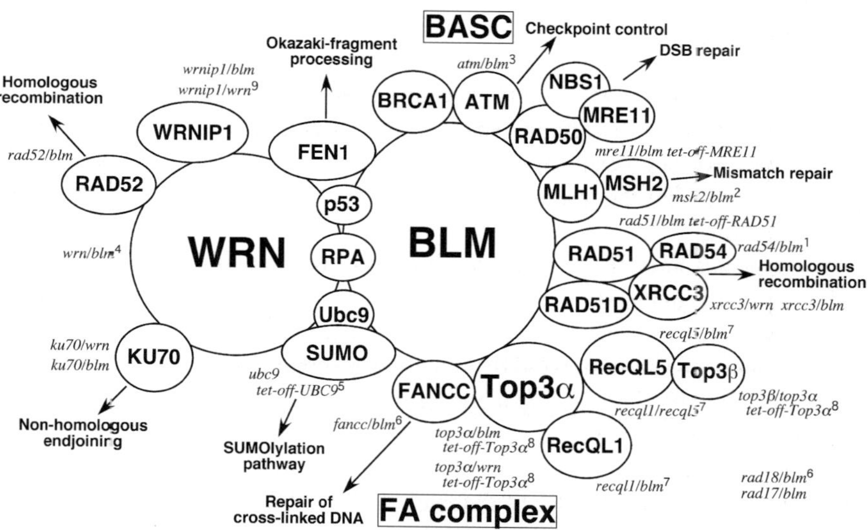

Figure 5-2. RECQL-interacting proteins and DT40 *recql* mutants. Proteins that interact with BLM and WRN are indicated. To address functional relationships for each interaction, we and others have generated and characterized the DT40 mutant cell lines indicated in the Figure. Not all mutants have been published. Because RAD51, MRE11, and Top3α are essential genes, conditional mutants of the *blm* background were constructed using the tet-off system (Sonoda et al., 1998). References: 1. Wang et al., 2000a; 2. Onoda et al., 2004; 3. Wang et al., 2004; 4. Imamura et al., 2002; 5. Hayashi et al., 2002; 6. Hirano et al., 2005; 7. Wang et al., 2003; 8. Seki et al., 2006; 9. Kawabe et al., 2006.

2.4.1   BLM-interacting proteins

BLM interacts with several proteins that participate in the maintenance of genome integrity (Figure 5-2). BLM is a component of BASC (BRCA1-Associated genome Surveillance Complex), which includes BRCA1, MLH1, MSH6, MSH2, replication factor C, RAD50 and ATM (Wang et al., 2000b). Indeed, BRCA1, which is defective in many cases of hereditary breast cancer, associates with BLM in human cells, and both proteins co-localize in cells exposed to hydroxyurea. MLH1, a eukaryotic homologue of the *E. coli* MutL protein, participates in mismatch repair with MSH6 and MSH2 and directly interacts with BLM (Langland et al., 2001). *ATM* is a causative gene for the genetic disorder ataxia telangiectasia, which is characterized by a defect in a damage-dependent checkpoint. The protein encoded by *ATM*, which has protein kinase activity, interacts with and phosphorylates BLM upon γ-irradiation (Ababou et al., 2000) although the functional significance of the phosphorylation remains to be elucidated.

Like yeast Sgs1, human BLM physically interacts with a higher eukaryotic Top3 orthologue, Top3α (Wu et al., 2000). Moreover, BLM dissociates double Holliday junctions in concert with Top3α (Figure 5-1) (Wu 2003). Interestingly, both RECQL1 and RECQL5β also interact with Top3α (Johnson et al., 2000; Shimamoto et al., 2000), and RECQL5β associates with Top3β (Shimamoto et al., 2000). The observation that human BLM lacking the N-terminal 1-133 amino acid region required for interaction with Top3α does not suppress the elevated levels of SCE in BS cells (Hu et al., 2001), suggests that the interaction between Top3α and BLM is important for BLM function.

Fanconi anaemia (FA) is a rare chromosomal instability disorder with a wide range of features, including predisposition to certain cancers. Cells derived from FA patients are hypersensitive to DNA cross-linking agents such as MMC and cis-platinum (CDDP). Over ten genetic complementation groups have been identified for FA (Meetei et al., 2005), and BLM is present in a complex containing five of the known FA proteins (termed FANC A, C, E, F and G) (Meetei et al., 2003). It is noteworthy that this FA sub-complex contains Top3α as well as BLM (Meetei et al., 2003), suggesting that there is a defect in a molecular pathway common to these apparently distinct cancer predisposition disorders.

RAD51, a key player in homologous recombination, interacts with BLM (Wu et al., 2001). Moreover, a RAD51 paralog, RAD51D, directly interacts with BLM (Braybrooke et al., 2003). Taking into account that *S. cerevisiae* and *S. pombe* RecQ proteins are involved in homologous recombination, the interaction between BLM and RAD51/RAD51D is likely important for BLM function.

2.4.2   WRN-interacting proteins

It has been reported that the N-terminal 50 amino acid region of WRN interacts with Ku80, and Ku70/80 stimulates the exonuclease activity of WRN (Cooper et al., 2000). Since the Ku70/80 complex is known to participate in DNA double-strand break repair with DNA-dependent protein kinase, WRN may function in the repair of double-strand breaks occurring during replication. DNA polymerase β (Harrigan et al., 2003) and poly-(ADP-ribose) polymerase-1 (PARP-1) (von Kobbe C et al., 2003), both of which are involved in base excision repair, also interact with WRN, suggesting the involvement of WRN in this repair mode.

RAD52, which has single-strand annealing activity, interacts and co-localizes with WRN in the nucleus when replication forks arrest following DNA damage (Baynton et al., 2003). RAD52 inhibits the exonuclease activity of WRN but stimulates its helicase activity on D-loop substrates (Figure 5-1). These data suggest that WRN and RAD52 act together to rescue stalled replication forks (Baynton et al., 2003).

An interaction of WRN with DNA polymerase δ has been indicated by immunoprecipitation and further shown to be mediated through the p50 subunit. Ectopically expressed WRN recruits p50 and p125, the catalytic subunit of DNA polymerase δ, into nucleolus (Szekely et al., 2000). WRN stimulates DNA polymerase δ activity in the absence of proliferating cell nuclear antigen (PCNA) but does not enhance the activity of eukaryotic DNA polymerases α and ε (Kamath-Loeb et al., 2000). In addition to DNA polymerase δ, PCNA and DNA topoisomerase I have been suggested to associate with WRN. All of these proteins are found in the replication-related 17S multiprotein complex (Lebel et al., 1999).

WRNIP1 (Werner Interacting Protein 1) interacts with the N-terminal portion of WRN, which contains the exonuclease domain (Kawabe et al., 2001). WRNIP1 has homology to replication factor C family proteins, which are required for loading PCNA onto template-primer junctions, and it is conserved from *E. coli* to humans. Interestingly, human WRNIP1 directly interacts with three of the four subunits of DNA polymerase δ and facilitates loading of the complex onto a template-primer substrate, which stimulates polymerase activity (Tsurimoto et al., 2005). Thus, recent data indicate that both WRN and WRNIP1 function to modulate DNA polymerase δ activity at replication forks.

2.4.3   Proteins that interact with both BLM and WRN

Direct interaction between WRN and BLM has been shown *in vitro*, and BLM inhibits the exonuclease activity of WRN (von Kobbe et al., 2002).

However, the amount of WRN that co-immunoprecipitates with BLM is low, suggesting that only a limited proportion of these proteins are engaged in complexes. RPA, a single-stranded DNA binding protein that is required for replication, repair and recombination, binds to and stimulates the helicase activity of BLM and WRN (Constantinou et al., 2000; Brosh et al., 2000). FEN1 (flap structure-specific endonuclease 1), which is required for Okazaki fragment processing during DNA replication, interacts with WRN, and its flap endonuclease activity is greatly stimulated by WRN (Brosh et al., 2001a). Stimulation of FEN1 activity by BLM has also been reported (Wang et al., 2005). The tumor suppressor p53 interacts with BLM and WRN and inhibits the unwinding of Holliday junctions by BLM or WRN (Figure 5-1) (Yang et al., 2002). The p53 protein alone binds to Holliday junctions, suggesting that it regulates homologous recombination through its ability to modulate interactions of BLM with recombination intermediates. The exonuclease activity of WRN is also inhibited by p53 in a dose-dependent manner (Brosh et al., 2001b). Finally, Ubc9 and SUMO are involved in a biochemical pathway by which Ubc9 conjugates SUMO to target proteins. Both BLM and WRN are sumoylated (Kawabe et al., 2000; Eladad et al., 2005), but the physiological significance of these modifications is unclear.

3.   ANALYSIS OF *RECQL* MUTANTS OF DT40 CELLS

3.1   DT40 cells as a model for the functional analysis of vertebrate RECQL proteins

There are both clear advantages and disadvantages in using DT40 cells to analyze the functions of vertebrate RECQLs at the cellular level. Although p53 interacts with both BLM and WRN in human cells, we cannot address these relationships functionally, since DT40 cells lack p53 (Takao et al., 1999). In contrast, the major advantages are as follows.

1. In contrast to cell lines derived from patients of diverse genetic backgrounds, DT40 cells provide a standard background that allows the phenotypes of RECQL-deficient derivatives (see Figure 5-2) to be more rigorously compared.
2. Analyses of yeasts have provided information about RecQ functions (see 2.3), but data obtained for the single RecQ proteins of *S. cerevisiae* and *S. pombe* are not sufficient for understanding the functional relationships among the five vertebrate RECQLs. The use of DT40 cells has allowed us to obtain information concerning these relationships, since we can

construct and characterize cells with any combination of double and triple *recql* mutations.

3. Data from yeast studies (see 2.3) has also suggested that RecQ functions with Top3, and is involved in homologous recombination and checkpoint controls. To extend these results, we have constructed cells with various combinations of mutations affecting *RECQL* genes, *RAD52* epistasis group genes, genes encoding checkpoint proteins, *TOP3α*, and *TOP3β*.

4. Numerous proteins are known to interact with RECQLs (see 2.4). As most of these are involved in the maintenance of genome stability, researchers have already constructed many DT40 cell lines disrupted for genes encoding such proteins (Takata et al., this issue, chapter 17; Takeda, this issue, chapter 18). Therefore, it is straightforward to generate double mutant cells with disruptions of genes encoding a RECQL protein and an interacting protein of interest. As shown in Figure 5-2, we have constructed and characterized dozens of DT40 mutant cell lines defective for *RECQL* and other genes.

3.2 Functional relationships among vertebrate RECQLs

3.2.1 Involvement of RECQLs in cell growth

DT40 cells singly disrupted for four *RECQL* genes (*recql1, blm, wrn,* and *recql5*) but not *RECQL4* (*recql4*) have been constructed (Wang et al., 2000a; Imamura et al., 2001; Imamura et al., 2002; Wang et al., 2003). Cells with *recql1* and *recql5* mutations grow normally, although *blm* cells grow more slowly than wild type cells (Wang et al., 2003). To study the functional relationship among RECQLs, we and others generated *recql1/recql5, recql1/blm, recql5/blm,* and *blm/wrn* double mutant cell lines (Imamura et al., 2002; Wang et al., 2003). Growth defects were not observed for *recql1/recql5* cells, but the growth of *recql1/blm* and *blm/recql5* cells is much more impaired than that of any single mutant, suggesting that RecQL1 or RecQL5 functions related to cell growth are important only in the *blm* background (Wang et al., 2003). Although *wrn* cells grow slightly more slowly than wild type cells, the growth rate of *wrn* cells is considerably exacerbated by disruption of the *BLM* gene (Imamura et al., 2002). Taken together, BLM is the most important of the RECQLs tested for cell growth, and RECQL1, WRN, and RECQL5 play complementary or supplementary roles in this respect. An accumulation of DNA damage in mutant cells does not explain the slow growth of double mutants, because there is no apparent increase in chromosomal aberrations in *recql1/blm, recql5/blm* and *blm/wrn* cells (Wang et al., 2003; Imamura et al., 2002). Moreover, the frequency of spontaneous SCE observed in several RECQL-defective cell lines (see 3.2.2) does not correlate with the extent of growth defects. Since all RECQLs tested efficiently unwind fork-like structures and at least BLM and WRN can unwind G4 DNA (Figure 5-1), it is likely that RECQLs function redundantly in DNA synthesis to support the smooth progression of replication forks.

3.2.2   Contributions of RECQLs to SCE

Like what is seen for cells derived from BS patients, an about 10-fold increase in SCE frequency is observed in DT40 *blm* cells (Wang et al., 2000a; Imamura et al., 2001). The *recql1, wrn,* and *recql5* single disruptants and *recql1/recql5* double disruptants have an apparently normal frequency of SCE, suggesting that BLM alone efficiently regulates SCE (Wang et al., 2003). It is noteworthy that BLM (but not RECQL1, WRN, or RECQL5) together with Top3α dissolves double Holliday junctions to produce noncrossovers *in vitro* (Figure 5-1), suggesting that this activity of BLM/Top3α is involved in SCE suppression *in vivo*. In contrast, for *recql1/blm, blm/wrn,* and *blm/recql5* cells, the frequency of SCE is unaltered, reduced, and enhanced, respectively, as compared to *blm* cells (Table 5-1) (Wang et al., 2003; Imamura et al., 2002). Although RECQL5 interacts with Top3α and Top3β, RECQL5 cannot dissolve double Holliday junctions *in vitro* in the presence of Top3α (Figure 5-1). Nonetheless, in the absence of BLM, RECQL5 seems to adopt the role of BLM in suppressing SCE *in vivo*. Interestingly, mouse *RECQL5$^{-/-}$* ES cells exhibit an elevated SCE frequency even in the presence of BLM, and a further increase in SCE frequency is observed in mouse *BLM$^{-/-}$/RECQL5$^{-/-}$* double mutant ES cells (Hu et al., 2005), as also seen for DT40 *blm/recql5* cells (Wang et al., 2003). Thus, it is suggested that, like BLM, RECQL5 is required for the suppression of SCE.

Although deletion of the *RECQL1* gene has no effect on the frequency of spontaneous SCE in wild type or *blm* cells, the frequency of MMC-induced SCE is increased in *recql1/blm* cells as compared to *blm* cells, while *recql1* cells are similar to wild type cells in this respect (Wang et al., 2003). These results suggest that RECQL1 plays a supplemental role for BLM, one that is detected only in its absence and following MMC exposure. Curiously, disruption of *WRN* partially reduces the elevated SCE frequency of DT40 *blm* cells, suggesting that WRN is involved in SCE formation in the absence of BLM (Imamura et al., 2002).

3.2.3 Sensitivity of RECQL-disrupted cell lines to genotoxic agents

Cells with *recql1*, *recql5* and *recql1/recql5* mutations have the same sensitivity as wild-type cells towards reagents and treatments we tested (Wang et al., 2003). In contrast, *blm* and *wrn* cells are moderately sensitive to 4NQO, CPT, MMS, CDDP, and UV (Imamura et al., 2002). Although the

Table 5-1. Effect of depletion of specific gene products on spontaneous SCE in *blm* cells.

	Genotype *x* gene	Wild-type (A)	*blm* (B)	*x* (C)	*x/blm* (D)	%	ref.
BASC	*mre11*	2.2	24.0	2.0[1]	19.3[1]	20.6↓	Unpublished data
	atm	2.1	25.3	2.0	22.7	10.8↓	Wang et al., 2004
	msh2	2.2	23.9	2.8	24.3	0.9↓	Onoda et al., 2004
FA	*fancc*	3.7	28.6	9.1	28.2	0[3]	Hirano et al., 2005
RECQLs	*recql1*	2.1	26.4	2.0	27.5[4]	4.9↑	Wang et al., 2003
	wrn	1.8	20.7	2.3	14.6	34.9↓	Imamura et al., 2002
	recql5	2.1	26.4	2.4	45.2	76.1↑	Wang et al., 2003
Homologous recombination	*rad54*	2.1	26.4	1.7	8.2	73.3↓	Wang et al., 2000a
	rad51	2.5	22.9	1.5[2]	15.2[2]	32.8↓	Fig. 5-3

The average numbers of spontaneous sister chromatid exchanges (SCE) per metaphase nucleus observed in DT40 mutant cell lines are indicated. From two to three, and 20 to 29 SCE are observed in wild-type cells and *blm* cells, respectively. Under conditions of Rad51 depletion or in RAD54 disruptants, the level of spontaneous SCEs is slightly decreased compared to that of wild-type cells. SCE is moderately elevated in *fancc* cells. Effects of the indicated gene disruptions on SCE levels in *blm* cells are calculated by the formula (B-A)-(D-C)/(B-A) X 100 (%). Arrows next to percentages indicate an increase or decrease in SCE levels relative to that of *blm* single mutant cells caused by the indicated gene disruption. [1,2] SCEs are measured under conditions of MRE11 or RAD51 depletion (Figure 5-3). [3] This formula cannot be simply applied in the case of *fancc* and *fancc/blm* cells. [4] Mitomycin C-induced but not spontaneous SCE is synergistically increased in *recql1/blm* cells compared to that in *blm* single mutant cells (Wang et al., 2003).

growth rates of *recql1/blm* and *blm/recql5* cells are considerably reduced compared to the corresponding single disruptants, *recql1/blm* and *blm/recql5* cells are as sensitive as *blm* cells to various DNA damaging agents (Wang et al., 2003). Thus, whereas RECQL1 and RECQL5 seem to partially complement BLM function in cell growth and in the suppression of SCE, it is not the case for DNA damage tolerance or repair. In contrast, *blm/wrn* cells exhibit increased sensitivities to 4NQO, CPT, MMS, CDDP, and UV-irradiation as compared with either single disruptant (Imamura et al., 2002), suggesting that WRN and BLM are involved in DNA repair or damage tolerance in a complementary fashion.

3.3   Functional relationships between BLM and Top3α

Since TOP3α is essential for cell growth, we generated cell lines disrupted for the TOP3α gene by introducing a plasmid carrying mouse *TOP3α*, the expression of which can be turned off by tetracycline (tet-off m*TOP3α*). We have generated *top3α* tet-off m*TOP3α*, *top3β*, *top3α/−top3β* tet-off m*TOP3α*, *top3α/blm* tet-off m*TOP3α*, *top3β/blm,* and *top3α/wrn* tet-off m*TOP3α* cell lines (Seki et al., 2006). Like *blm* cells, Top3α-depleted cells show an elevated SCE frequency, implying that the suppression of SCE requires Top3α as well as BLM functions (unpublished result). Moreover, Top3α depletion causes cellular accumulation in G_2 phase, nuclear enlargement, and chromosome gaps and breaks that occur at the same positions in sister chromatids. Although these phenotypes observed in Top3α-depleted cells are not affected by disruption of either the *WRN* or *TOP3β* gene, all these phenotypes are suppressed by disruption of the *BLM* gene. These results clearly indicate a functional interaction between BLM and Top3α *in vivo* (Seki et al., 2006). The main conclusion of these studies is that BLM, together with Top3α, mediates the dissolution of sister chromatids at a late stage of DNA replication.

3.4   RECQLs and proteins involved in repair
and genome maintenance pathways

3.4.1   WRN and WRNIP1

The WRNIP1 protein interacts with WRN. To examine their *in vivo* functional relationships, we generated and characterized *wrnip1, wrn,* and *wrnip1/wrn* cells. While *wrnip1/wrn* cells grow at a rate similar to that of wild-type cells, they exhibit an increase in spontaneous SCE compared to either single mutant. In addition, whereas *wrnip1* and *wrn* cells are moderately sensitive to CPT, *wrnip1/wrn* cells show higher sensitivity (Kawabe et al., 2006). These results suggest that WRNIP1 functions in a repair or damage tolerance pathway in a WRN-independent manner, although the two proteins interact *in vivo*.

3.4.2   BASC and BLM

To address the relationship between BLM and each BASC component, we generated *mre11/blm* tet-off *hMRE11*, *atm/blm*, and *msh2/blm* cell lines. MRE11 is essential for cell growth, and its depletion results in a high incidence of chromosomal aberrations (Yamaguchi-Iwai et al., 1999). Disruption of *BLM* does not affect the phenotype of MRE11-depleted cells, because there is no further increase in chromosomal aberrations in *mre11/blm* tet-off *hMRE11* compared to *mre11* tet-off *hMRE11* cells in the presence of doxycycline (Dox) (unpublished data). However, the elevated SCE frequency of *blm* cells is moderately reduced by MRE11 depletion, suggesting that MRE11 partially participates in SCE formation in *blm* cells (Table 5-1).

Although ATM interacts with BLM and phosphorylates BLM following UV- or γ-ray irradiation, there is no evidence for a functional interaction between ATM and BLM (Wang et al., 2004; Fukuo et al., 2004). Cells with *atm* and *blm* mutations are moderately sensitive to X-irradiation and MMS, respectively. The X-ray sensitivity of *atm/blm* cells is almost the same as that of *atm* cells, and the MMS sensitivity of *atm/blm* cells is the same as that of *blm* cells (Wang et al., 2004), indicating that ATM and BLM function in different pathways upon induction of DNA damage by these agents. In addition, disruption of the *ATM* gene only slightly affects the elevated frequency of SCE in *blm* cells (Table 5-1) (Wang et al., 2004). Moreover, this frequency is not affected by the additional disruption of the *MSH2* gene (Onoda et al., 2004). Thus among the BASC components, at least ATM and MSH2 are not essential for SCE in *blm* cells (Table 5-1).

3.4.3   The FA complex and BLM

The level of spontaneous SCE is elevated approximately two-fold in *fancc* cells (Hirano et al., 2005), an effect that requires Xrcc3, whereas *fancc/rad18* double disruptants exhibit a higher incidence of SCE than does either single mutant. Unexpectedly, the frequency of SCE in *fancc/blm* cells is similar to that of *blm* cells (Hirano et al., 2005). Moreover, the MMC-induced formation of GFP-BLM nuclear foci is severely compromised in both human and chicken *fancc* and *fancd2* cells (Hirano et al., 2005). These results are suggesting a functional relationship between FANCC and BLM. Since one of the FA multi-protein complexes includes FANCC, BLM, and Top3α (Meetei et al., 2003), the suppression of excess SCE by FANCC appears to depend on the BLM/Top3α function.

3.4.4 Homologous recombination and BLM/WRN

Genetic analysis of RecQ in both budding and fission yeast revealed that its function is closely associated with homologous recombination, prompting us to examine the relationship between vertebrate RECQLs and homologous recombination proteins by generating *xrcc3/blm, xrcc3/wrn, rad52/blm, rad54/blm,* and *rad51/blm* tet-off *hRAD51* cells. Our previous analysis of *rad54/blm* cells revealed that the elevated levels of SCE in *blm* cells are greatly reduced by disruption of *RAD54* (Table 5-1) (Wang et al., 2000a). Moreover, the elevated targeted integration frequency in *blm* cells is almost completely abolished by *RAD54* disruption (Wang et al., 2000a). Thus, we provided evidence indicating that the elevated levels of SCE and targeted integration in *blm* cells depend on homologous recombination. As expected, both XRCC3 (unpublished data) and RAD51 (Figure 5-3 and Table 5-1) are also required for SCE in *blm* cells. In contrast, RAD52 does not contribute to SCE in *blm* cells (unpublished data). Based on previous reports and unpublished data, we summarize in Table 5-1 the effects of various gene disruptions on the frequency of SCE in *blm* cells.

An examination of the sensitivity of single and double mutants (*xrcc3/ blm, rad52/blm, rad54/blm, ku70/blm, rad18/blm, atm/blm,* and *wrnip1/blm*) to various genotoxic agents revealed that BLM is epistatic to XRCC3, suggesting that BLM is involved in XRCC3-mediated homologous recombination. In contrast, the sensitivity of *xrcc3/wrn* cells to various DNA damaging agents is greater than that of either single mutant (unpublished data). Taking into account that BLM suppresses SCE mediated by XRCC3/RAD51/RAD54 (Table 5-1), it may play a role in the processing of homologous recombination intermediates. If the sensitivity of *blm* cells to various DNA damaging agents is mainly due to a failure to process such intermediates, they should also be sensitive to X-irradiation, which generates DNA double strand breaks (DSBs). *blm* cells, however, are sensitive to MMS and UV-irradiation but not to X-irradiation compared to wild type cells (Wang et al., 2000a; Imamura et al., 2001, 2002; Wang et al., 2004). Thus, we predict that lack of BLM affects the quality of homologous recombination products but not viability following DSB formation.

The MMS sensitivity of *blm* cells could be explained by a catastrophic collapse of replication forks. Interestingly, the spectrum of genotoxic agents to which *blm* cells are sensitive is remarkably similar to that of *rad17* cells

(Kobayashi et al., 2004), which are defective in the replication checkpoint. Indeed, both *blm* and *rad17* cells are sensitive to MMS, CDDP, CPT, 4NQO, and UV-irradiation but not to X-irradiation (Imamura et al., 2002; Kobayashi et al., 2004). Moreover, *rad17/blm* double mutants are as sensitive to MMS as *rad17* cells (unpublished data), suggesting that the BLM function is controlled by the Rad17 checkpoint protein.

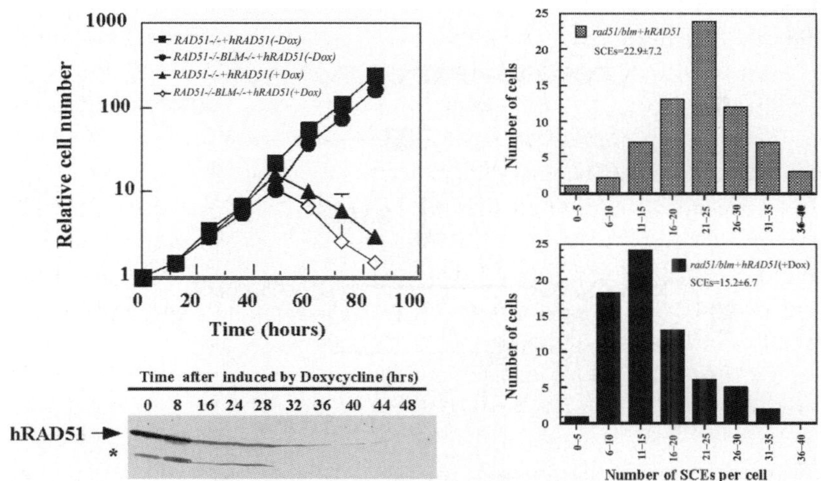

Figure 5-3. RAD51 is required for the elevated levels of SCE in *blm* cells. The *rad51/blm* tet-off hRAD51 cells are derivatives of *rad51* tet-off hRAD51 cells (Sonoda et al., 1998, 1999). (A) Growth curves of *rad51* tet-off hRAD51 and *rad51/blm* tet-off hRAD51 cells in the presence or absence of doxycycline (Dox). (B) After Dox addition, both cell lines cease growth just after the disappearance of hRAD51 protein, as previously described (Sonoda et al., 1998, 1999). The arrow indicates hRAD51 protein in *rad51/blm* tet-off hRAD51 cells after Dox addition, and an asterisk shows hRAD51 degradation products. (C) Sister chromatid exchange (SCE) in *rad51/blm* tet-off hRAD51 cells in the absence (upper panel) or presence (lower panel) of Dox (see plate 7).

Recently, an attractive model has been proposed for the error-free bypass of DNA lesions at replication forks in budding yeast, which depends on proteins involved in homologous recombination, Rad51 and the RecQ orthologue Sgs1 (Figure 5-4) (Liberi et al., 2005). Since all of the data we have obtained with MMS-treated DT40 disruptants is compatible with this model, we thus propose that XRCC3 and BLM, like budding yeast Rad51 and Sgs1, are involved in the error-free bypass of DNA lesions at replication forks. BLM interacts with RAD51 and RAD51D (Braybrooke et al., 2003) which like XRCC3 is one of five RAD51 paralogs, further supporting our proposal. If this is the case, mutations will accumulate in the absence of BLM, resulting in a high cancer risk. Taking into account that WRN functions in DNA repair or damage tolerance in a fashion complementary to that of BLM (Imamura et al., 2002), it is likely that WRN is also involved in error-free lesion bypass independent of XRCC3 and BLM. It is noteworthy that WRN interacts with RAD52 at replication forks following DNA damage (Baynton et al., 2003), and simultaneous deletion of XRCC3 and RAD52 results in lethality (Fujimori et al., 2001). Taken together, it is probable that the BLM/XRCC3 and WRN/RAD52 pathways regulate the restart of replication forks in a complementary fashion. This notion will be tested experimentally in the near future.

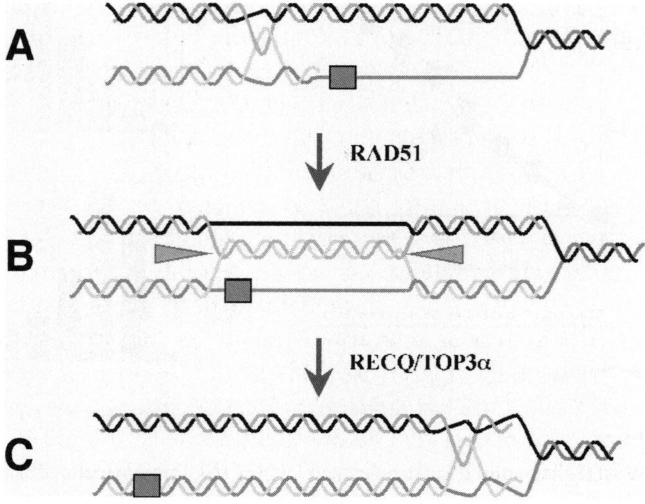

Figure 5-4. Template switching model for replication fork restart on damaged templates. The model is adapted from a published report (Liberi et al., 2005). Replication forks on a damaged template bypass DNA lesions through template switching mediated by hemicatenanes (structure A). A Rad51 homologous recombination-dependent pathway converts/stabilizes the intermediates as a pseudo double Holliday junction by extending pairing between the newly synthesized strands (structure B). The RECQ/Top3α complex (arrow heads) may mediate the conversion of this structure to a replication fork containing hemicatenanes (structure C).

4. CONCLUDING REMARKS AND FUTURE PERSPECTIVES

Of the five vertebrate RECQLs, only RECQL1 and RECQL5 seem to compensate for the absence of BLM (Wang et al., 2003). In contrast, WRN has roles complementary to those of BLM (Imamura et al., 2002). Thus, an assignment of the cellular functions of BLM is crucial for understanding the roles of all vertebrate RECQLs. Four major findings related to BLM functions can be summarized. First, BLM is required for efficient cell growth and this function is assured by RECQL1, WRN, and RECQL5 in its absence (see 3.2.1). Second, BLM suppresses SCE, and disruption of *RECQL1, WRN,* or *RECQL5* affects SCE formation in *blm* cells in different ways. Since SCE in *blm* cells is mediated by the XRCC3/RAD51/Rad54-dependent homologous recombination pathway (Figure 5-3 and Table 5-1), all RECQLs tested are somehow linked to homologous recombination (see 3.2.2). Third, BLM together with Top3α appears to be involved in sister chromatid dissolution at a late stage of DNA replication (see 3.3). Fourth, the MMS sensitivity of *blm* cells can be explained by a model that BLM is involved in recombination-dependent replication fork restart (Figure 5-4, see 3.4,4). This process is independent of RECQL1 and RECQL5 but may involve WRN in a different manner. Further analyses of *wrn* DT40 cells will clarify the role of WRN in replication fork reactivation.

Finally, a recent systematic synthetic lethal screen of the budding yeast genome provided extensive information about genes that genetically interact with RecQ (Tong et al., 2004). For example, simultaneous deletion of Sgs1 and Sod1 (superoxide dismutase 1), which detoxifies superoxide, confers lethality. Since oxidative stress is thought to be a major source of DNA damage, it is of interest to analyze DT40 *recql* cells in the absence of enzymes that counteract such stress. This study is now in progress in our laboratory. Recently, we have constructed a targeting vector for *RECQL4* gene disruption. We expect that future analysis of *recql4* DT40 cells will provide new insights into the function of *RECQL4*, mutations of which cause Rothmund-Thomson syndrome. We expect that the continued generation and analysis of DT40 *recql* mutants will contribute to defining the cellular functions of vertebrate RECQLs.

REFERENCES

Ababou M, Dutertre S, Lecluse Y, Onclercq R, Chatton B, Amor-Gueret M. 2000. ATM-dependent phosphorylation and accumulation of endogenous BLM protein in response to ionizing radiation. Oncogene. 19; 5955-5963.

Aurias A, Antoine JL, Assathiany R, Odievre M, Dutrillaux B. 1985. Radiation sensitivity of Bloom's syndrome lymphocytes during S and G2 phases. Cancer Genet. Cytogenet. 16; 131-136.

Bachrati CZ, Hickson ID. 2003. RecQ helicases: suppressors of tumorigenesis and premature aging. Biochem J. 374; 577-606.

Baynton K, Otterlei M, Bjoras M, von Kobbe C, Bohr VA, Seeberg E. 2003. WRN interacts physically and functionally with the recombination mediator protein RAD52. J Biol Chem. 278; 36476-36486.

Beamish H, Kedar P, Kaneko H, Chen P, Fukao T, Peng C, Beresten S, Gueven N, Purdie D, Lees MS. et al., 2002. Functional link between BLM defective in Bloom's syndrome and the ataxia-telangiectasia-mutated protein, ATM. J. Biol. Chem. 277; 30515-30523.

Braybrooke JP, Li JL, Wu L, Caple F, Benson FE, Hickson ID. 2003. Functional interaction between the Bloom's syndrome helicase and the RAD51 paralog, RAD51L3 (RAD51D). J Biol Chem. 278; 48357-48366.

Brosh RM Jr, Li JL, Kenny MK, Karow JK, Cooper MP, Kureekattil RP, Hickson ID, Bohr VA. 2000. Replication protein A physically interacts with the Bloom's syndrome protein and stimulates its helicase activity. J Biol Chem. 275; 23500-23508.

Brosh RM Jr, von Kobbe C, Sommers JA, Karmakar P, Opresko PL, Piotrowski J, Dianova I, Dianov GL, Bohr VA. 2001a. Werner syndrome protein interacts with human flap endonuclease 1 and stimulates its cleavage activity. EMBO J. 20; 5791-5801.

Brosh RM Jr, Karmakar P, Sommers JA, Yang Q, Wang XW, Spillare EA, Harris CC, Bohr VA. 2001b. p53 Modulates the exonuclease activity of Werner syndrome protein. J Biol Chem. 276; 35093-35102.

Cheok CF, Wu L, Garcia PL, Janscak P, Hickson ID.2005. The Bloom's syndrome helicase promotes the annealing of complementary single-stranded DNA. Nucleic Acids Res. 33; 3932-3941.

Cooper MP, Machwe A, Orren DK, Brosh RM, Ramsden D, Bohr VA. 2000. Ku complex interacts with and stimulates the Werner protein. Genes Dev. 14; 907-912.

Constantinou A, Tarsounas M, Karow JK, Brosh RM, Bohr VA, Hickson ID, West SC. 2000. Werner's syndrome protein (WRN) migrates Holliday junctions and co-localizes with RPA upon replication arrest. EMBO Rep. 1; 80-84.

Davey S, Han CS, Ramer SA, Klassen JC, Jacobson A, Eisenberger A, Hopkins KM, Lieberman HB, Freyer GA. 1998. Fission yeast rad12+ regulates cell cycle checkpoint control and is homologous to the Bloom's syndrome disease gene. Mol Cell Biol. 18; 2721-2728.

Eladad S, Ye TZ, Hu P, Leversha M, Beresten S, Matunis MJ, Ellis NA. 2005. Intra-nuclear trafficking of the BLM helicase to DNA damage-induced foci is regulated by SUMO modification. Hum Mol Genet. 14; 1351-1365.

Ellis NA, Groden, J, Ye TZ, Straughen J, Lennon DJ, Ciocci S, Proytcheva M, German J. 1995. The Bloom's syndrome gene product is homologous to RecQ helicases. Cell 83; 655-666.

Enomoto T. 2001. Functions of RecQ family helicases: possible involvement of Bloom's and Werner's syndrome gene products in guarding genome integrity during DNA replication. J Biochem. 129; 501-507.

Epstein CJ, Martin G M, Schultz AL, Motulsky A G. 1966. Werner's syndrome : a review of its symptomatology, natural history, pathologic features, genetics and relationship to the natural aging process. Medicine 45; 177-221.

Frei C, Gasser SM. 2000. The yeast Sgs1p helicase acts upstream of Rad53p in the DNA replication checkpoint and colocalizes with Rad53p in S-phase-specific foci. Genes Dev. 14; 81-96.

Fukao T, Chen P, Ren J, Kaneko H, Zhang GX, Kondo M, Yamamoto K, Furuichi Y, Takeda S, Kondo N, Lavin MF. 2004. Disruption of the BLM gene in ATM-null DT40 cells does not exacerbate either phenotype. Oncogene. 23; 1498-1506.

Fukuchi K, Martin GM, Monnat RJ Jr. 1989. Mutator phenotype of Werner syndrome is characterized by extensive deletions. Proc Natl Acad Sci U S A. 86; 5893-5897.

Fujimori A, Tachiiri S, Sonoda E, Thompson LH, Dhar PK, Hiraoka M, Takeda S, Zhang Y, Reth M, Takata M. 2001. Rad52 partially substitutes for the Rad51 paralog XRCC3 in maintaining chromosomal integrity in vertebrate cells. EMBO J. 20; 5513-5520.

Gangloff S, McDonald JP, Bendixen C, Arthur L, Rothstein R. 1994. The yeast type I topoisomerase Top3 interacts with Sgs1, a DNA helicase homolog: a potential eukaryotic reverse gyrase. Mol Cell Biol. 14; 8391-8398.

Gangloff S, Soustelle C, Fabre F. 2000. Homologous recombination is responsible for cell death in the absence of the Sgs1 and Srs2 helicases. Nat Genet. 25; 192-194.

Garcia PL, Liu Y, Jiricny J, West SC, Janscak P. 2004. Human RECQ5beta, a protein with DNA helicase and strand-annealing activities in a single polypeptide. EMBO J. 23; 2882-2891.

German J. 1993. Bloom's syndrome: a mendelian prototype of somatic mutational disease. Medicine 72; 393-406.

Goodwin A, Wang SW, Toda T, Norbury C, Hickson ID. 1999. Topoisomerase III is essential for accurate nuclear division in Schizosaccharomyces pombe. Nucleic Acids Res. 27; 4050-4058.

Hand R, German J. 1975. A retarded rate of DNA chain growth in Bloom's syndrome. Proc. Natl. Acad. Sci. USA 72; 758-762.

Harrigan JA, Opresko PL, von Kobbe C, Kedar PS, Prasad R, Wilson SH, Bohr VA. 2003. The Werner syndrome protein stimulates DNA polymerase beta strand displacement synthesis via its helicase activity. J Biol Chem. 278; 22686-22695.

Hayashi T, Seki M, Maeda D, Wang W, Kawabe Y, Seki T, Saitoh H, Fukagawa T, Yagi H, Enomoto T. 2002. Ubc9 is essential for viability of higher eukaryotic cells. Exp Cell Res. 280; 212-221.

Hirano S, Yamamoto K, Ishiai M, Yamazoe M, Seki M, Matsushita N, Ohzeki M, Yamashita YM, Arakawa H, Buerstedde JM, Enomoto T, Takeda S, Thompson LH, Takata M. 2005. Functional relationships of FANCC to homologous recombination, translesion synthesis, and BLM. EMBO J. 24; 418-427.

Hu P, Beresten SF, van Brabant AJ, Ye TZ, Pandolfi PP, Johnson FB, Guarente L, Ellis NA. 2001. Evidence for BLM and topoisomerase IIIalpha interaction in genomic stability. Hum Mol Genet. 10; 1287-1298.

Hu Y, Lu X, Barnes E, Yan M, Lou H, Luo G. 2005. Recql5 and Blm RecQ DNA helicases have nonredundant roles in suppressing crossovers. Mol Cell Biol. 25; 3431-3442.

Huang S, Li B, Gray MD, Oshima J, Mian IS, Campisi J. 1998. The premature ageing syndrome protein, WRN, is a 3'→5' exonuclease. Nat Genet. 20; 114-116.

Imamura O, Fujita K, Shimamoto A, Tanabe H, Takeda S, Furuichi Y, Matsumoto T. 2001. Bloom helicase is involved in DNA surveillance in early S phase in vertebrate cells. Oncogene. 20; 1143-1151.

Imamura O, Fujita K, Itoh C, Takeda S, Furuichi Y, Matsumoto T. 2002. Werner and Bloom helicases are involved in DNA repair in a complementary fashion. Oncogene. 21; 954-963.

Ira G, Malkova A, Liberi G, Foiani M, Haber JE. 2003. Srs2 and Sgs1-Top3 suppress crossovers during double-strand break repair in yeast. Cell. 115; 401-411.

Johnson FB, Lombard DB, Neff NF, Mastrangelo MA, Dewolf W, Ellis NA, Marciniak RA, Yin Y, Jaenisch R, Guarente L. 2000. Association of the Bloom syndrome protein with topoisomerase IIIalpha in somatic and meiotic cells. Cancer Res. 60; 1162-1167.

Kamath-Loeb AS, Johansson E, Burgers PM, Loeb LA. 2000. Functional interaction between the Werner Syndrome protein and DNA polymerase delta. Proc Natl Acad Sci USA. 97; 4603-4608.

Kawabe Y, Seki M, Yoshimura A, Nishino K, Hayashia T, Takeuchi T, Iguchi S, Kusa Y, Ohtsuki M, Tsuyama T, Imamura O, Matsumoto T, Furuichi Y, Tada S, Enomoto T. 2006 Analyses of the interaction of WRNIP1 with Werner syndrome protein (WRN) in vitro and in the cell. DNA Repair 5; 816-828.

Kawabe Y, Seki M, Seki T, Wang WS, Imamura O, Furuichi Y, Saitoh H, Enomoto T. 2000. Covalent modification of the Werner's syndrome gene product with the ubiquitin-related protein, SUMO-1. J Biol Chem. 275; 20963-20966.

Kawabe Y, Branzei D, Hayashi T, Suzuki H, Masuko T, Onoda F, Heo SJ, Ikeda H, Shimamoto A, Furuichi Y, Seki M, Enomoto T. 2001. A novel protein interacts with the Werner's syndrome gene product physically and functionally. J Biol Chem. 276; 20364-20369.

Kitao S, Ohsugi I, Ichikawa K, Goto M, Furuichi Y, and Shimamoto A. 1998. Cloning of two new human helicase genes of the RecQ family: Biological significance of multiple species in higher eukaryotes. Genomics 54; 443-452.

Kitao S, Shimamoto A, Goto M, Miller RW, Smithson WA, Lindor NM, Furuichi Y. 1999. Mutations in RECQL4 cause a subset of cases of Rothmund-Thomson syndrome. Nat. Genet. 22; 82-84.

Kobayashi M, Hirano A, Kumano T, Xiang SL, Mihara K, Haseda Y, Matsui O, Shimizu H, Yamamoto K. 2004. Critical role for chicken Rad17 and Rad9 in the cellular response to DNA damage and stalled DNA replication. Genes Cells. 9; 291-303.

Kuhn EM. 1980. Effects of X-irradiation in G1 and G2 on Bloom's Syndrome and normal chromosomes. Hum. Genet. 54; 335-341.

Langland G, Kordich J, Creaney J, Goss KH, Lillard-Wetherell K, Bebenek K, Kunkel TA, Groden J. 2001. The Bloom's syndrome protein (BLM) interacts with MLH1 but is not required for DNA mismatch repair. J Biol Chem. 276; 30031-30035.

Lebel M, Leder P. 1998. A deletion within the murine Werner syndrome helicase induces sensitivity to inhibitors of topoisomerase and loss of cellular proliferative capacity. Proc Natl Acad Sci USA. 95; 13097-13102.

Lebel M, Spillare EA, Harris CC, Leder P. 1999. The Werner syndrome gene product co-purifies with the DNA replication complex and interacts with PCNA and topoisomerase I. J Biol Chem. 274; 37795-37799.

Liberi G, Maffioletti G, Lucca C, Chiolo I, Baryshnikova A, Cotta-Ramusino C, Lopes M, Pellicioli A, Haber JE, Foiani M. 2005. Rad51-dependent DNA structures accumulate at damaged replication forks in sgs1 mutants defective in the yeast ortholog of BLM RecQ helicase. Genes Dev. 19; 339-350.

Lindor NM, Furuichi Y, Kitao S, Shimamoto A, Arndt C, Jalal S. 2000. Rothmund-Thomson syndrome due to RECQ4 helicase mutations: report and clinical and molecular comparisons with Bloom syndrome and Werner syndrome. Am J Med Genet. 90; 223-228.

Lonn U, Lonn S, Nylen U, Winblad G, German J. 1990. An abnormal profile of DNA replication intermediates in Bloom's syndrome. Cancer Res. 50; 3141-3145.

Martin GM. 1977. Cellular aging-clonal senescence. Am. J. Pathol. 89; 484-511.

Meetei AR, Medhurst AL, Ling C, Xue Y, Singh TR, Bier P, Steltenpool J, Stone S, Dokal I, Mathew CG, Hoatlin M, Joenje H, de Winter JP, Wang W. 2005. A human ortholog of archaeal DNA repair protein Hef is defective in Fanconi anemia complementation group M.Nat Genet. 37; 958-963.

Meetei AR, Sechi S, Wallisch M, Yang D, Young MK, Joenje H, Hoatlin ME, Wang W. 2003. A multiprotein nuclear complex connects Fanconi anemia and Bloom syndrome. Mol Cell Biol. 23; 3417-3426.

Miyajima A, Seki M, Onoda F, Shiratori M, Odagiri N, Ohta K, Kikuchi Y, Ohno Y, Enomoto T. 2000. Sgs1 helicase activity is required for mitotic but apparently not for meiotic function. Mol. Cell. Biol. 20; 6399-6409.

Murray JM, Lindsay HD, Munday CA, Carr AM. 1997. Role of Schizosaccharomyces pombe RecQ homolog, recombination, and checkpoint genes in UV damage tolerance. Mol Cell Biol. 17; 6868-6875.

Nakayama H, Nakayama K, Nakayama R, Irino N, Nakayama Y, Hanawalt PC. 1984. Isolation and genetic characterization of a thymineless-death resistant mutant of *Escherichia coli* K12: identification of a new mutation (*recQ1*) that blocks the RecF recombination pathway. Mol. Gen. Genet. 195; 474-480.

Onoda F, Seki M, Miyajima A, Enomoto T. 2001. Involvement of SGS1 in DNA damage-induced heteroallelic recombination that requires RAD52 in Saccharomyces cerevisiae. Mol Gen Genet. 264; 702-708.

Onoda F, Seki M, Wang W, Enomoto T. 2004. The hyper unequal sister chromatid recombination in an sgs1 mutant of budding yeast requires MSH2. DNA Repair 3; 1355-1362.

Poot M, Hoehn H, Runger TM, Martin GM. 1992. Impaired S-phase transit of Werner syndrome cells expressed in lymphoblastoid cell lines. Exp. Cell Res. 202; 267-273.

Poot M, Yom JS, Whang SH, Kato JT, Gollahon KA, Rabinovitch PS. 2001. Werner syndrome cells are sensitive to DNA cross-linking drugs. FASEB J. 15; 1224-1226.

Poot M, Gollahon KA, Emond MJ, Silber JR, Rabinovitch PS. 2002. Werner syndrome diploid fibroblasts are sensitive to 4-nitroquinoline-N-oxide and 8-methoxypsoralen: implications for the disease phenotype. FASEB J. 16; 757-758.

Prince PR, Ogburn CE, Moser MJ, Emond MJ, Martin GM, Monnat RJ Jr. 1999. Cell fusion corrects the 4-nitroquinoline 1-oxide sensitivity of Werner syndrome fibroblast cell lines. Hum Genet. 105; 132-138.

Puranam KL, Blackshear PJ. 1994. Cloning and characterization of RECQL, a potential human homologue of the *Escherichia coli* DNA helicase RecQ. J. Biol. Chem. 269; 29838-29845.

Sangrithi MN, Bernal JA, Madine M, Philpott A, Lee J, Dunphy WG, Venkitaraman AR. 2005. Initiation of DNA replication requires the RECQL4 protein mutated in Rothmund-Thomson syndrome. Cell 121; 887-898.

Seki M, Miyazawa H, Tada S, Yanagisawa J, Yamaoka T, Hoshino S, Ozawa K, Eki T, Nogami M, Okumura K, Taguchi H, Hanaoka F, Enomoto T. 1994a. Molecular cloning of cDNA encoding human DNA helicase Q1 which has homology to *Escherichia coli* RecQ helicase and localization of the gene at chromosome 12p12. Nucleic Acids Res. 22; 4566-4573.

Seki M, Nakagawa T, Seki T, Kato G, Tada S, Takahashi Y, Yoshimura A, Kobayashi T, Aoki A, Otsuki M, Habermann FA, Tanabe H, Ishii Y, Enomoto T. 2006 Bloom helicase and DNA topoisomerase III α are involoved in the dissolution of sister chromatids. Mol. Cell. Biol. In press.

Seki M, Yanagisawa J, Kohda T, Sonoyama T, Ui M, Enomoto T. 1994b. Purification of two DNA-dependent adenosinetriphosphatases having DNA helicase activity from HeLa cells and comparison of the properties of the two enzymes. J. Biochem. 115; 523-531.

Sharma S, Sommers JA, Choudhary S, Faulkner JK, Cui S, Andreoli L, Muzzolini L, Vindigni A, Brosh RM Jr. 2005. Biochemical analysis of the DNA unwinding and strand annealing activities catalyzed by human RECQ1. J Biol Chem. 280; 28072-28084.

Shimamoto A, Nishikawa K, Kitao S, Furuichi Y. 2000. Human RecQ5beta, a large isomer of RecQ5 DNA helicase, localizes in the nucleoplasm and interacts with topoisomerases 3alpha and 3beta. Nucleic Acids Res. 28; 1647-1655.

Shinya A, Nishigori C, Moriwaki S, Takebe H, Kubota M, Ogino A, Imamura S. 1993. A case of Rothmund-Thomson syndrome with reduced DNA repair capacity. Arch Dermatol. 129; 332-336.

Smith PJ, Paterson MC. 1982. Enhanced radiosensitivity and defective DNA repair in cultured fibroblasts derived from Rothmund Thomson syndrome patients. Mutat Res. 94; 213-228.

Sonoda E, Sasaki MS, Buerstedde JM, Bezzubova O, Shinohara A, Ogawa H, Takata M, Yamaguchi-Iwai Y, Takeda S. 1998. Rad51-deficient vertebrate cells accumulate chromosomal breaks prior to cell death. EMBO J. 17 598-608.

Sonoda E, Sasaki MS, Morrison C, Yamaguchi-Iwai Y, Takata M, Takeda S. 1999. Sister chromatid exchanges are mediated by homologous recombination in vertebrate cells. Mol Cell Biol.19; 5166-5169.

Szekely AM, Chen YH, Zhang C, Oshima J, Weissman SM. 2000. Werner protein recruits DNA polymerase delta to the nucleolus. Proc Natl Acad Sci USA. 97; 11365-11370.

Takao N, Kato H, Mori R, Morrison C, Sonada E, Sun X, Shimizu H, Yoshioka K, Takeda S, Yamamoto K. 1999. Disruption of ATM in p53-null cells causes multiple functional abnormalities in cellular response to ionizing radiation. Oncogene. 18; 7002-7009.

Takeuchi F, Hanaoka F, Goto M, Akaoka I, Hori T, Yamada M, Miyamoto T. 1982. Altered frequency of initiation sites of DNA replication in Werner's syndrome cells. Hum. Genet. 60; 365-368.

Tong AH, et al., 2004. Global mapping of the yeast genetic interaction network. Science 303; 808-813.

Tsurimoto T, Shinozaki A, Yano M, Seki M, Enomoto T. 2005. Human Werner helicase interacting protein 1 (WRNIP1) functions as a novel modulator for DNA polymerase delta. Genes Cells. 10; 13-22.

Ui A, Satoh Y, Onoda F, Miyajima A, Seki M, Enomoto T. 2001. The N-terminal region of Sgs1, which interacts with Top3, is required for complementation of MMS sensitivity and suppression of hyper-recombination in sgs1 disruptants. Mol Genet Genomics. 265; 837-850.

von Kobbe C, Karmakar P, Dawut L, Opresko P, Zeng X, Brosh RM Jr, Hickson ID, Bohr VA. 2002. Colocalization, physical, and functional interaction between Werner and Bloom syndrome proteins. J Biol Chem. 277; 22035-22044.

von Kobbe C, Harrigan JA, May A, Opresko PL, Dawut L, Cheng WH, Bohr VA. 2003. Central role for the Werner syndrome protein/poly(ADP-ribose) polymerase 1 complex in the poly(ADP-ribosyl)ation pathway after DNA damage. Mol Cell Biol. 23; 8601-8613.

Wang W, Seki M, Narita Y, Sonoda E, Takeda S, Yamada K, Masuko T, Katada T, Enomoto T. 2000a. Possible association of BLM in decreasing DNA double strand breaks during DNA replication. EMBO J. 19; 3428-3435.

Wang W, Seki M, Narita Y, Nakagawa T, Yoshimura A, Otsuki M, Kawabe Y, Tada S, Yagi H, Ishii Y, Enomoto T. 2003. Functional relation among RecQ family helicases RecQL1, RecQL5, and BLM in cell growth and sister chromatid exchange formation. Mol Cell Biol. 23; 3527-3535.

Wang W, Seki M, Otsuki M, Tada S, Takao N, Yamamoto K, Hayashi M, Honma M, Enomoto T. 2004. The absence of a functional relationship between ATM and BLM, the components of BASC, in DT40 cells. Biochim Biophys Acta. 1688; 137-144.

Wang W, Bambara RA. 2005. Human Bloom protein stimulates flap endonuclease 1 activity by resolving DNA secondary structure. J Biol Chem. 280; 5391-5399.

Wang XW, Tseng A, Ellis NA, Spillare EA, Linke SP, Robles AI, Seker H, Yang Q, Hu P, Beresten S. et al., 2001. Functional interaction of p53 and BLM DNA helicase in apoptosis. J. Biol. Chem. 276; 32948-32955.

Wang Y, Cortez D, Yazdi P, Neff N, Elledge SJ, Qin J. 2000b BASC, a super complex of BRCA1-associated proteins involved in the recognition and repair of aberrant DNA structures. Genes Dev. 14; 927-939.

Wu L, Davies SL, North PS, Goulaouic H, Riou JF, Turley H, Gatter KC, Hickson ID. 2000 The Bloom's syndrome gene product interacts with topoisomerase III. J Biol Chem. 275; 9636-9644.

Wu L, Davies SL, Levitt NC, Hickson ID. 2001 Potential role for the BLM helicase in recombinational repair via a conserved interaction with RAD51. J Biol Chem. 276; 19375-19381.

Wu L, Hickson ID. 2003 Bloom's syndrome helicase suppresses crossing over during homologous recombination. Nature 426; 870-874.

Wu L, Lung Chan K, Ralf C, Bernstein DA, Garcia PL, Bohr VA, Vindigni A, Janscak P, Keck JL, Hickson ID. 2005 The HRDC domain of BLM is required for the dissolution of double Holliday junctions. EMBO J. 24; 2679-2687.

Yamaguchi-Iwai Y, Sonoda E, Sasaki MS, Morrison C, Haraguchi T, Hiraoka Y, Yamashita YM, Yagi T, Takata M, Price C, Kakazu N, Takeda S. 1999 Mre11 is essential for the maintenance of chromosomal DNA in vertebrate cells. EMBO J. 18; 6619-6629.

Yang Q, Zhang R, Wang XW, Spillare EA, Linke SP, Subramanian D, Griffith JD, Li JL, Hickson ID, Shen JC, Loeb LA, Mazur SJ, Appella E, Brosh RM Jr, Karmakar P, Bohr VA, Harris CC. 2002 The processing of Holliday junctions by BLM and WRN helicases is regulated by p53. J Biol Chem. 277; 31980-31987.

Yannone SM, Roy S, Chan DW, Murphy MB, Huang S, Campisi J, Chen DJ. 2001 Werner syndrome protein is regulated and phosphorylated by DNA-dependent protein kinase. J Biol Chem. 276; 38242-38248.

Yu CE, Oshima J, Fu YH, Wijsman EM, Hisama F, Alisch R, Matthews S, Nakura J, Miki T, Ouais S, Martin GM, Mulligan J, Schellenberg GD. 1996 Positional cloning of the Werner's syndrome gene. Science 272; 258-262.

Chapter 6

GENETIC ANALYSIS OF APOPTOTIC EXECUTION

Sandrine Ruchand[1], Kumiko Samejima[1], Damien Hudson[1], Scott H. Kaufmann[2] and William C. Earnshaw[1]
[1]Wellcome Trust Centre for Cell Biology, ICMB, Swann Building, University of Edinburgh, Mayfield Road, Edinburgh EH9 3JR, Scotland, UK
[2]Division of Oncology Research, Mayo Clinic, 200 First St., S.W., Rochester, MN 55905

Abstract: Chicken DT40 cells are a very favourable system to use for the study of apoptosis. These cells undergo apoptosis readily in response to a variety of physiological and experimental stimuli. Their response to chemotherapeutic agents such as etoposide is rapid and efficient, facilitating biochemical analysis of apoptotic execution. Because many of the genes involved in apoptotic execution are not essential for cellular life, DT40 cells also provide a ready system for the genetic analysis of apoptosis, in which true null mutations may be isolated. Here we describe standard procedures for the induction and analysis of apoptosis in DT40 cells. We also describe a few of the conclusions from our studies in which several components of the apoptotic execution pathway have been knocked out.

Key words: apoptotic execution; caspases

1.   INTRODUCTION

Cell death by apoptosis is an essential mechanism by which unwanted cells are eliminated from numerous biological systems. It is characterized by a series of distinct morphological and biochemical changes (Wyllie et al., 1980). The apoptotic process can be divided into two different phases: a death commitment phase followed by an execution phase during which the

J.-M. Buerstedde and S. Takeda (eds.), Reviews and Protocols in DT40 Research, 75–90.

cell undergoes dramatic alterations (Takahashi and Earnshaw, 1996). These alterations include chromatin condensation, plasma membrane blebbing, decrease in cell volume, nuclear fragmentation and formation of apoptotic bodies. At the nuclear level, the chromatin is fragmented and the DNA is cleaved at internucleosomal regions giving a typical ladder pattern on agarose gel electrophoresis (Wyllie et al., 1980). The major nuclease responsible for this distinct DNA cleavage has been identified as CAD/CPAN/DFF40 (Enari et al., 1998; Halenbeck et al., 1998; Liu et al., 1998).

Various processes that occur during the execution phase of apoptosis, including CAD activation, require the catalytic activity of a family of highly specific <u>c</u>ysteine-dependent <u>asp</u>artate specific prote<u>ases</u> called caspases (Cohen, 1997; Thornberry, 1998; Earnshaw et al., 1999). Caspases involved in the apoptotic process have been grouped into two main classes according to their primary structure and direct experimental evidence. The first class, initiator caspases, have long prodomains that enable them to transduce various signals into proteolytic activity, thereby initiating a caspase cascade leading to the activation of the downstream effector caspases. The second class, effector caspases, lack extended prodomains that sense various signals and are instead activated *en masse* by the initiator caspases. Once activated, these effector caspases are responsible for the bulk of the proteolytic cleavages observed during apoptosis (for reviews see (Budihardjo et al., 1999; Earnshaw et al., 1999; Hengartner, 2000)).

Many different apoptotic signaling pathways have been studied in DT40 cells. This system has been a useful model for studying gamma irradiation and anti-IgM cross-linking of B cells surface receptors (Takata et al., 1994; Uckun et al., 1996). More recently we have taken advantage of DT40 cells to study the apoptotic execution process (Samejima et al., 2001; Ruchaud et al., 2002; Korfali et al., 2004).

In this chapter we describe different assay systems used to study apoptosis in DT40 cells. Then we present specific examples of how the DT40 gene knockout system helped us dissect the apoptotic execution process and in particular the events taking place at the nuclear level.

2.   ASSAY SYSTEMS TO STUDY APOPTOSIS IN DT40 CELLS

2.1   Inducing apoptosis in DT40 cells

A wide variety of apoptosis-inducing agents have been used on DT40 cells. These apoptotic stimuli range from gamma/UV irradiation, anti-IgM cross-linking of surface B cell receptors, oxidative stress and chemotherapeutic agents (Takata et al., 1994; Uckun et al., 1996; Maruo et al., 1999; Ding et al., 2000; Hawkins et al., 2002). These treatments have mostly been used to study the involvement of specific genes (as DT40 gene knockouts) in particular apoptotic-signaling pathways.

2.2 Apoptotic stimuli used in signaling studies

A wide variety of apoptotic stimuli have been employed in previous studies of apoptosis in DT40 cells. For example, a number of DNA damaging agents, including gamma irradiation, methyl methanesulfonate (MMS) and the DNA cross linking agent cisplatin, have been used to analyze the involvement of RAD54, RAD52 and RAD51 in DNA repair and homologous recombination in DT40 cells (Bezzubova et al., 1997; Sonoda et al., 1998; Yamaguchi-Iwai et al., 1998). In DT40 cells with targeted disruption of genes encoding the protein tyrosine kinases (PTK) Lyn, Syk and BTK, as well as phospholipase C-γ2, the effect of gene disruption was studied using anti-IgM cross-linking or ionizing radiation to induce apoptosis (Takata et al., 1994; Takata et al., 1995; Yang et al., 1995; Uckun et al., 1996). Further studies examined the effects of chemotherapeutic agents such as cytarabine, etoposide, doxorubicin, campto-thecin, and cisplatin on several PTK knockout lines (*Lyn$^{-/-}$, Syk$^{-/-}$* and *BTK$^{-/-}$*) and demonstrated the involvement of Lyn in the apoptotic signaling pathway induced by topoisomerase II inhibitors (Maruo et al., 1999). More recently, Ding et al., (Ding et al., 2000) induced apoptosis in DT40 cells using oxidative stress (H_2O_2) and reported an essential role of Syk in the activation of the serine-threonine kinase Akt survival pathway after exposure to this stimulus. In other studies, the microtubule poisons vinblastine and nocodazole were utilized to demonstrate a critical role of the *MEKK1* gene in the response to mitotic spindle dysfunction (Kwan et al., 2001). Finally, antibodies to the putative B cell surface death receptor chB6 have been utilized to explore the apoptotic response upon cross-linking of this molecule cells (Funk et al., 2003).

2.2.1 Agents used to study apoptotic execution

When the emphasis has been on the analysis of apoptotic pathways per se, the topoisomerase II inhibitor etoposide, the DNA damaging agent bleomycin and the protein kinase inhibitor staurosporine have been used to study apoptotic execution events in DT40 cells (Samejima et al., 2001; Ruchaud et al., 2002; Korfali et al., 2004). Etoposide has been the drug of choice in our lab for several reasons: rapid apoptotic induction (<2h), synchrony of apoptosis and highly reproducible results.

2.3 Detecting and quantifying apoptotic cells

Apoptotic cells can be detected based on their distinct morphological and biochemical features. These include chromatin condensation, DNA fragmentation, phosphatidylserine (PS) exposure, caspase activation, and mitochondrial dysfunction.

2.3.1 Chromatin condensation

Chromatin condensation can be visualized by fluorescence microscopy using dyes that stain nucleic acid (DAPI, Hoechst, etc.) (Samejima et al., 2001; Ruchaud et al., 2002; Korfali et al., 2004). Apoptotic cell nuclei exhibit chromatin condensation as shown in Figure 1.

2.3.2 DNA fragmentation

Oligonucleosomal DNA cleavage occurring in the later stages of apoptotic execution can be detected by conventional agarose gel electrophoresis (Hughes and Cidlowski, 2000; Kaufmann et al., 2000; Samejima et al., 2001). HMW DNA cleavage occurring in the early stages can be detected by pulsed-field gel electrophoresis. Another method commonly used to detect DNA fragmentation is the Terminal transferase-mediated dUTP nick end-labeling (TUNEL) reaction. This labels the 3'hydroxy ends of the DNA breaks. TUNEL-labeled cells can be analyzed either under the microscope or by flow cytometry. Furthermore, since apoptotic cells lose part of their fragmented DNA upon fixation and washing under non-cross-linking conditions, they can be analyzed according to their DNA content by flow cytometry and detected as a sub G1 peak.

2.3.3 Phosphatidylserine exposure

Phosphatidylserine (PS) exposure on the outer leaflet of the plasma membrane is a hallmark of apoptosis (Fadok et al., 1992). PS binds with high affinity to the polypeptide annexin V in a Ca^{2+}-dependent manner. Therefore, Annexin V conjugated to a variety of dyes is commonly used to detect apoptotic cells by either microscopy or flow cytometry (Green, 2000; Samejima et al., 2001; Ruchaud et al., 2002; Korfali et al., 2004).

2.3.4   Caspase activation

Caspase activation is another hallmark of apoptosis (Cohen, 1997; Budihardjo et al., 1999; Earnshaw et al., 1999). Thus, identifying cells in which caspases are active and quantifying their activity is a good method to detect and quantify apoptotic cells (see Section 2.4 for more details) (Stennicke and Salvesen, 2000).

2.3.5   Mitochondrial dysfunction and cytochrome c release

Mitochondria contain a variety of pro- and anti-apoptotic factors. Thus mitochondrial dysfunction and the release of factors such as cytochrome c can trigger cell death. The presence of these factors in the cytosol can therefore serve as an apoptotic marker (Zamzami et al., 2000; Hawkins et al., 2002; Brown et al., 2004; Green and Kroemer, 2004).

2.4   Isolating apoptotic DT40 cells

A typical healthy DT40 cell culture contains approximately 5% of dying or dead cells. Upon the induction of apoptosis, the bulk of the cells usually undergo apoptosis in asynchronous manner. Therefore, isolation of apoptotic cells from a mixed population is sometimes required for biochemical analysis of apoptotic pathways.

2.4.1   Elutriation

A cell population can be fractionated on the basis of cell shape and density using elutriation. As described in Method 7, elutriation is a powerful method to obtain cells synchronized in the different phases of cell cycle. We have used this technique to successfully isolate apoptotic cells using conditions to collect cells of 8 μm in diameter (Beckmann Rotor JE5 4000 rpm 40 ml/min; our unpublished data).

2.4.2   Isolation of apoptotic cells using annexin V-biotin and streptavidin-magnetic beads

The high affinity between PS and annexin V can be also used for the isolation of apoptotic cells. Binding of annexin V-biotin to apoptotic cells followed by the binding of biotin to streptavidin-magnetic beads enables separation of apoptotic cells from living ones. We successfully separated apoptotic cells from a mixed population in this way using a commercially available kit. This procedure is easy and quick (less than 30 min). The biggest advantage of this method is that it does not require special equipment such as an elutriator or a cell sorter. The disadvantage is that apoptotic cells retain external annexin V-biotin.

2.4.3   Other methods

It is possible to separate cells using a fluorescent cell sorter. Apoptotic cells labelled with annexin V conjugated to a fluorescent dye or labeled with the TUNEL reaction can be separated by cell sorting. Moreover, cells containing lower amounts of DNA than normal (sub-G1 peak) can be collected as apoptotic cells. The biggest advantage of this method is that analysis and sorting of cells can be preformed in the same time.

Another way to separate cells utilizes the difference in density between living and apoptotic cells. The isolation of apoptotic Molt4 human leukemia cells by ficoll gradient has been reported (Smith et al., 1992) and it should be possible to apply similar methodology to DT40 cells.

2.5   Caspase affinity labeling and activity analysis in DT40 apoptotic cell extracts

Each caspase recognizes a very specific peptide sequence (for review see (Cohen, 1997; Earnshaw et al., 1999)). By adding a chromogenic or fluorogenic group at the C-terminus of the preferred caspase recognition sequences, it has been possible to design somewhat specific synthetic peptide substrates to measure caspase activities (for a comprehensive review see (Stennicke and Salvesen, 2000). A wide variety of caspase inhibitors have been engineered similarly. Most of them are aldehyde, fluoromethyl ketone or chloromethyl ketone derivatives of each caspase cleavage peptide sequence. While the aldehydes are reversible inhibitors, the fluoro- and chloromethyl ketones are irreversible as result of their ability to alkylate the active cysteine residue present in the caspase active site. These compounds have been further modified by the addition of detectable moieties such as fluorochromes or biotin that allow detection of specific active caspases in cell lysates or localization at a single cell level.

These synthetic substrates have proven to be very useful tools to analyze the role of caspases during the apoptotic process of a variety of cell lines derived from a large number of species. The fact that the cleavage recognition sequence of each caspase is conserved between species has made these tools available to apoptotic studies in chicken DT40 cells. (Takahashi et al., 1996a; Martins et al., 1997; Ruchaud et al., 2002; Korfali et al., 2004).

2.6 In vitro apoptosis induction using DT40 apoptotic cell extracts on isolated nuclei

In order to dissect apoptotic events occurring at the nuclear level, our laboratory has developed an *in vitro* assay in which isolated interphase nuclei (Wood and Earnshaw, 1990) are incubated in cytosolic extracts prepared from apoptotic cells (Lazebnik et al., 1995b). This technique has the advantage that apoptotic changes are induced in a highly synchronous manner in the nuclei. Nuclei isolated from HeLa, Jurkat and DT40 cells have been successfully used in this assay. The cytosolic extracts can be prepared from any cell type (i.e., DT40 wild type or knockout cells) induced to undergo apoptosis by a variety of treatments. Furthermore, the apoptotic extracts can be immuno-depleted for a certain protein or supplemented with specific inhibitors enzymes (e.g., caspases, kinases, phosphatases) implicated in apoptosis. This flexible system, which allows a wide variety of combinations, turned out to be essential to analyze the phenotype of one of our DT40 knockout (see Section 3.4 below).

In these experiments, apoptosis is induced in isolated nuclei after addition of apoptotic cell extracts and an ATP regenerating system. The nuclei are harvested during a time course for up to 5 h and then subjected to one or more different types of analysis, including immunoblotting, immunofluorescence, agarose gel electrophoresis to detect DNA fragmentation, or morphological analysis (i.e., for chromatin condensation) after DNA. As illustrated in Fig. 1, the apoptotic morphology after DNA staining can be classified into sequential stages according to the level of chromatin condensation.

3. DT40 KNOCKOUTS RELATED TO APOPTOTIC EXECUTION

In order to analyze the events taking place in the nucleus during apoptotic execution, we took advantage of the DT40 system to generate several knockout cell lines. To further analyze the DNA fragmentation and chromatin condensation process, we generated a DT40 cell line deficient for the CAD/CPAN/DFF40 nuclease (Samejima et al., 2001). We also generated cell lines deficient for caspase-6 and -7, two downstream effector caspases whose functions were poorly understood (Ruchaud et al., 2002; Korfali et al., 2004).

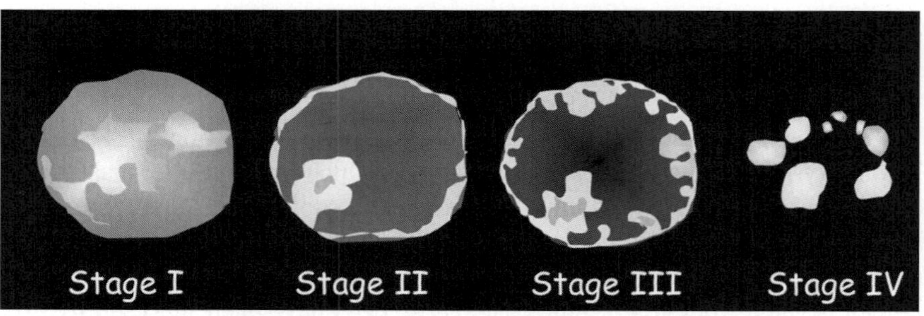

Figure 6-1. Schematic representation of nuclei incubated in DT40 apoptotic extracts or control DT40 extracts following DAPI staining. Sequential stages of chromatin condensation occurring during the incubation are represented. Nuclei incubated in control extracts define stage I. After addition of apoptotic extracts, the chromatin begins to condense against the nuclear periphery and in the nucleoli (stage II). Then, the peripheral chromatin ring condenses into discrete masses that separate from one another (stage III). Finally, the chromatin masses form discrete apoptotic bodies and the nuclear shape is lost (stage IV), ending with the nuclei disintegration.

3.1   Vector design

Targeting vectors were designed for *CAD, Caspase-6* and *Caspase-7* using DNA from a lambdaFix II DT40 genomic library screened with chicken cDNA probes. The genomic library was made in our lab, as no commercial DT40 genomic library was available. The targeting arms were approximately 2 and 3kb in length for each targeting vector and were either cut out directly from the phage DNA using restriction enzymes or were amplified by PCR. Because these apoptotic genes are not essential, no rescue construct was required.

It was possible to delete the entire open reading frame for CAD. For Caspase 6, 888 bp out of 915 bp were deleted from the open reading frame, leaving only nine amino acids of the prodomain. From the Caspase 7 open reading frame, 786 bp of 930 bp were deleted, leaving 13 amino acids of the prodomain and 3 amino acids of the large subunit.

In general, vectors are designed to disrupt an N-terminal part of the protein, as small peptides are usually not functional or expressed. It is preferable to delete the entire open reading frame but this depends on the availability of cloned upstream sequence. Also, for bigger genes with many introns, large deletions can lead to complications such as the possibility of deleting or interfering with another gene embedded within the sequence.

3.2   *CAD* null DT40 cells

DNA fragmentation and chromatin condensation are two major nuclear features of apoptosis. DNA fragmentation generally occurs at two levels: Early in the process 50-200 kbp fragments known as high molecular weight (HMW) fragments are generated; and later on oligonucleosomal DNA fragments become evident (Samejima and Earnshaw, 2005). Chromatin condensation can also be divided into different phases: chromatin condensed against nuclear periphery (stage II-III) followed by the formation of discrete ball-like masses of chromatin (stage IV) (see Figure 1 above).

CAD/CPAN/DFF40 is a major apoptotic nuclease. In living cells, CAD exists as an inactive complex with ICAD/DFF45. In apoptotic cells, cleavage of ICAD by caspase-3 and/or -7 activates CAD. Purified CAD is capable of inducing both oligonucleosomal DNA fragmentation and apoptotic chromatin condensation in isolated nuclei. It has been suggested that CAD might be responsible for HMW DNA cleavage (Zhang et al., 1998; McIlroy et al., 1999).

3.2.1   CAD is required for oligonucleosomal DNA fragmentation but dispensable for HMW DNA cleavage in etoposide-treated DT40 cells

To clarify the role of CAD in apoptosis, we generated a chicken DT40 subline in which both alleles of the gene were disrupted (Samejima et al., 2001).

Etoposide (a topoisomerase II inhibitor) induced apoptosis in $CAD^{-/-}$ cells as efficiently as in wild type cells. However, oligonucleosomal DNA fragmentation was not detected in $CAD^{-/-}$ cells, indicating that CAD is essential for oligonucleosomal DNA fragmentation in these cells. Surprisingly, HMW DNA cleavage occurs normally in this subline, suggesting that factors other than CAD could generate these fragments. It is possible that in this instance HMW cleavage was caused by topo II, the target of etoposide, as topo II has been shown to cause HMW cleavage under certain (but not all) conditions (Li et al., 1999).

3.2.2   CAD is required for completion of nuclear disassembly but dispensable for initial stages of chromatin condensation

Upon the induction of apoptosis, $CAD^{-/-}$ cells exhibited stage II-III chromatin condensation but failed to proceed to stage IV. This result suggested that CAD is required for completion of nuclear disassembly during apoptosis but that other factors are responsible for the initial stages of chromatin condensation.

3.3 *Caspase-7* null DT40 cells

Caspase-7 proenzyme contains a short prodomain and is believed to be a downstream effector caspase. Caspase-7 shares a high degree of homology with caspase-3 and shows very similar activities against synthetic tetra-peptide-based substrates *in vitro*, though caspase-7 might have some unique roles during apoptosis (Fernandes-Alnemri et al., 1995; Talanian et al., 1997). Caspase-7 appears to be essential for the clonal deletion of auto-reactive B cells by B-cell receptor cross-linking (Nitta et al., 2001). IgM treatment of immature B-lymphocytes causes growth arrest and caspase-7 activation independently of caspase-8 activation and cytochrome c release (Ruiz-Vela et al., 1999). In addition, B-cell receptor cross-linking induces the selective activation of caspase-7 but not caspase-3 (Bras et al., 1999).

3.3.1 Delayed CAD activation in Caspase-7-deficient cells

To further clarify the role of caspase-7 during apoptotic execution we generated a chicken DT40 cell line in which both alleles of the gene were disrupted (Korfali et al., 2004).

In analyzing the caspase-7 deficiency phenotype we first observed that the cells were more resistant to the commonly used apoptosis-inducing drugs etoposide and staurosporine. We then showed that *caspase-7$^{-/-}$* cells exhibited a delay in both TUNEL labeling and DNA fragmentation after apoptotic induction, suggesting a delay in the activation of CAD/CPAN/DFF40 (Samejima et al., 2001). These observations complement our results obtained with *CAD$^{-/-}$* cells and suggest that caspase-7 may be directly or indirectly involved in ICAD/DFF45 cleavage and therefore in the activation of CAD.

3.3.2 A possible upstream role for caspase-7

Because caspase-7 is not thought to cleave ICAD directly, further studies were performed to examine a possible regulatory role for chicken caspase-7. WhileIt is believed that caspase-7 is an effector caspase playing a role in the later stages of apoptotic execution, our results also revealed that *caspase-7$^{-/-}$* cells show a delay in PS externalization during apoptosis induced by drugs such as etoposide and staurosporine, an event that is believed to be an early apoptotic marker.

We then analyzed the cleavage of various caspase-7 substrates such as PARP, lamin B1 and B2. To analyze the cleavage of PARP, a classical substrate of both caspases -3 and -7, we used a cell-free system (see Section

2.6). *Caspase-7 $^{-/-}$* cells showed a delay in cleaving PARP1 as well as lamin B1 and B2.

The fact that these substrates were cleaved eventually in *caspase-7 $^{-/-}$* cells implied that caspase-7 is not essential for their cleavage, but instead suggested that the enzyme might be involved in activating other caspases that are responsible for processing these substrates. Consistent with this hypothesis, we demonstrated that the activation of several caspases (-3, -6) was delayed in the null cells undergoing apoptosis, suggesting that caspase-7 might be upstream of these other effector caspases in DT40 cells during apoptosis.

All our results strongly suggested that caspase-7 functions close to the apex of the apoptotic cascade in DT40 B lymphocytes, earlier than other effector caspases.

3.4   *Caspase-6* null DT40 cells

The specific role played by caspase-6 during apoptosis was much less well understood prior to our study. Nuclear lamins were the first caspase-6 substrates identified (Orth et al., 1996; Takahashi et al., 1996a). Studies in which caspase-6 activity was selectively inhibited with the serpin SPI-2 suggested that lamin cleavage might be required for nuclear disassembly during apoptotic execution (Lazebnik et al., 1995a; Takahashi et al., 1996b). Other substrates have been reported to be cleaved by caspase-6, but these are generally also cleaved by other caspases such as caspase-3. In fact, the only substrate presently thought to be cleaved exclusively by caspase-6 is lamin A/C (Takahashi et al., 1996a; Slee et al., 2001). Because expression of uncleavable mutant lamin A or B causes significant delays in the onset of chromatin condensation and nuclear shrinkage during apoptosis (Rao et al., 1996), lamin cleavage appears to be an important event in the nuclear apoptotic process.

3.4.1   Caspase-6 deficient DT40 cells showed no apoptotic phenotype

To examine the specific role of caspase-6 during nuclear disassembly, we isolated a DT40 clone in which both alleles of the *Caspase-6* gene were disrupted (Ruchaud et al., 2002). To our surprise, no obvious morphological differences were observed in the apoptotic process of those deficient cells.

Because lamin A is thought to be cleaved only by caspase-6 and many lymphoid cells do not express lamin A (Guilly et al., 1987; Kaufmann, 1989), we checked lamin A/C status in DT40 cells. We were unable to detect any lamin A/C in DT40 cells.

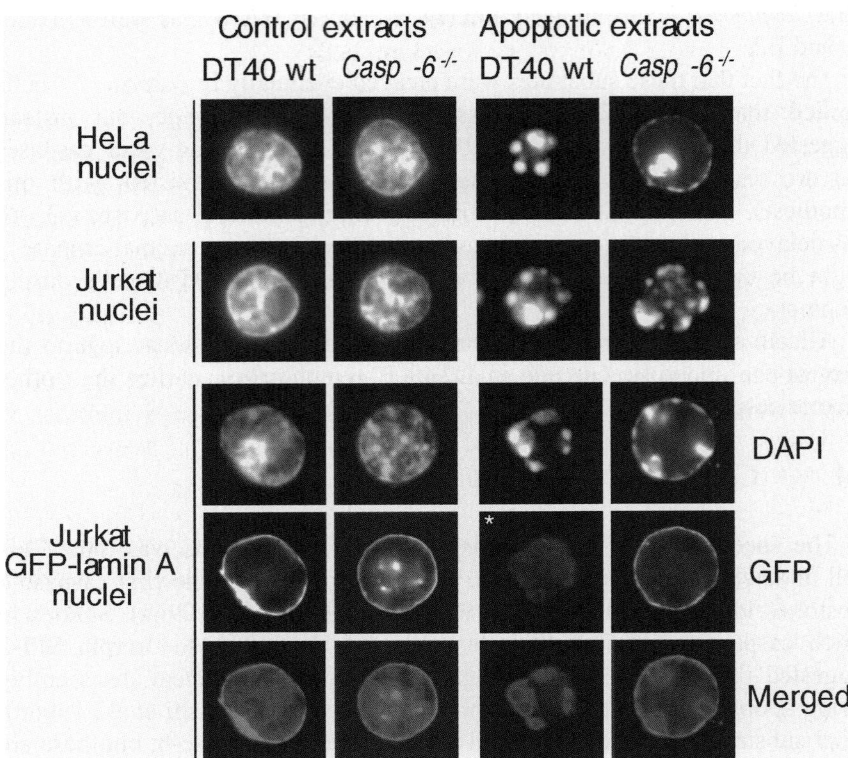

Figure 6-2. Isolated HeLa, Jurkat or Jurkat:GFP-lamin A nuclei were incubated for 2 h in extracts from wild type and *caspase-6⁻ᐟ⁻* DT40 cells treated with 1 μM staurosporine or diluent for 8 h. A nucleus representative of the major population is shown for each condition (DAPI) along with the GFP in the Jurkat:GFP-lamin A nuclei. The signal in the panel indicated by a white star was enhanced relative that in the other panels in order to see the residual GFP-lamin A fluorescence (From Ruchaud, 2002, EMBO J.).

3.4.2 What we can learn from the use of cell-free systems prepared from genetic knockout cells

We had to turn to a cell-free system in order to examine the role of lamin cleavage in nuclear disassembly. In this system, HeLa cell nuclei (containing Lamin A) were induced to undergo apoptotic morphological changes in the presence of extracts from wild type and *caspase-6*$^{-/-}$ apoptotic DT40 cells (see Section 2.6). The use of this system allowed us to detect a block in chromatin condensation and apoptotic body formation in HeLa cell nuclei in the absence of caspase-6 (see Fig. 2). Similar results were obtained when nuclei from lamin A-transfected lymphoid Jurkat or DT40 cells were incubated in caspase-6 deficient apoptotic extracts (Ruchaud et al., 2002); (unpublished data) (see Fig. 2). Furthermore, the caspase-6 inhibitor z-VEID-fmk mimicked the effects of caspase-6 deficiency and prevented the cleavage of lamin A.

The use of the cell-free system enabled us to show that, in cells expressing lamin A, caspase-6-mediated lamin A cleavage is essential for the chromatin to undergo complete condensation during apoptosis and therefore for the formation of apoptotic bodies. Taken together our results strongly suggest that the only essential function of this particular caspase during apoptosis is restricted to the cleavage of Lamin-A.

CONCLUSION

DT40 cells have turned out to be a highly useful tool to study the apoptotic execution process. The results we described here for three independent DT40 knockouts (three enzymes) demonstrate the power of this model for obtaining a much better understanding of the role played by all the different factors involved in the apoptotic process. The fact that the system allows sequential knockouts of multiple genes makes it even more powerful for studying cell death events where most of the factors involved are not essential for life.

REFERENCES

Bezzubova O, Silbergleit A, Yamaguchi-Iwai Y, Takeda S, Buerstedde JM. 1997. Reduced X-ray resistance and homologous recombination frequencies in a RAD54-/- mutant of the chicken DT40 cell line. Cell 89:185-193.

Bras A, Ruiz-Vela A, Gonzalez de Buitrago G, Martinez AC. 1999. Caspase activation by BCR cross-linking in immature B cells: differential effects on growth arrest and apoptosis. Faseb J 13:931-944.

Brown CY, Bowers SJ, Loring G, Heberden C, Lee RM, Neiman PE. 2004. Role of Mtd/Bok in normal and neoplastic B-cell development in the bursa of Fabricius. Dev Comp Immunol 28:619-634.

Budihardjo I, Oliver H, Lutter M, Luo X, Wang X. 1999. Biochemical pathways of caspase activation during apoptosis. Annu Rev Cell Dev Biol 15:269-290.

Cohen GM. 1997. Caspases: the executioners of apoptosis. Biochem. J. 326:1-16.

Ding J, Takano T, Gao S, Han W, Noda C, Yanagi S, Yamamura H. 2000. Syk is required for the activation of Akt survival pathway in B cells exposed to oxidative stress. J Biol Chem 275:30873-30877.

Earnshaw WC, Martins LM, Kaufmann SH. 1999. Mammalian caspases: Structure, activation, substrates and functions during apoptosis. Ann. Rev. Biochem. 68:383-424.

Enari M, Sakahira H, Yokoyama H, Okawa K, Iwamatsu A, Nagata S. 1998. A caspase-activated DNase that degrades DNA during apoptosis, and its inhibitor ICAD. Nature 391:43-50.

Fadok VA, Voelker DR, Campbell PA, Cohen JJ, Bratton DL, Henson PM. 1992. Exposure of phosphatidylserine on the surface of apoptotic lymphocytes triggers specific recognition and removal by macrophages. J Immunol 148:2207-2216.

Fernandes-Alnemri T, Takahashi A, Armstrong R, Krebs J, Fritz L, Tomaselli K, Wang L, Yu Z, Croce CM, Salvesen G, Earnshaw WC, Litwack G, Alnemri ES. 1995. *Mch3*, a novel human apoptotic cysteine protease highly related to CPP32. Cancer Res. 55:6045-6052.

Funk PE, Pifer J, Kharas M, Crisafi G, Johnson A. 2003. The avian chB6 alloantigen induces apoptosis in DT40 B cells. Cell Immunol 226:95-104.

Green DR. 2000. Apoptosis and sphingomyelin hydrolysis. The flip side. J Cell Biol 150: F5-7.

Green DR, Kroemer G. 2004. The pathophysiology of mitochondrial cell death. Science 305:626-629.

Guilly MN, Bensussan A, Bourge JF, Bornens M, Courvalin JC. 1987. A human T lymphoblastic cell line lacks lamins A and C. Embo J 6:3795-3799.

Halenbeck R, MacDonald H, Roulston A, Chen TT, Conroy L, Williams LT. 1998. CPAN, a human nuclease regulated by the caspase-sensitive inhibitor DFF45. Curr Biol 8:537-540.

Hawkins TE, Das D, Young B, Moss SE. 2002. DT40 cells lacking the Ca^{2+}-binding protein annexin 5 are resistant to Ca^{2+}-dependent apoptosis. Proc Natl Acad Sci U S A 99:8054-8059.

Hengartner MO. 2000. The biochemistry of apoptosis. Nature 407:770-776.

Hughes FM, Jr., Cidlowski JA. 2000. Apoptotic nuclease assays. Methods Enzymol 322: 47-62.

Kaufmann SH. 1989. Additional members of the rat liver lamin polypeptide family. Structural and immunological characterization. J Biol Chem 264:13946-13955.

Kaufmann SH, Mesner PW, Jr., Samejima K, Tone S, Earnshaw WC. 2000. Detection of DNA cleavage in apoptotic cells. Methods Enzymol 322:3-15.

Korfali N, Ruchaud S, Loegering D, Bernard D, Dingwall C, Kaufmann SH, Earnshaw WC. 2004. Caspase-7 gene disruption reveals an involvement of the enzyme during the early stages of apoptosis. J Biol Chem 279:1030-1039.

Kwan R, Burnside J, Kurosaki T, Cheng G. 2001. MEKK1 is essential for DT40 cell apoptosis in response to microtubule disruption. Mol Cell Biol 21:7183-7190.

Lazebnik YA, Takahashi A, Moir R, Goldman R, Poirier GG, Kaufmann SH, Earnshaw WC. 1995a. Studies of the lamin proteinase reveal multiple parallel biochemical pathways during apoptotic execution. Proc. Nat. Acad. Sci. (USA) 92:9042-9046.

Lazebnik YA, Takahashi A, Poirier GG, Kaufmann SH, Earnshaw WC. 1995b. Characterization of the Execution Phase of Apoptosis *in vitro* Using Extracts From Condemned-Phase Cells. Journal of Cell Science S19:41-49.

Li TK, Chen AY, Yu C, Mao Y, Wang H, Liu LF. 1999. Activation of topoisomerase II-mediated excision of chromosomal DNA loops during oxidative stress. Genes Dev 13:1553-1560.

Liu X, Li P, Widlak P, Zou H, Luo X, Garrard WT, Wang X. 1998. The 40-kDa subunit of DNA fragmentation factor induces DNA fragmentation and chromatin condensation during apoptosis. Proc Natl Acad Sci USA 95:8461-8466.

Martins LM, Kottke T, Mesner PW, Basi GS, Sinha S, Frigon N, Jr., Tatar E, Tung JS, Bryant K, Takahashi A, Svingen PA, Madden BJ, McCormick DJ, Earnshaw WC, Kaufmann SH. 1997. Activation of multiple interleukin-1beta converting enzyme homologues in cytosol and nuclei of HL-60 cells during etoposide-induced apoptosis. J Biol Chem 272:7421-7430.

Maruo A, Oishi I, Sada K, Nomi M, Kurosaki T, Minami Y, Yamamura H. 1999. Protein tyrosine kinase Lyn mediates apoptosis induced by topoisomerase II inhibitors in DT40 cells. Int Immunol 11:1371-1380.

McIlroy D, Sakahira H, Talanian RV, Nagata S. 1999. Involvement of caspase 3-activated DNase in internucleosomal DNA cleavage induced by diverse apoptotic stimuli. Oncogene 18:4401-4408.

Nitta T, Igarashi K, Yamashita A, Yamamoto M, Yamamoto N. 2001. Involvement of polyamines in B cell receptor-mediated apoptosis: spermine functions as a negative modulator. Exp Cell Res 265:174-183.

Orth K, Chinnaiyan AM, Garg M, Froelich CJ, Dixit VM. 1996. The CED-3/ICE-like protease Mch2 is activated during apoptosis and cleaves the death substrate lamin A. J. Biol. Chem. 271:16443-16446.

Rao L, Perez D, White E. 1996. Lamin proteolysis facilitates nuclear events during apoptosis. J. Cell Biol. 135:1441-1455.

Ruchaud S, Korfali N, Villa P, Kottke TJ, Dingwall C, Kaufmann SH, Earnshaw WC. 2002. Caspase-6 gene disruption reveals a requirement for lamin A cleavage in apoptotic chromatin condensation. Embo J 21:1967-1977.

Ruiz-Vela A, Gonzalez de Buitrago G, Martinez AC. 1999. Implication of calpain in caspase activation during B cell clonal deletion. Embo J 18:4988-4998.

Samejima K, Earnshaw WC. 2005. Trashing the genome: the role of nucleases during apoptosis. Nat Rev Mol Cell Biol.

Samejima K, Tone S, Earnshaw WC. 2001. CAD/DFF40 nuclease is dispensable for high molecular weight DNA cleavage and stage I chromatin condensation in apoptosis. J Biol Chem 276:45427-45432.

Slee EA, Adrain C, Martin SJ. 2001. Executioner caspase-3, -6, and -7 perform distinct, non-redundant roles during the demolition phase of apoptosis. J. Biol. Chem. 276:7320-7326.

Smith GK, Duch DS, Dev IK, Kaufmann SH. 1992. Metabolic effects and kill of human T-cell leukemia by 5-deazaacyclotetrahydrofolate, a specific inhibitor of glycineamide ribonucleotide transformylase. Cancer Res 52:4895-4903.

Sonoda E, Sasaki MS, Buerstedde JM, Bezzubova O, Shinohara A, Ogawa H, Takata M, Yamaguchi-Iwai Y, Takeda S. 1998. Rad51-deficient vertebrate cells accumulate chromosomal breaks prior to cell death. Embo J 17:598-608.

Stennicke HR, Salvesen GS. 2000. Caspase assays. Methods Enzymol 322:91-100.

Takahashi A, Alnemri E, Lazebnik YA, Fernandes-Alnemri T, Litwack G, Moir RD, Goldman RD, Poirier GG, Kaufmann SH, Earnshaw WC. 1996a. Cleavage of lamin A by Mch2α but not CPP32: Multiple ICE-related proteases with distinct substrate recognition properties are active in apoptosis. Proc. Nat. Acad. Sci. (USA) 93:8395-8400.

Takahashi A, Earnshaw WC. 1996. ICE-related proteases in apoptosis. Curr Opin Genet Dev 6:50-55.

Takahashi A, Musy P-Y, Martins LM, Poirier GG, Turner PC, Moyer RW, W.C. E. 1996b. CrmA/SPI-2 inhibition of an endogenous ICE-related protease responsible for lamin A cleavage and apoptotic nuclear fragmentation. J. Biol. Chem. 271:32487-32490.

Takata M, Homma Y, Kurosaki T. 1995. Requirement of phospholipase C-gamma 2 activation in surface immunoglobulin M-induced B cell apoptosis. J Exp Med 182: 907-914.

Takata M, Sabe H, Hata A, Inazu T, Homma Y, Nukada T, Yamamura H, Kurosaki T. 1994. Tyrosine kinases Lyn and Syk regulate B cell receptor-coupled Ca^{2+} mobilization through distinct pathways. Embo J 13:1341-1349.

Talanian RV, Quinlan C, Trautz S, Hackett MC, Mankovich JA, Banach D, Ghayur T, Brady KD, Wong WW. 1997. Substrate specificities of caspase family proteases. J. Biol. Chem. 272:9677-9682.

Thornberry NA. 1998. Caspases: key mediators of apoptosis. Chem. Biol. 5:R97-R103.

Uckun FM, Waddick KG, Mahajan S, Jun X, Takata M, Bolen J, Kurosaki T. 1996. BTK as a mediator of radiation-induced apoptosis in DT-40 lymphoma B cells. Science 273:1096-1100.

Wood ER, Earnshaw WC. 1990. Mitotic chromatin condensation in vitro using somatic cell extracts and nuclei with variable levels of endogenous topoisomerase II. J Cell Biol 111:2839-2850.

Wyllie AH, Kerr JFR, Currie AR. 1980. Cell death: The significance of apoptosis. Int. Rev. Cytol. 68:251-305.

Yamaguchi-Iwai Y, Sonoda E, Buerstedde JM, Bezzubova O, Morrison C, Takata M, Shinohara A, Takeda S. 1998. Homologous recombination, but not DNA repair, is reduced in vertebrate cells deficient in RAD52. Mol Cell Biol 18:6430-6435.

Yang C, Maruyama S, Yanagi S, Wang X, Takata M, Kurosaki T, Yamamura H. 1995. Syk and Lyn are involved in radiation-induced signaling, but inactivation of Syk or Lyn alone is not sufficient to prevent radiation-induced apoptosis. J Biochem (Tokyo) 118:33-38.

Zamzami N, Metivier D, Kroemer G. 2000. Quantitation of mitochondrial transmembrane potential in cells and in isolated mitochondria. Methods Enzymol 322:208-213.

Zhang J, Liu X, Scherer DC, van Kaer L, Wang X, Xu M. 1998. Resistance to DNA fragmentation and chromatin condensation in mice lacking the DNA fragmentation factor 45. Proc. Natl. Acad. Sci. (USA) 95:12480-12485.

Chapter 7

THE DT40 SYSTEM AS A TOOL FOR ANALYZING KINETOCHORE ASSEMBLY

Masahiro Okada, Tetsuya Hori and Tatsuo Fukagawa
Department of Molecular Genetics, National Institute of Genetics and The Graduate University for Advanced Studies, Mishima, Shizuoka 411-8540, Japan

Abstract: The kinetochore is structure composed of many proteins that assembles on centromeric DNA to mediate the binding of spindle microtubules to chromosomes and chromosome movement. Budding yeast has been used as a model organism to investgate the mechanisms underlying kinetochore assembly. Although the basic features of chromosome segregation are thought to be common between all eukaryotes, it is difficult to identify components of vertebrate kinetochores by sequence homology with budding yeast kinetochore proteins. Therefore, we must use vertebrate systems to understand the mechanisms of kinetochore assembly and function. Several experimental strategies, including RNA interference (RNAi) in cultured human cells, knockout mice, *Drosophila* genetics, *C. elegans* with RNAi, and immunodepletion in *Xenopus* egg extracts, have been used to examine the mechanisms of kinetochore assembly. Our Lab. is using DT40 cells to identify and characterize kinetochore components. The DT40 system is a powerful and reliable tool for study of kinetochore assembly. Herein, we review recent advances in our understanding of the mechanisms underlying kinetochore assembly in DT40 cells.

Key words: kinetochore, centromere, CENPs, mitosis, DT40

1. INTRODUCTION

The centromere plays a fundamental role in proper chromosome segregation during mitosis and meiosis in eukaryotic cells. Its functions include sister chromatid adhesion and separation, microtubule attachment, chromosome movement, establishment of heterochromatin, and mitotic checkpoint control. These functions are important for chromosome segregation, and thus, the centromere has been widely studied and reviewed (Pluta

J.-M. Buerstedde and S. Takeda (eds.), Reviews and Protocols in DT40 Research, 91–106.
© 2006 *Springer.*

et al., 1995; Craig et al., 1999; Choo, 2001; Mellone and Allshire, 2003; Cleveland et al., 2003; Fukagawa, 2004; Maiato et al., 2004). The kinetochore, a multi-protein structure essential for microtubule binding, is formed at centromeres, and formation of the kinetochore is one major function of the centromeres. The mechanisms underlying kinetochore assembly have been well studied in budding yeast. In budding yeast, more than 60 kinetochore proteins have been identified with a combination of genetic and biochemical approaches, and the detailed protein-protein interactions have been clarified (Cheeseman et al., 2002; DeWulf et al., 2003, Westermann et al., 2003). Although the basic features of chromosome segregation are thought to be common between all eukaryotes, it is difficult to identify vertebrate kinetochore components by sequence homology with budding yeast kinetochore proteins. Strategies including RNA interference (RNAi) in cultured humn cells, knockout mice, *Drosophila* genetics, RNAi in *C. elegans*, and immunodepletion with *Xenopus* egg extracts, have been used to study the process of kinetochore assembly (Wood et al., 1997; Howman et al., 2000; Blower and Karpen, 2001; Oegema et al., 2001; Goshima et al., 2003). We used a conditional knockout approach in the chicken B lymphocyte cell line DT40 (Buerstedde and Takeda, 1991) to study kinetochore assembly. The high level of homologous recombination in DT40 cells allows efficient targeted disruption of genes of interest. Because the entire chicken genome has been sequenced, geneomic approaches, such as DNA microarray, and proteomics approaches can be used with these cells. In this review, we describe recent advances in our understanding of the general mechanism of kinetochore assembly and compare our results with the DT40 system with the results of investigators using different experimental systems.

1.1   Traditional kinetochore proteins in vertebrate cells

Observations of chromosomes by traditional electron microscopy revealed that the kinetochore of vertebrate cells is a trilaminar button-like structure on the surface of the centromeric heterochromatin (Pluta et al., 1995; Craig et al., 1999; Choo, 2001; Fukagawa, 2004; Maiato et al., 2004). The inner kinetochore plate is closely apposed to the centromeric DNA and is essential for kinetochore assembly. Components of the inner kinetochore protein complex in vertebrate cells have been identified primarily by immunological methods. Antibodies against centromere proteins have been isolated from patients with autoimmune diseases (Moroi et al., 1980) and from immunized animals (Compton et al., 1991), and this has led to the identification of several centromere-associated proteins known as CENPs (Pluta et al., 1995; Craig et al., 1999). Sera of patients with autoimmune diseases mainly recognize three peptides, CENP-A, CENP-B, and CENP-C (Table 7-1).

Table 7-1. Vertebrate proteins that constitutively localize to the centromere throughout the cell cycle and their counterparts in the yeasts S. cerevisiae and S. pombe.

Vertebrate proteins	Yeast Homologs
CENP-A	Cse4p (S. cerevisiae)
	spCENP-A as Cnp1/Sim2 (S. pombe)
CENP-B	Abp1 (S. pombe)
	Cbh1 (S. pombe)
	Cbh2 (S. pombe)
CENP-C	Mif2p (S. cerevisiae)
	spCENP-C as Cnp3 (S. pombe)
CENP-H	? (S. cerevisiae)
	Sim4 ? (S.pombe)
CENP-I	Ctf3p (S. cerevisiae)
	Mis6 (S. pombe)
Mis12	Mtw1p (S. cerevisiae)
	Mis12 (S. pombe)
DC8/DC31	Nsl1 (S. cerevisiae)
	Mis14 (S. pombe)
PMF1	Nnf1p(S. cerevisiae)
	Nnf1 (S. pombe)
Af15q14/hKNL1	Spc105p (S. cerevisiae)
	Spc7 (S. pombe)
Q9H410/c20orf172	Dsn1p (S. cerevisiae)
	Mis13 (S. pombe)

CENP-A is a 17-kDa centromere-specific protein, and the C-terminal 90 amino acids are 60% identical to those of histone H3 (Palmer and Margolis, 1987; Shelby et al., 1997). CENP-A is found only at active centromeres (Warburton et al., 1997) and co-purifies with nucleosomes. CENP-A is highly conserved and is contained in the centromeric chromatin of all nucleated eukaryotic cells (Cleveland et al., 2003). These features of CENP-A with respect to centromere organization suggest that it is a key factor in kinetochore assembly (Hayashi et al., 2004; Regnier et al., 2005).

CENP-B is a specific binding protein for α-satellite sequences (Masumoto et al., 1989). α-satellite sequences may be involved in centromere function (see below); however, available data suggest that CENP-B and CENP-B binding sites do not play significant roles in kinetochore assembly. Kinetochores of human neocentromeres and the Y chromosome do not contain CENP-B (Craig et al., 1999). Analysis of stable dicentric chromosomes in which one centromere has been inactivated revealed that α-satellite arrays bind CENP-B but are not associated with assembly of a functional kinetochore (Sullivan and Schwartz, 1995). Mice that lack CENP-B do not have significant difficulty with chromosome segregation (Hudson et al., 1998). Taken together, the data suggest that vertebrate CENP-B is involved in formation of the heterochromatin rather than in assembly of the kinetochore similar to the *S. pombe* homologs of CENP-B (Nakagawa et al., 2002).

CENP-C is an evolutionarily conserved protein. The *S. cerevisiae* homolog of CENP-C, Mif2p, binds to centromeric DNA (Meluh and Koshland, 1995), and mutation or overexpression of *Mif2* causes errors in chromosome segregation (Brown et al., 1993). CENP-C is localized to the inner kinetochore plates adjacent to the centromeric DNA (Saitoh et al., 1992) and is known to bind DNA directly (Yang et al., 1996).

1.2   Constitutive centromere proteins identified recently

CENP-A, -B, and -C were originally identified as antigens recognized by antisera from patients with autoimmune disease. These three proteins associate constitutively with the centromere throughout the vertebrate cell cycle. After the identification of CENP-A, -B, and -C, several constitutive centromere components were identified and characterized (Table 7-1) with cell biology methods, homology searches with yeast proteins, and novel bioinformatics and proteomics approaches.

CENP-H was first identified as a centromere component in mouse cells (Sugata et al., 1999), and the human and chicken homologs were then isolated and characterized (Sugata et al., 2000; Fukagawa et al., 2001). CENP-H localizes to the centromere throughout the cell cycle, presents at the inner kinetochore plate, and is found only in active centromeres, including neocentromeres (Sugata et al., 2000).

CENP-I was originally identified due to weak similarity with Mis6, a fission yeast centromere protein (Saitoh et al., 1997; Nishihashi et al., 2002). We reported that CENP-I localizes constitutively to the centromere throughout the cell cycle in chicken DT40 cells (Nishihashi et al., 2002). The human homolog of CENP-I was identified and was also found to localize to the centromere throughout the cell cycle in human cells (Liu et al., 2003).

Mis12 was originally identified as a fission yeast centromere protein (Goshima et al., 1999), but ordinary BLAST searches did not identify homologous sequences in the genomes of higher vertebrates. hMis12, a human homolog of Mis12, was recently identified through a novel bioinformatics approach (Goshima et al., 2003). Although hMis12 protein shows low homology with fission yeast Mis12, localization analysis revealed that hMis12 is a constitutive inner centromere protein (Goshima et al., 2003). Recently Mis12-associated proteins were identified from human HeLa cells through proteomics approaches (Cheeseman et al., 2004; Obuse et al., 2004). Among Mis12-associated proteins, AF15q14/hKNL1, DC8/DC31, Q9H410, and PMF1 were confirmed to be constitutive kinetochore components (Table 7-1).

1.3   DT40 knockouts to analyze kinetochore assembly

To analyze kinetochore assembly in vertebrate cells, we have used chicken DT40 cells because they have a high level of homologous recombination and condensed mitotic chromosomes that are easily visualized under light microscopy and because several conditional knockout systems have been established (Brown et al., 2003). In addition, structures such as the mitotic spindle, centrosome, and centromere in DT40 cells are much larger and more elaborate than the analogous structures in yeast, and the stages of mitosis (prometaphase chromosome congression, metaphase, and transition to anaphase) are easily distinguished under a light microscope. The generation and analysis of DT40 conditional knockouts of the constitutive centromere proteins CENP-A, -C, -H, and -I and for transient centromere-associated proteins, including Nuf2, Hec1, and ZW10 have been reported (Fukagawa and Brown, 1997; Fukagawa et al., 1999; Fukagawa et al., 2001; Nishihashi et al., 2002; Hori et al., 2003; Regnier et al., 2005).

Disruption of the *CENP-C* gene in DT40 cells showed that CENP-C is essential for cell proliferation (Fukagawa and Brown, 1997). Analysis of a conditional knockout of CENP-C in chicken DT40 cells revealed that the absence of CENP-C caused mitotic delay, chromosome missegregation, and apoptosis (Fukagawa and Brown, 1997; Fukagawa et al., 1999). The phenotypes of conditional knockouts of CENP-H and CENP-I in DT40 cells (Fukagawa et al., 2001; Nishihashi et al., 2002) are quite similar, suggesting that these proteins may form a complex. Disruption of CENP-A in DT40 cells caused mislocalization of CENP-C, CENP-H, CENP-I, and Nuf2, indicating that CENP-A is upstream of these proteins in the process of kinetochore assembly (Regnier et al., 2005).

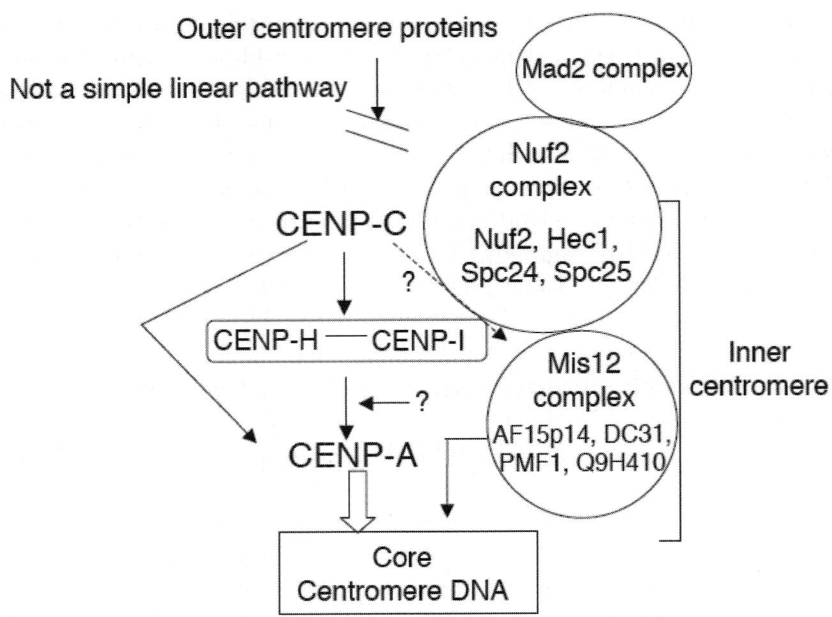

Figure 7-1. Current model of kinetochore assembly in vertebrate cells. In chicken DT40 cells, CENP-I and CENP-H are interdependent, and both proteins are required for kinetochore localization of CENP-C, but not CENP-A. RNAi experiments in human cells revealed that CENP-C is recruited by CENP-A and is not dependent on CENP-H and CENP-I and that Mis12 is independent of the CENP-A loading pathway (Goshima et al. 2003). The Nuf2 complex may act as a connector between the inner and outer kinetochores. Outer kinetochore proteins, including several checkpoint proteins, are not assembled via a simple linear pathway.

Our data from DT40 knockouts and data from other RNAi analyses in human HeLa cells (Goshima et al., 2003; Obuse et al., 2004) contributed to the current model of kinetochore assembly in vertebrate cells (Fig. 7-1). In fission yeast, CENP-A localization is altered in Mis6/CENP-I mutants, suggesting that Mis6/CENP-I may be involved in CENP-A localization to the centromere (Takahashi et al., 2000). The situation in vertebrate cells is different. Depletion of CENP-I in DT40 cells does not alter the normally discrete centromeric CENP-A signals, whereas the discrete centromeric CENP-C and CENP-H signals become diffuse or abolished (Nishihashi et al., 2002). Furthermore, the normally discrete CENP-I signals are abolished in CENP-H-deficient cells (Fukagawa et al., 2001), indicating that CENP-H and CENP-I are interdependent for targeting to centromeres. Because CENP-H is necessary for recruitment of CENP-C to form a functional centromere (Fukagawa et al., 2001), formation of CENP-A-containing nucleosomes, which is mediated by unknown factors, occurs first, and CENP-H and CENP-I are then targeted to these structures, possibly as a macromolecular complex. CENP-C then interacts with the CENP-A/CENP-H/CENP-I chromatin to form the kinetochore plate structure (Fig. 7-1). RNAi analyses of hMis12 in HeLa cells caused chromosome misalignment but not mitotic delay, whereas gene knockouts of CENP-A, -C, -H, and -I resulted in mitotic delay (Fukagawa et al., 1999; Howman et al. 2000; Fukagawa et al., 2001; Nishihashi et al., 2002). hMis12 may not be involved in the CENP-A pathway (Fig. 7-1) because CENP-A localization is not altered by RNAi of hMis12 in HeLa cells (Goshima et al., 2003). The differences in mitotic phenotypes between RNAi based knockdown of hMis12 and DT40 knockouts of CENP-A, -C, -H, and -I suggest that there may be several pathways for kinetochore assembly. However, we recently observed severe mitotic delay in DT40 knockouts of Mis12 and KNL1, a component of the Mis12 complex (Kline et al., 2006). Because RNAi analysis sometimes causes incomplete knockdown, further studies are needed to obtain certain conclusions.

When the functions of inner kinetochore proteins are disrupted, checkpoint proteins still localize to kinetochores, and cell cycle progression is delayed during mitosis. Mitotic checkpoint proteins, including members of the Mad and Bub families, localize to the outer kinetochore (Craig et al., 1999; Maiato et al., 2004). We also observed chromosomes in which kinetochores attached to microtubules in CENP-I-deficient cells (Nishihashi et al., 2002). The fact that microtubule binding and mitotic checkpoint signaling proceed even when a number of components of the inner kinetochore plate are absent indicates that kinetochore assembly does not follow a simple linear pathway in vertebrates. Even if kinetochore assembly does not occur via a simple linear pathway in vertebrate cells, there should be molecular connectors between the outer and inner kinetochores, and we would like to propose that the Nuf2 complex acts as such a connector. The Nuf2 complex may be targeted to the CENP-A/CENP-H/CENP-I/CENP-C complex during interphase (Fig. 7-1) because the Nuf2 complex starts to localize to the centromere during interphase (G2), and centromere localization of the complex is not altered in Nuf2- or Hec1-deficient DT40 cells (Hori et al. 2003). Hec1 also interacts with the Mad complex, which localizes to the outer kinetochore (Martin-Lluesma et al., 2002; Hori et al., 2003). When associated with the outer kinetochore, the Nuf2 complex acts as a structural component in vertebrate kinetochores (Hori et al., 2003). Our yeast two-hybrid analysis revealed that Hec1 interacts with CENP-H (Mikami et al., 2005). On the basis of these data, we believe that the Nuf2 complex may act to connect the outer and inner kinetochores (Fig. 7-1). Recently, purification of the Mis12 complex revealed that the Mis12 complex interacts with the Nuf2 complex (Cheeseman et al., 2004; Obuse et al., 2004). This data also supports our hypothesis that the Nuf2 complex functions as connector between the outer and inner kinetochores.

1.4　　Identification new kinetochore components with DT40 cells

More than 60 kinetochore proteins have been identified in budding yeast (Cheeseman et al., 2002; De Wulf et al., 2003; Westermann et al., 2003). Therefore, we predict that other components of the vertebrate inner

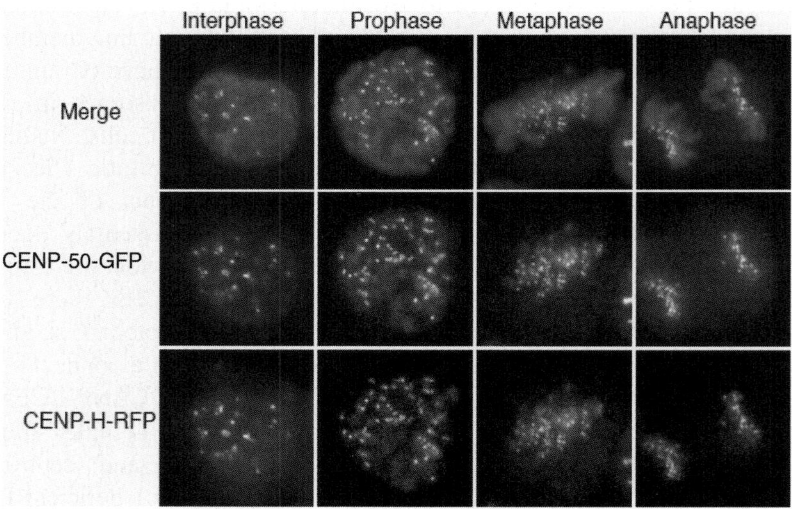

Figure 7-2. Localization of CENP-50, a member of the CENP-H/I complex proteins. GFP tagged CENP-50 protein (green) was expressed in DT40 cells that express chicken CENP-H-RFP (red). CENP-50 protein co-localizes to CENP-H throughout the cell cycle (Minoshima et al., 2005).

kinetochore remain to be identified. To identify vertebrate proteins that are required for kinetochore assembly, we used a combination of genetically modified DT40 cells and proteomics. For example, we generated DT40 cell lines in which expression of CENP-H was completely replaced with CENP-H-FLAG or CENP-H-GFP (Mikami et al., 2005) or expression of CENP-I was completely replaced with CENP-I-GFP. We then immunoprecipitated chromatin lysates of these cells with anti-GFP or anti-FLAG antibodies. Proteins co-precipitated with antibodies were analyzed by mass spectrometry, and we identified several polypeptides and cloned he corresponding cDNAs (Okada et al., 2006). We confirmed that these proteins constitutively localize to centromeres (Fig. 7-2) and named them the CENP-H/I complex proteins. We are sure that this is a powerful approach for identification of new proteins and will be widely used. Knockout analyses Knockout analyses of the CENP-H/I complex proteins in DT40 cells must further our understanding of the molecular mechanisms of kinetochore assembly.

2. USE OF HUMAN-CHICKEN HYBRID CELLS TO UNDERSTAND THE SIGNIFICANCE OF CENTROMERE DNAS

Centromere DNAs of vertebrate cells are large (one- to several-Mb) and consist of highly repetitive tandem sequence repeats (Fig. 7-3). Furthermore, large portions of these tandem repeats can be deleted and normal chromosome segregation still occurs (Sullivan et al., 2001). Deletions of centromere DNA can occur naturally (Choo, 2001) or can be induced experimentally (Brown et al., 2000). If specific core sequences that are thought to be centromere DNAs are removed from chromosomes, functional centromeres may develop at other sites along the chromosome, and the centromeric activity is transmitted via an epigenetic mechanism (Karpen and Allshire, 1997). This phenomenon of neocentromere formation occurs when a euchromatic chromosome locus acquires centromere activity (Choo, 2001). Detailed sequence analyses of one neocentromere revealed that the primary DNA sequence was unchanged in comparison with the corresponding region of the normal chromosome from which it arose (Choo, 2001). This finding supports the idea that centromere assembly is not dependent on primary DNA sequence but is instead regulated by epigenetic forces. However, neocentromere formation is a rare event, and normal human centromeres have common repetitive sequences (α-satellite repeats), suggesting that such sequences have some significance for centromere formation and function.

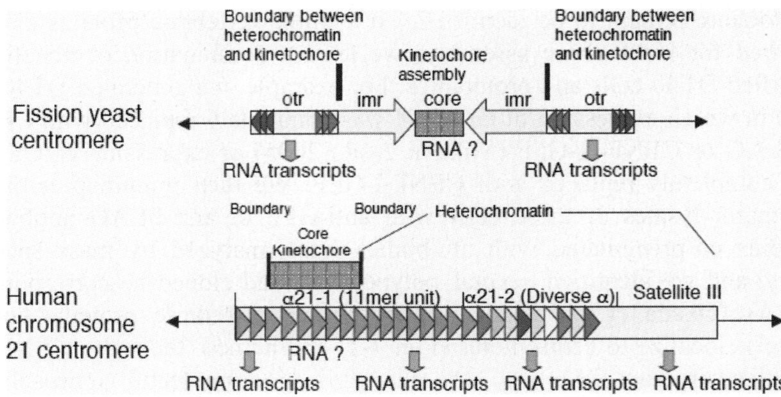

Figure 7-3. Schematic representations of the human chromosome 21 centromere region in chicken-human hybrid DT40 cells and of the fission yeast centromere. α-satellite arrays are composed of the α21-I array that comprises repeating 11-mer units and the α21-II array that contains diverged α satellite repeats. RNA transcripts were detected from α21-I, α21-II and satellite III arrays. We are not sure that RNA molecules are trasscribed from core kinetochore region. We also suppose that the boundary between the core kinetochore and heterochromatin structure is not determined by the primary DNA sequence in vertebrate cells (Fukagawa et al., 2004) (see plate 11).

To study the significance of centromere DNAs in DT40 cells, human-chicken hybrid cells have been widely used. The sequence information that comprise chicken centromeric DNA is not known, whereas the α-satellite repeats of human chromosomes are well characterized and function as centromeric DNA in DT40 cells. Spence et al. (2002) transferred the human X chromosome into DT40 cells and created a series of mini-chromosome derivatives containing α-satellite arrays of different sizes by telomere-directed breakage. Detailed analyses of these derivatives revealed that the site of kinetochore assembly could be localized to a subdomain of the major α-satellite array (DXZ1 locus) and coincided with a cleavage site for DNA topoisomerase II (topo II). α-satellite arrays of the human Y chromosome have also been analyzed in DT40 cells (Yang et al., 2000). DT 40 cell line is a useful tool to analyze human centromeric repeat.

We used human-chicken hybrid DT40 cells that contain human chromosome 21 to examine the role of RNAi machinery in vertebrate centromeres (Fukagawa et al., 2004). Although there were reports that the RNAi machinery is related to chromosome segregation in fission yeast (Volpe et al., 2002; Provost et al., 2002; Hall et al., 2003), it is unclear whether the RNAi machinery is associated with chromosome segregation in vertebrate cells. Therefore, we generated a conditional loss-of-function mutant of *Dicer* in a human-chicken hybrid DT40 cell line that contains human chromosome 21. The molecular structure of the chromosome 21 centromere is well characterized (Ikeno et al., 1998; Ohzeki et al., 2002), and it is possible to characterize human centromere repeats in Dicer-deficient cells. Loss of Dicer leads to cell death with accumulation of abnormal mitotic cells with premature sister chromatid separation (Fukagawa et al., 2004). We also observed abnormalities in centromere localization of heterochromatin proteins, and we concluded that Dicer-related RNAi machinery is involved in formation of the heterochromatin structure in the centromere region of higher vertebrate cells (Fukagawa et al., 2004). Most importantly, we detected aberrant accumulation of transcripts from all domains with α-satellite sequences in Dicer-deficient human-chicken hybrid DT40 cells. The centromere region is thought to be composed of two distinct domains (Partridge et al., 2000). One is the central region that is required for kinetochore assembly, and the other is a heterochromatic region that is essential for establishment of centromere cohesion (Fig. 7-3). In fission yeast RNAi mutants, the central core region, which contains centromere-specific CENP-A and is essential for kinetochore assembly, did not produce RNA transcripts, whereas the outer repeat region, which associates with Swi6, produced long RNA transcripts (Volpe et al., 2002). In contrst, we detected transcripts from a core region containing α21-I sequence and heterochromatic region with α21-II sequences (Fukagawa et al., 2004). There are two possible explanations for the differences between fission yeast and vertebrate cells. One possibility is that the RNA transcript degraded by Dicer may be involved in the kinetochore assembly pathway in vertebrate cells but not in fission yeast. Although we cannot exclude this possibility, it is unlikely, because localization of inner kinetochore proteins CENP-A and CENP-C is not altered in Dicer-deficient cells (Fukagawa et al., 2004). The second possibility is that part of the α21-I region is essential for kinetochore assembly and the remainder of the α21-I region is composed of heterochromatic structures similar to the α21-II region (Fig. 7-3). Analysis of the human X chromosome centromere suggested that only parts of the long α-satellite array of the X chromosome centromere were used for assembly of the core kinetochore (Spence et al., 2002). Recently, Sullivan and Karpen (2004) reported that chromatin of core kinetochore contains blocks of histone H3 nucleosomes interspersed with blocks of CENP-A nucleosomes. Therefore, we hypothesized that the site of the core kinetochore on human chromosome 21 comprises an approximately 100-kb region within the approximately 2-Mb α21-I region and contains CENP-A (Fukagawa et al., 2004). The other region of the α21-I array, which contains histone H3, produces RNA transcripts and is composed of heterochromatic structures (Fig. 7-3). The boundary that delineates core kinetochore and heterochromatin regions in vertebrate centromeres is not clear from the nucleotide sequence. However, this boundary in the fission yeast centromere is clear (Fig. 7-3). These studies indicated that the human-chicken hybrid DT40 cell system is a useful tool for studies of *cis* DNA elements required for centromere function.

CONCLUSION

In *S. cerevisiae*, at least 60 proteins are involved in centromere function. We believe that more than 100 proteins are involved in vertebrate centromere functions, which include kinetochore assembly, sister chromatid cohesion, microtubule attachment, and mitotic checkpoint control. Although we have identified new centromere proteins in recent years, we have not yet identified many of the proteins that are necessary for proper kinetochore assembly in vertebrate cells. In addition, RNA molecules may be involved in kinetochore assembly and function in vertebrate cells. We must clarify the mechanisms that determine when and how a site for kinetochore assembly is chosen and transmitted. It is likely that there are common principles by which centromeres function in all eukaryotes. As mentioned above, the chicken DT40 cell line is a powerful tool for studies of kinetochore assembly. At present, the RNAi knockdown system is being used widely to examine cellular functions in cultured vertebrate cells. However, there are several problems with this technique, including incomplete knockdown or non-specific knockdown of expression. In contrast, phenotype analyses of DT40 cells are relatively accurate. Because the entire chicken genome has been sequenced, it is easy to generate mutants and to conduct genome-wide studies. Furthermore, it is useful to apply proteomics approach. We are confident that the common principles by which centromeres function in all eukaryotic cells will be clarified by studies of DT40 cells.

ACKNOWLEDGEMENTS

We thank members of the Fukagawa laboratory for discussion and comments. Experiments in the Fukagawa laboratory are supported by Grants-in-Aid for Scientific Research on Priority Areas "Genome Biology", "Cancer Cell Biology", "Cell Cycle", and "Nuclear Dynamics" from the Ministry of Education, Science, Sports and Culture of Japan.

REFERENCES

Blower MD, Karpen GH. 2001. The role of Drosophila CID in kinetochore formation, cell-cycle progression and heterochromatin interactions. Nat Cell Biol. 3: 730-739.

Brown MT, Goetsch L, Hartwell LH. 1993. *MIF2* is required for mitotic integrity during anaphase spindle elongation in *Sacchromyces cerevisiae*. J Cell Biol. 123: 387-403.

Brown WR, Mee PJ, Shen MH. 2000. Artificial chromosomes: ideal vectors? Trends Biotechnol. 18: 218-223.

Brown WR, Hubbard SJ, Tickle C, Wilson SA. 2003. The chicken as a model for large-scale analysis of vertebrate gene function. Nat Rev Genet. 4: 87-98.

Buerstedde JM, Takeda S. 1991. Increased ratio of targeted to random integration after transfection of chicken B cell lines. Cell. 67: 179-188.

Choo KH. 2001. Domain organization at centromere and neocentromere. Dev Cell. 1: 165-177.

Cheeseman IM, Anderson S, Jwa M, Green EM, Kang J, Yates JR III, Chan CS, Drubin DG, Barnes G. 2002. Phospho-regulation of kinetochore-microtubule attachments by the Aurora kinase Ipl1p. Cell. 111: 163-172.

Cheeseman IM, Niessen S, Anderson S, Hyndman F, Yates, JR III, Oegema K, Desai, A. 2004. A conserved protein network controls assembly of the outer kinetochore and its ability to sustain tension. Genes Dev. 18: 2255-2268.

Cleveland DW, Mao Y, Sullivan KF. 2003. Centromeres and kinetochores: from epigenetics to mitotic checkpoint signaling. Cell. 112: 407-421.

Compton DA, Yen TJ, Cleveland DW. 1991. Identification of novel centromere/kinetochore-associated proteins using monoclonal antibodies generated against human mitotic chromosome scaffolds. J Cell Biol. 112: 1083-1097.

Craig JM, Earnshaw WC, Vagnarelli P. 1999. Mammalian centromeres: DNA sequence, protein composition, and role in cell cycle progression. Exp Cell Res. 246: 249-262.

De Wulf P, McAinsh AD, Sorger, PK. 2003. Hierarchical assembly of the budding yeast kinetochore from multiple subcomplexes. Genes Dev. 17: 2902-2921.

Fukagawa T, Brown WRA. 1997. Efficient conditional mutation of the vertebrate CENP-C gene. Hum Mol Genet. 6: 2301-2308.

Fukagawa T, Pendon C, Morris J, Brown W. 1999. CENP-C is necessary but not sufficient to induce formation of functional centromere. EMBO J. 18: 4196-4209.

Fukagawa T, Mikami Y, Nishihashi A, Regnier V, Haraguchi T, Hiraoka Y, Sugata N, Todokoro K, Brown W, Ikemura T. 2001. CENP-H, a constitutive centromere component, is required for centromere targeting of CENP-C in vertebrate cells. EMBO J. 20: 4603-4617.

Fukagawa T. 2004. Assembly of kinetochores in vertebrate cells. Exp Cell Res. 296: 21-27.

Fukagawa T, Nogami M, Yoshikawa M, Ikeno M, Okazaki T, Takami Y, Nakayama T, Oshimura M. 2004. Dicer is essential for formation of the heterochromatin structure in vertebrate cells. Nat Cell Biol. 6: 784-791.

Goshima G, Saitoh S, Yanagida M. 1999. Proper metaphase spindle length is determined by centromere proteins Mis12 and Mis6 required for faithful chromosome segregation. Genes Dev. 13: 1664-1677.

Goshima G, Kiyomitsu T, Yoda K, Yanagida M. 2003. Human centromere chromatin protein hMis12, essential for equal segregation, is independent of CENP-A loading pathway. J Cell Biol. 160: 25-39.

Hall IM, Noma K, Grewal SI. 2003. RNA interference machinery regulates chromosome dynamics during mitosis and meiosis in fission yeast. Proc Natl Acad Sci USA. 100: 193-198.

Hayashi T, Fujita Y, Iwasaki O, Adachi Y, Takahashi K, Yanagida M. 2004. Mis16 and Mis18 are required for CENP-A loading and histone deacetylation at centromeres. Cell. 118: 715-729.

Hori T, Haraguchi T, Hiraoka Y, Kimura H, Fukagawa T. 2003. Dynamic behavior of Nuf 2-
    Hec1 complex that localizes to the centrosome and centromere and is essential for mitotic
    progression in vertebrate cells. J Cell Sci. 116: 3347-3362.
Howman EV, Fowler KJ, Newson AJ, Redward S, MacDonald AC, Kalitsis P, Choo KHA.
    2000. Early disruption of centromeric chromatin organization in centromere protein A
    (Cenpa) null mice. Proc Natl Acad Sci USA. 97: 1148-1153.
Hudson DF, Fowler KJ, Earle E, Saffery R, Kalitsis P, Trowell H, Hill J, Wreford NG, de
    Kretser DM, Cancilla MR, Howman E, Hii L, Cutts SM, Irvine DV, Choo KH. 1998.
    Centromere protein B null mice are mitotically and meiotically normal but have lower
    body and testis weights. J Cell Biol. 141: 309-319.
Ikeno M, Grimes B, Okazaki T, Nakano M, Saitoh K, Hoshino H, McGill NI, Cooke H,
    Masumoto H. 1998. Construction of YAC-based mammalian artificial chromosomes. Nat
    Biotechnol. 16: 431-439.
Karpen GH, Allshire RC. 1997. The case for epigenetic effects on centromere identity and
    function. Trends Genet. 13: 489-496.
Kline SL, Cheeseman IM, Hori T, Fukagawa T, Deasai A. 2006. The human Mis12 complex
    is required for kinetochore assembly and proper chromosome segreagation. J Cell Biol.
    173: 9-17.
Liu ST, Hittle JC, Jablonski SA, Campbell MS, Yoda K, Yen TJ. 2003. Human CENP-I
    specifies localization of CENP-F, MAD1 and MAD2 to kinetochores and is essential for
    mitosis. Nat Cell Biol. 5: 341-345.
Maiato H, DeLuca J, Salmon ED, Earnshaw WC. 2004. The dynamic kinetochore-microtubule
    interface. J Cell Sci. 117: 5461-5477.
Martin-Lluesma S, Stucke VM, Nigg EA. 2002. Role of Hec1 in spindle checkpoint signaling
    and kinetochore recruitment of Mad1/Mad2. Science. 297: 2267-2270.
Masumoto H, Masukata H, Muro Y, Nozaki N, Okazaki, T. 1989. A human centromere
    antigen (CENP-B) interacts with a short specific sequence in alphoid DNA, a human
    centromeric satellite. J Cell Biol. 109: 1963-1973.
Measday V, Hailey DW, Pot I, Givan SA, Hyland KM, Cagney G, Fields S, Davis TN, Hieter
    P. 2002. Ctf3, the Mis6 budding yeast homolog, interacts with Mcm22p and Mcm16p at
    the yeast outer kinetochore. Genes Dev. 16: 101-113.
Mellone BG, Allshire RC. 2003. Stretching it: putting the CEN(P-A) in centromere. Curr
    Opin Cell Biol. 13: 191-198.
Meluh PB, Koshland D. 1995. Evidence that the *Mif2* gene of *Saccharomyces cerevisiae*
    encodes a centromere protein with homology to the mammalian centromere protein
    CENP-C. Mol Biol Cell. 6: 793-807.
Mikami Y, Hori T, Kimura H, Fukagawa T. 2005. The functional region of CENP-H interacts
    with the Nuf 2 complex that localizes to centromere during mitosis. Mol Cell Biol. 25:
    1958-1970.
Minoshima Y, Hori T, Okado M, Kimura H, Haraguchi T, Hiraoka Y, Bao Y-C, Kawashima T,
    Kitamura T, Fukagawa T. 2005. The constitutive centromere component CENP-50 is
    required for recovery from spindle damage. Mol Cell Biol. 25: 10315-10328.
Moroi Y, Peebles C, Fritzler MJ, Steigerwald J, Tan EM. 1980. Autoantibody to centromere
    (kinetochore) in scleroderma sera. Proc Natl Acad Sci USA. 77: 1627-1631.
Nakagawa H, Lee JK, Hurwitz J, Allshire RC, Nakayama J, Grewal SI, Tanaka K, Murakami
    Y. 2002. Fission yeast CENP-B homologs nucleate centromeric heterochromatin by
    promoting heterochromatin-specific histone tail modifications. Genes Dev. 16: 1766-1778.
Nishihashi A, Haraguchi T, Hiraoka Y, Ikemura T, Regnier V, Dodson H, Earnshaw WC,
    Fukagawa T. 2002. CENP-I is essential for centromere function in vertebrate cells. Dev
    Cell. 2: 463-476.

Obuse C, Iwasaki O, Kiyomitsu T, Goshima G, Toyoda Y, Yanagida M. 2004. A conserved Mis12 centromere complex is linked to heterochromatic HP1 and outer kinetochore protein Zwint-1. Nat Cell Biol. 6: 1135-1141.

Oegema K, Desai A, Rybina S, Kirkham M, Hyman, AA. 2001. Functional analysis of kinetochore assembly in Caenorhabditis elegans. J. Cell Biol. 153: 1209-1225.

Ohzeki J, Nakano M, Okada T, Masumoto H. 2002. CENP-B box is required for de novo centromere chromatin assembly on human alphoid DNA. J Cell Biol. 159: 765-775.

Okada M, Cheeseman IM, Hori T, Okawa K, McLeod IX, Yates III JR, Desai A, Fukagawa T. 2006. The CENP-H-I complex is required for efficient incorporation of newly synthesized CENP-A into centromeres. Nat Cell Biol. 8: 446-457.

Palmer DK, Margolis RL. 1987. A 17-kD centromere protein (CENP-A) copurifies with nucleosome core particles and with histones. J Cell Biol. 104: 805-815.

Partridge JF, Borgstrom B, Allshire RC. 2000. Distinct protein interaction domains and protein spreading in a complex centromere. Genes Dev. 14: 783-791.

Pluta AF, Mackay AM, Ainsztein AM, Goldberg IG, Earnshaw WC. 1995. The centromere: Hub of chromosomal activities. Science. 270: 1591-1594.

Provost P, Silverstein RA, Dishart D, Walfridsson J, Djupedal I, Kniola B, Wright A, Samuelsson B, Radmark O, Ekwall K. 2002. Dicer is required for chromosome segregation and gene silencing in fission yeast cells. Proc Natl Acad Sci USA. 99: 16648-16653.

Regnier V, Vagnarelli P, Fukagawa T, Zerjal T, Burns E, Trouche D, Earnshaw W, Brown W. 2005. CENP-A is required for accurate chromosome segregation and sustained kinetochore association of BubR1. Mol Cell Biol. 25: 3967-3981.

Saitoh H, Tomkiel J, Cooke CA, Ratrie H, Maurer M, Rothfield NF, Earnshaw WC. 1992. CENP-C, an autoantigen in scleroderma, is a component of the human inner kinetochore plate. Cell. 70: 115-125.

Saitoh S, Takahashi K, Yanagida M. 1997. Mis6, a fission yeast inner centromere protein, acts during G1/S and forms specialized chromatin required for equal segregation. Cell. 90: 131-143.

Shelby RD, Vafa O, Sullivan KF. 1997. Assembly of CENP-A into centromere chromatin requires a cooperative array of nucleosomal DNA contact sites. J Cell Biol. 136: 501-513.

Spence JM, Critcher R, Ebersole TA, Valdivia MM, Earnshaw WC, Fukagawa T, Farr CJ. 2002. Co-localization of centromere activity, proteins and topoisomerase II within a subdomain of the major human X alpha-satellite array. EMBO J. 21: 5269-5280.

Sugata N, Munekata E, Todokoro K. 1999. Characterization of a novel kinetochore protein, CENP-H. J Biol Chem. 274: 27343-27346.

Sugata N, Li S, Earnshaw WC, Yen TJ, Yoda K, Masumoto H, Munekata E, Warburton PE, Todokoro, K. 2000. Human CENP-H multimers colocalize with CENP-A and CENP-C at active centromere-kinetochore complexes. Hum Mol Genet. 9: 2919-2926.

Sullivan BA, Schwartz S. 1995. Identification of centromeric antigens in dicentric Robertsonian translocations: CENP-C and CENP-E are necessary components of functional centromeres. Hum Mol Genet. 4: 2189-2197.

Sullivan BA, Blower MD, Karpen GH. 2001. Determining centromere identity: cyclical stories and forking path. Nat Rev Genet. 2: 584-596.

Sullivan BA, Karpen GH. 2004. Centromeric chromatin exhibits a histone modification pattern that is distinct from both euchromatin and heterochromatin. Nat Struct Mol Biol. 11: 1076-1083.

Takahashi K, Chen ES, Yanagida M. 2000. Requirement of Mis6 centromere connector for localizing a CENP-A-like protein in fission yeast. Science. 288: 2215-2219.

Volpe TA, Kidner C, Hall IM, Teng G, Grewal SI, Martienssen RA. 2002. Regulation of heterochromatic silencing and histone H3 lysine-9 methylation by RNAi. Science. 297: 1833-1837.

Warburton PE, Cooke CA, Bourassa S, Vafa O, Sullivan BA, Stetten G, Gimelli G, Warburton D, Tyler-Smith C, Sullivan KF, Poirier GG, Earnshaw WC. 1997. Immunolocalization of CENP-A suggests a novel nucleosome structure at the inner kinetochore plate of active centromeres. Curr Biol. 7: 901-904.

Westermann S, Cheeseman IM, Anderson S, Yates JR III, Drubin DG, Barnes G. 2003. Architecture of the budding yeast kinetochore reveals a conserved molecular core. J Cell Biol. 163: 215-222.

Wood KW, Sakowicz R, Goldstein LS, Cleveland DW. 1997. CENP-E is a plus end-directed kinetochore motor required for metaphase chromosome alignment. Cell. 91: 357-366.

Yang CH, Tomkiel J, Saitoh H, Johnson DH, Earnshaw WC. 1996. Identification of overlapping DNA-binding and centromere-targeting domains in the human kinetochore protein CENP-C. Mol Cell Biol. 16: 3576-3586.

Yang JW, Pendon C, Yang J, Haywood N, Chand A, Brown WR. 2000. Human minichromosomes with minimal centromeres. Hum Mol Genet. 9: 1891-1902.

Chapter 8

ANALYSING THE DNA DAMAGE AND REPLICATION CHECKPOINTS IN DT40 CELLS

Michael D. Rainey, George Zachos and David A.F. Gillespie
CR-UK Beatson Laboratories, Beatson Institute for Cancer Research, Garscube Estate, Bearsden, Glasgow G61 1BD

Abstract: Eukaryotic cells respond to DNA damage or blocks to DNA replication by triggering a variety of "checkpoint" responses which delay cell cycle progression, modulate DNA replication, and facilitate DNA repair. Checkpoints play a vital role in maintaining genome integrity, particularly under conditions of genotoxic stress, and mutations in checkpoint genes can predispose to cancer and aging. Checkpoints are best understood at the molecular level in model organisms such as fission yeast, where the presence of aberrant DNA structures is sensed and relayed via signal transduction pathways to activate the checkpoint effector kinases, Chk1 and Cds1/ Chk2, which implement appropriate responses. Many of the yeast checkpoint sensor, transducer, and effector proteins are conserved in vertebrate cells, raising the question of whether they function in a similar or analogous way. DT40 cells provide a particularly tractable experimental system for genetic and biochemical dissection of checkpoints in vertebrates. Thus far, gene knockouts in DT40 have revealed that the Chk1 and Chk2 checkpoint effector kinases control a very different range of checkpoint responses in vertebrates compared to yeast. In future, these and other DT40 mutants will provide powerful tools for understanding the molecular basis of these unexpected differences and detailed studies of checkpoint mechanisms.

Key words: Cell cycle, checkpoints, DNA damage, DNA replication, Chk1, Chk2.

J.-M. Buerstedde and S. Takeda (eds.), Reviews and Protocols in DT40 Research, 107–117.
© 2006 *Springer.*

1.   THE DNA DAMAGE AND REPLICATION CHECKPOINTS

The accurate replication and division of large genomes presents eukaryotic cells with a considerable challenge, particularly when DNA suffers damage or replication is impeded. As a result, cells have evolved surveillance mechanisms, termed checkpoints, whose purpose is to detect such problems and to trigger cellular responses which minimize the probability of lethal or permanent genetic damage (Hartwell & Weinert, 1989).

Checkpoints are probably activated transiently in the course of most unperturbed cell cycles in response to low levels of spontaneous DNA damage and minor replication problems, however it is technically difficult to detect such stochastic events in individual cells. As a result, checkpoints are generally studied using radiation or other genotoxic agents to inflict DNA damage or inhibit DNA synthesis in most or all cells in a population simultaneously. Such treatments trigger multiple, functionally distinct responses which are collectively referred to under the umbrella terms "DNA damage" and "Replication" checkpoints respectively (Fig. 1). A brief operational definition of the best-characterized of these is given below.

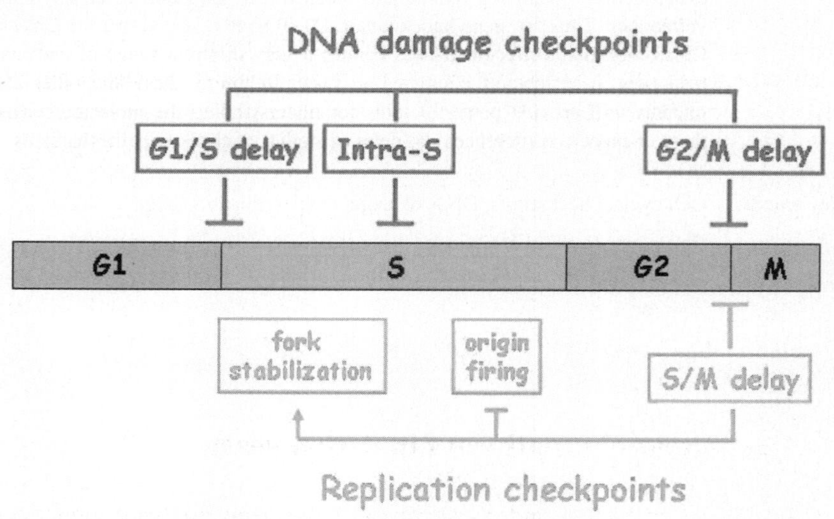

Figure 8-1. Overview of the various branches of the DNA damage (red) and Replication (green) checkpoints which exist in vertebrate cells. See text for a detailed explanation (see plate 12).

DNA damage can trigger delays or arrests at multiple points in the cell cycle (Fig. 1). Genetically normal cells can often arrest in G1 and thus avoid replicating damaged DNA through p53-mediated induction of the p21CIP1 cyclin-dependent kinase (CDK) inhibitor, which blocks entry to S-phase (Bates & Vousden, 1996). This G1/ S checkpoint however is frequently compromised or defective in tumour cells as a result of mutation or loss of p53 function, and indeed appears to be absent in DT40 cells for this reason (Takao et al., 1999). This particular response therefore will not be considered further here. In comparison, the G2/ M checkpoint serves to prevent cells with damaged genomes, and in particular with DNA double strand breaks (DSBs), from attempting mitosis and thus risking permanent loss of genetic information (O'Connell et al., 2000). Arrest in G2 is imposed primarily through suppression of mitotic cyclin B/ CDK1 activity and is presumed to allow time for DNA damage to be repaired prior to mitosis (O'Connell et al., 2000). A functional G2/M checkpoint is retained in virtually all tumour cell lines, including DT40. Finally, the intra-S checkpoint can slow the rate at which damaged DNA is replicated, at least in part by moderating the rate of replication origin firing (Bartek et al., 2004).

Whereas many genotoxic agents can damage DNA in all phases of the cell cycle, drugs which inhibit DNA synthesis primarily affect cells in S-phase. Agents used to block replication experimentally include aphidicolin, which inhibits DNA polymerase directly, and the ribonucleotide reductase inhibitor hydroxyurea, which inhibits DNA polymerase by depleting nucleotide precursor pools. Both agents block replication fork progression (often referred to as replication fork "stalling") and trigger three mechanistically distinct checkpoint responses. As shown in Fig. 1, two of these act on the replication machinery itself; the first is required to actively stabilize stalled replication forks and protect them from a poorly defined process of functional inactivation termed "collapse", whilst the second suppresses latent replication origin firing to prevent the accumulation of further stalled forks (Lambert & Carr, 2005). These functions are essential if DNA replication is to resume and reach completion after a prolonged period of interruption. Finally, a distinct branch of the replication checkpoint delays the onset of mitosis until genome duplication is achieved. This mitotic delay is generally termed the S/M checkpoint to distinguish it from the G2/ M checkpoint triggered by DNA damage (Fig. 1).

1.1   Checkpoint pathways in fission yeast

Checkpoint pathways and mechanisms have been analysed genetically and biochemically in detail in the fission yeast, *S. pombe* (Furuya & Carr, 2003; Rhind & Russell, 2000). Since it is generally believed that many of the concepts derived from this model can be extrapolated to vertebrates, a simplified view of the functional organization of the fission yeast DNA damage and replication checkpoint pathways is shown in Fig. 2.

In yeast, the presence of damaged or incompletely-replicated DNA is detected through the action of multiple "Checkpoint Rad" proteins which act as sensors. Some of these likely form PCNA-like sliding clamps on DNA (Rad9-Hus1-Rad1), play a role in loading such clamps on DNA (Rad17), or associate with extended tracts of single-stranded DNA (Rad3-Rad26), although exactly how different kinds of lesions are distinguished is unclear. At all events, when DNA damage or stalled replication forks are detected, the PI-3 kinase-like kinase (PIKK) Rad3 becomes activated and mediates selective activation of one or other of the downstream Chk1 and Cds1 (the homologue of vertebrate Chk2) effector kinases in conjunction with the Mrc1 and Crb2 adaptor proteins (Rhind & Russell, 2000). Crucially, Chk1 and Cds1 control distinct checkpoint responses; whereas Chk1 is selectively activated by DNA damage to enforce the G2/M checkpoint delay (O'Connell et al., 2000), Cds1 is selectively activated by DNA synthesis inhibition and acts to stabilize stalled replication forks, suppress origin firing, and delay mitosis whilst replication is blocked (Fig. 2: (Lambert & Carr, 2005; Rhind & Russell, 2000)).

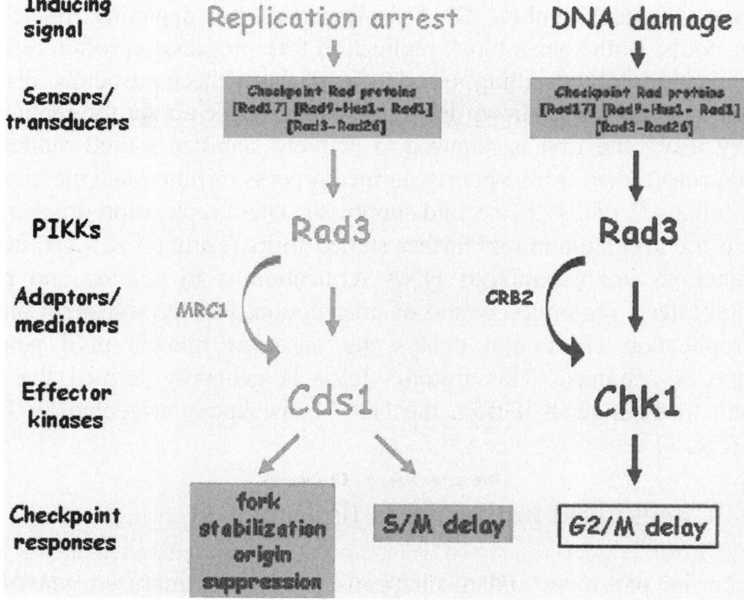

Figure 8-2. Overview of the principal components and functional organization of checkpoint pathways in the fission yeast, *S. pombe.* See text for a detailed description (see plate 13).

1.2   Studying checkpoints in DT40 cells

With the obvious exception of the p53-mediated G1/S arrest (Takao et al., 1999), all of the other DNA damage and replication checkpoint responses discussed above can be studied in DT40 cells using combinations of flow cytometry, centrifugal elutriation, and other techniques.

If equipment is available, ionizing radiation (IR) provides a precise and convenient means of inducing DNA damage in cultured cells. We use an Alcyon II CGR MeV cobalt source to administer doses of between 0.1 and 10 Grays to DT40 cultures up to 50 mls in volume. As shown in Fig. 3A, moderate doses of IR induce a robust G2 delay in DT40 cells as revealed by DNA content flow cytometry (Method 20). The scale (ie proportion of cells affected) and duration of this delay is markedly dose dependent, however in a culture exposed to 4 Grays most cells arrest in G2 after 8 or 10 hours followed by a gradual release at later times (Fig. 3A).

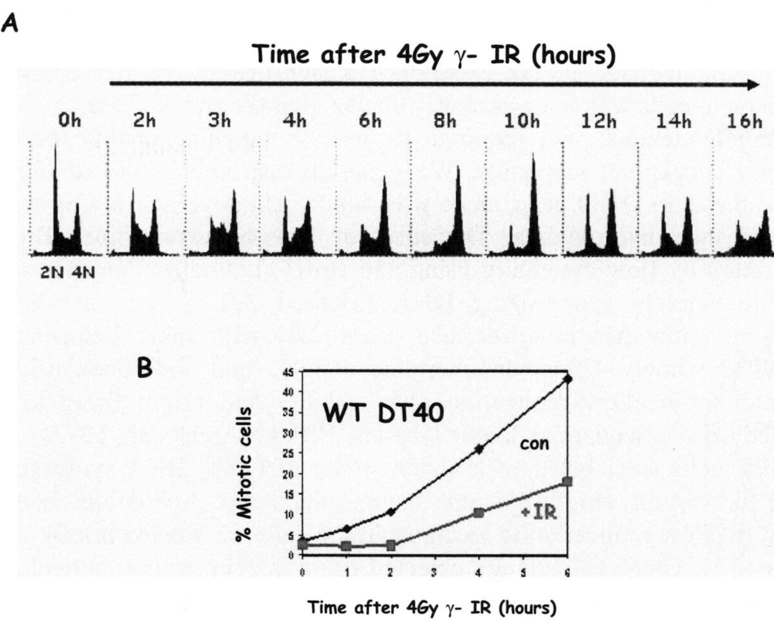

Figure 8-3. Visualizing ionizing radiation (IR)-induced G2/ M checkpoint arrest in WT DT40 cells by A) DNA content flow cytometry (Method 20), or B) quantifying the rate at which mitotic cells accumulate in the presence of nocodazole with or without prior irradiation using anti-pH3 antibodies (Method 6). Cultures were exposed to 4 Grays of IR in each case (see plate 14).

One problem with this method is that it does not permit the fate of that subset of cells that are in G2 at the time of irradiation to be followed unambiguously. This may be important for certain experimental purposes, since evidence suggests that the delay induced in G2 cells may be mechanistically distinct from that experienced by cells which sustain damage in G1 or S-phase but arrest only on reaching G2 (Xu et al., 2002).

One way to study such cells specifically is to use "nocodazole trapping". By disrupting microtubule assembly nocodazole traps cells in mitosis, thereby allowing the rate at which cells are entering M-phase from G2 to be estimated by flow cytometry using fluorescent antibodies specific for histone H3 phosphorylated on serine 10 (anti-pH3; see Method 6). During the first two hours or so of nocodazole treatment most or all of the cells which become trapped in mitosis will have been in G2-phase when drug was added. As shown in Fig. 3B, mitotic accumulation in this protocol is completely blocked for at least 2 hours in DT40 cultures exposed to 4 Grays of IR immediately prior to addition of nocodazole, demonstrating that G2 cells which sustain damage arrest rapidly. Note however that mitotic accumulation resumes again after 4 hours, significantly before the maximal G2 arrest is detected by DNA content at around 8 or 10 hours (Fig. 3A, B). This illustrates the complexity of the G2/M arrest response and suggests that cells which sustain damage in G2 experience a significantly shorter delay in comparison to cells which are similarly damaged in G1 and S-phase.

Different methods are required to assess the functionality of the replication checkpoint responses. We generally use aphidicolin to inhibit DNA synthesis in DT40 cells, since it is rapidly effective and in our hands less toxic than hydroxyurea. Detection of bromo-deoxyuridine (BrdU) incorporation by flow cytometry using anti-BrdU antibodies identifies cells which are actively synthesizing DNA (Method 20). Certain anti-BrdU monoclonal antibodies however also cross-react with other halogenated nucleotides, namely chloro-deoxyuridine (CldU) and iodo-deoxyuridine (IdU), and these allow replication fork viability and origin firing to be monitored using a sequential double labeling method (Aten et al., 1992).

Briefly, cells are labeled with a short pulse of CldU, DNA synthesis is blocked for varying lengths of time using aphidicolin. Aphidcolin is then removed to allow replication to resume and the cells are labeled briefly with IdU (Fig. 4A). CldU and IdU are detected using discriminating monoclonal antibodies and species-specific secondary antibodies coupled to either FITC (green) or Texas Red (red). Sites of incorporation are then visualized in individual labeled nuclei by confocal microscopy.

Replication forks which remain viable throughout the period of DNA synthesis inhibition and which can resume replication incorporate both CldU and IdU and therefore appear as yellow. In contrast, sites which incorporate only CldU (red) represent replication forks which have collapsed, whereas incorporation of IdU (green) alone signifies the presence of a new replication fork generated by the firing of a latent origin of replication during the period when elongation was blocked by aphidicolin (sometimes described as "futile" origin firing; Fig. 4B). Using this method, essentially all stalled replication forks in WT DT40 cells are observed to remain viable during at least 8 hours of aphidicolin treatment (Fig. 4B, far right). In contrast, Chk1-/- DT40 cells exhibit a combination of progressive replication fork collapse and futile origin firing which eventually exhausts their replication capacity completely (Fig. 4B; (Zachos et al., 2003).

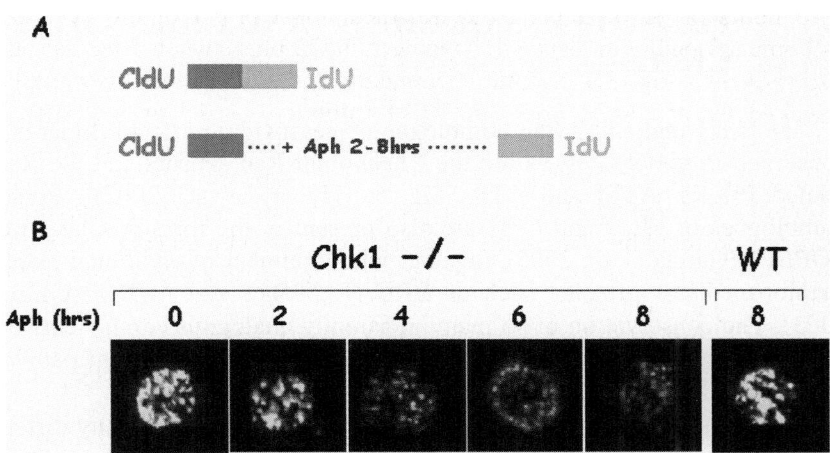

Figure 8-4. Visualizing replication fork stability and origin firing during aphidicolin-induced replication arrest in WT and Chk1-/- DT40 cells using dual labelling with CldU and IdU. A) Principle of the method; cells are exposed to a short pulse of CldU, replication is then blocked with aphidicolin for various periods of time. The inhibitor is then removed and cells are exposed to a short pulse of IdU. Incorporation of CldU and IdU into discrete replication foci in individual nuclei is then detected using monoclonal antibodies specific for each halogenated nucleotide and visualized by confocal microscopy. Foci which appear as yellow are interpreted to represent replication forks which were able to restart after aphidicolin was removed (ie viable), red alone indicates previously active forks which became incapable of resuming replication (collapsed), whilst green signifies new forks generated by origin firing during the period when elongation was blocked (futile origin firing). A more detailed description of the protocol employed can be found in (Zachos et al., 2003) (see plate 15).

The S/M checkpoint ensures that mitosis does not occur until DNA replication is completed. This checkpoint response can also be studied using nocodazole to trap mitotic calls cultured in the presence of aphidicolin to block DNA synthesis. However, because S/M delay occurs only in S-phase cells, it is desirable to monitor the onset of mitosis in a population of cells synchronized in this phase of the cell cycle. This can be accomplished using centrifugal elutriation to purify a living population of G1 and early S-phase cells (Method 7; (Zachos et al., 2005)). The elutriated cells are then returned to culture in the presence or absence of aphidicolin plus nocodazole and cells which enter mitosis prematurely with unreplicated DNA quantified by flow cytometry using anti-pH3 antibodies (Method 6). In principle this kind of synchronization approach could also be used to study the effect of irradiation or other DNA damaging agents on the rate at which cells progress through S-phase (the intra-S checkpoint) by DNA content flow cytometry.

2.   PHENOTYPES OF CHK1-/- AND CHK2-/- DT40 CELLS

The Chk1 and Chk2 (the homologue of yeast Cds1) effector kinases are conserved in vertebrates, as are the Checkpoint Rad sensors and the Rad3-related PIKKs, ATM and ATR (Fig. 5; (Nyberg et al., 2002)). Putative homologues of Mrc1 and Crb2 are also present in the form of Claspin and TOPBP1 (Garcia et al., 2005), together with a number of additional putative adaptor/mediator proteins such as BRCA1, 53BP1 and MDC1 (Canman, 2003). The conservation of so many apparently analogous components begs the question of whether, or to what extent, vertebrate checkpoint pathways conform to the fission yeast paradigm.

This question is currently under intensive investigation in many different organisms and using many different experimental approaches, however we reasoned that DT40 cells might provide a particularly advantageous system in view of the relative ease of genetic modification and their tractability in terms of biochemical and cell cycle analysis.

To dissect the molecular functions of Chk1 and Chk2 we decided to use gene targeting to "knockout" each kinase in DT40 cells. Loss of Chk1 function is lethal in mouse blastocytes and embryonic stem cells, however Chk1-deficient DT40 cells proved to be viable (Zachos et al., 2003), as did DT40 cells devoid of functional Chk2 (Zachos et al., 2005). By comparing how these mutant lines respond to agents that damage DNA or block replication using the methods outlined above, it is possible to build up a picture of which checkpoints are controlled by Chk1 and Chk2.

As one might predict, Chk1 is absolutely essential for G2/M arrest in response to DNA damage induced by IR in DT40 cells (Zachos et al., 2003). In comparison, Chk2-deficient DT40 cells exhibit only a very subtle defect in G2/M arrest after irradiation. This partial defect is confined to the small number of cells that are in G2 at the time of damage (MDR unpublished results) and is obviously epistatic to Chk1, although the molecular basis for this is unknown. More surprisingly, Chk1, rather than Chk2, proves to be essential for replication fork stability and replication origin suppression when DNA synthesis is blocked (Fig. 4B; (Zachos et al., 2003)). This was rather unexpected, since in fission yeast these functions are controlled by Cds1 (i.e., the Chk2 homologue) rather than Chk1 (Fig. 2). In comparison, the replication checkpoint responses are retained and appear to operate normally in DT40 cells which lack functional Chk2 (Zachos et al., 2005).

Chk1 is also required to maintain the S/M checkpoint when DNA synthesis is blocked (Zachos et al., 2005). In marked contrast to the situation with DNA damage however, where no mitotic delay at all can be detected, premature mitosis during replication arrest in Chk1-/- cells appears to occur only after the collapse of all viable replication forks. These experiments lead us to suspect that the process of DNA replication *per se* may be inherently incompatible with mitosis. If this idea is correct, the principal role of the replication checkpoint in vertebrates at least (and thus of Chk1) may be to ensure that replication forks and origins are preserved until genome duplication is achieved (Zachos et al., 2005).

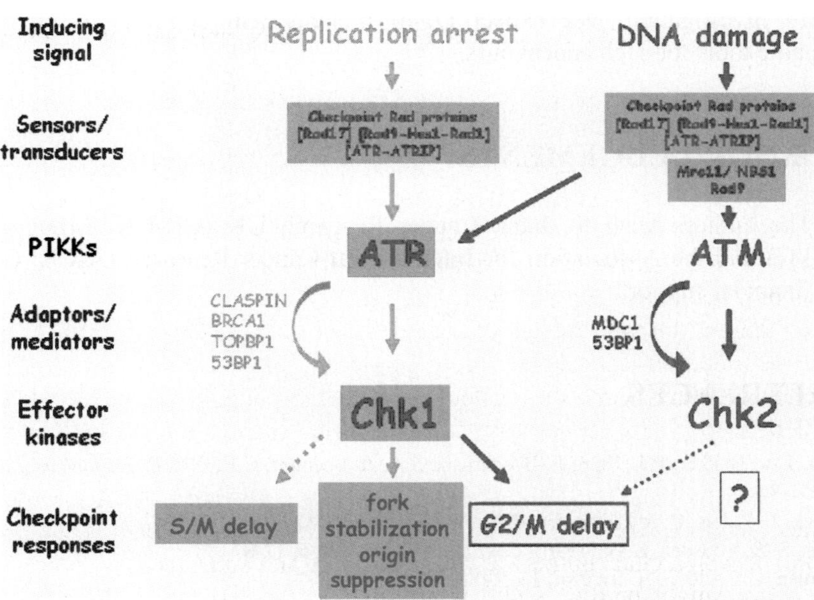

Figure 8-5. Overview of the principal components and proposed functional organization of checkpoint pathways in vertebrate cells, focusing on the functions controlled by the effector kinases, Chk1 and Chk2. See text for a detailed description (see plate 16).

CONCLUSIONS

Observations from many different experimental organisms and systems, including DT40 Chk1 and Chk2 knockout mutants, has provided a basic outline of how the DNA damage and Replication checkpoint pathways are organized in vertebrates (Fig. 5). It becomes evident that whereas the basic mechanisms of certain responses have been substantially conserved from fission yeast to vertebrates, others may operate in quite different ways. In particular, Chk1 has become the lynchpin of both the DNA damage and Replication checkpoints in vertebrates (Fig. 5), whereas fission yeast dedicate Chk1 to damage-induced G2/ M arrest (Fig. 2). This raises the question of how Chk1 is activated and directed to appropriate downstream targets and substrates in response to such a wide and diverse range of genomic problems. Indeed, what are the crucial targets of Chk1? At present only dual specificity phosphatases of the Cdc25 family have been identified with any certainty. And what of the role of Chk2? Has is it really shrunk to relative insignificance or are we missing something? Much therefore remains to be learned about how checkpoints work in vertebrates, whilst connections between checkpoints and other processes such as DNA repair, recombination, and apoptosis also merit future investigation. We expect DT40 mutants will continue to provide valuable tools for such endeavours.

ACKNOWLEDGEMENTS

The authors wish to thank Cancer Research UK (CR-UK; MDR and DAFG) and the Association for International Cancer Research (AICR; GZ) for financial support.

REFERENCES

Aten, J.A., Bakker, P.J., Stap, J., Boschman, G.A. & Veenhof, C.H. (1992). *Histochem J*, **24,** 251-9.

Bartek, J., Lukas, C. & Lukas, J. (2004). *Nat Rev Mol Cell Biol*, **5,** 792-804.

Bates, S. & Vousden, K.H. (1996). *Curr Opin Genet Dev*, **6,** 12-8.

Canman, C.E. (2003). *Curr Biol*, **13,** R488-90.

Furuya, K. & Carr, A.M. (2003). *J Cell Sci*, **116,** 3847-8.

Garcia, V., Furuya, K. & Carr, A.M. (2005). *DNA Repair (Amst)*, **4,** 1227-39.

Hartwell, L.H. & Weinert, T.A. (1989). *Science*, **246,** 629-34.

Lambert, S. & Carr, A.M. (2005). *Biochimie*, **87,** 591-602.

Nyberg, K.A., Michelson, R.J., Putnam, C.W. & Weinert, T.A. (2002). *Annu Rev Genet*, **36,** 617-56.

O'Connell, M.J., Walworth, N.C. & Carr, A.M. (2000). *Trends Cell Biol*, **10,** 296-303.

Rhind, N. & Russell, P. (2000). *J Cell Sci*, **113 (Pt 22),** 3889-96.

Takao, N., Kato, H., Mori, R., Morrison, C., Sonada, E., Sun, X., Shimizu, H., Yoshioka, K., Takeda, S. & Yamamoto, K. (1999). *Oncogene*, **18**, 7002-9.
Xu, B., Kim, S.T., Lim, D.S. & Kastan, M.B. (2002). *Mol Cell Biol*, **22**, 1049-59.
Zachos, G., Rainey, M.D. & Gillespie, D.A. (2003). *Embo J*, **22**, 713-23.
Zachos, G., Rainey, M.D. & Gillespie, D.A. (2005). *Mol Cell Biol*, **25**, 563-74.

Chapter 9

USING DT40 TO STUDY CLATHRIN FUNCTION

Frank R. Wettey and Antony P. Jackson
Department of Biochemistry, University of Cambridge, Tennis Court Road, Cambridge, CB2 1QW, UK.

Abstract: Clathrin-coated pits and vesicles play a major role in eukaryotic membrane trafficking pathways. We have used the DT40 system to delete endogenous clathrin genes selectively from DT40 and replace them with clathrin under the control of a tetracycline-regulatable promoter. This enabled clathrin expression to be manipulated, and the functional consequences of clathrin depletion to be studied in a stable vertebrate context. Here we describe the background to the work on clathrin, our practical experience with using the tetracycline-regulatable expression system in DT40 and some novel insights into membrane trafficking pathways obtained using the cell-lines generated during the course of this work.

Key words: clathrin, DT40, receptor-mediated endocytosis, tetracycline-regulatable expression system

1. INTRODUCTION

A characteristic and defining feature of eukaryotic cells is the possession of a complex suite of internal membrane compartments. The composition, size, shape and polarity of these membrane boundaries are maintained by an elaborate vesicular transport system in which selected proteins are transported from one compartment to another via distinct classes of coated transport vesicles (Figure 9-1) (Mellman and Warren, 2000). The clathrin-coated vesicle (CCV) was the first such example to be discovered (Roth and Porter, 1964; Kanaseki and Kadota, 1969; Pearse, 1975), and continues to provide

J.-M. Buerstedde and S. Takeda (eds.), Reviews and Protocols in DT40 Research, 119–143.
© 2006 *Springer.*

important insights into the problems of intracellular membrane transport. The CCVs have been implicated in several distinct membrane transport steps. In particular, they are responsible for the internalisation of plasma membrane proteins by receptor-mediated endocytosis (RME) and the transport of lysosomal enzymes from the *trans*-Golgi network (TGN) (Figure 9-1). Clathrin lattices are also found on endosomes, where they play a role in the recycling of proteins to the plasma membrane and in the targeting of some ubiquitinated proteins for lysosomal degradation. In neurones and neuroendocrine cells, clathrin is required for the maturation of regulated secretory granules (Kirchhausen, 2000; Brodsky et al., 2001).

1.1   Receptor-mediated endocytosis (RME)

RME is probably the best understood of all clathrin-mediated transport steps. Proteins internalised by RME include a wide variety of receptors carrying ligands required for growth (for example transferrin, low-density lipoprotein etc.) (Mousavi et al., 2004). But RME plays other roles as well. For example, by removing hormone receptor complexes, RME can act to attenuate the signal. In other cases, by delivering bound receptors to intracellular compartments for optimum signalling, RME may enhance and sustain signal transduction (Sorkin and Von Zastrow, 2002). At the inner-surface of the plasma membrane coat components sequentially assemble to form a clathrin-coated pit. Receptors interact with a class of molecules called adaptors and become clustered into this growing structure. The adaptors link the membrane proteins with the clathrin that forms the outer layer of the coat. Together with accessory and regulatory molecules, cargo proteins, adaptors and clathrin co-assemble, and the growing coated pit invaginates. Finally, the membrane neck is severed to form a closed coated vesicle (Figure 9-2A).

1.2   Adaptors and accessory proteins

Adaptors represent a structurally diverse group of proteins, whose role is to recognise many different classes of cargo receptors. The 'classic' family of adaptor proteins (APs) are closely related and comprise AP1, AP2, AP3 and AP4. Each of these four classes is localised to different intracellular compartments and varies in its receptor specificity. APs are heterotetramers. There are two 110-130 kDa large subunits (a β-class adaptin, and one of either an α, γ, δ or ε adaptin), together with a 50 kDa medium chain (μ adaptin) and a 15-20 kDa small chain (σ adaptin). The two large subunits contain a C-terminal appendage domain connected to the N-terminal core

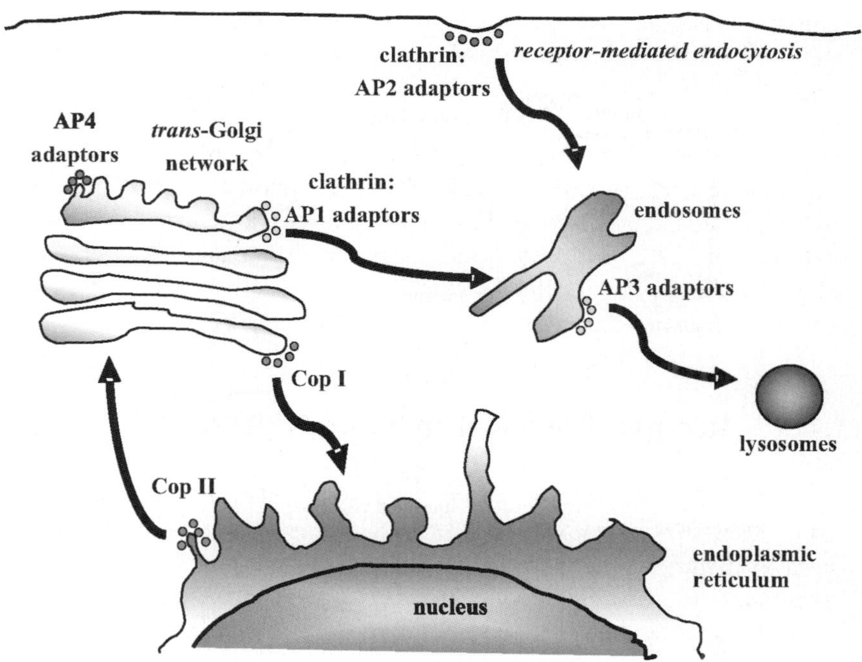

Figure 9-1. Major trafficking pathways in a typical eukaryotic cell. Proteins move from the endoplasmic reticulum (ER) through the secretory pathway via distinct classes of coated transport vesicles. For example, transport between the ER and the Golgi is largely carried out by Cop I and Cop II vesicles. Clathrin-coated vesicles are required for receptor-mediated endocytosis from the plasma membrane and transport of lysosomal proteins from the *trans*-Golgi network to endosomes/lysosomes. Clathrin-coated vesicles at different cellular locations are associated with different classes of adaptor proteins (APs) (see text). The diagram is highly simplified. For example, only one endosome compartment is shown and multiple pathways to and from each compartment are omitted for clarity (see plate 17).

domain via an unstructured flexible linker sequence (Owen et al., 2004). The AP2 adaptor (comprising α, β2, μ2 and σ2 adaptin subunits) is associated with RME from the plasma membrane (Figure 9-2B). It interacts initially with the plasma membrane via membrane docking proteins and binding sites for phosphatidylinositol (4,5)-bisphosphate (PIP2), a lipid particularly concentrated in the plasma membrane (Collins et al., 2002). Internalised receptors contain short sequence motifs in their cytoplasmic domains that are recognised by APs. There are different classes of internalisation signals that bind to different components of the AP structure. One important class of signal is of the form YXXΦ in which Φ is a bulky hydrophobic amino-acid, but X is

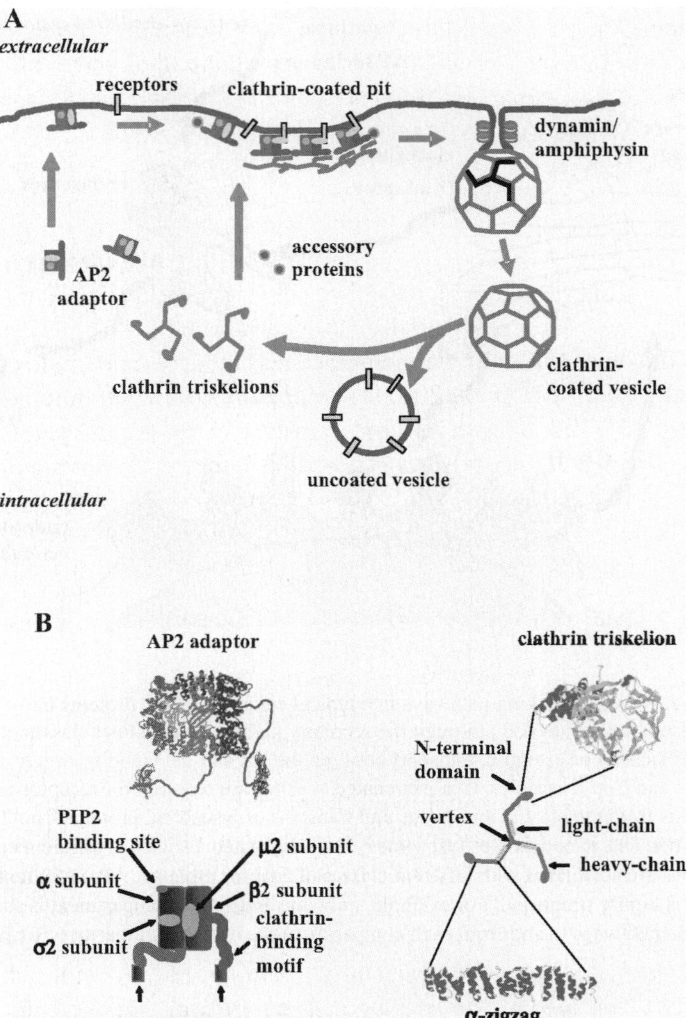

Figure 9-2. A. Outline of the main events in receptor-mediated endocytosis. The AP2 adaptor attaches to the inner-surface of the plasma membrane, initially via its PIP2 binding site. A conformational change allows the μ2 subunit to bind the cytoplasmic domain of the receptors (see section 3.5). Clathrin, together with other accessory proteins, binds the adaptors and drives membrane curvature to form the coated pit. The final conversion of the coated pit into the coated vesicle is catalysed by dynamin, which assembles with amphiphysin on the membrane neck of the highly invaginated pit. The black outline on the coated vesicle shows the location of an individual triskelion on the assembled coat. B. Structures of the AP2 adaptor and clathrin triskelion. Structures were downloaded from http://www.pdb.org/pdb/Welcome.do (see plate 18).

variable. This sequence motif is recognised by the AP μ subunit (Ohno et al., 1995; Ohno et al., 1998). More recently, it is becoming appreciated that other adaptor families are important in extending the range of proteins incorporated into coated pits. For example, β-arrestins act as specialised adaptors for G protein-coupled receptors (Robinson, 2004).

A variety of accessory proteins help to co-ordinate the coat assembly in RME. Proteins such as CALM and its brain-specific isoform AP180 possess multiple short binding sites for PIP2, clathrin and the AP2 α subunit-appendage domain. It can promote clathrin and AP2-binding to the coated pit as flat lattices (Ford et al., 2001). Epsin1 is implicated in membrane deformation that occurs with pit formation. Here the protein inserts an α-helix into the outer leaflet of the membrane and induces membrane curvature (Chen et al., 1998; Itoh et al., 2001). The protein Eps15 can bind both epsin and clathrin/AP2 and is localised predominantly to the growing edges of the coated pit. As such it may organise the spatial location of epsins and similar molecules (Tebar et al., 1996). The accessory protein endophilin I has lysophosphatidic acid acyl transferase activity, which is thought to facilitate invagination by altering the local lipid composition of the membrane (Ringstad et al., 1999; Schmidt et al., 1999).

1.3   Clathrin assembly

Clathrin assembles onto the adaptors to form the outer layer of the coat. It probably acts to stabilise the curvature introduced into the growing pit, whilst increasing its deformation until the entire region invaginates to form a closed vesicle. In its unassembled cytosolic form, clathrin consists of three identical heavy chains (CHC) with a molecular weight of approximately 192 kDa, each bound to one of two 23-27 kDa light chains (LCa or LCb) to form a three-legged structure called a triskelion (Kirchhausen, 2000). The triskelions provide the basic structural building block, with which the characteristic hexagonal and pentagonal polyhedral cage of the coated vesicle is assembled. Each 'leg' of the triskelion can be subdivided into discrete segments. The C-terminal domain forms the triskelion vertex, where the three heavy chains stably interact. The proximal and distal regions of the leg are formed from a repeated sequence of short α-helices in a so called α-zigzag conformation that provides rigidity to the structure (Ybe et al., 1999). The vertex is puckered and the angle between the proximal and distal regions of the leg is flexible, so as the triskelions assemble, they tend to form closed cages with hexagonal and pentagonal faces. The globular N-terminal domain comprises a seven-bladed β-propeller structure. In the assembled coat, this N-terminal domain protrudes inwards. The clefts formed from adjacent blades of the propeller act as binding sites for short motifs on the adaptors such as the linker region of the AP β subunit, as well as other accessory proteins (ter Haar et al., 2000; Fotin et al., 2004) (Figure 9-2B). The associated light chains bind to the proximal region of the triskelion leg in an extended conformation and help to regulate assembly competence (Brodsky et al., 1991). In addition to the ubiquitously expressed CHC gene responsible for membrane traffic, most vertebrates including chicken possess a second CHC-like gene, but expression of this isoform is largely confined to muscle and it does not bind clathrin light chains, suggesting a specialised and perhaps different role (Towler et al., 2004).

1.4   Membrane fission and uncoating

As the triskelions assemble, the pit invaginates until only a thin neck connects the forming CCV to the plasma membrane. The protein amphiphysin dimerises onto the neck and provides a platform for the assembly of the GTPase dynamin (Peter et al., 2004) (Figure 9-2A). In a way that is not fully understood, the hydrolysis of GTP drives the final scission of the membrane neck (Sever, 2002). Following detachment from the plasma membrane, clathrin is quickly uncoated by the combined action of the ATPase Hsp70 and the coat component auxilin (which provides the DnaJ domain for the Hsp70) (Lemmon, 2001). The APs and accessory proteins are probably uncoated separately. Here the protein synaptojanin may play a role. It is recruited onto the pit by its interaction with amphiphysin and endophilin. Synaptojanin possesses a lipid phosphatase activity, which converts PIP2 to phosphatidylinositol (PtdIns), thus weakening the attachment of coat components with the vesicle membrane (Cremona et al., 1999; Verstreken et al., 2003).

1.5   Clathrin coat formation at the *trans*-Golgi network (TGN) and other intracellular compartments

Coated vesicle formation at the TGN is likely to be fundamentally similar to that at the plasma membrane. However, there are also some important differences. For example, the coat specificity is not the same. The major cargo carried by TGN-derived coated vesicles are the cation-dependent and cation-independent mannose 6-phosphate receptors (MPR) (Griffiths et al., 1988). Their ligands are lysosomal enzymes that are tagged with mannose 6-phosphate as they proceed through the Golgi stacks. The major destination for these CCVs is the late endosome compartment for ultimate delivery to the lysosome (Luzio et al., 2003). The differing specificity of TGN-derived compared to plasma membrane-derived CCVs is accounted for by the presence of a different adaptor, AP1. This adaptor comprises γ, β1, μ1 and σ1 adaptin subunits (Owen et al., 2004). The selective targeting of AP1 to the TGN may be due partly to its inability to bind PIP2 and its requirement for the small molecular GTPase Arf1, which is localised to the TGN membrane by an unknown mechanism (Austin et al., 2002; Bonifacino and Lippincott-Schwartz, 2003). In addition to AP1, clathrin at the TGN can interact with a number of monomeric adaptors called GGAs (Golgi-localised γ-ear containing Arf-binding protein). The precise role of the GGAs is not clear, although they are believed to exhibit an analogous and partly overlapping role to AP complexes. They are implicated in recruitment of mannose 6-phosphate receptors and other accessory proteins (Dell'Angelica et al., 2000; Hirst et al., 2000; Robinson, 2004). The adaptor AP4 is also present at the TGN (Dell'Angelica et al., 1999; Hirst et al., 1999). As opposed to yeast, vertebrate AP3 is localised to a more peripheral endosomal compartment rather than the TGN and may play a role in endosome to lysosome traffic. In melanocytes, the AP3 adaptors are required for the transport of material to melanosomes, but it is not clear whether this adaptor requires clathrin in mammals (Dell'Angelica et al., 1997; Robinson and Bonifacino, 2001).

1.6 Experimental approaches to understanding clathrin function

Our current understanding of clathrin function rests on a variety of experimental approaches. At one extreme, there has been a recent outpouring of detailed atomic resolution structures of almost all the major coated vesicle proteins and these have provided remarkable molecular insights into key aspects of coat assembly and regulation (Owen and Luzio, 2000; Fotin et al., 2004). The use of simplified *in vitro* assays has been essential to dissect the temporal sequence of events during coated pit formation and the trans-formation into coated vesicles (Smythe et al., 1992). These approaches also need to be complimented with studies in living cells to place the results in proper physiological context. Early attempts to interfere with CCV formation *in vivo* exploited the inhibition of coated pit assembly by cytosolic acidification, potassium depletion and cytoskeletal disruption (Larkin et al., 1983; Cosson et al., 1989). These approaches do indeed block coated pit formation and some insights into receptor trafficking can be inferred by their use, but inevitably they are highly disruptive and non-physiological and risk widespread non-specific effects that are difficult to control for. Expression of a dominant-negative truncated clathrin has been described and provides a cleaner method of interfering, but does not remove the wild-type protein (Bennett et al., 2001). Gene targeting has been used to inactivate clathrin in single-celled eukaryotes such as bakers' yeast *S. cerevisiae* (Lemmon and

Jones, 1987; Payne et al., 1987) and the slime mould *D. discoideum* (Ruscetti et al., 1994). However, there are significant differences in the membrane trafficking pathways in vertebrates and these distantly related organisms making detailed comparisons sometimes difficult. The recent development of antisense technology and siRNA inhibition has greatly extended experimental options (Hinrichsen et al., 2003; Iversen et al., 2003; Motley et al., 2003), although transient transfection of siRNA has its own problems such as variable levels of inhibition.

It is within this context that the use of DT40 to study clathrin is particularly appealing. Firstly, the ability to target genes in DT40 with high efficiency offers an ease of genetic manipulation unique among vertebrate cell-lines and ensures that the endogenous gene can be permanently and completely removed in a cloned homogeneous background (Buerstedde and Takeda, 1991). Secondly, as detailed below, the endogenous genes can be replaced by a tightly inducible expression system. Thirdly, although in comparison to cell-lines such as HeLa and CHO that are used extensively in work on membrane traffic, the relatively small size of DT40 may seem unprepossessing, it does contain a well-defined and active secretory and endocytic pathway (Figure 9-3). Indeed, in our experience DT40 is one of the most endocytically active cell-lines we have examined. Furthermore, DT40 grows quickly and can easily attain high cell densities allowing bulk biochemical assays to be performed with relative ease. Fourthly, many of the key proteins of interest in membrane traffic are strongly conserved within vertebrates – the amino-acid sequence of CHC itself for example shows 96% identity between chicken and human. Hence there is a good chance that many antibodies raised against the mammalian proteins can be used directly in DT40-related work.

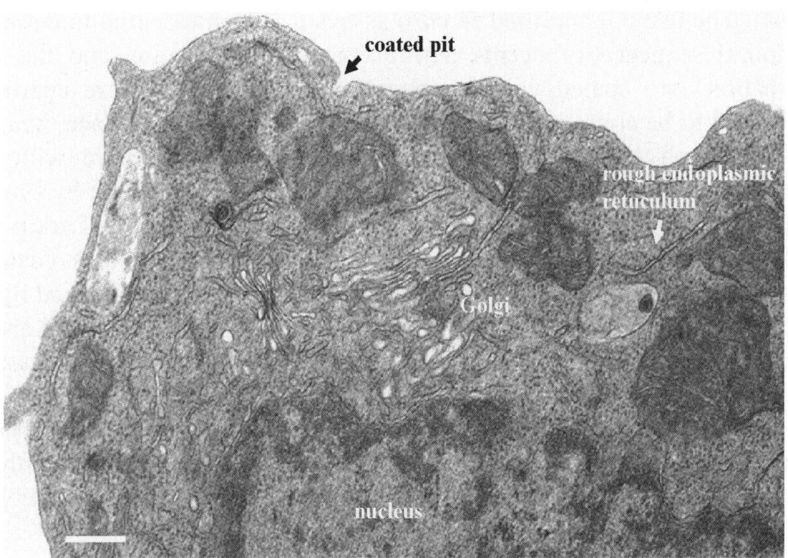

Figure 9-3. Electron micrograph of a typical DT40 cell showing well-developed secretory pathway organelles. Bar = 1 μm.

2. GENERATING A CLATHRIN KNOCKOUT IN DT40

By comparison to its mammalian homologue, the chicken CHC gene is relatively compact with small introns. It is localised on chromosome 19, one of the microchromosomes (http://www.ensembl.org/Gallus_gallus/geneview?gene=ENSGALG00000005139).There are only two alleles of this gene in DT40 making it an attractive target for gene knockout. However, previous work suggests caution in attempting to inactivate clathrin even in a vertebrate tissue-culture cell-line. For example, in *S. cerevisiae*, the viability of clathrin-deleted strains depends on the genetic background of the yeast (Lemmon and Jones, 1987; Payne et al., 1987). In *D. discoideum*, clathrin deletion is not normally lethal, but becomes so if the cells are grown in suspension culture due to a block on cytokinesis (Ruscetti et al., 1994; Niswonger and O'Halloran, 1997). Inactivating mutations in the CHC gene of the fruit fly *D. melanogaster* generally cause lethality during embryogenesis (Bazinet et al., 1993). Consequently, we decided to combine the advantages of efficient gene targeting in DT40 with the Tetracycline-regulatable gene expression system (Tet) to allow the controllable expression of clathrin (Gossen and Bujard, 1992).

2.1 The Tet expression system in DT40

The Tet based regulation has several advantages over other inducible expression systems such as those regulated by metal ions, heat shock or steroid hormones. Tet systems have very high maximal expression levels, with low to undetectable background transcription in many cases and they allow precise control over the desired expression level by well-characterised, specific inducers that do not have any pleiotropic effects. They have fast kinetics and a wide dynamic range of induction levels ranging over several orders of magnitude (Yin et al., 1996). In the Tet system, a cDNA to be controllably expressed is ligated downstream of a hybrid promoter comprising the Tn10 encoded regulatory elements (TRE) of the *E. coli* tetracycline-resistance operon and the minimal promoter of the human cytomegalovirus (P_{minCMV}). Transcription is controlled by a chimeric transactivator, either rtTA (Tet-On) or tTA (Tet-Off). Each transactivator consists of the tetracycline repressor (TetR) fused to the activation domain of the virion protein 16 (VP16) from herpes simplex virus. The transactivators stimulate transcription upon binding the TRE of the otherwise silent promoter. However, the Tet-On transactivator only binds in the presence of tetracycline or the more potent analogue doxycycline, whilst the Tet-Off transactivator only binds in its absence (Baron and Bujard, 2000) (Figure 9-4A and B).

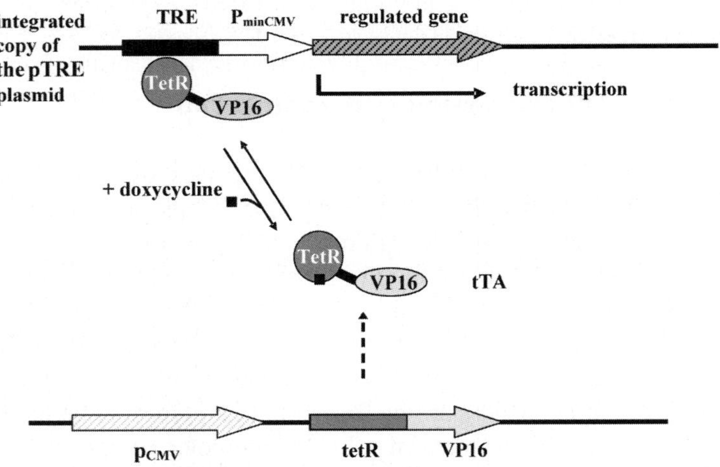

Figure 9-4. Schematic showing the two types of Tet-responsive expression systems. A. Tet-On and B, Tet-Off. An integrated plasmid expressing either the Tet-On (rtTA) or Tet-Off (tTA) transactivator drives expression of the regulated gene from a TRE promoter. The Tet-On transactivator binds TRE in the presence of tetracycline or doxycycline, the Tet-Off transactivator binds in its absence.

Before choosing either of these systems, one has to think carefully about the issues to be addressed (knockdown versus regulated expression or overexpression, the importance of residual expression etc.). Difficulties in expressing a gene of interest using the Tet system have been reported for some cell-types (Yin et al., 1996), although it works well in DT40 (Wang et al., 1996). It appears that a perfect Kozak initiation sequence is not necessary for protein expression from the TRE/P_{minCMV} promoter (our own data and others). A problem that has been observed by many researchers for other cell types and has also been our experience with DT40 is that cell growth at high densities (i.e., above 2×10^6 cells/ml) tends to repress expression of the Tet-Off regulated gene, even in the absence of doxycycline. The reason is not clear. One factor that has a major influence on the basal activity and maximal expression is the genome integration site of both the transactivator plasmid and the plasmid encoding the regulated gene (Damke et al., 1995). This is illustrated in Table 9-1, which shows data from our own results generated during the construction of the CHC knockouts. In our initial experiments, we generated separate DT40 clones each stably expressing a randomly integrated copy of the pTet-On or pTet-Off plasmid. When these were separately and transiently transfected with a plasmid expressing luciferase under the control of the TRE/P_{minCMV} promoter, there was a wide range of quantitative responses in enzyme activity with respect to doxycycline addition. The most likely explanation is that this variation in transactivator efficiency reflects differences in the random integration of the transactivator within the DT40 genome. Another point to note is the tendency of the Tet-On system to produce a higher background than Tet-Off, a feature that has been noticed in other cell-lines (Yin et al., 1996). In light of these results, we chose the Tet-Off system for further development. The Tet-Off DT40 clone giving the largest fold response to doxycycline addition (G526, Table 9-1) was used to generate a heterozygous line deleted for the first CHC allele using standard DT40 protocols (Winding and Berchtold, 2001). From this, we generated several lines heterozygous for the endogenous CHC gene and stably transfected with the full-length human CHC cDNA that had been integrated into the plasmid pBI-L (Baron et al., 1995). This plasmid allows the bidirectional expression of the gene of interest (human clathrin) and luciferase, both under the control of the TRE/P_{minCMV} promoter. The advantage of this approach is that by measuring luciferase activity with and without doxycycline, it affords an easy and convenient way to identify the clones most strongly responsive to doxycycline addition. Once again there is a wide variation in the degree of enzyme repression (Table 9-1). The important message is that although the Tet system can give very tight responses in DT40, it is still worth the extra effort to use suitable preliminary screening assays so as to optimise the chance of obtaining the best responders. From our experience, the luciferase reporter system is quantitative, reproducible and easy to use in DT40 screens with both transient and stable transfections.

Table 9-1. Comparison of Tet-On and Tet-Off system and transient vs. stable lines: individual DT40 clones stably expressing pTet-On (clones G438 to G448) or pTet-Off (clones G526 to G577) were established and each was transiently transfected by standard electroporation with the reporter plasmid pUHC13-3 containing the luciferase gene under the control of the TRE/P_{minCMV} promoter. Cells were grown either with or without 1µg/ml doxycycline. At 48hrs post transfection, the cells were harvested, lysed and assayed for luciferase activity. There was a wide variation from clone to clone in response to doxycycline. The best responder from the Tet-Off experiment (clone G526) was used to derive a heterozygous endogenous CHC knockout. This was then double transfected with a puromycin selection marker and a bi-directional luciferase reporter plasmid also expressing human CHC. Stable transfected, puromycin-resistant cell clones (H508-D5 to H526-G1) were assayed for luciferase activity (see methods section) after 48hrs of growth in either the presence or absence of 1µg/ml doxycycline. Once again there is a wide range of responses from individual clones. The best responders were used to generate the final $CHC^{-/-}$ DT40 cell. Representative experiments of only a small number of clones are shown for illustration.

clone	luciferase activity (RLU/µg)		activation factor
	- doxycycline	+ doxycycline	
Tet-On: transient			
G438	36	1,665	46
G439	21	718	34
G440	10	595	60
G441	11	709	64
G442	8	866	108
G443	13	2,207	170
G444	9	2,496	277
G445	7	2,208	315
G446	23	2,088	91
G447	22	2,356	107
G448	8	1,555	194
Tet-Off: transient			
G526	4,320	7	617
G528	2,095	6	349
G533	2,581	8	323
G534	3,202	9	356
G553	1,676	6	279
G555	3,489	9	388
G563	1,373	5	275
G576	203	11	18
G577	424	3	141
Tet-Off: stable			
H508-D5	23,536,645	12,090	1,947
H519-C1	15,398,349	29,547	521
H520-C5	164,424,753	2,370	69,384
H521-C6	174,983,543	1,441	121,432
H523-D6	113,968,789	1,223	93,188
H525-E3	10,472,089	1,250	8,378
H526-G1	4,172,100	1,257	3,319

As a result of these preliminary experiments, we were able to select a Tet-Off clone heterozygous for endogenous clathrin that generated a five orders of magnitude luciferase reporter response to doxycycline (Table 9-1), similar if not better than that reported from other cell-lines (Gossen and Bujard, 1992). This clone was used as the basis for the elimination of the final endogenous clathrin allele by targeted gene disruption to generate a CHC$^{-/-}$ DT40 cell-line conditionally deficient in clathrin expression (Wettey et al., 2002). When grown in the presence of 50-100 ng/ml doxycycline, clathrin expression was efficiently repressed (Figure 9-5A). Note the slower decline in clathrin expression compared to luciferase activity, a result that is consistent with the long half-life of the clathrin protein (Acton and Brodsky, 1990) and which typically necessitates waiting three days following doxycycline addition before conducting experiments. The use of the Tet-Off system offers another unique advantage. Growing cells in varying concentrations of doxycycline allows the expression of the regulatable gene to be set at predetermined levels so as to allow *in vivo* titration experiments (see section 3.2). This is shown in Figure 9-5B for both CHC and luciferase. In this experiment, the CHC results were obtained with the DT40-CHC$^{-/-}$ conditionally deficient cells, whilst the luciferase example was obtained with a DT40 Tet-Off cell line transiently transfected with the response plasmid. The effective range of doxycycline for both cells is comparable, reproducible and similar to that determined for the Tet system in other cell types (Yin et al., 1996). Hence, preliminary transient experiments designed to identify the most responsive pTet-Off-integrated clones can, if needed, also be used to accurately predict the quantitative response to doxycycline titration shown by the final stable construct.

3.   FUNCTIONAL STUDIES

Here we summarise how clathrin depletion in DT40 has produced some unexpected insights and how experiments using these cells can be applied to a variety of questions both general to vertebrate membrane traffic and specific to lymphocyte biology.

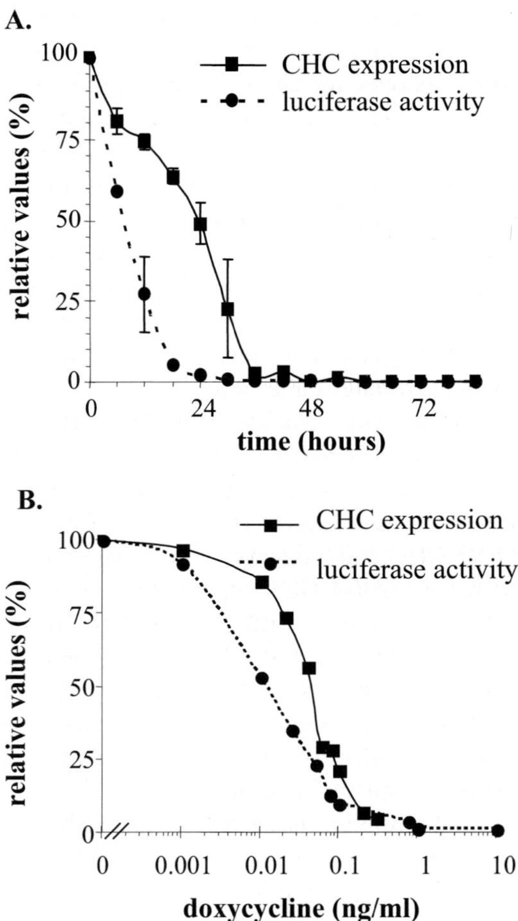

Figure 9-5. Tight regulation of reporter expression with the Tet-Off system in DT40. A. Time-course. Transgenic DT40-CHC$^{-/-}$ cells stably expressing human CHC and luciferase from a bi-directional promoter were incubated in the presence of 50ng/ml doxycycline for the indicated times. Cell lysates were prepared and analysed in duplicates for luciferase activity (see methods section) and for human CHC expression by immunoblotting (Wettey et al., 2002). Luciferase activity and CHC expression are expressed as a percentage relative to their level at the beginning of the experiment. B. Titration of Tet-regulated genes in DT40. For CHC, transgenic DT40 - CHC$^{-/-}$ cells (squares) were incubated for 72 hrs with the indicated concentrations of doxycycline, cell lystes prepared, assayed for human CHC content by immunoblotting (Wettey et al., 2002) and the signals quantified by densitometry. For luciferase activity, cells of a DT40 Tet-Off cell line (circles) were transiently transfected by electroporation with the plasmid pUHC13-3 containing the luciferase gene under the control of the TRE/P$_{minCMV}$ promoter and incubated with the indicated concentrations of doxycycline for 48 hours. Luciferase activity was measured in duplicate as described in materials. Values are means of duplicates with a standard error ≤15%.

3.1   Lethality in clathrin-depleted DT40 cells

An early finding was that clathrin depletion rapidly led to cell death by apoptosis (Wettey et al., 2002). Following this observation, the originally derived clathrin knockout line was redesignated DKO-S (*d*ouble *k*nockout, *s*ensitive to clathrin-depletion). During the course of the work however, we isolated a variant derived from DKO-S cells that did not die when clathrin expression was repressed. This variant was recloned and its properties have remained stable. It is designated DKO-R (*d*ouble *k*nockout, *r*esistant to clathrin depletion). Both DKO-S and DKO-R cells repress clathrin to the same extent and with the same kinetics in the presence of doxycycline. Both cell-lines activate caspase when challenged with the DNA-damaging drug cytosine arabinoside. Hence the DKO-R cells have retained their apoptotic pathway, but they no longer respond to the apoptotic signal from clathrin depletion. We normally grow our DT40 cells in media supplemented with 10% foetal calf serum and 1% chicken serum (Buerstedde and Takeda, 1991). DKO-S cells grown in media with 10% foetal calf serum but 0.25% chicken serum grew normally when expressing clathrin, but died sooner when clathrin expression was repressed (Wettey et al., 2002), suggesting that chicken serum contains a factor (or factors) required for survival and is not normally growth limiting, but becomes so when clathrin is removed. This view is supported by the fact that DKO-S cells can survive clathrin depletion if the media is supplemented with higher than normal levels of chicken serum (2% or greater) and some batches of chicken serum are better than others at supporting the growth of clathrin-depleted DKO-S cells (Stoddart et al., 2005). The identity of the factor is not yet clear, but serum is a rich supply of growth and survival factors that are known to be internalised by CCVs. It may be significant for example that some lymphocyte cell-lines undergo apoptosis when transferrin, uptake is selectively blocked (Moura et al., 2004).

3.2   Effects of regulated clathrin depletion
on receptor-mediated endocytosis

The rate of receptor-mediated endocytosis of transferrin was significantly reduced by about 84% in the absence of clathrin. Although even here, transferrin uptake was not completely eliminated (Wettey et al., 2002). Clathrin-independent pathways of endocytosis have been described previously (Nichols and Lippincott-Schwartz, 2001), but the ability to cleanly remove clathrin in a cell-line that is normally highly endocytically active allows these pathways to be experimentally examined in detail (see section 3.3). Growing DKO-R cells in varying concentrations of doxycycline allows clathrin expression to be set at predetermined levels *in vivo* (Figure 9-6A and B) and hence the dose-dependent effect of clathrin reduction on a particular pathway - in this case receptor-mediated endocytosis - can be quantitatively investigated. Since the DKO cells grown without doxycycline express a higher level of clathrin than wild-type DT40, the titration

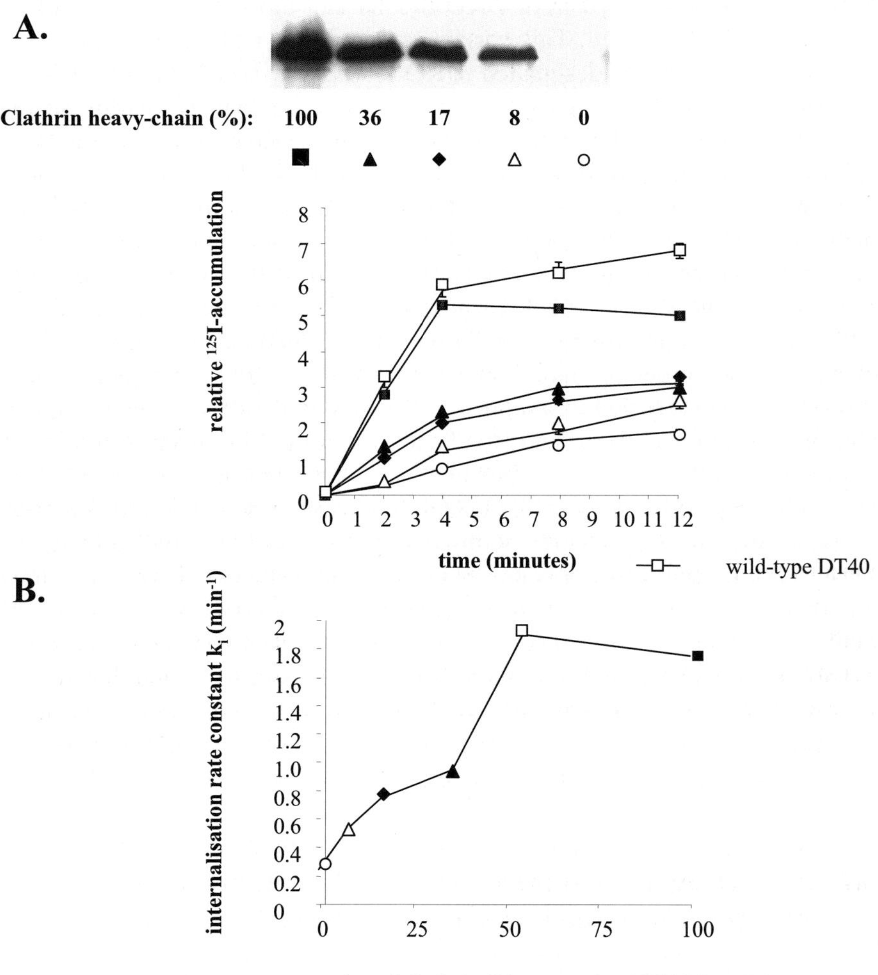

Figure 9-6. Titration of receptor-mediated endocytosis (RME) with pre-set levels of clathrin in DKO-R cells generated by incubating DKO-R cells with increasing concentrations of doxycycline from 0 to 50 ng/ml. A. Clathrin expression was measured by immunoblotting (Wettey et al., 2002). A serial dilution of lysate from untreated cells was run on the same gel for quantification by densitometry. RME of ^{125}I-labelled chicken transferrin (conalbumin) and curve fitting to obtain values of internalisation rate constant k_i are as described in the methods section. B. Dependence of internalisation rate on expressed clathrin level.

experiments can explore a range of clathrin levels both above and below that of wild-type. It is notable for example that the internalisation rate for wild-type DT40 is similar to the level for unrepressed DKO-R cells, suggesting that at the level of clathrin normally expressed in these cells, factors other than clathrin are limiting for its endocytic capacity.

3.3   Clathrin and B cell receptor endocytosis

Antigen binding to the B cell antigen receptor (BCR) activates both BCR signalling (Shinohara and Kurosaki, this issue, Chapter 10) and internalisation. The mechanism by which the receptor is endocytosed and the effect of endocytosis on signal transduction are controversial. In particular, the relative importance of clathrin, actin and lipid-rafts on these pathways has been difficult to disentangle (Reth and Wienands, 1997). As a pre-B cell-line, DT40 expresses the BCR and its trafficking and signalling has recently been studied in DKO-S cells grown with and without clathrin (Stoddart et al., 2005). Clathrin-depletion significantly, but not entirely inhibited antigen-stimulated BCR endocytosis. The residual clathrin-independent endocytosis of the BCR was blocked by agents that inhibited actin polymerization and raft integrity and also affected the clathrin-dependent pathway. Rafts or actin on their own were not sufficient, since depletion of clathrin together with either a raft or actin antagonist completely abolished internalisation. Taken together, the data suggest that a major pathway of internalisation required clathrin acting in concert with lipid rafts and cytoskeletal components and that a minor alternative pathway exists, which is dependent on lipids rafts and actin, but not clathrin. Clathrin-depletion also led to a significant enhancement of BCR signalling via MAP kinase indicating that, unlike some receptors, delivery to the endosome is not a pre-requisite for BCR signalling.

3.4   Endocytosis of avian leucosis virus

In addition to receptors and other cell-surface molecules, certain viruses have exploited endocytic pathways to enter cells. The DT40 cell-line was originally generated by transformation with avian leucosis virus (ALV), and DT40 sheds low levels of ALV into the medium (Baba et al., 1985). These viral particles can often be seen in electron micrographs of DT40 cells both in the medium and inside the cells in membrane-bounded compartments. In some cases ALV particles can be seen clearly associated with clathrin-coated pits (Figure 9-7). In the absence of clathrin, ALV particles were no longer associated with the internal compartments and strikingly accumulated at the plasma membrane (Wettey et al., 2002).Further work using DKO-S and DKO-R cells has quantitatively confirmed this picture for the case of GFP-tagged ALV subgroup B and established a central role for clathrin in the uptake of this virus into cells (Diaz-Griffero et al., 2005).

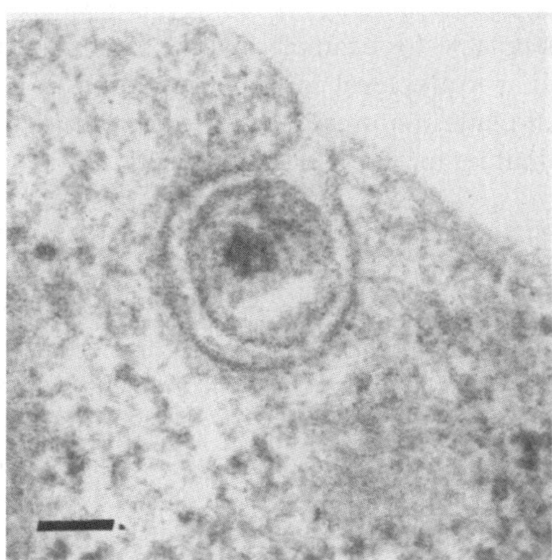

Figure 9-7. Transmission electron micrograph of an endogenous avian leucosis virus secreted from DT40 being re-internalised by a clathrin-coated pit. Here the electron dense coat around the pit can be clearly seen. Bar = 100 nm.

3.5　Clathrin as a regulator for receptor-mediated endocytosis: its role in AP2 phosphorylation

As noted above (section 1.2), the plasma membrane-associated adaptor AP2 captures receptors containing YXXΦ motifs in their cytoplasmic tails via interaction with the AP2 μ2 subunit. Strikingly however, the atomic resolution structure of AP2 clearly shows that the binding site of the μ2 subunit is occluded, implying that a conformational change is required to allow μ2 to bind receptors (Collins et al., 2002). This conformational change is driven by the phosphorylation of threonine 156 of the μ2 subunit (Olusanya et al., 2001). A strong candidate for the enzyme that carries out this reaction is adaptor-associated kinase 1 (AAK1) (Conner and Schmid, 2002). An inactive form of AAK1 associates at low stoichiometry with cytosolic AP2 and is thus delivered to the growing coated pit along with AP2. Clathrin-depletion in DKO-R cells lead to a significant reduction in the steady-state level of μ2 phosphorylation *in vivo*, suggesting a role for clathrin in the regulation of this step. Purified clathrin activated AAK1 activity and in permeabilised cells, ligand sequestration into the growing coated pits was driven by clathrin, concomitantly with enhanced µ2 phosphorylation (Jackson et al., 2003). In DKO-R cells, immunofluorescence studies show that in the absence of clathrin, the AP2 adaptor still bound its target membrane. In fact, the binding was increased in the absence of clathrin (Wettey et al., 2002). This can be explained if initial binding of AP2 to the membrane occurs independently of receptor tails, probably via its lipid PIP2-binding site (Collins et al., 2002). In this way, AP2 is pre-targeted to the plasma membrane, but its functional interaction with receptors requires activation of µ2 phosphorylation in a clathrin-dependent manner (see Figure 9-2A). The increased membrane binding in the absence of clathrin suggests that AP2 requires clathrin for its detachment from the plasma membrane. This is the first indication that clathrin is not just a structural coat component, but also plays an important regulatory role, ensuring that AP2 becomes competent to bind receptors only within the context of the growing coated pit.

3.6   Perturbation of trafficking at the TGN by regulated clathrin depletion

Similar to the effect on AP2, clathrin removal in DKO-R cells caused an increase in membrane binding of the adaptors AP1 and GGA1 at the TGN (Wettey et al., 2002). Clathrin depletion also caused a dramatic reorganisation in the intracellular distribution of the MPR, from a faint but tight perinuclear region to a bright and more peripheral localisation (Figure 9-8), clearly indicating a connection between the MPR trafficking and clathrin. There is some evidence that CCVs may be particularly important for the recycling of the MPR from late endosome compartments to TGN. For example, in fibroblast from mice lacking the AP1 µ1 subunit, the MPR accumulates in post TGN compartments (Meyer et al., 2000), a result similar to the effect shown with clathrin-deficient DKO-R cells. Surprisingly however, clathrin depletion failed to affect the steady-state morphology of lysosomes. Moreover, the trafficking of the soluble lysosomal enzyme β-glucuronidase and of the membrane-bound lysosomal protein LEP100 was unaffected by clathrin depletion (Wettey et al., 2002). Although subtle effects on lysosomal function cannot yet be ruled out, the results do constrain current models that place a central role for clathrin in lysosomal transport. It has been noted previously that lymphocytes from I-cell disease patients, whose soluble lysosomal enzymes lack mannose 6-phosphate, nonetheless retain viable lysosomes (Gabel and Kornfeld, 1984; Glickman and Kornfeld, 1993; Dittmer et al., 1999). This suggests that lymphocytes posses a MPR-independent pathway of lysosomal traffic. DT40 is a pre B-lymphocyte and may well express this pathway. But if so, the pathway may also be clathrin-independent.

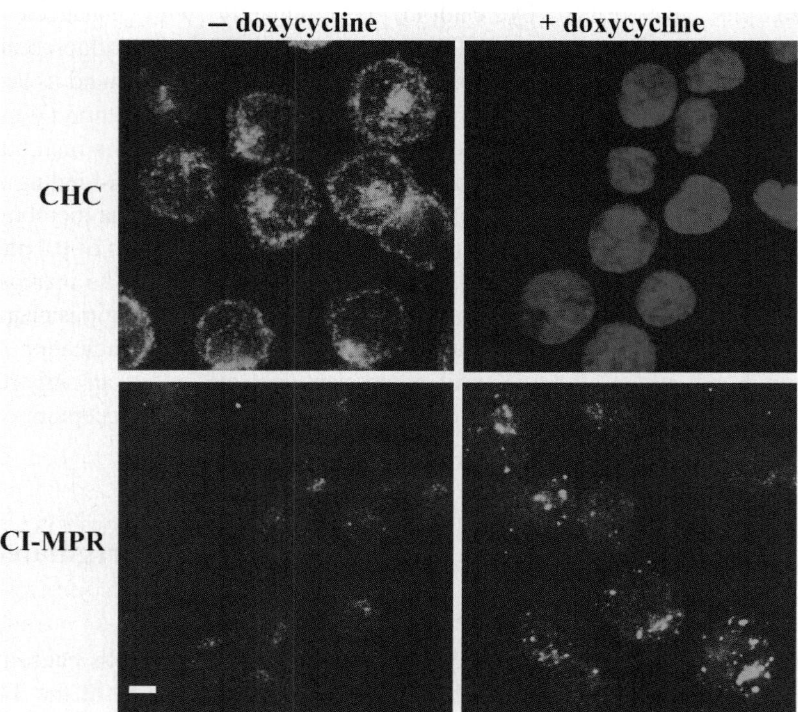

Figure 9-8. Effect of CHC depletion on the distribution of the cation-independent mannose 6-phosphate receptor (CI-MPR). DKO-R cells were incubated for 96 hrs either with or without 50 ng/ml doxycycline and separately assayed by indirect immunofluorescence microscopy as described (see methods section) using specific antibodies against clathrin heavy-chain (monoclonal antibody TD1) and CI-MPR (rabbit polyclonal antibody PL603-CI) followed by Alexa 488-conjugated secondary antibodies. No specific staining is visible in doxycycline-treated cells stained for CHC. In this panel, the nuclei have been stained with Hoechst 33342 dye to reveal the cells. Bar = 5 µm (see plate 19).

4.   FUTURE PROSPECTS

DT40 has already been well exploited in a variety of fields within cell-biology such as signal transduction, apoptosis, the cell cycle and gene conversion (Kurosaki et al., 2000; Sonoda et al., 2001; Winding and Berchtold, 2001; Arakawa et al., 2002; Yamazoe et al., 2004). By contrast, studies on membrane trafficking pathways in DT40 have been relatively under researched. Our experience with clathrin shows that this deserves to change. Our results have provided further evidence for the redundancy built into membrane trafficking pathways and revealed a number of unexpected avenues of enquiry, for example connecting clathrin-mediated pathways with cell death.

As noted above, clathrin is a long lived protein, which takes three to four days to disappear once doxycycline is added. It is an open question whether the clathrin-depleted cells can compensate during this time by up-regulating other membrane trafficking pathways or whether a re-routing into already functioning alternative pathways such as fluid-phase endocytosis occurs. In order to prevent the effects of alternative pathways that may obscure or complicate experimental analysis, we need to develop approaches that allow for a rapid inactivation of clathrin (within minutes). One attractive possibility is to develop a temperature-sensitive derivative of chicken clathrin, based perhaps on the sequence of the temperature-sensitive allele of yeast clathrin (Pishvaee et al., 1997). This work could easily be extended to include other specific mutations to allow a detailed structure/function analysis of clathrin in its physiological context. For this, the easiest approach would be to transfect the constructs into the DKO-R cell-line and then repress wild-type clathrin with doxycycline. However, direct targeted integration of mutant constructs to replace the endogenous clathrin genes is more elegant and likely to generate a cleaner experimental model, because the mutant completely replaces the wild-type protein (Fukagawa and Brown, 1997; Fukagawa et al., 2001) (Arakawa and Buerstedde, this issue, Chapter 1). Such an approach in vertebrate cells can only be easily attempted using the remarkable properties of DT40.

ACKNOWLEDGEMENT

The work described in Table 9-1 and Figure 9-5 was carried out in the laboratory of Prof. J. C Howard, University of Cologne.

REFERENCES

Acton SL, Brodsky FM (1990) Predominance of clathrin light chain LCb correlates with the presence of a regulated secretory pathway. J Cell Biol 111:1419-1426.

Arakawa H, Hauschild J, Buerstedde JM (2002) Requirement of the activation-induced deaminase (AID) gene for immunoglobulin gene conversion. Science 295:1301-1306.

Austin C, Boehm M, Tooze SA (2002) Site-specific cross-linking reveals a differential direct interaction of class 1, 2, and 3 ADP-ribosylation factors with adaptor protein complexes 1 and 3. Biochemistry 41:4669-4677.

Baba TW, Giroir BP, Humphries EH (1985) Cell lines derived from avian lymphomas exhibit two distinct phenotypes. Virology 144:139-151.

Baron U, Bujard H (2000) Tet repressor-based system for regulated gene expression in eukaryotic cells: principles and advances. Methods Enzymol 327:401-421.

Baron U, Freundlieb S, Gossen M, Bujard H (1995) Co-regulation of two gene activities by tetracycline via a bidirectional promoter. Nucleic Acids Res 23:3605-3606.

Bazinet C, Katzen AL, Morgan M, Mahowald AP, Lemmon SK (1993) The Drosophila clathrin heavy chain gene: clathrin function is essential in a multicellular organism. Genetics 134:1119-1134.

Bennett EM, Lin SX, Towler MC, Maxfield FR, Brodsky FM (2001) Clathrin hub expression affects early endosome distribution with minimal impact on receptor sorting and recycling. Mol Biol Cell 12:2790-2799.

Bonifacino JS, Lippincott-Schwartz J (2003) Coat proteins: shaping membrane transport. Nat Rev Mol Cell Biol 4:409-414.

Brodsky FM, Chen CY, Knuehl C, Towler MC, Wakeham DE (2001) Biological basket weaving: formation and function of clathrin-coated vesicles. Annu Rev Cell Dev Biol 17:517-568.

Brodsky FM, Hill BL, Acton SL, Nathke I, Wong DH, Ponnambalam S, Parham P (1991) Clathrin light chains: arrays of protein motifs that regulate coated-vesicle dynamics. Trends Biochem Sci 16:208-213.

Buerstedde JM, Takeda S (1991) Increased ratio of targeted to random integration after transfection of chicken B cell lines. Cell 67:179-188.

Chen H, Fre S, Slepnev VI, Capua MR, Takei K, Butler MH, Di Fiore PP, De Camilli P (1998) Epsin is an EH-domain-binding protein implicated in clathrin-mediated endocytosis. Nature 394:793-797.

Collins BM, McCoy AJ, Kent HM, Evans PR, Owen DJ (2002) Molecular architecture and functional model of the endocytic AP2 complex. Cell 109:523-535.

Conner SD, Schmid SL (2002) Identification of an adaptor-associated kinase, AAK1, as a regulator of clathrin-mediated endocytosis. J Cell Biol 156:921-929.

Cosson P, de Curtis I, Pouyssegur J, Griffiths G, Davoust J (1989) Low cytoplasmic pH inhibits endocytosis and transport from the trans-Golgi network to the cell surface. J Cell Biol 108:377-387.

Cremona O, Di Paolo G, Wenk MR, Luthi A, Kim WT, Takei K, Daniell L, Nemoto Y, Shears SB, Flavell RA, McCormick DA, De Camilli P (1999) Essential role of phosphoinositide metabolism in synaptic vesicle recycling. Cell 99:179-188.

Damke H, Gossen M, Freundlieb S, Bujard H, Schmid SL (1995) Tightly regulated and inducible expression of dominant interfering dynamin mutant in stably transformed HeLa cells. Methods Enzymol 257:209-220.

Dell'Angelica EC, Mullins C, Bonifacino JS (1999) AP-4, a novel protein complex related to clathrin adaptors. J Biol Chem 274:7278-7285.

Dell'Angelica EC, Ohno H, Ooi CE, Rabinovich E, Roche KW, Bonifacino JS (1997) AP-3: an adaptor-like protein complex with ubiquitous expression. Embo J 16:917-928.

Dell'Angelica EC, Puertollano R, Mullins C, Aguilar RC, Vargas JD, Hartnell LM, Bonifacino JS (2000) GGAs: a family of ADP ribosylation factor-binding proteins related to adaptors and associated with the Golgi complex. J Cell Biol 149:81-94.

Diaz-Griffero F, Jackson AP, Brojatsch J (2005) Cellular uptake of avian leukosis virus subgroup B is mediated by clathrin. Virology 337:45-54.

Dittmer F, Ulbrich EJ, Hafner A, Schmahl W, Meister T, Pohlmann R, von Figura K (1999) Alternative mechanisms for trafficking of lysosomal enzymes in mannose 6-phosphate receptor-deficient mice are cell type-specific. J Cell Sci 112 (Pt 10):1591-1597.

Ford MG, Pearse BM, Higgins MK, Vallis Y, Owen DJ, Gibson A, Hopkins CR, Evans PR, McMahon HT (2001) Simultaneous binding of PtdIns(4,5)P2 and clathrin by AP180 in the nucleation of clathrin lattices on membranes. Science 291:1051-1055.

Fotin A, Cheng Y, Sliz P, Grigorieff N, Harrison SC, Kirchhausen T, Walz T (2004) Molecular model for a complete clathrin lattice from electron cryomicroscopy. Nature 432:573-579.

Fukagawa T, Brown WR (1997) Efficient conditional mutation of the vertebrate CENP-C gene. Hum Mol Genet 6:2301-2308.

Fukagawa T, Regnier V, Ikemura T (2001) Creation and characterization of temperature-sensitive CENP-C mutants in vertebrate cells. Nucleic Acids Res 29:3796-3803.

Gabel CA, Kornfeld S (1984) Targeting of beta-glucuronidase to lysosomes in mannose-6 phosphate receptor-deficient MOPC 315 cells. J Cell Biol 99:296-305.

Glickman JN, Kornfeld S (1993) Mannose 6-phosphate-independent targeting of lysosomal enzymes in I-cell disease B lymphoblasts. J Cell Biol 123:99-108.

Gossen M, Bujard H (1992) Tight control of gene expression in mammalian cells by tetracycline-responsive promoters. Proc Natl Acad Sci USA 89:5547-5551.

Griffiths G, Hoflack B, Simons K, Mellman I, Kornfeld S (1988) The mannose 6-phosphate receptor and the biogenesis of lysosomes. Cell 52:329-341.

Hinrichsen L, Harborth J, Andrees L, Weber K, Ungewickell EJ (2003) Effect of clathrin heavy chain- and alpha-adaptin-specific small inhibitory RNAs on endocytic accessory proteins and receptor trafficking in HeLa cells. J Biol Chem 278:45160-45170.

Hirst J, Bright NA, Rous B, Robinson MS (1999) Characterization of a fourth adaptor-related protein complex. Mol Biol Cell 10:2787-2802.

Hirst J, Lui WW, Bright NA, Totty N, Seaman MN, Robinson MS (2000) A family of proteins with gamma-adaptin and VHS domains that facilitate trafficking between the trans-Golgi network and the vacuole/lysosome. J Cell Biol 149:67-80.

Itoh T, Koshiba S, Kigawa T, Kikuchi A, Yokoyama S, Takenawa T (2001) Role of the ENTH domain in phosphatidylinositol-4,5-bisphosphate binding and endocytosis. Science 291:1047-1051.

Iversen TG, Skretting G, van Deurs B, Sandvig K (2003) Clathrin-coated pits with long, dynamin-wrapped necks upon expression of a clathrin antisense RNA. Proc Natl Acad Sci USA 100:5175-5180.

Jackson AP, Flett A, Smythe C, Hufton L, Wettey FR, Smythe E (2003) Clathrin promotes incorporation of cargo into coated pits by activation of the AP2 adaptor micro2 kinase. J Cell Biol 163:231-236.

Kanaseki T, Kadota K (1969) The "vesicle in a basket". A morphological study of the coated vesicle isolated from the nerve endings of the guinea pig brain, with special reference to the mechanism of membrane movements. J Cell Biol 42:202-220.

Kirchhausen T (2000) Clathrin. Annu Rev Biochem 69:699-727.

Kurosaki T, Maeda A, Ishiai M, Hashimoto A, Inabe K, Takata M (2000) Regulation of the phospholipase C-gamma2 pathway in B cells. Immunol Rev 176:19-29.

Larkin JM, Brown MS, Goldstein JL, Anderson RG (1983) Depletion of intracellular potassium arrests coated pit formation and receptor-mediated endocytosis in fibroblasts. Cell 33:273-285.

Lemmon SK (2001) Clathrin uncoating: Auxilin comes to life. Curr Biol 11:R49-52.

Lemmon SK, Jones EW (1987) Clathrin requirement for normal growth of yeast. Science 238:504-509.

Luzio JP, Poupon V, Lindsay MR, Mullock BM, Piper RC, Pryor PR (2003) Membrane dynamics and the biogenesis of lysosomes. Mol Membr Biol 20:141-154.

Mellman I, Warren G (2000) The road taken: past and future foundations of membrane traffic. Cell 100:99-112.

Meyer C, Zizioli D, Lausmann S, Eskelinen EL, Hamann J, Saftig P, von Figura K, Schu P (2000) mu1A-adaptin-deficient mice: lethality, loss of AP-1 binding and rerouting of mannose 6-phosphate receptors. Embo J 19:2193-2203.

Motley A, Bright NA, Seaman MN, Robinson MS (2003) Clathrin-mediated endocytosis in AP-2-depleted cells. J Cell Biol 162:909-918.

Moura IC, Lepelletier Y, Arnulf B, England P, Baude C, Beaumont C, Bazarbachi A, Benhamou M, Monteiro RC, Hermine O (2004) A neutralizing monoclonal antibody (mAb

A24) directed against the transferrin receptor induces apoptosis of tumor T lymphocytes from ATL patients. Blood 103:1838-1845.

Mousavi SA, Malerod L, Berg T, Kjeken R (2004) Clathrin-dependent endocytosis. Biochem J 377:1-16.

Nichols BJ, Lippincott-Schwartz J (2001) Endocytosis without clathrin coats. Trends Cell Biol 11:406-412.

Niswonger ML, O'Halloran TJ (1997) A novel role for clathrin in cytokinesis. Proc Natl Acad Sci U S A 94:8575-8578.

Ohno H, Aguilar RC, Yeh D, Taura D, Saito T, Bonifacino JS (1998) The medium subunits of adaptor complexes recognize distinct but overlapping sets of tyrosine-based sorting signals. J Biol Chem 273:25915-25921.

Ohno H, Stewart J, Fournier MC, Bosshart H, Rhee I, Miyatake S, Saito T, Gallusser A, Kirchhausen T, Bonifacino JS (1995) Interaction of tyrosine-based sorting signals with clathrin-associated proteins. Science 269:1872-1875.

Olusanya O, Andrews PD, Swedlow JR, Smythe E (2001) Phosphorylation of threonine 156 of the mu2 subunit of the AP2 complex is essential for endocytosis *in vitro* and *in vivo*. Curr Biol 11:896-900.

Owen DJ, Luzio JP (2000) Structural insights into clathrin-mediated endocytosis. Curr Opin Cell Biol 12:467-474.

Owen DJ, Collins BM, Evans PR (2004) Adaptors for clathrin coats: structure and function. Annu Rev Cell Dev Biol 20:153-191.

Payne GS, Hasson TB, Hasson MS, Schekman R (1987) Genetic and biochemical characterization of clathrin-deficient Saccharomyces cerevisiae. Mol Cell Biol 7:3888-3898.

Pearse BM (1975) Coated vesicles from pig brain: purification and biochemical characterization. J Mol Biol 97:93-98.

Peter BJ, Kent HM, Mills IG, Vallis Y, Butler PJ, Evans PR, McMahon HT (2004) BAR domains as sensors of membrane curvature: the amphiphysin BAR structure. Science 303:495-499.

Pishvaee B, Munn A, Payne GS (1997) A novel structural model for regulation of clathrin function. Embo J 16:2227-2239.

Reth M, Wienands J (1997) Initiation and processing of signals from the B cell antigen receptor. Annu Rev Immunol 15:453-479.

Ringstad N, Gad H, Low P, Di Paolo G, Brodin L, Shupliakov O, De Camilli P (1999) Endophilin/SH3p4 is required for the transition from early to late stages in clathrin-mediated synaptic vesicle endocytosis. Neuron 24:143-154.

Robinson MS (2004) Adaptable adaptors for coated vesicles. Trends Cell Biol 14:167-174.

Robinson MS, Bonifacino JS (2001) Adaptor-related proteins. Curr Opin Cell Biol 13:444-453.

Roth TF, Porter KR (1964) Yolk Protein Uptake in the Oocyte of the Mosquito Aedes Aegypti. L. J Cell Biol 20:313-332.

Ruscetti T, Cardelli JA, Niswonger ML, O'Halloran TJ (1994) Clathrin heavy chain functions in sorting and secretion of lysosomal enzymes in Dictyostelium discoideum. J Cell Biol 126:343-352.

Schmidt A, Wolde M, Thiele C, Fest W, Kratzin H, Podtelejnikov AV, Witke W, Huttner WB, Soling HD (1999) Endophilin I mediates synaptic vesicle formation by transfer of arachidonate to lysophosphatidic acid. Nature 401:133-141.

Sever S (2002) Dynamin and endocytosis. Curr Opin Cell Biol 14:463-467.

Smythe E, Carter LL, Schmid SL (1992) Cytosol- and clathrin-dependent stimulation of endocytosis in vitro by purified adaptors. J Cell Biol 119:1163-1171.

Sonoda E, Morrison C, Yamashita YM, Takata M, Takeda S (2001) Reverse genetic studies of homologous DNA recombination using the chicken B-lymphocyte line, DT40. Philos Trans R Soc Lond B Biol Sci 356:111-117.

Sorkin A, Von Zastrow M (2002) Signal transduction and endocytosis: close encounters of many kinds. Nat Rev Mol Cell Biol 3:600-614.

Stoddart A, Jackson AP, Brodsky FM (2005) Plasticity of B cell receptor internalization upon conditional depletion of clathrin. Mol Biol Cell 16:2339-2348.

Tebar F, Sorkina T, Sorkin A, Ericsson M, Kirchhausen T (1996) Eps15 is a component of clathrin-coated pits and vesicles and is located at the rim of coated pits. J Biol Chem 271:28727-28730.

ter Haar E, Harrison SC, Kirchhausen T (2000) Peptide-in-groove interactions link target proteins to the beta-propeller of clathrin. Proc Natl Acad Sci U S A 97:1096-1100.

Towler MC, Gleeson PA, Hoshino S, Rahkila P, Manalo V, Ohkoshi N, Ordahl C, Parton RG, Brodsky FM (2004) Clathrin isoform CHC22, a component of neuromuscular and myotendinous junctions, binds sorting nexin 5 and has increased expression during myogenesis and muscle regeneration. Mol Biol Cell 15:3181-3195.

Verstreken P, Koh TW, Schulze KL, Zhai RG, Hiesinger PR, Zhou Y, Mehta SQ, Cao Y, Roos J, Bellen HJ (2003) Synaptojanin is recruited by endophilin to promote synaptic vesicle uncoating. Neuron 40:733-748.

Wang J, Takagaki Y, Manley JL (1996) Targeted disruption of an essential vertebrate gene: ASF/SF2 is required for cell viability. Genes Dev 10:2588-2599.

Wettey FR, Hawkins SF, Stewart A, Luzio JP, Howard JC, Jackson AP (2002) Controlled elimination of clathrin heavy-chain expression in DT40 lymphocytes. Science 297:1521-1525.

Winding P, Berchtold MW (2001) The chicken B cell line DT40: a novel tool for gene disruption experiments. J Immunol Methods 249:1-16.

Yamazoe M, Sonoda E, Hochegger H, Takeda S (2004) Reverse genetic studies of the DNA damage response in the chicken B lymphocyte line DT40. DNA Repair (Amst) 3:1175-1185.

Ybe JA, Brodsky FM, Hofmann K, Lin K, Liu SH, Chen L, Earnest TN, Fletterick RJ, Hwang PK (1999) Clathrin self-assembly is mediated by a tandemly repeated superhelix. Nature 399:371-375.

Yin DX, Zhu L, Schimke RT (1996) Tetracycline-controlled gene expression system achieves high-level and quantitative control of gene expression. Anal Biochem 235:195-201.

Chapter 10

GENETIC ANALYSIS OF B CELL SIGNALING

Hisaaki Shinohara and Tomohiro Kurosaki
Laboratory for Lymphocyte Differentiation, RIKEN Research Center for Allergy and Immunology, Tsurumi-ku, Yokohama, Kanagawa 230-0045

Abstract: Recent evidence indicates that B lymphocytes are instructed continuously by B cell receptor (BCR) signals to make crucial cell-fate decisions at several checkpoints during their development and humoral immune responses, reinforcing the importance of studies of the BCR signals. One of the best cell lines for these studies is a chicken DT40 B-cell line, because a high propensity of homologous recombination in this cell line allows us to easily take both genetic and biochemical approaches. Here, based upon the recent data, mainly obtained from the DT40 system, we discuss several new aspects on mechanisms by which BCR signals are propagated and modified.

Key words: BCR, signal, DT40, protein kinase, adaptor molecule, calcium mobilization.

1.   INTRODUCTION

The B-cell antigen receptor (BCR) is characterized by a complex hetero-oligomeric structure in which antigen binding and signal transduction are compartmentalized into distinct receptor subunits (Fig. 1 a). For effective humoral immune responses, mature B cells must respond to foreign antigens and generate antigen-specific effector cells. So, it is easy to imagine that the BCR complex is required for the later stages of B cell maturation as well as effector phases of mature B cells. However, the BCR is required also during early B-cell ontogeny (antigen-independent phase). Therefore, the classic distinction between antigen-dependent and antigen-independent stages of B cell development is no longer so clear-cut. For instance, targeted disruption of the BCR in mature B cells has shown that the BCR is required for the maintenance of B cells before they encounter foreign antigens (Kraus et al., 2004). In addition to BCR expression *per se*, signals through the BCR are required for

J.-M. Buerstedde and S. Takeda (eds.), Reviews and Protocols in DT40 Research, 145–187.

selecting a correct B cell fate. Indeed, disruption of genes encoding downstream components of the BCR signaling pathways leads to various B cell anomalies including early developmental arrest or late humoral dysfunction.

The purpose of this review is to summarize the major advances made in elucidation of the biological roles of the BCR and its signaling components and to discuss the mechanisms of how the BCR signal is fine-tuned by these components, thereby being translated into biological outcomes - with a particular focus on PLC-γ2-PKCβ-calcium pathway.

1.1   Function of the BCR components

The BCR signal transduction component comprises a disulfide-bounded heterodimer of the Igα (CD79α) and Igβ (CD79β). Igα and Igβ contain tyrosine residues in their cytoplasmic domain that are imbedded in immunoreceptor tyrosine-based activation motifs (ITAMs) (Fig. 1 a)(Reth, 1989). The functional importance of the ITAMs was first revealed in experiments in which chimeric receptors containing ITAMs in their cytoplasmic domain were constructed and expressed in lymphocytes. Cross-linking these chimeric receptors using anti-receptor antibodies induces signaling events including protein tyrosine kinase (PTK) activation and calcium mobilization (Law et al., 1993; Sanchez et al., 1993). This signaling function is abolished by mutations that change either of the two conserved tyrosine residues of the ITAM to phenylalanine, indicating the importance of phosphorylation on these ITAM tyrosine residues.

The effects of Igα/Igβ and their ITAM tyrosines on B cell development and activation were more directly examined by in vivo mouse targeting experiments. B cells with complete V(D)J$_H$ rearrangements fail to accumulate in mice entirely lacking Igβ (Gong and Nussenzweig, 1996). Since Igα and Igβ are required for cell surface expression of the BCR, the phenotype in the Ig$\beta^{-/-}$ mice could be due to loss of cell surface expression of pre-BCR similar to the lack of BCR expression that occurs after Igβ ablation, and/or to a lack of signaling ability of the pre-BCR. The pre-BCR is composed of membrane Igμ, surrogate light chains (SLCs; made of VpreB and λ5 proteins), and Igα/Igβ. Therefore, the necessity of the signaling ability of the pre-BCR has been more directly addressed by using mice that harbor mutations only in the cytoplasmic domains of Igα and/or Igβ, thereby keeping the normal cell surface expression of the pre-BCR. B cell development was arrested at the pro-B stage in mice that carry a deletion of the cytoplasmic domains of both Igα and Igβ (Ig$\alpha^{\Delta c\,/\Delta c}Ig\beta^{\Delta c/\Delta c}$) (Reichlin et al., 2001) or a mutated Igα ITAM with a turuncation of Igβ (Ig$\alpha^{FF/FF}$Ig$\beta^{\Delta c/\Delta c}$) (Kraus et al., 2001). Because a decreased, but significant, level of expression of Igμ was observed in mice, this arrest is more likely to be due to a lack of clonal expansion of pre-B cells with sucessuful IgH rearrangements, rather than to inefficient V-to-DJ$_H$ recombination. In contrast to Igα$^{\Delta c/\Delta c}$Igβ$^{\Delta c/\Delta c}$ or Igα$^{FF/FF}$Igβ$^{\Delta c/\Delta c}$, the milder

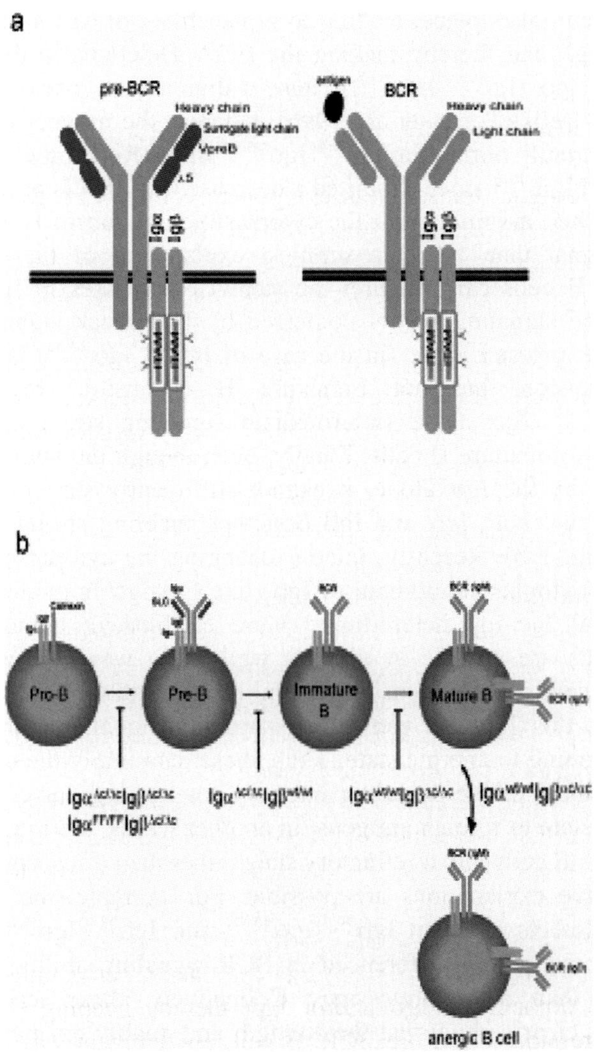

Figure 10-1. a) Simplified scheme of pre-BCR and BCR. Immunoreceptor tyrosine-based activation motif (ITAM) contains two functional tyrosines. b) Selection steps during B cell development and the various mutated IgαIgβ involved in the developmental steps, as revealed by genetic studies. Vertical 'T-bars' indicate blocks in development. The Igα$^{wt/wt}$Igβ$^{\alpha c/\alpha c}$ mice were exchanged the cytoplasmic domain of Igα for the cytoplasmic domain of Igβ by gene targeting (see plate 20).

developmental defects were observed in mice harboring one intact Igα or Igβ (Igβ$^{wt/wt}$Igα$^{\Delta c/\Delta c}$ or Igα$^{wt/wt}$Igβ$^{\Delta c/\Delta c}$) (Fig. 1 b). Given the importance of pre-BCR, in expansion of pre-B cells, the above data demonstrate that one of the intact cytoplasmic domain of either Igα or Igβ is sufficient for pre-BCR-mediated expansion and subsequent generation of pre-B cells.

After clonal expansion of pre-B cells, successful light-chain gene rearrangement takes place, leading to replacement of SLCs of the pre-BCR by Igκ or Igλ and thereby making the BCR. Deletions in the cytoplasmic domains of Igα (Igβ$^{wt/wt}$Igα$^{\Delta c/\Delta c}$) cause a dramatic decrease in numbers of immature B cells (Torres et al., 1996), whereas the number of immature B cells is apparently normal in Igβ$^{\Delta c/\Delta c}$Igα$^{\omega\tau/\omega\tau}$ mice (Reichlin et al., 2001). But these Igβ$^{\Delta c/\Delta c}$Igα$^{\omega\tau/\omega\tau}$ mice exhibited a decrease in numbers of mature B cells. Taken together, assuming that the cytoplasmic domain of Igα can induce a stronger signal than Igβ, the simplest explanation of these data is that developing B cells can interpret the stepwise increases in signals through Igα-Igβ. The signaling ability conferred by Igβ (weak signaling strength) when it is expressed alone (in the case of Igβ$^{wt/wt}$Igα$^{\Delta c/\Delta c}$) is sufficient for pre-B expansion, but not immature B expansion. In the case of Igβ$^{\Delta c/\Delta c}$Igα$^{\omega\tau/\omega\tau}$, Igα alone (intermediate signaling strength) can support expansion of immature B cells. Finally, even though the intermediate signal is provided by the Igα alone, it cannot sufficiently support expansion of mature B cells; both Igα and Igβ (together enabling strong signaling) are required (Fig. 1 b). Recently, mice exchanging the cytoplasmic domain of Igα for the cytoplasmic domain of Igβ (like Igα-Igα homodimer instead of physiological Igα-Igβ heterodimer) were established, demonstrating that these B cells are anergic in vivo as well as *in vitro*, despite apparently normal B cell development (Reichlin et al., 2004). According to our assumption, this Igα-Igα can induce stronger signal, rather than normal situation, leading to anergic state. Thus, these data imply the existence of an optimal window of the signaling intensity for making naïve B cells into a susceptible state to foreign antigens; in another words, too much constitutive signal brings B cells into a refractory state rather than a susceptible state.

Alternative explanations are possible. For example, the distinct developmental defects between Igβ$^{\Delta c/\Delta c}$Igα$^{\omega\tau/\omega\tau}$ and Igβ$^{wt/wt}$Igα$^{\Delta c/\Delta c}$ mice could result from qualitative differenced in BCR signaling abilities of Igα amd Igβ, rather than quantitative ones. Collectively, these series of elegant experiments clearly show that the strength and quality emanating from pre-BCR as well as BCR affect the B cell fate in many different ways, depending upon maturation stages of B cells.

1.2   Activation mechanisms of PTKs

As the BCR has no intrinsic PTK activity, this receptor utilizes several distinct families of cytoplasmic PTKs and PTPases for initiating signals. Indeed, three distinct types of PTKs have been found to be activated upon BCR engagement: the Src family kinases (SFKs) (Lyn, Blk, and Fyn), Syk, and Btk (Fig. 2) (Gauld and Cambier, 2004; Kurosaki et al., 2000; Xu et al., 2005). Gene-targeting experiments using mice and cell lines have dissected the functions of these PTKs and PTPases in BCR signaling as well as B cell development. A summary of the DT40 cell lines deficient in various signaling molecules in our laboratory are presented in Table 1.

Table 10-1. A summary of DT40 knock out cells.

Description	reference
1. Syk$^{-/-}$	(Takata et al., 1994)
2. Lyn$^{-/-}$	(Takata et al., 1994)
3. PLC-γ2$^{-/-}$	(Takata et al., 1995)
4. Btk$^{-/-}$	(Takata and Kurosaki, 1996)
5. Lyn$^{-/-}$/Syk$^{-/-}$	(Takata and Kurosaki, 1996)
6. IP3R-type1$^{-/-}$/ -Type2$^{-/-}$	(Sugawara et al., 1997)
7. IP3R-type1$^{-/-}$/ -Type3$^{-/-}$	(Miyakawa et al., 1999)
8. IP3R-type2$^{-/-}$/ -Type3$^{-/-}$	(Miyakawa et al., 1999)
9. IP3R-type1$^{-/-}$/ -Type2$^{-/-}$/ -Type3$^{-/-}$	(Sugawara et al., 1997)
10. SHIP1$^{-/-}$	(Okada et al., 1998)
11. Grb2$^{-/-}$	(Hashimoto et al., 1998)
12. SHP2$^{-/-}$	(Maeda et al., 1998)
13. SHP1$^{-/-}$/ SHP2$^{-/-}$	(Maeda et al., 1998)
14. Shc$^{-/-}$	(Hashimoto et al., 1998)
15. BLNK$^{-/-}$	(Ishiai et al., 1999a)
16. BLNK$^{-/-}$/ Syk$^{-/-}$	(Ishiai et al., 2000)
17. Grap$^{-/-}$	(Liou et al., 2000)
18. Grb2$^{-/-}$/ Grap$^{-/-}$	(Liou et al., 2000)
19. Cbl$^{-/-}$	(Yasuda et al., 2000)
20. BCAP$^{-/-}$	(Okada et al., 2000)
21. Vav3$^{-/-}$	(Inabe et al., 2002)
22. PI3K (p110α)$^{-/-}$	(Inabe et al., 2002)
23. TRPC1$^{-/-}$	(Mori et al., 2002)
24. Cbl-b$^{-/-}$	(Yasuda et al., 2002)
25. BLNK$^{-/-}$/ Grb2$^{-/-}$	(Johmura et al., 2003)
26. rasGRP1$^{-/-}$	(Oh-hora et al., 2003)
27. rasGRP3$^{-/-}$	(Oh-hora et al., 2003)
28. rasGRP1$^{-/-}$/ rasGRP3$^{-/-}$	(Oh-hora et al., 2003)
29. Sos1$^{-/-}$	(Oh-hora et al., 2003)
30. Sos2$^{-/-}$	(Oh-hora et al., 2003)
31. Sos1$^{-/-}$/ Sos2$^{-/-}$	(Oh-hora et al., 2003)
32. TAK1$^{-/-}$	(Shinohara et al., in press)
33. CARMA1$^{-/-}$	(Shinohara et al., in press)
34. PKC$\beta^{-/-}$	(Shinohara et al., in press)

A targeted gene disruption of Lyn in a chicken B cell line DT40, resulted in a profound decrease in tyrosine phosphorylation of cellular proteins. Since this particular cell line expresses dominantly Lyn among SFKs, this residual receptor-induced phosphorylation is presumably attributable to Syk (Takata et al., 1994). In fact, the complementary pattern of the receptor-induced phosphosphorylation in the absence of Syk supports this contention. Lyn/Syk double deficient DT40 cells exhibit no tyrosine phosphorylation induced by BCR cross-linking (Takata and Kurosaki, 1996). Thus, two important conclusions can be drawn by these data; 1) Lyn and Syk are initially activated PTKs in the context of BCR signaling; 2) these PTKs can be activated independently of each other at least to some extent, although kinetic experiments have shown that the increased activity of the SFKs precedes Syk activation in BCR signaling context. The latter conclusion that Syk is able to be activated independently of Lyn is in sharp contrast to TCR signaling in which Syk homologue Zap-70 is stringently regulated by SFKs such as Lck, implicating the existence of differential activation modes between Syk and Zap-70 (Kurosaki et al., 1994).

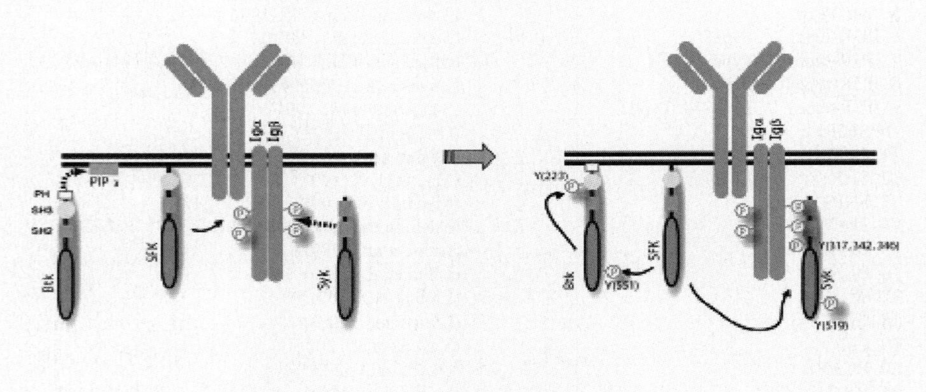

Figure 10-2. Activation mechanisms of SFK, Syk, and Btk in BCR signaling. After BCR aggregation, activated SFK phosphorylates ITAMs within Igα-Igβ. The activity of SFK at resting condition is determined by the phosphorylation status of its carboxyterminal tyrosine, which is regulated by Csk and CD45. Carboxyterminal phosphorylation renders SFK inactive. Btk is recruited to the plasma membrane by interaction of its PH domain with PI-3,4,5-P3 (PIP3), a product of PI3K, while Syk is recruited to the doubly-phosphorylated ITAMs within Igα-Igβ by its SH2 domains. Activated SFK phosphorylates the tyrosine residue in the activation loop of Btk and Syk (Tyr551 of Btk and Tyr520 of Syk) so leading to their activation. DAG: diacylglycerol (see plate 21).

1.2.1   Activation of Lyn

The exact mechanism of how Lyn is activated upon BCR ligation is still not totally clear. But, a clue came from the characterization of cholesterol- and sphingolipid-enriched membrane microdomains, which are referred to here as lipid rafts (alternatively known as glycolipid-enriched microdomains; GEMs). One of the most important properties of lipid rafts is their ability to include or exclude different proteins to variable extents (Cherukuri et al., 2001; Simons and Toomre, 2000). Indeed, monomeric BCR at resting level seems to have a weak affinity for rafts, which heavily skews the equilibrium distribution of the receptor towards the non-raft regions of the membrane, whereas Lyn resides constitutively in rafts. It has been observed that the BCR, once oligomerized by multivalent ligand binding, moves into rafts; this indicates a model whereby the movement of BCR into rafts would lead to co-localize the BCR with Lyn (Cheng et al., 1999). Since Lyn has been shown to interact with Igα-Igβ through its N-terminal domain (Pleiman et al., 1994), this specific interaction occurring in the rafts may contribute to the initial activation of Lyn.

The activities of SFKs are regulated by tyrosine phosphoryloation status; autophosphorylation of Tyr416 in their catalytic domain is stimulatory, and carboxy-terminal phosphorylation (Tyr527) is inhibitory. Phosphorylation of Tyr527 mediates an intramolecular association with the kinase's own src homolofy (SH)2 domain, leading to a repressed conformation of its kinase activity (Huse and Kuriyan, 2002). The phosphorylation of the inhibitory carboxy-terminal tyrosine is regulated by a tyrosine kinase, Csk, and a tyrosine phosphatase, CD45. In accord with this concept, C-terminal tyrosine of Lyn is hyperphosphorylated in CD45-deficient and hypophosphorylated in Csk-deficient DT40 cells (Hata et al., 1994; Yanagi et al., 1996). As the extent of receptor-induced tyrosine phosphorylation of cellular substrates is significantly increased by loss of Csk and decreased by loss of CD45, Csk and CD45 function like a rheostat to tune BCR signal by determining the phosphorylation status of the inhibitory tyrosine of Lyn.

The existence of a positive-feedback regulatory loop for Lyn activation has been suggested by recent data (Fig. 4 a) (Singh et al., 2005). A clue to this positive loop came from the observation that the phosphorylation and activation of Lyn is dependent upon calcium mobilization, particularly from the endoplasmic reticulum (ER) pool. Given that this BCR-mediated calcium mobilization requires upstream PTKs, the generated calcium downstream of Lyn is thought to function as a potent re-stimulator for Lyn. Then, the question arises about the connection between calcium and Lyn activation. Stimulation of the BCR leads to a rapid production of reactive oxygen species (ROS), in particular H_2O_2, which is dependent on PTK activity, PI3K function, and calcium (Tonks, 2005). Thus, an idea is proposed that calcium influences the levels of ROS, which in turn positively regulates activation of Lyn. In fact, H_2O_2 scavengers inhibited BCR-induced Lyn activation and this was not rescued by ionomycin-induced increases in Ca^{2+} (Singh et al., 2005). Although how ROS activates Lyn is unclear, the potential candidates are PTPases such as SHP-1. Hence, according to this scenario, oxidation of PTPases by a Ca^{2+}-ROS pathway leads to inactivation of the phosphatase activity, thereby keeping the activation status of Lyn.

1.2.2   Activation of Syk

Weiss and Littman initially proposed a sequential activation model, based on the cooperativity between Lck and Zap-70 in T cells (Weiss and Littman, 1994). Most of the observations to date in B cells support that this mechanism also operates in B cells. According to this model, BCR engagement leads to phosphorylation of the Igα-Igβ ITAM tyrosine residues by SFKs. Then Syk is recruited to the doubly-phosphorylayed ITAMs and is activated by SFK phosphorylation (Fig. 2 left panel) (Kurosaki et al., 2000). Detailed biochemical and genetic studies, however, have underscored the existence of an additional activation mode of Syk that is independent of SFKs (Kurosaki et al., 1994).

Then, what is the underlying mechanism that explains the SFK-independent activation of Syk? Because association of Syk with BCR complex is observed before receptor stimulation (although the molecular details are still unclear), receptor ligation might directly stimulate the activity of the pre-associated Syk (Hutchcroft et al., 1992). One of the physiological substrates of the activated Syk might be Igα-Igβ ITAMs. Since Syk SH2 domains bind to the phosphorylated ITAMs, this phosphorylation could bring additional Syk into close proximity with the pre-bound Syk, thereby allowing transphosphorylation and/or autophsosphorylation of each other even in the absence of SFKs. Consistent with this notion, the kinase activity of Syk, but not Zap-70, is activated when bound to doubly-phosphorylated ITAMs of Igα-Igβ (Johnson et al., 1995; Rolli et al., 2002). As the Syk mutant (Tyr520 to Phe) is not able to transmit signals, one target of the auto/transphosphorylation is Tyr520 located in the Syk activation loop (Fig. 3) (Kurosaki et al., 1995). On the other hand, in the case of Zap-70, phosphorylation of the homologous reside Tyr493 is stringently regulated by SFKs (Chan et al., 1995). Given that CD4 and CD8 bring SFKs into close proximity with TCR, this stringent requirement for SFKs in Zap-70 activation might contribute to preventing spontaneous TCR activation without recognition of MHC by CD4 and CD8. Previously, the Tyr493 phosphorylation (corresponding to Tyr520 in Syk) in Zap-70 is though to be mediated directly through transphosphorylation by SFKs. However, recent data has provided the revised model in which SFKs such as Lck phosphorylate Tyr315 and Tyr319 located in the linker region connecting the SH2 domains to the kinase domain (interdomain B) first, rather than Tyr493, thereby leading to a structural rearrangement of Zap-70 (Brdicka et al., 2005). This rearrangement could result in increased accessibility of the Tyr493 to SFKs and/or elevated kinase activity of Zap-70 (Fig. 3).

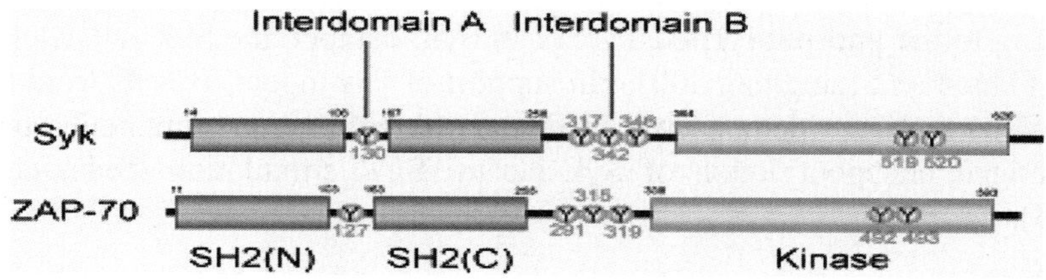

Figure 10-3. Schematic structure of Syk/ ZAP-70 family protein tyrosine kinases. The tandem SH2 domains are shown as blue boxes and the kinase domains as green boxes. Black bars depict the interdomains connecting the SH2-SH2 (A) and SH2-kinase domains (B). The tyrosines in Syk which have been shown to undergo phosphorylation are indicated. These sites may be important in regulating enzymatic activity or recruiting other signalling proteins. Phosphorylation of Y342 and Y346 is dependent on the SFK; phosphorylation of Y130, Y317, Y519 and Y520 is dependent on Syk itself and thus these may be sites of auto-phosphorylation. Homologous tyrosines (see plate 22).

Although Syk is activated in the absence of Lyn, the positive effects of Lyn on Syk are clear. In fact, BCR-mediated activation of Syk is dramatically inhibited by loss of Lyn in DT40 B cells (Kurosaki et al., 1994). Hence, under physiological conditions, BCR ligation activates SFKs as well as pre-associated Syk, leading to phosphorylation of ITAMs in Igα-Igβ, and the subsequent recruitment of Syk to the phosphorylated ITAMs. Then, recruited Syk is activated by SFK-dependent transphosphorylation and by Syk-mediated auto/transphosphorylation (Fig. 2).

In addition to Tyr520, Syk also undergoes phosphorylation on several other tyrosine residues that are located in the linker region connecting the two SH2 domains (interdomain A), or the SH2 domains to the kinase domain (interdomain B) including Tyr317, Tyr342, and Tyr346 (Keshvara et al., 1998) (Fig. 3). Tyr342 and Tyr346 correspond to Tyr315 and Tyr319 in Zap-70 aforementioned. Phosphorylation of these sites (Tyr317, Tyr342, and Tyr346) occurs as a consequence of Syk activation, and these phosphorylation, in turn, could function as an initiator to amplify or attenuate the activation of

Syk, thereby contributing to formation of a positive- or a negative-feedback regulation loop for Syk activation, respectively. Assuming that the similar mechanism operates in the case of Tyr342 and Tyr346 in Syk, like Zap-70, phosphorylation of these sites could further enhance its enzymatic activity. On the other hand, Tyr317 in Syk is reported to provide a binding site for Cbl and Cbl-b (Deckert et al., 1998). Considering that the ring finger domains of Cbl family act as E3 ubiquitin ligases that downregulate the activities of signaling molecules through protein ubiquitination/degradation (Joazeiro et al., 1999; Levkowitz et al., 1999; Yokouchi et al., 1999), a promising model has emerged that proposes that Cbl and/or Cbl-b, after binding to the phosphorylated Tyr317 in Syk, dampen the Syk activity (Fig. 4 b) (Thien and Langdon, 2001). In support of this model, B cells from Cbl-b-deficient mice demonstrate reduced ubiquitination and subsequent prolonged phosphorylation of Syk during BCR stimulation (Sohn et al., 2003).

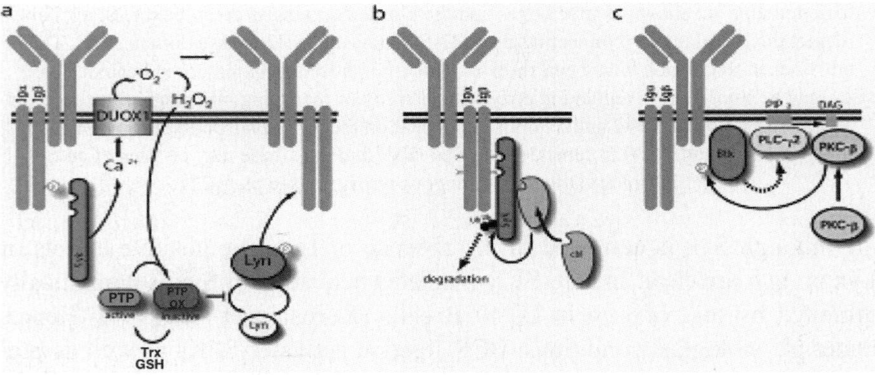

Figure 10-4. Positive and negative feed-back loops in BCR-signaling. a). Redox regulation of Protein tyrosine phosphatases (PTPs) determines the activation status of the protein tyrosine kinase, Lyn, which in turn controls signal output emanating from BCR. Thioredoxin (Trx) and glutathione (GSH) systems control PTP oxidation status. Recently, an annotated murine cDNA sequence identified as DUOX1 was discovered that displayed 87% sequence identity with that of its human DUOX (dual oxidases) counterpart. DUOX represents the second family of nonphagocytic NADPH oxidases that possesses two EF hand motifs. And DUOX1 is thought to be a connector between calcium and supreoxide (H2O2) generation. b). Activated Syk autophosphorylates Tyr317, in addition to its substrates such as BLNK. This Tyr317 phosphorylation provides a binding site for Cbl, which acts as E3 ubiquitin ligase. Ubiquitin-mediated degradation of Syk results in the attenuation of BCR signaling. c). In this model, PLC-γ2 is activated by virtue of Btk action, leading to generation of DAG. Then, the generated DAG recruits PKCβ to the plasma membrane, wherein PKCβ is activated. Phosphorylation of Btk by PKCβ down-regulates Btk kinase activity and decreases its membrane localization, both of which in turn inhibit PLC-γ2 activity (see plate 23).

1.2.3   Activation of Btk

As increased activity of Btk follows the increased activity of SFKs upon receptor cross-linking, it was proposed that Btk is activated through phosphorylation of SFKs (Saouaf et al., 1994). To verify this proposal, the requirement for SFKs in Btk activation was reexamined in the context of BCR signaling. In Lyn-deficient DT40 B cells, the BCR-induced tyrosine phosphorylation of Btk still occurred despite delayed time kinetics. In contrast, susutained Btk phosphorylation was clearly inhibited in Syk-deficient DT40 cells (Kurosaki and Kurosaki, 1997). Thus, the simple explanation for these results is the existence of two phases of Btk activation in BCR signaling: the initial phase and the sustained phase, mediated by Lyn and Syk, respectively (Fig. 2). Supporting involvement of Syk in Btk activation, BCR-mediated tyrosine phosphoryulation of Btk in Blk/Lyn double-deficient splenic B cells was comparable to that in wild-type cells (Tarakhovsky, 1997). Furthermore, the physical association of Btk with Syk was reported (Morrogh et al., 1999).

In fibroblasts and COS cells, Btk is activated through SFK-dependent transphosphorylation of Tyr551, which is located in the activation loop of the catalytic domain of Btk (Mahajan et al., 1995; Rawlings et al., 1996). This results in a five- to ten-fold increase in Btk enzymatic activity, subsequently leading to autophosphorylation at Tyr223 in the SH3 domain (Fig. 2). Although it still remains unresolved which PTK(s) phosphorylates Tyr551 in BCR signaling context, transphosphorylation of Tyr551 is a prerequisite for participation of Btk in BCR signal transduction (Kurosaki and Kurosaki, 1997). Autophosphorylation of Tyr223 in the Btk SH3 domain has been suggested to participate in Btk function. Crystallographic analysis of Itk (a Btk/Tec family PTK expressed in T cells) shows the existence of an intramolecular interaction between the proline-rich region and the SH3 domain of Itk (Berg et al., 2005). Interestingly, Tyr223 of Btk is located within the interface of the interaction between these two regions. Thus, an intriguing possibility is that phosphorylation of Tyr223 may disrupt this intramolecular interaction, thereby exposing binding sites for association with other signaling molecules. Indeed, Tyr223 phosphorylation was reported to prevent binding to the Wiskott Aldrich syndrome protein (WASP) and increase the affinity to Syk (Fig. 2) (Morrogh et al., 1999).

In addition to regulation by phosphorylation, other mechanisms of Btk regulation have been identified. Btk interacts with phospatidylinositol-3,4,

5-trisphosphate (PI-3,4,5-P$_3$) through its PH domain,an interaction that is required for recruitment of Btk to the membrane (Kojima et al., 1997; Rameh et al., 1997; Salim et al., 1996). Supporting this mechanism, the PH domain gain-of-function mutant E41K shows increased membrane localization and phosphorylation of Btk in resting cells (Li et al., 1995), bringing Btk in close proximity of other BCR signaling components. Since PI-3,4,5-P$_3$ is generated by PI3-K, Btk is targeted to the plasma membrane after PI3-K activation, thereby rendering Btk susceptible to transphosphorylation at Tyr551 by Lyn and Syk.

The existence of a negative-feedback regulatory loop for Btk activation has been suggested (Fig. 4 c). This came from the observation that BCR-mediated tyrosine phosphorylation of Btk is increased and prolonged in PKCβ-deficient B cells, which indicates PKCβ functions as a potent inhibitor of Btk (Kang et al., 2001). As Btk has a positive role in the PLC-γ2-calcium-PKCβ pathway, a model is proposed in which PKCβ, after being activated by a Btk-dependent mechanism, phosphorylates Btk, which in turn, blocks membrane recruitment of Btk and subsequent PLC-γ2 activation. Indeed, Ser180 in the TH domain of Btk has been identified as a PKCβ phosphorylation site, and the Ser180Ala Btk mutant is hyperactive, presumably owing to enhanced membrane localization as well as subsequent enhanced enzymatic activity (Kang et al., 2001). So, this negative-feedback mechanism could contribute to controlling the duration of Btk residency in the plasma membrane and subsequent PLC-γ2 activation.

1.3   Activation mechanisms of lipid-metabolizing enzymes

BCR ligation leads to the rapid stimulation of phosphoinositide metabolism and the generation of multiple second messengers, which influence B cell development and function. In fact, deletion of an enzyme that is responsible for this metabolism (PI3K, PLC-γ2, SHIP, PTEN), results in aberrant B cell development and immune responses (Fruman et al., 1999; Hashimoto et al., 2000; Helgason et al., 2000; Liu et al., 1998; Suzuki et al., 2003; Suzuki et al., 1999; Wang et al., 2000). The balance of phosphoinositide metabolism is mediated at least five enzymes: phosphatidylinositol phosphate kinase (PIPK), PI3K, PLC-γ2, SHIP, and PTEN (Fig. 5). Hence, the generation of lipid second messenger is dependent upon the balance between these enzymes, the activities of which are regulated by several mechanisms, including tyrosine phosphorylation and subcellular localization.

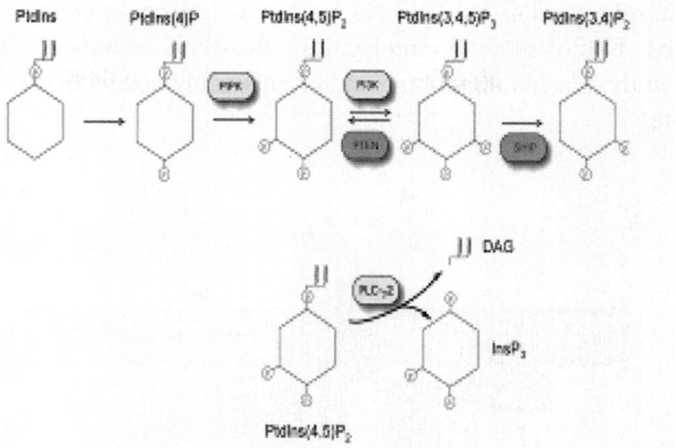

Figure 10-5. Phosphoinositide metabolism mediated by PIPK, PI3K, SHIP, PTEN and PLC-γ2. Both Phosphoinositide 3-kinase (PI3K) and phospholipase Cγ2 (PLC-γ2) share the common substrate, phosphoinositide 4,5 bis-phosphate (PtdIns(4,5)P2), the product of phosphatidylinositol-4 phosphate 5-kinase (PIPK). PI3K phosphorylates PtdIns(4,5)P2 to give rise to PtdIns(3,4,5)P3; (PLC-γ2) hydrolyses PtdIns(4,5)P2 to produce InsP3 and diacylglycerol (DAG); SRC-homology-2-domain-containing inositol 5-phosphatase (SHIP) and phosphatase and tensin homologue (PTEN) hydrolyse PtdIns(3,4,5)P3 to PtdIns(3,4)P2 and PtdIns(4,5)P2, respectively (see plate 24).

1.3.1 Regulation of PIP$_2$ hydrolysis

Adaptor molecules serve as a substrate of BCR-induced tyrosine kinases such as Syk, and phosphorylation of these molecules is essential for subsequent PLC-γ2 activation. One of the adaptor molecules, BLNK (also known as SLP-65, BASH, and BCA), after being phosphorylated by Syk, binds to Btk, Vav, and PLC-γ2 in a SH2-phosphotyrosine-dependent manner (Chiu et al., 2002; Fu et al., 1998; Hashimoto et al., 1999; Ishiai et al., 1999a; Ishiai et al., 1999b; Johmura et al., 2003; Takata et al., 1995). As BLNK is translocated to the plasma membrane after BCR stimulation, these binding causes two consequences; 1) these partners (Btk, Vav, and PLC-γ2) also go to the plasma membrane where PIP$_2$ is located; 2) Btk and PLC-γ2 are located into close proximity with each other. Thus, the first mechanism makes PLC-γ2 gain access to its substrate PIP$_2$, while the second one facilitates phosphorylation of Tyr753 and Tyr759 on PLC-γ2 by Btk, which is essential for enhancing PLC-γ2 enzymatic activity (Humphries et al., 2004; Kim et al., 2004; Watanabe et al., 2001). In addition to Tyr753 and Tyr759, PLC-γ2 undergoes phosphorylation at Tyr1217 in a Btk-independent manner (Fig. 6 a) (Kim et al., 2004). Although the importance of this phosphorylation in the BCR signaling context is clear, this is probably not caused by direct activation of the PLC-γ2 lipase activity.

Rather, phosphorylation on Tyr1217 seems to induce association with as-yet unidentified SH2-containing molecules, possibly stabilizing the PLC-γ2 residency in the plasma membrane and/or enhancing accessibility of PLC-γ2 to its substrate.

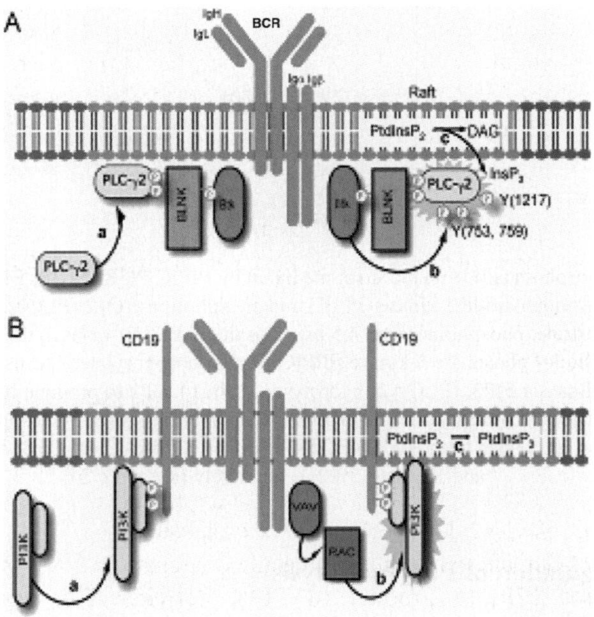

Figure 10-6. Two-step models for PLC-γ2 and PI3K activation. The first step of activation of (A) phospholipase Cγ2 (PLC-γ2) and (B) phosphoinositide 3-kinase (PI3K) requires their recruitment to the rafts (step a), presumably because their substrate, phosphatidylinositol-4, 5-bisphosphate (PtdInsP2) is enriched in the rafts. Tyrosine phosphorylation of (A) B-cell linker (BLNK) and (B) CD19 participates in this process. Then, in the case of PLC-γ2 activation, this enzyme undergoes tyrosine phosphorylation by Btk (step b in A), which is important for PLC-γ2 activation (step c, in A). In the case of PI3K, the recruited PI3K is then activated by activated RAC (step b, in B). BCR, B-cell receptor; DAG, diacylglycerol; IgH, immunoglobulin heavy chain; IgL, immunoglobulin light chain; P, phosphate (see plate 25).

Although the importance of BLNK in PLC-γ2 activation is clear, there remain several questions that need to be addressed, including how BLNK is recruited to the plasma membrane, and what other mechanism, in addition to tyrosine phosphorylation on Tyr753, Tyr759, and Tyr1217 of PLC-γ2, contributes to PLC-γ2 activation during BCR signaling. In the first issue, an idea has been brought by two lines of recent observations. A region containing highly conserved leucine zipper in the BLNK N-terminus is reported to be responsible for targeting BLNK to the plasma membrane

(Kohler et al., 2005), suggesting that the binding proteins to this leucine zipper region could contribute to recruitment of BLNK to the plasma membrane. On the other hand, it has been proposed that binding of the BLNK SH2 domain to the phosphorylated Igα could recruit BLNK to the BCR signaling complex. Indeed, Igα undergoes phosphorylation of three tyrosine residues on BCR ligation, two of which are crucial for binding to Syk SH2 domains and the remaining one for binding to the BLNK SH2 domain (Engels et al., 2001; Pike and Ratcliffe, 2005; Siemasko et al., 2002). Collectively, BLNK might be localized in the plasma membrane, constitutively, by virtue of its leucin zipper region, and once stimulated, BLNK moves to the BCR complex by its SH2 domain, thereby undergoing tyrosine phosphorylation by Syk.

PIPK and PI3K are thought to contribute to PLC-γ2 activation. As mentioned above, the importance of the Btk kinase activity in PLC-γ2 activation is clear, however, new data suggest that Btk plays an additional role in a kinase-independent manner as well (Saito et al., 2003). The Btk PH/TH domain associates with PIPK and participates in recruitment of PIPK to the plasma membrane. Because PIP_2, a substrate for both PI3K and PLC-γ2, is synthesized from PIP by PIPK, this PIPK recruitment results in generation of more PIP_2. Thus, Btk plays two roles in PLC-γ2 activation, phosphorylating Tyr753 and Tyr759 on PLC-γ2 and recruiting the PIPK to the membrane, thereby ensuring that the activated PLC-γ2 does not run out of its substrate.

Various PH-domain-containing molecules, including Btk and PLC-γ2, selectively bind PIP_3, a product of PI3K activity. This indicates two mechanisms by which the PI3K and PLC-γ2 pathways are connected. First, PIP_3 is involved indirectly in PLC-γ2 activation through its binding to the Btk PH domain and subsequent Btk activation. Second, the interaction of the PLC-γ2 PH domain with PIP_3 could be required to promote and/or stabilize PLC-γ2 localization in the plasma membrane, thereby leading to the complete activation of PLC-γ2. Hence, the PI3K pathway would function as an amplifier of the PLC-γ2 pathway (Kurosaki, 1999).

Attenuator of the BCR-induced PLC-γ2 pathway is also known (Fig. 7). Tyrosine phosphorylation status of the inhibitory receptor PIRB affects the activity of the PLC-γ2 pathway (Hayami et al., 1997; Kubagawa et al., 1997). Since the ligand for PIRB is known to be MHC class I molecules (Nakamura et al., 2004; Takai, 2005), two modes of PIRB operation can be envisaged. First, when a target cell expresses both a BCR-binding antigen and class I, PIRB on B cells might be brought into close proximity to antigen-bound BCR. Mimicking this situation, the colligation of PIRB and BCR results in the recruitment of SHP1, a tyrosine phosphatase, to the phosphorylated immunoreceptor tyrosine-based inhibitory motifs (ITIMs) in the cytoplasmic domain of PIRB (Blery et al., 1998; Maeda et al., 1998).

The associated SHP1 then induces dephosphorylation of Syk, Btk, and BLNK, which in turn, downregulates PLC-γ2 activation (Maeda et al., 1999). Second, assuming that a target cell or a B cell by itself expresses only the class I molecule, but not the antigen, the interaction between class I and PIRB might sequester PIRB away from the vicinity of the BCR on B cells. According to this model, together with the evidence that PIRB is, to some extent, phosphorylated and associated with SHP1 in resting B cells (Ho et al., 1999), PIRB could function as a positive regulator of BCR signaling by excluding SHP1 that could, otherwise, inhibit BCR signaling. Hence, regardless of which mode is operating, the tyrosine phosphorylation status of PIRB and its localization relative to the BCR could be one of crucial determinants for fine-tuning the PLC-γ2 activity.

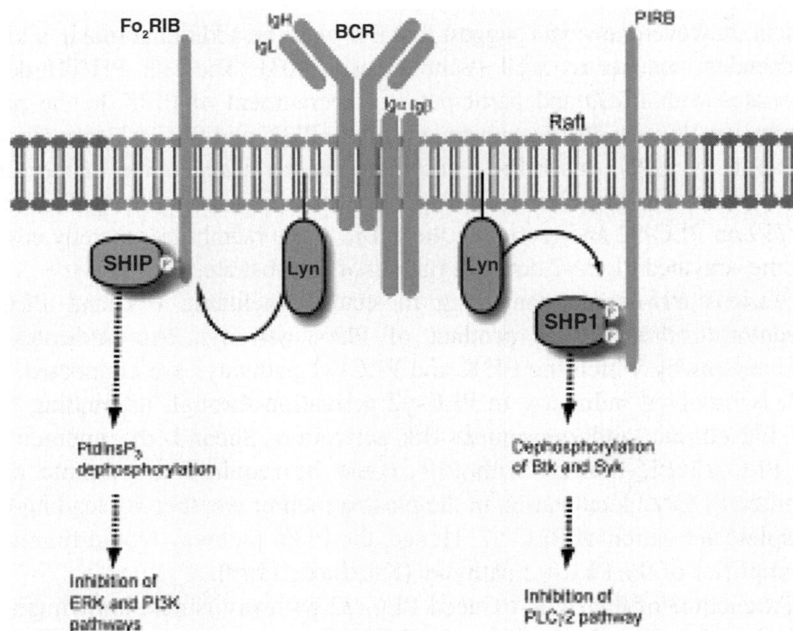

Figure 10-7. Negative-regulatory loops mediated by inhibitory receptors on B cells. Once the immunoreceptor tyrosine-based inhibitory motif (ITIM) of FcγRIIB is phosphorylated, SRC-homology-2 (SH2)-domain-containing inositol 5-phosphatase (SHIP) is recruited, which, in turn, have negative influences on phosphoinositide 3-kinase (PI3K) and extracellular signal-regulated kinase (ERK) pathways. By contrast, tyrosine phosphorylation of paired immunoglobulin-like receptor B (PIRB) ITIMs recruites SH2-domain-containing protein tyrosine phosphatase 1 (SHP1), which, in turn, dephosphorylates various protein-tyrosine kinases, including Syk and Bruton's tyrosine kinase (Btk). BCR, B-cell receptor; IgH, immunoglobulin heavy chain; IgL, immunoglobulin light chain; PLC-γ2, phospholipase Cγ2; PtdInsP3, phosphatidylinositol-3,4,5-triphosphate (see plate 26).

1.3.2 Regulation of PIP$_3$ generation

On BCR ligation, the cytoplasmic tail of CD19 is phosphorylated by Lyn, and provides binding sites for the SH2 domains of the p85 subunit of PI3K and Vav (Fearon and Carroll, 2000; Tedder et al., 1997). Through the interaction between CD19 and PI3K, PI3K could be translocated to the plasma membrane, whereby PI3K gains access to its substrate, PIP$_2$ (Fig. 6 b). Recent results indicate that this targeting mechanism is necessary, but not sufficient, for PI3K activation on BCR engagement. An initial clue to an additional activation mechanism came from the observation that FcεRI-mediated Akt activation, a readout of PI3K activity, is clearly inhibited in mast cells from Vav1$^{-/-}$ mice (Manetz et al., 2001). A role for Vav is further supported by the fact that Akt activation upon BCR engagement is also attenuated in Vav3$^{-/-}$ DT40 B cells. Despite the reduced Akt activity, targeting of PI3K to the plasma membrane appears to be normal in Vav3$^{-/-}$ DT40 cells, suggesting that Vav3 participates in a post-PI3K- translocational process. Moreover, an activated Rac1, a downstream target of Vav family, is able to increase the enzymatic activity of PI3K *in vitro,* suggesting that Vav acts on enzymatic activation of PI3K through Rac1 (Fig. 6b) (Inabe et al., 2002). So, a model is proposed in which PI3K, after being recruited to the plasma membrane by CD19, undergoes a conformational change, thereby leading to the increased its lipid kinase activity, by Vav. This model, however, opposes the previous view that PI3K functions as an upstream regulator for Vav-family members, on the basis of in vitro evidence that PIP$_3$, a product of PI3K, binds to the Vav PH domain, thereby enhancing the GEF activity of Vav (Turner and Billadeau, 2002). One interpretation to reconcile these two models is that Vav and PI3K function as both upstream and downstream of one another, creating a positive-feed forward amplifycation loop; Vav regulates PI3K, producing PIP$_3$, which in turn positively regulates the Vav function.

The translocation of SHIP to the plasma membrane seems to be a prerequisite for its involvement in dampening the PI3K pathway, because enforced recruitment of SHIP to the plasma membrane results in a reduction of the level of PIP$_3$ through its conversion to PIP$_2$ (Bolland et al., 1998). This mechanism is typified by the example of colligation of BCR and Fcγ RIIB (Fig. 7), in which SHIP is recruited, through its SH2 domain, to the phosphorylated ITIM in the cytoplasmic domain of Fcγ RIIB after colligation of these receptors (Ono et al., 1996; Ono et al., 1997). The generation of PIP$_3$ on BCR ligation alone is also enhanced in B cells from SHIP$^{-/-}$ mice and SHIP$^{-/-}$ DT40 cells (Brauweiler et al., 2000; Helgason et al., 2000; Okada et al., 1998), which shows the importance of SHIP in BCR signaling. Although

BCR-mediated SHIP translocation to the membrane has been observed, as for colligation of the BCR and Fcγ RIIB, a target molecule that drives SHIP to the membrane still remains to be determined in the BCR signaling.

PTEN acts as a tumor suppressor, and its encoding gene is mutated or deleted in various tumors (Di Cristofano and Pandolfi, 2000). PTEN functions as an 'off' switch in B cells, because Akt hyperactivation is observed in B cells from B-cell-specific mutation of PTEN (Suzuki et al., 2003). Although the function of PTEN is clear, there is little information regarding the regulation of PTEN expression, localization, or activity.

1.4   Activation mechanisms of Serine-Threonine Kinases (IKK, JNK, and ERK)

One event downstream of the activation of lipid metabolizing enzymes is activation of serine-threonine kinases. Although involvement of PLC-γ2 in activation of IKK, JNK, and ERK upon BCR engagement has been well known, until recently, the molecular connections between PLC-γ2 and IKK/JNK/ERK remained unclear. But, recent data has begun to shed new light on these connections.

1.4.1   Activation of IKK and JNK

PLC-γ2 activation leads to hydrolysis of PIP_2, yielding inositol-1,4, 5-trisphosphate (IP_3) and diacylglycerol (DAG). Soluble IP_3 and membrane-bound DAG leads to calcium release and activation of PKC isoforms, respectively. The PKC family comprises at least eleven members, which are categorized into three classes on the basis of their structure and activation requirements (Guo et al., 2004). The classical PKC isoforms (α, β, and γ) are are regulated by both Ca^{2+} and DAG, the novel PKC isoforms (δ, ε, θ, and η) are regulated by DAG, but not Ca^{2+}; and the atypical PKC isoforms (ζ, τ, and λ) require neither Ca^{2+} nor DAG. Among these PKC isoforms, mice deficient in PKCβ have reduced numbers of mature B cells in the periphery, a loss of peritoneal B1 B cells, and reduced T-cell-independent antibody responses (Saijo et al., 2002; Su et al., 2002). Moreover, these PKCβ$^{-/-}$ B cells fail to induce IKK activation and subsequent NF-κB activation upon BCR engagement. In contrast, CD40-dependent NF-κB activation is apparently normal in PKCβ$^{-/-}$ B cells, suggesting involvement of PKCβ in IKK activation specifically in BCR signaling context.

In addition to PKCβ, BCR stimulation requires adaptor molecules, CARMA1, Bcl10, and MALT1 to activate IKK (Fig. 8) (Lin and Wang,

2004; Thome, 2004). Indeed, B cells lacking any of these adaptor proteins are defective in IKK activation (Egawa et al., 2003; Hara et al., 2003; Jun et al., 2003; Newton and Dixit, 2003; Ruefli-Brasse et al., 2003; Ruland et al., 2003; Xue et al., 2003). In overexpression studies, CARMA1 directly binds to Bcl10 through CARD-CARD interactions and Bcl10 interacts with MALT1 (Gaide et al., 2002; Pomerantz et al., 2002; Uren et al., 2000). So, the question arises as to the connection between PKCβ and the adaptor complex CARMA1/Bcl10/MALT1. Given the evidence that BCR-mediated serine-threonine phosphorylation of CARMA1 and subsequent inducible association of CARMA1 with Bcl10 are drastically decreased in PKCβ-deficient DT40 B cells, this PKCβ-mediated phosphorylation of CARMA1 is likely to be a prerequisite for the inducible association of CARMA1 and Bcl10, probably contributing to formation of a stable CARMA1/Bcl10/-MALT1 complex. Then, the IKK complex appears to be recruited to this trimeric complex (Shinohara H et al., in press). It was thought that PKCβ might directly phosphorylate IKKα/IKKβ, being responsible for subsequent IKK activation. However, this idea seems improbable. Instead, another serine-threonine kinase, TAK1, appears to be involved in phosphorylation on IKKα/IKKβ. Indeed, in vivo and in vitro experiments support this idea; i) BCR-mediated IKK activation is abolished in TAK1-deficient DT40 B cells; ii) activation loop serine residues of IKKβ is phosphorylated by TAK1 in *in vitro* conditions (Shinohara H et al., in press). Since TAK1 is associated with the phosphorylated CARMA1, a model has been proposed in which after being phosphorylated by PKCβ, CARMA1 facilitates scaffolding TAK1 and IKK, thereby allowing TAK1 to phosphorylate the activation loop serine residues of IKKβ (Fig. 8). Moreover, since both CARMA1- and TAK1-deficient DT40 B cells fail to activate JNK after BCR ligation, this interaction also makes TAK1 to function as a MAP3K for JNK activation in BCR signaling context (Shinohara H et al., in press).

Although the functional importance of TAK1 is clear, a question remains about how TAK1 is activated after BCR stimulation. Loss of CARMA1 appears to decrease BCR-mediated TAK1 activation status, suggesting that this association may induce TAK1 enzymatic activity, like that phosphorylated Igα enhances Syk enzymatic activity, mentioned above. It is also possible that a more upstream kinase may transphosphorylate critical residues in TAK1 that is important for full activation of TAK1. These two possibilities are not mutually exclusive; probably, both mechanisms are operating in the BCR signaling context.

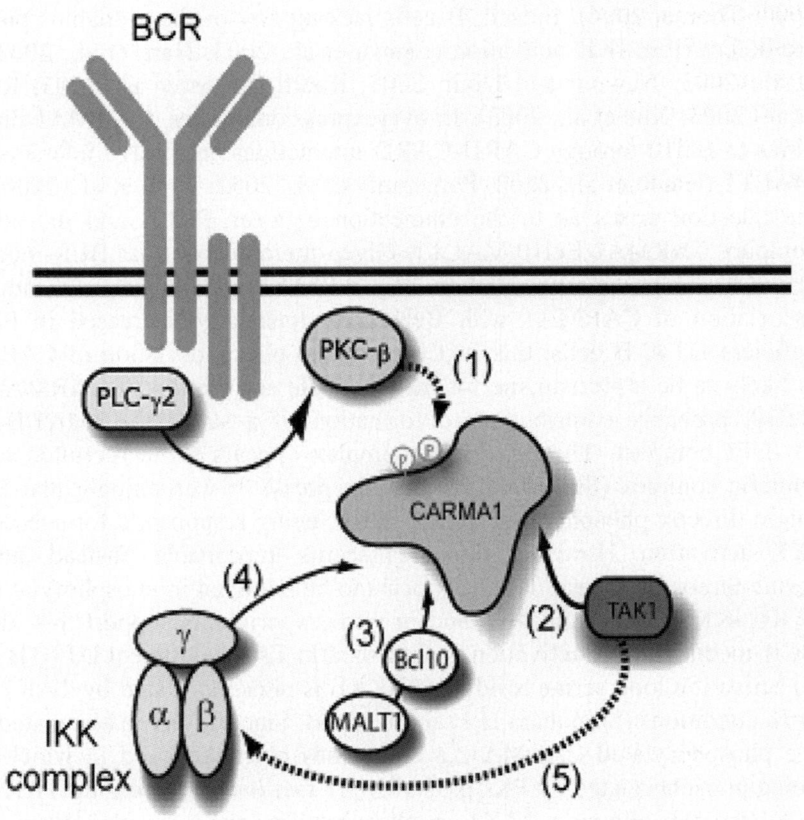

Figure 10-8. A model for BCR-mediated NF-κB activation. Stimulation of BCR leads to activation of proximal protein tyrosine kinases including Syk and Btk. Btk phosphorylate several tyrosine residues on phospholipase Cγ2 (PLC-γ2), and subsequent activation of protein kinase Cβ (PKCβ). Activated PKCβ phosphorylates CARMA1 (1), directly or indirectly, which is able to recruit TAK1 to the phosphorylated CARMA1 (2). Meanwhile, the IKK complex probably through the Bcl10/ MALT1 complex is recruited to the phosphorylated CARMA1 (3 and 4). These interactions (CARMA1-IKK and CARMA1-TAK1) contribute to access of two key protein kinases, TAK1 and IKK, leading to activation of the IKK complex (5) (see plate 27).

1.4.2   Activation of ERK

Like other cell types, Ras is activated by BCR stimulation and introduction of dominant-negative Ras inhibits BCR-mediated ERK activation in DT40 B cells as well as primary B cells (Hashimoto et al., 1998; Iritani et al., 1997; Nagaoka et al., 2000), demonstrating that Ras functions as an upstream regulator for subsequent ERK activation. Thus, the question raises about how Ras is activated upon BCR stimulation. It has been for long time thought that Sos, one type of nucleotide exchange factors for Ras, participates in BCR-mediated Ras activation (Campbell et al., 1998). However, data emerging from several laboratories indicate that RasGRP3, another type of exchange factors for Ras, rather than Sos, plays a more dominant role for coupling BCR to Ras activation (Ehrhardt et al., 2004; Oh-hora et al., 2003; Teixeira et al., 2003). Ras activation occurred normally even in Sos1/Sos2 double-deficient DT40 B cells, while this activation was, although not completely, inhibited in RasGRP3-deficient DT40 B cells. Necessity of RasGRP3 in Ras activation appears to well explain the previous findings that PLC-γ2 was required for BCR-mediated Ras activation (Hashimoto et al., 1998). Since RasGRP3, but not Sos1/Sos2, has the C1 domain that binds to DAG, a product of PLC-γ2 action, the idea has been evoked that PLC-γ2 participates in recruitment of RasGRP3 in a DAG-C1 dependent manner. Indeed, deletion of the C1 domain of RasGRP3 fails to move RasGRP3 to the plasma membrane and to activate Ras in BCR signaling. Moreover, wild-type RasGRP3 cannot move to the plasma membrane in PLC-γ2-deficeint DT40 B cells (Fig. 9) (Oh-hora et al., 2003). In addition to recruitment of RasGRP3 to the membrane, phosphorylation on Thr133 on RasGRP3 is also important for full activation of RasGRP3 (Aiba et al., 2004; Zheng et al., 2005). This is mediated by activated PKCβ, thereby leading to increased enzymatic activity of Ras GRP3. Collectively, a new model is proposed in which DAG generated by PLC-γ2 facilitates recruitment of both PKCβ and RasGRP3 to the plasma membrane, wherein PKCβ phsophorylates Thr133 in RasGRP3, being crucial for full activation of RasGRP3 (Fig. 9).

Ras is downregulated following interaction with the Ras-GTPase activating protein, RasGAP (Bar-Sagi and Hall, 2000; Campbell et al., 1998; Reuther and Der, 2000). BCR stimulation results in tyrosine phosphorylation of RasGAP and the GTPase activating activity of RasGAP appears to be decreased by this tyrosine phosphorylation (Tamir et al., 2000; Yamanashi et al., 2000). Thus, assuming that BCR activates Ras through operating two mechanisms concomitantly, by activating RasGRP3 and by inhibiting RasGAPs, the residual Ras activation in RasGRP3-deficient DT40 B cells could be due to ongoing inhibition of RasGAP during BCR stimulation. Once Ras is activated, this GTP-bond Ras binds directly to Raf-1, the MAP3K in the ERK pathway. Activated Raf-1 and B-Raf phosphoryalte and

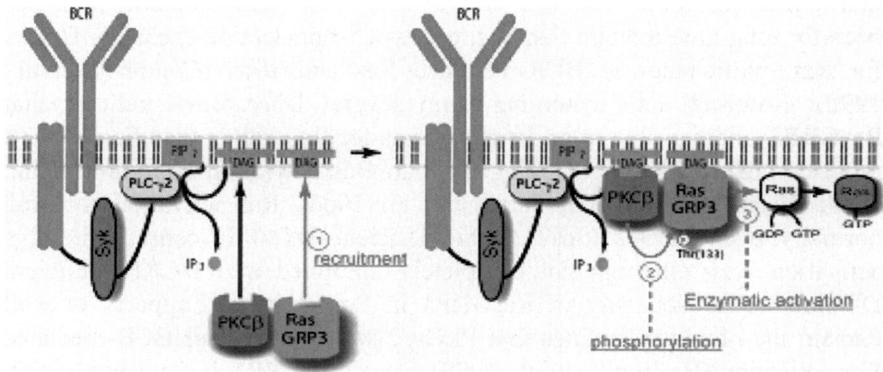

Figure 10-9. A model of Ras regulation in B cells by a Ras-guanyl nucleotide exchange factor, RasGRP3. After BCR stimulation, RasGRP3 is recruited to plasma membrane by binding to DAG, a PLC-γ2 product (1). PKC is also recruited to membrane and phosphorylate RasGRP3 at Thr-133, thereby resulting in its activation (2). Then, fully activated RasGRP3 turns RasGDP into its active RasGTP form (3) (see plate 28).

activate MEK1/MEK2, which both in turn phosphorylate ERK1/ERK2. Phosphorylated ERKs form dimmers, a step required for nuclear translocation and the subsequent phosphorylation of transcriptional regulatory proteins, including Fos, Jun, and Ets family members (Dong et al., 2002).

1.5   Regulation of calcium mobilization

After PLC-γ2 activation, the generated IP_3 binds to IP_3 receptors located in the ER, thereby stimulating the release of calcium from internal stores. It is thought that, once these calcium stores are emptied, the entry of external calcium is triggered through so-called store-operated calcium channels (SOCs) in the plasma membrane (Fig. 10) (Parekh and Penner, 1997).
Regulation of IP_3 receptors

Three distinct IP_3 receptor genes are identified with differential expression patterns, depending upon distinct cell types (Dent et al., 1996; Miyakawa et al., 1999; Newton et al., 1994; Wojcikiewicz et al., 1994). Triple knockout of all three IP_3 receptor isoforms in DT40 B cells abolishes the BCR-induced calcium mobilization both from internal stores and from extracellular stores (Sugawara et al., 1997), whereas this calcium mobilization still occurs by a single knockout of these three receptors. Thus, overall, three IP3 receptors are essential and functional redundant molecules for BCR-induced calcium mobilization (Miyakawa et al., 1999). And these data appears to support the store-operated coupling model, in which calcium release occurs first, then leading to the entry of external calcium.

Although expression of only one isoform does not abolish calcium mobilization, detailed calcium signaling patterns differ significantly among these three IP3 receptor isoforms. For instance, DT40 B cells expressing only type 2 IP_3 receptor showed regular and robust calcium oscillations upon BCR ligation, whereas monophasic calcium transient or rapidly dampened calcium oscillations were observed in mutant B cells expressing type 3 or type 1 alone, respectively (Miyakawa et al., 1999). Hence, differential and combinatorial expression of these three IP_3 receptors is one of the critical determinants for creating transient or oscillatory calcium signals, which in turn could regulate the selectivity of transcriptional factors in B cells. For instance, NF-κB is activated by a large transient calcium rise, whereas NFAT requires a slow, but sustained calcium plateau (Dolmetsch et al., 1997).

Then, the question arises about the mechanisms underlying such differential calcium oscillation patterns. Although being not entirely clear, a clue came from mutational analysis of type 1 IP_3 receptor (Miyakawa et al., 2001). Binding of IP_3 to the IP_3 receptors is essential, but not sufficient to open the calcium release channel embedded in these receptors, and calcium by itself is thought to be required as a co-agonist for the IP_3 receptors. Asp2100 is responsible for such calcium sensor, because its replacement with Glu causes a 10-fold decrease in the calcium sensitivity, but bears normal IP_3 sensitivity. This Asp2100Glu mutant exhibits a dramatic decrease in calcium oscillation activity. Furthermore, consistent with the ability of type 2 IP_3 receptor to create robust calcium oscillations, this receptor, compared with type 1 and type 3, possesses higher calcium sensitivity (Miyakawa et al., 1999). Therefore, these data provide the compelling evidence that calcium initially released by IP_3 receptors feeds back to augment further calcium release in a positively cooperative fashion, thereby contributing to generation of calcium oscillations. Coupling between calcium release and calcium influx.

The mechanism by which the store-operated channels (SOCs) are activated by intracellular calcium release still remains mystery. SOCs are named functionally, particularly in terms of electrophysiological criteria, but not molecularly. Hence, for solving this important question, two interrelated issues should be addressed; the molecular identity of SOCs, and the nature of the signal that activates them. In the first issue, much recent interest has been directed towards the TRP-family of ion channels and they have been considered likely to encode components of SOCs, as discussed in a more detail in the next section. Because of a limited amount of information about these two issues in lymphocytes, it is important to incorporate more from other cell types in this section.

There are currently three (Fig. 10), but not mutually exclusive, models for activation of SOCs, and recent studies have suggetsed that such three models could coexist in the same cells and cooperate even in a single receptor system. The diffusible messenger model is the oldest one (Randriamampita and Tsien, 1993). According to this model, calcium influx factor (CIF) is produced by depleted internal stores and it triggers activation of SOCs (Fig. 10 a, left pannel). After initial excitement, the CIF model was strongly criticized because of a continuous uncertainty about the molecular nature, but a few groups have continued their struggle, attempting to identify native CIF and determine the CIF-mediated pathway (Randriamampita and Tsien, 1995; Rzigalinski et al., 1999; Takemura and Ohshika, 1999; Thomas and Hanley, 1995; Trepakova et al., 2000). Although the original CIF model hypothesized that CIF directly binds and activates SOCs, an indirect activation model has been recently proposed (Smani et al., 2004; Smani et al., 2003). In this model, CIF-induced displacement of inhibitory calmodulin from iPLA2 is a key event, leading to iPLA2 activation. Once iPLA2 is activated by CIF, lysophospholipids are generated, which in turn activate SOCs. Hence, according to this variant model (Fig. 10 a right pannel), a more direct activator for SOCs is considered to be lysophospholipids rather than CIF.

The conformational coupling model postulates that a direct interaction between the ER-resident proteins such as IP3 receptors and SOCs is required for opening SOCs (Fig. 10b left pannel). This mechanism has been particularly appealing, because structural and functional coupling of transient receptor protein C3 (TRPC3) channel, possibly one of SOCs, and IP3 receptor was demonstrated in *in vitro* system (Kiselyov et al., 2000). However, in contradiction with this model, triple knock-out of all three IP3 receptor isoforms in DT40 B cells still exhibits proper calcium influx after pharmacological emptying of the ER calcium stores such as thapsigargin or ionomycin treatment (Sugawara et al., 1997). One interpretation that reconciles these two data is that opening mechanisms for SOCs might differ in between BCR signaling and pharamcological contexts, although some overlapping exists. In support of this explanation, in a triple knock-out background, BCR-mediated calcium influx can be restored by type 1 IP_3 receptor mutants that do have IP_3 binding, but not channel activity (van Rossum et al., 2004). Conversely, the IP_3 binding mutants cannot restore, simply suggesting that IP_3-mediated conformational change in the IP_3 receptor is required for opening SOCs in the BCR signaling context.

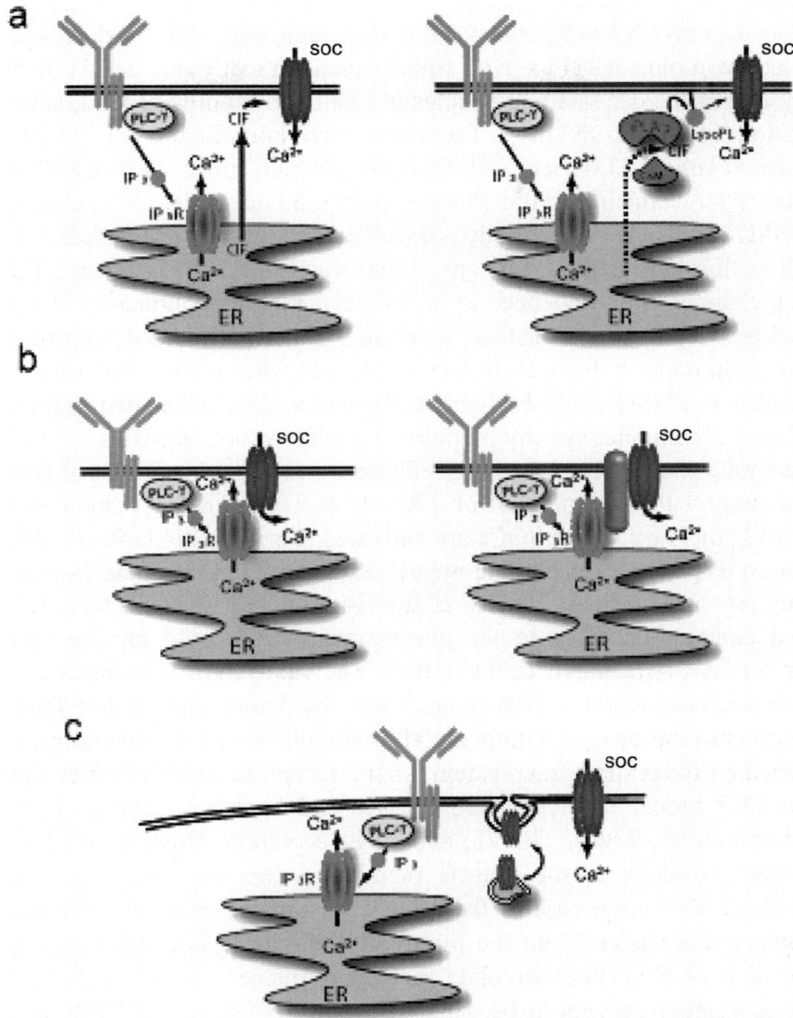

Figure 10-10. Models of SOCs activation. (A) Left, The direct activation model by the diffusible-messenger. In this model, it is postulated that depletion of calcium from intracellular store induces production of calcium influx factor (CIF). CIF diffuses to the plasma membrane and activates SOCs directly. Right, The indirect activation model by the diffusible-messenger. CIF produced by depletion of the internal calcium pool diffuses and activates iPLA2 by displacement of inhibitory calmodulin (CaM). Once iPLA2 gets activated, it generates lysophospholipids, which in turn activate SOCs. (B) Left, The direct conformational-coupling model. In this model, it is postulated that conformational change of the IP3 receptor takes place by it opening and this change is transmitted to SOCs by direct interaction and activation of SOCs. Right, The indirect conformational-coupling model. Instead of the IP3 receptor by itself, other ER-resident molecules sense the conformational change of the IP3 receptor and directly interact to the SOC, thereby activating them. (C) The secretion model. This model suggests a mechanism by which depletion of calcium from intracellular store initiates the vesicular translocation and insertion of calcium channels to the plasma membrane. Then, delivery of these channels to the membrane might be a trigger for cation influx (see plate 29).

As an obvious extension of this model, the conformational change in the IP$_3$ receptor can be transmitted to SOCs through other ER-resident and/or intermediate molecules (Fig. 10 b right pannel) (Yuan et al., 2003). In this regard, one of the candidate molecules is a calcium-binding protein, stromal interaction molecule (STIM). There are two homologues of STIM in mammalian cells, STIM1 and STIM2, both of which appear to be distributed ubiquitously (Marchant, 2005; Putney, 2005). Knockdown of STIM1, but not STIM2, by RNAi substantially reduced activity of SOCs in Jurkat T cells as well as fibroblast cells (Liou et al., 2005; Roos et al., 2005). Since STIM1 has no channel-like sequence, it is unlikely that this molecule has the channel activity by itself. Instead, since this protein is located both on the plasma membrane and intracellular membranes, presumably the ER, it is proposed that STIM1 could bridge two channel molecules located in the ER and plasma membrane, thereby participating in activation of SOCs.

Assuming that SOCs comprise TRP-family members, a series of recent studies suggest that recruitment of TRPs to the plamsma membrane is the third mechanism by which SOCs are activated (Fig. 10 c) (Clapham, 2003). Such a concept has considerable appeal as many TRP channels expressed *in vitro* are constitutively active. If this is also true *in vivo*, then BCR-induced delivery of TRPs to the plasma membrane could be one of the trigger for SOCs-mediated cation influx. The first report corroborating an exocytic mechanism for a TRP channel was the demonstration that TRPV2 translocates to the plasma membrane after stimulation of the insulin growth factor using overexpression system (Kanzaki et al., 1999). Since then, several TRP family members, TRPC3 (Xu and Sternberg, 2003), TRPC5 (Bezzerides et al., 2004), TRPV5, and TRPV6 (van de Graaf et al., 2003), are shown to move to the plasma membrane after receptor stimulation (Montell, 2004). In the case of the TRPC5 in hippocampal neurons, incorporation of this channel into the plasma membrane is initiated by growth factors such as NGF that stimulate receptor tyrosine kinases (RTKs), and this incorporation appears to be dependent on PI3K, Rac, and PIPK in the NGF signaling context (Bezzerides et al., 2004). Because these enzymes and GTP-binding proteins are also activated in the BCR signaling context, the previous findings that PI3K is required for BCR-mediated calcium influx (Clayton et al., 2002; Jou et al., 2002) might be accounted for by its involvemnt in translocation of SOCs to the plasma membrane.

1.5.1   Looking for Ca^{2+} entry channels

Proteins of the TRP-family are presently the best candidates for the pore-forming subunit of SOCs. The TRP family can be classified into six subfamilies (TRPC, TRPV, TRPM, TRPA, TRPP, and TRPML), and total number of these genes is now turned out to be 28 in mice (see Fig. 11 for their structures) (Clapham, 2003). Among them, TRPV5 and TRPV6 seem to be the stronger candidates, but their physiological properties, when overexpressed each alone in heterologous cells such as HEK293 cells, do not fully resemble those of endogenous SOCs (Hoenderop et al., 2001; Vennekens et al., 2000; Voets et al., 2001). Thus, the best speculation at present is that SOC channels in B lymphocytes are heteromultimeric complexes of TRP components, and additional regulatory or adapter subunits, needed to promote channel assembly and subsequent coupling to store depletion mentioned above.

After calcium influx occurs, it is proposed that PLC-γ2 is activated again, thereby forming a positive-forward loop between PLC-γ2 and calcium, based upon the existence of the C2 calcium binding domain in PLC-γ2 (Nishida et al., 2003). Supporting this model, the PLC-γ2 mutant devoid of this C2 domain exhibited no more such amplification. Given the physical association between PLC-γ2 and TRP channels such as TRPC3 (Nishida et al., 2003), this model also suggests that the PLC-γ2 C2 domain could quickly sense the alteration of the calcium concentration just induced by TRPC3, possibly making a positive loop in a very restricted region inside B lymphocytes (Fig. 11).

In contrast to significant progress about how calcium signals are generated, how calcium signals are terminated is largely unknown. Several mechanisms, including inhibition of SOCs, SERCA-mediated uptake of Ca^{2+} into intracelullular Ca^{2+} stores, and the action of Ca^{2+} pumps (PMCAs) which pump Ca^{2+} out the cells are proposed (Berridge et al., 2003). Indeed, the importance of PMCAs was recently shown in B and T lymphocytes (Bautista et al., 2002; Chen et al., 2004).

1.6   NFAT and NF-κB

Increased intracellular calcium promotes nuclear translocation of two important sets of transcription factors, namely the NFAT and the NF-κB family (Fig. 12) (Dolmetsch et al., 1997). Activation of the NF-κB transcription factors NF-κB1, NF-κB2, c-Rel, RelA, and RelB is essential for B cell development as well as immune responses (Henderson and

Calame, 1998; Li and Verma, 2002). NF-κB is released from inhibitors of the IκB family by phosphorylation and ubiqutination-mediated proteolysis.

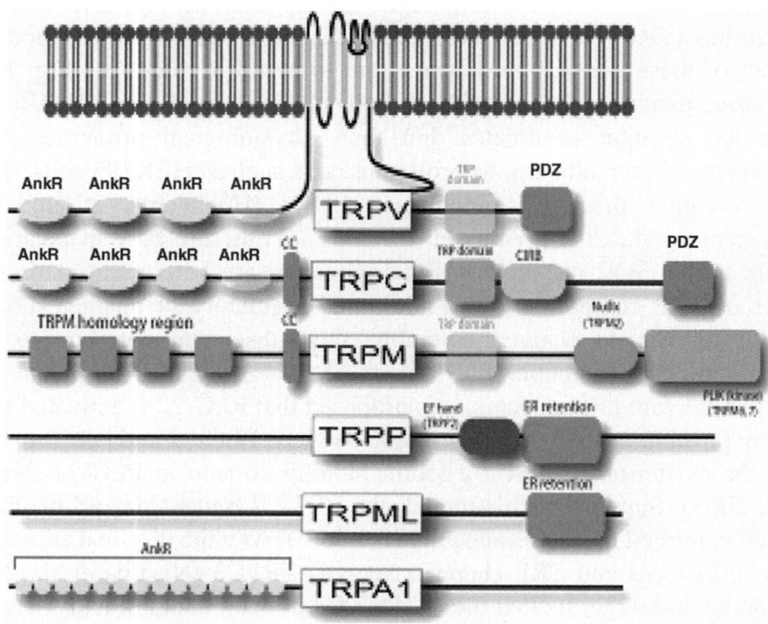

Figure 10-11. Schematic structure of TRP-family ion channels. Schematic structure of TRP-ion channels. All the members of the family consist of six transmembrane domain-containing ion channels flanked with two cytoplasmic tails that are characteristic to each member. The TRP box is EWKFAR in TRPC, but is less conserved in TRPV and TRPM. CC indicates coiled-coil domain. Ankyrin repeats (AnkR) range from 0 to 14 in number. CIRB stands for putative calmodulin- and IP3 receptor-binding domain. EF hand, canonical helix-loop-helix Ca2þ-binding domain; PDZ, amino acids-binding PDZ domains; PLIK, phospholipase C-interacting kinase; Nudix, NUDT9 hydrolase protein homologue-binding ADP ribose (see plate 30).

and this IκB phosphorylation is triggered by a multimolecular complex comprising two serine threonine kinases, IKKα and IKKβ, and a regulatory subunit IKKγ (Ghosh and Karin, 2002). Thus, activation of the IKK complex as mentioned above is essential for subsequent NF-κB activation. There are four calcium-regulated members of the NFAT family (Feske et al., 2003; Rao et al., 1997). The activity of these proteins is determined by their phosphorylation status, which is tightly regulated by the interplay between calcineurin and opposing kinases. When calcinurin is activated through an increase in calcium levels, NFAT is dephosphorylated at a large number of phosphorylated serine residues and rapidly enters the nucleus (Okamura et al., 2000; Shibasaki et al., 1996). Conversely in stimulated cells where NFAT is already localized to the nucleus, termination of calcium signaling

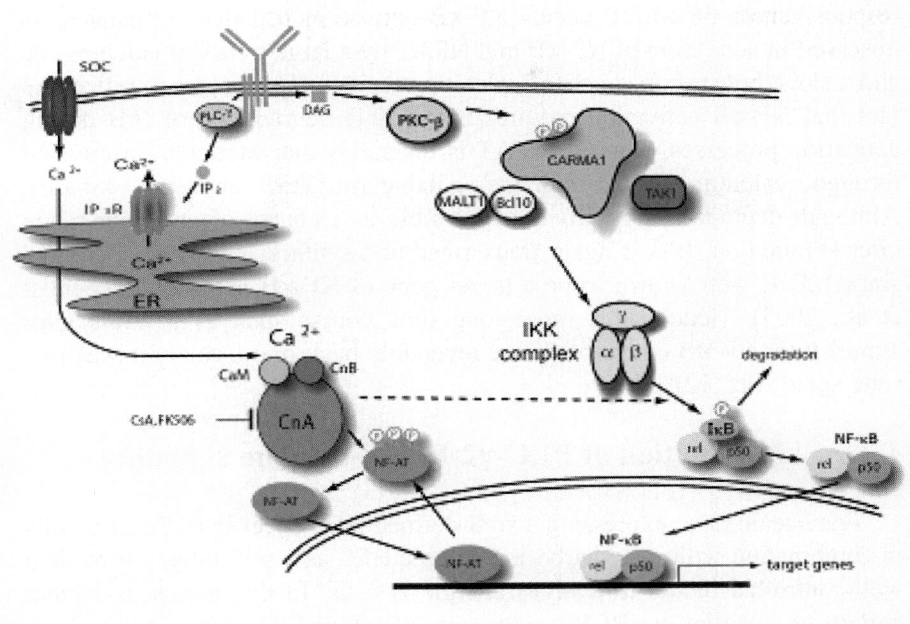

Figure 10-12. Regulation of NF-κB and NFAT by PLC-γ2. BCR-mediated PLC-γ2 activation causes elevation of cytoplasmic concentration of Ca2+, which leads to the activation of calcineurin. NFAT is dephosphorylated by calcineurin and translocates to the nucleus. DAG, generated by PIP2hydrolysis, activates PKCβ, which then activates IKK complex presumably by promoting membrane recruitment and aggregation of the CARMA1/Bcl10/MALT1 complex. This complex activates IKKs, resulting in the phosphorylation and ubiquitin-mediated degradation of IκB. Released rel/p50 complex translocates to the nucleus. It has been reported that inhibition of calcineurin blocks the activation of NF-κB suggesting that calcineurin is involved in NF-κB activation pathway (Biswas et al., 2003) (see plate 31).

results in re-phosphorylation of NFAT, exposure of a nuclear export signal that is bound by the nuclear export receptor Crm1, and transport of NFAT back to the cytoplasm (Kehlenbach et al., 1998; Klemm et al., 1997). Hence, an implication of these findings is that in order to maintain NFAT in a dephosphorylated state for subsequent gene regulation, calcineurin should be present in the nucleus of stimulated cells. Furthermore, calcineurin-mediated dephosphorylation must be capable of being reversed by NFAT kinases (Hogan et al., 2003; Kiani et al., 2000). More importantly, the property of reversible activation and deactivation by calcineurin and NFAT kinase, respectively, confers on NFAT a remarkable activity to sense dynamic changes in Ca^{2+} in lymphocytes (Dolmetsch et al., 1997; Dolmetsch et al., 1998; Li et al., 1998; Tomida et al., 2003). Thus, the differential responsiveness of NFAT versus NF-κB activation in calcium changes, as observed in activation of NF-κB and NFAT by a large transient calcium rise and a slow, but sustained calcium plateau, respectively, appears to reflect the fact that NF-κB activation requires irreversible degradation of IκB during activation processes, whereas NFAT is reversibly activated and deactivated through calcium-sensing enzyme calcineurin and opposing kinases. Although degradation of IκB is irreversible in a course of one hour range after stimulation, IκB is again transcribed and synthesized after a lag time, since IκB is well known to be a target gene of NF-κB pathway (Siebenlist et al., 2005). Hence, in a more long time course such as 5 hours after stimulation, NF-κB components are reversibly back into a calcium-sensitive state again (Fig. 12).

1.7   Modulation of PLC-γ2-PKCβ-calcium signaling

Transgenic mice expressing a well-defined HEL-specific BCR on B cells in combination with mice harboring soluble HEL as a self-antigen provide a well-controlled model to analyze anergic B cells. In this model, a distinct pattern of signaling by BCRs, activating NFAT or ERK, but not NF-κB or JNK, characterizes anergic B cells (Glynne et al., 2000; Healy et al., 1997). These results indicate that chronic, suboptimal stimulation induces an inhibitory feedback to selectively uncouple the receptor from activation signaling pathways. In anergic B cells, basal calcium level is high and antigen-mediated calcium oscillations are dampened (Healy et al., 1997). This differential calcium signaling, given the evidence that NF-κB, but not NFAT, requires a large transient calcium rise, could explain why NF-κB pathway is selectively uncoupled in anergic cells.

Therefore, the above findings raise the question of how PLC-γ2-calcium signaling properties are modulated by the inhibitory feedback operating in anergic B cells. Several data suggest ubiquitin E3 ligases (E3s) as an potential candidate mediator for this feedback (Davis and Ben-Neriah, 2004; Liu, 2004). In the case of T cells, T cells are induced to the anergy state through anergy-inducing stimuli such as TCR stimulation without costimulation and treatment of the calcium ionophore ionomycin (Heissmeyer et al., 2004; Macian et al., 2002). Indeed, after ionomycin stimulation, at least three E3s, Itch, Cbl-b, and GRAIL are upregulated at the mRNA and protein levels (Heissmeyer et al., 2004). When anergic T cells are re-stimulated, Itch and closely related HECT-domain E3 Nedd4 becomes localized to the membrane, whereby they target key signaling molecules, including PLC-γ1 and PKCθ for degradation, thereby diminishing calcium mobilization. Hence, if similar mechanism also operates in B cells, signaling molecules involved in PLC-γ-calcium pathway, such as Syk, BLNK, PLC-γ2, $_3$IP R, or TRPs, could be a target for ubiquitin-mediated downregulation, thereby dampening calcium signaling observed in anergic B cells. As discussed As discussed above, Syk was indeed shown to be a potential target for Cbl and/or Cbl-b, thereby undergoing poly-ubiqitination-induced degradation (Sohn et al., 2003).

In addition to poly-ubiquitin-mediated degradation, studies in yeast and in mammalian cells have shown that mono-ubiquitin tags are necessary and sufficient to trigger endocytosis (Dupre et al., 2004). For instance, binding of epidermal growth factor (EGF) to its receptor (EGFR) stimulates receptor internalization, and EGF-dependent mono-ubiquitination of the EGFR by the E3 ligase Cbl triggers a sorting event on the early endosome (Levkowitz et al., 1998). Sorting involves recognition of the mono-ubiquitinated receptors by proteins, such as Hrs and Tsg101, that contain ubiquitin-binding domains (Chin et al., 2001; Lu et al., 2003). Hence, if Cbl and/or Cbl-b are upregulated in anergic B cells, like ionomycin-induced T cell anergy, enhanced cycles of endocytic sequetration and recycling would be expected. This enhanced endocytosis is likely to terminate BCR-mediated calcium signaling prematurely by sequestering the early signaling complexes from the plasma membrane supply of PIP$_2$ and PIP$_3$ needed for Btk and PLC-γ2 activity. Supporting this possibility, endocytosis of BCRs to a large pool of recycling intracellular receptors is enhanced in anergic B cells compared with that of their naive counterparts (Heltemes-Harris et al., 2004; Morris et al., 2000). Thus, E3s could modulate calcium signaling by, at least, two ways; degradation of signaling proteins and enhancing endocytosis.

1.8   Concluding remarks

Significant progress has been made defining the role of BCR signaling components in PLC-γ2 activation, and deciphering the means by which BCR fine-tunes the PLC-γ2 activity by these components. In this regard, the in vitro DT40 system contributes significantly to this progress. However, the molecular identification of calcium entry channels and elucidation of biochemical sequence of events leading to turn on and off these channels still remain as a future challenge. Given that many channels, pumps, and adaptors may functionally connect to each other, which in turn contributes to spatiotemporal regulation of effective calcium signal, an answer to these questions will requires use of multiple fluorescent proteins whose location and expression can be monitored simultaneously inside DT40 cells, in addition to deletion or activation of wild-type and mutated genes under regulated conditions.

REFERENCES

Aiba, Y., Oh-hora, M., Kiyonaka, S., Kimura, Y., Hijikata, A., Mori, Y., and Kurosaki, T. (2004). Activation of RasGRP3 by phosphorylation of Thr-133 is required for B cell receptor-mediated Ras activation. Proc Natl Acad Sci U S A *101*, 16612-16617.

Bar-Sagi, D., and Hall, A. (2000). Ras and Rho GTPases: a family reunion. Cell *103*, 227-238.

Bautista, D. M., Hoth, M., and Lewis, R. S. (2002). Enhancement of calcium signalling dynamics and stability by delayed modulation of the plasma-membrane calcium-ATPase in human T cells. J Physiol *541*, 877-894.

Berg, L. J., Finkelstein, L. D., Lucas, J. A., and Schwartzberg, P. L. (2005). Tec family kinases in T lymphocyte development and function. Annu Rev Immunol *23*, 549-600.

Berridge, M. J., Bootman, M. D., and Roderick, H. L. (2003). Calcium signalling: dynamics, homeostasis and remodelling. Nat Rev Mol Cell Biol *4*, 517-529.

Bezzerides, V. J., Ramsey, I. S., Kotecha, S., Greka, A., and Clapham, D. E. (2004). Rapid vesicular translocation and insertion of TRP channels. Nat Cell Biol *6*, 709-720.

Biswas, G., Anandatheerthavarada, H. K., Zaidi, M., and Avadhani, N. G. (2003). Mitochondria to nucleus stress signaling: a distinctive mechanism of NFκB/Rel activation through calcineurin-mediated inactivation of IκBβ. J Cell Biol *161*, 507-519.

Blery, M., Kubagawa, H., Chen, C. C., Vely, F., Cooper, M. D., and Vivier, E. (1998). The paired Ig-like receptor PIR-B is an inhibitory receptor that recruits the protein-tyrosine phosphatase SHP-1. Proc Natl Acad Sci U S A *95*, 2446-2451.

Bolland, S., Pearse, R. N., Kurosaki, T., and Ravetch, J. V. (1998). SHIP modulates immune receptor responses by regulating membrane association of Btk. Immunity *8*, 509-516.

Brauweiler, A., Tamir, I., Dal Porto, J., Benschop, R. J., Helgason, C. D., Humphries, R. K., Freed, J. H., and Cambier, J. C. (2000). Differential regulation of B cell development, activation, and death by the src homology 2 domain-containing 5' inositol phosphatase (SHIP). J Exp Med *191*, 1545-1554.

Brdicka, T., Kadlecek, T. A., Roose, J. P., Pastuszak, A. W., and Weiss, A. (2005). Intramolecular regulatory switch in ZAP-70: analogy with receptor tyrosine kinases. Mol Cell Biol *25*, 4924-4933.

Campbell, S. L., Khosravi-Far, R., Rossman, K. L., Clark, G. J., and Der, C. J. (1998). Increasing complexity of Ras signaling. Oncogene *17*, 1395-1413.

Chan, A. C., Dalton, M., Johnson, R., Kong, G. H., Wang, T., Thoma, R., and Kurosaki, T. (1995). Activation of ZAP-70 kinase activity by phosphorylation of tyrosine 493 is required for lymphocyte antigen receptor function. EMBO J *14*, 2499-2508.

Chen, J., McLean, P. A., Neel, B. G., Okunade, G., Shull, G. E., and Wortis, H. H. (2004). CD22 attenuates calcium signaling by potentiating plasma membrane calcium-ATPase activity. Nat Immunol *5*, 651-657.

Cheng, P. C., Dykstra, M. L., Mitchell, R. N., and Pierce, S. K. (1999). A role for lipid rafts in B cell antigen receptor signaling and antigen targeting. J Exp Med *190*, 1549-1560.

Cherukuri, A., Dykstra, M., and Pierce, S. K. (2001). Floating the raft hypothesis: lipid rafts play a role in immune cell activation. Immunity *14*, 657-660.

Chin, L. S., Raynor, M. C., Wei, X., Chen, H. Q., and Li, L. (2001). Hrs interacts with sorting nexin 1 and regulates degradation of epidermal growth factor receptor. J Biol Chem *276*, 7069-7078.

Chiu, C. W., Dalton, M., Ishiai, M., Kurosaki, T., and Chan, A. C. (2002). BLNK: molecular scaffolding through 'cis'-mediated organization of signaling proteins. EMBO J *21*, 6461-6472.

Clapham, D. E. (2003). TRP channels as cellular sensors. Nature *426*, 517-524.

Clayton, E., Bardi, G., Bell, S. E., Chantry, D., Downes, C. P., Gray, A., Humphries, L. A., Rawlings, D., Reynolds, H., Vigorito, E., and Turner, M. (2002). A crucial role for the p110δ subunit of phosphatidylinositol 3-kinase in B cell development and activation. J Exp Med *196*, 753-763.

Davis, M., and Ben-Neriah, Y. (2004). Behind the scenes of anergy: a tale of three E3s. Nat Immunol *5*, 238-240.

Deckert, M., Elly, C., Altman, A., and Liu, Y. C. (1998). Coordinated regulation of the tyrosine phosphorylation of Cbl by Fyn and Syk tyrosine kinases. J Biol Chem *273*, 8867-8874.

Dent, M. A., Raisman, G., and Lai, F. A. (1996). Expression of type 1 inositol 1,4,5-trisphosphate receptor during axogenesis and synaptic contact in the central and peripheral nervous system of developing rat. Development *122*, 1029-1039.

Di Cristofano, A., and Pandolfi, P. P. (2000). The multiple roles of PTEN in tumor suppression. Cell *100*, 387-390.

Dolmetsch, R. E., Lewis, R. S., Goodnow, C. C., and Healy, J. I. (1997). Differential activation of transcription factors induced by Ca^{2+} response amplitude and duration. Nature *386*, 855-858.

Dolmetsch, R. E., Xu, K., and Lewis, R. S. (1998). Calcium oscillations increase the efficiency and specificity of gene expression. Nature *392*, 933-936.

Dong, C., Davis, R. J., and Flavell, R. A. (2002). MAP kinases in the immune response. Annu Rev Immunol *20*, 55-72.

Dupre, S., Urban-Grimal, D., and Haguenauer-Tsapis, R. (2004). Ubiquitin and endocytic internalization in yeast and animal cells. Biochim Biophys Acta *1695*, 89-111.

Egawa, T., Albrecht, B., Favier, B., Sunshine, M. J., Mirchandani, K., O'Brien, W., Thome, M., and Littman, D. R. (2003). Requirement for CARMA1 in antigen receptor-induced NF-κB activation and lymphocyte proliferation. Curr Biol *13*, 1252-1258.

Ehrhardt, A., David, M. D., Ehrhardt, G. R., and Schrader, J. W. (2004). Distinct mechanisms determine the patterns of differential activation of H-Ras, N-Ras, K-Ras 4B, and M-Ras by receptors for growth factors or antigen. Mol Cell Biol *24*, 6311-6323.

Engels, N., Wollscheid, B., and Wienands, J. (2001). Association of SLP-65/BLNK with the B cell antigen receptor through a non-ITAM tyrosine of Igα. Eur J Immunol *31*, 2126-2134.

Fearon, D. T., and Carroll, M. C. (2000). Regulation of B lymphocyte responses to foreign and self-antigens by the CD19/CD21 complex. Annu Rev Immunol *18*, 393-422.

Feske, S., Okamura, H., Hogan, P. G., and Rao, A. (2003). Ca^{2+}/calcineurin signalling in cells of the immune system. Biochem Biophys Res Commun *311*, 1117-1132.

Fruman, D. A., Snapper, S. B., Yballe, C. M., Davidson, L., Yu, J. Y., Alt, F. W., and Cantley, L. C. (1999). Impaired B cell development and proliferation in absence of phosphoinositide 3-kinase p85α. Science *283*, 393-397.

Fu, C., Turck, C. W., Kurosaki, T., and Chan, A. C. (1998). BLNK: a central linker protein in B cell activation. Immunity *9*, 93-103.

Gaide, O., Favier, B., Legler, D. F., Bonnet, D., Brissoni, B., Valitutti, S., Bron, C., Tschopp, J., and Thome, M. (2002). CARMA1 is a critical lipid raft-associated regulator of TCR-induced NF-κB activation. Nat Immunol *3*, 836-843.

Gauld, S. B., and Cambier, J. C. (2004). Src-family kinases in B-cell development and signaling. Oncogene *23*, 8001-8006.

Ghosh, S., and Karin, M. (2002). Missing pieces in the NF-κB puzzle. Cell *109 Suppl*, S81-96.

Glynne, R., Akkaraju, S., Healy, J. I., Rayner, J., Goodnow, C. C., and Mack, D. H. (2000). How self-tolerance and the immunosuppressive drug FK506 prevent B-cell mitogenesis. Nature *403*, 672-676.

Gong, S., and Nussenzweig, M. C. (1996). Regulation of an early developmental checkpoint in the B cell pathway by Igβ. Science *272*, 411-414.

Guo, B., Su, T. T., and Rawlings, D. J. (2004). Protein kinase C family functions in B-cell activation. Curr Opin Immunol *16*, 367-373.

Hara, H., Wada, T., Bakal, C., Kozieradzki, I., Suzuki, S., Suzuki, N., Nghiem, M., Griffiths, E. K., Krawczyk, C., Bauer, B., et al. (2003). The MAGUK family protein CARD11 is essential for lymphocyte activation. Immunity *18*, 763-775.

Hashimoto, A., Okada, H., Jiang, A., Kurosaki, M., Greenberg, S., Clark, E. A., and Kurosaki, T. (1998). Involvement of guanosine triphosphatases and phospholipase C-γ2 in extracellular signal-regulated kinase, c-Jun NH2-terminal kinase, and p38 mitogen-activated protein kinase activation by the B cell antigen receptor. J Exp Med *188*, 1287-1295.

Hashimoto, A., Takeda, K., Inaba, M., Sekimata, M., Kaisho, T., Ikehara, S., Homma, Y., Akira, S., and Kurosaki, T. (2000). Cutting edge: essential role of phospholipase C-γ2 in B cell development and function. J Immunol *165*, 1738-1742.

Hashimoto, S., Iwamatsu, A., Ishiai, M., Okawa, K., Yamadori, T., Matsushita, M., Baba, Y., Kishimoto, T., Kurosaki, T., and Tsukada, S. (1999). Identification of the SH2 domain binding protein of Bruton's tyrosine kinase as BLNK-functional significance of Btk-SH2 domain in B-cell antigen receptor-coupled calcium signaling. Blood *94*, 2357-2364.

Hata, A., Sabe, H., Kurosaki, T., Takata, M., and Hanafusa, H. (1994). Functional analysis of Csk in signal transduction through the B-cell antigen receptor. Mol Cell Biol *14*, 7306-7313.

Hayami, K., Fukuta, D., Nishikawa, Y., Yamashita, Y., Inui, M., Ohyama, Y., Hikida, M., Ohmori, H., and Takai, T. (1997). Molecular cloning of a novel murine cell-surface glycoprotein homologous to killer cell inhibitory receptors. J Biol Chem *272*, 7320-7327.

Healy, J. I., Dolmetsch, R. E., Timmerman, L. A., Cyster, J. G., Thomas, M. L., Crabtree, G. R., Lewis, R. S., and Goodnow, C. C. (1997). Different nuclear signals are activated by the B cell receptor during positive versus negative signaling. Immunity *6*, 419-428.

Heissmeyer, V., Macian, F., Im, S. H., Varma, R., Feske, S., Venuprasad, K., Gu, H., Liu, Y. C., Dustin, M. L., and Rao, A. (2004). Calcineurin imposes T cell unresponsiveness through targeted proteolysis of signaling proteins. Nat Immunol *5*, 255-265.

Helgason, C. D., Kalberer, C. P., Damen, J. E., Chappel, S. M., Pineault, N., Krystal, G., and Humphries, R. K. (2000). A dual role for Src homology 2 domain-containing inositol-5-phosphatase (SHIP) in immunity: aberrant development and enhanced function of b lymphocytes in ship -/- mice. J Exp Med *191*, 781-794.

Heltemes-Harris, L., Liu, X., and Manser, T. (2004). Progressive surface B cell antigen receptor down-regulation accompanies efficient development of antinuclear antigen B cells to mature, follicular phenotype. J Immunol *172*, 823-833.

Henderson, A., and Calame, K. (1998). Transcriptional regulation during B cell development. Annu Rev Immunol *16*, 163-200.

Ho, L. H., Uehara, T., Chen, C. C., Kubagawa, H., and Cooper, M. D. (1999). Constitutive tyrosine phosphorylation of the inhibitory paired Ig-like receptor PIR-B. Proc Natl Acad Sci U S A *96*, 15086-15090.

Hoenderop, J. G., Vennekens, R., Muller, D., Prenen, J., Droogmans, G., Bindels, R. J., and Nilius, B. (2001). Function and expression of the epithelial Ca^{2+} channel family: comparison of mammalian ECaC1 and 2. J Physiol *537*, 747-761.

Hogan, P. G., Chen, L., Nardone, J., and Rao, A. (2003). Transcriptional regulation by calcium, calcineurin, and NFAT. Genes Dev *17*, 2205-2232.

Humphries, L. A., Dangelmaier, C., Sommer, K., Kipp, K., Kato, R. M., Griffith, N., Bakman, I., Turk, C. W., Daniel, J. L., and Rawlings, D. J. (2004). Tec kinases mediate sustained calcium influx via site-specific tyrosine phosphorylation of the phospholipase Cγ Src homology 2-Src homology 3 linker. J Biol Chem *279*, 37651-37661.

Huse, M., and Kuriyan, J. (2002). The conformational plasticity of protein kinases. Cell *109*, 275-282.

Hutchcroft, J. E., Harrison, M. L., and Geahlen, R. L. (1992). Association of the 72-kDa protein-tyrosine kinase PTK72 with the B cell antigen receptor. J Biol Chem *267*, 8613-8619.

Inabe, K., Ishiai, M., Scharenberg, A. M., Freshney, N., Downward, J., and Kurosaki, T. (2002). Vav3 modulates B cell receptor responses by regulating phosphoinositide 3-kinase activation. J Exp Med *195*, 189-200.

Iritani, B. M., Forbush, K. A., Farrar, M. A., and Perlmutter, R. M. (1997). Control of B cell development by Ras-mediated activation of Raf. EMBO J *16*, 7019-7031.

Ishiai, M., Kurosaki, M., Inabe, K., Chan, A. C., Sugamura, K., and Kurosaki, T. (2000). Involvement of LAT, Gads, and Grb2 in compartmentation of SLP-76 to the plasma membrane. J Exp Med *192*, 847-856.

Ishiai, M., Kurosaki, M., Pappu, R., Okawa, K., Ronko, I., Fu, C., Shibata, M., Iwamatsu, A., Chan, A. C., and Kurosaki, T. (1999a). BLNK required for coupling Syk to PLC-γ2 and Rac1-JNK in B cells. Immunity *10*, 117-125.

Ishiai, M., Sugawara, H., Kurosaki, M., and Kurosaki, T. (1999b). Cutting edge: association of phospholipase C-γ2 Src homology 2 domains with BLNK is critical for B cell antigen receptor signaling. J Immunol *163*, 1746-1749.

Joazeiro, C. A., Wing, S. S., Huang, H., Leverson, J. D., Hunter, T., and Liu, Y. C. (1999). The tyrosine kinase negative regulator c-Cbl as a RING-type, E2-dependent ubiquitin-protein ligase. Science *286*, 309-312.

Johmura, S., Oh-hora, M., Inabe, K., Nishikawa, Y., Hayashi, K., Vigorito, E., Kitamura, D., Turner, M., Shingu, K., Hikida, M., and Kurosaki, T. (2003). Regulation of Vav localization in membrane rafts by adaptor molecules Grb2 and BLNK. Immunity *18*, 777-787.

Johnson, S. A., Pleiman, C. M., Pao, L., Schneringer, J., Hippen, K., and Cambier, J. C. (1995). Phosphorylated immunoreceptor signaling motifs (ITAMs) exhibit unique abilities to bind and activate Lyn and Syk tyrosine kinases. J Immunol *155*, 4596-4603.

Jou, S. T., Carpino, N., Takahashi, Y., Piekorz, R., Chao, J. R., Carpino, N., Wang, D., and Ihle, J. N. (2002). Essential, nonredundant role for the phosphoinositide 3-kinase p110δ in signaling by the B-cell receptor complex. Mol Cell Biol *22*, 8580-8591.

Jun, J. E., Wilson, L. E., Vinuesa, C. G., Lesage, S., Blery, M., Miosge, L. A., Cook, M. C., Kucharska, E. M., Hara, H., Penninger, J. M., et al. (2003). Identifying the MAGUK protein Carma-1 as a central regulator of humoral immune responses and atopy by genome-wide mouse mutagenesis. Immunity *18*, 751-762.

Kang, S. W., Wahl, M. I., Chu, J., Kitaura, J., Kawakami, Y., Kato, R. M., Tabuchi, R., Tarakhovsky, A., Kawakami, T., Turck, C. W., et al. (2001). PKCβ modulates antigen receptor signaling via regulation of Btk membrane localization. EMBO J *20*, 5692-5702.

Kanzaki, M., Zhang, Y. Q., Mashima, H., Li, L., Shibata, H., and Kojima, I. (1999). Translocation of a calcium-permeable cation channel induced by insulin-like growth factor-I. Nat Cell Biol *1*, 165-170.

Kehlenbach, R. H., Dickmanns, A., and Gerace, L. (1998). Nucleocytoplasmic shuttling factors including Ran and CRM1 mediate nuclear export of NFAT *In vitro*. J Cell Biol *141*, 863-874.

Keshvara, L. M., Isaacson, C. C., Yankee, T. M., Sarac, R., Harrison, M. L., and Geahlen, R. L. (1998). Syk- and Lyn-dependent phosphorylation of Syk on multiple tyrosines following B cell activation includes a site that negatively regulates signaling. J Immunol *161*, 5276-5283.

Kiani, A., Rao, A., and Aramburu, J. (2000). Manipulating immune responses with immunosuppressive agents that target NFAT. Immunity *12*, 359-372.

Kim, Y. J., Sekiya, F., Poulin, B., Bae, Y. S., and Rhee, S. G. (2004). Mechanism of B-cell receptor-induced phosphorylation and activation of phospholipase C-γ2. Mol Cell Biol *24*, 9986-9999.

Kiselyov, K. I., Shin, D. M., Wang, Y., Pessah, I. N., Allen, P. D., and Muallem, S. (2000). Gating of store-operated channels by conformational coupling to ryanodine receptors. Mol Cell *6*, 421-431.

Klemm, J. D., Beals, C. R., and Crabtree, G. R. (1997). Rapid targeting of nuclear proteins to the cytoplasm. Curr Biol *7*, 638-644.

Kohler, F., Storch, B., Kulathu, Y., Herzog, S., Kuppig, S., Reth, M., and Jumaa, H. (2005). A leucine zipper in the N terminus confers membrane association to SLP-65. Nat Immunol *6*, 204-210.

Kojima, T., Fukuda, M., Watanabe, Y., Hamazato, F., and Mikoshiba, K. (1997). Characterization of the pleckstrin homology domain of Btk as an inositol polyphosphate and phosphoinositide binding domain. Biochem Biophys Res Commun *236*, 333-339.

Kraus, M., Alimzhanov, M. B., Rajewsky, N., and Rajewsky, K. (2004). Survival of resting mature B lymphocytes depends on BCR signaling via the Igα/β heterodimer. Cell *117*, 787-800.

Kraus, M., Pao, L. I., Reichlin, A., Hu, Y., Canono, B., Cambier, J. C., Nussenzweig, M. C., and Rajewsky, K. (2001). Interference with immunoglobulin Igα immunoreceptor tyrosine-based activation motif (ITAM) phosphorylation modulates or blocks B cell development, depending on the availability of an Igb cytoplasmic tail. J Exp Med *194*, 455-469.

Kubagawa, H., Burrows, P. D., and Cooper, M. D. (1997). A novel pair of immunoglobulin-like receptors expressed by B cells and myeloid cells. Proc Natl Acad Sci U S A *94*, 5261-5266.

Kurosaki, T. (1999). Genetic analysis of B cell antigen receptor signaling. Annu Rev Immunol *17*, 555-592.

Kurosaki, T., Johnson, S. A., Pao, L., Sada, K., Yamamura, H., and Cambier, J. C. (1995). Role of the Syk autophosphorylation site and SH2 domains in B cell antigen receptor signaling. J Exp Med *182*, 1815-1823.

Kurosaki, T., and Kurosaki, M. (1997). Transphosphorylation of Bruton's tyrosine kinase on tyrosine 551 is critical for B cell antigen receptor function. J Biol Chem *272*, 15595-15598.

Kurosaki, T., Maeda, A., Ishiai, M., Hashimoto, A., Inabe, K., and Takata, M. (2000). Regulation of the phospholipase C-γ2 pathway in B cells. Immunol Rev *176*, 19-29.

Kurosaki, T., Takata, M., Yamanashi, Y., Inazu, T., Taniguchi, T., Yamamoto, T., and Yamamura, H. (1994). Syk activation by the Src-family tyrosine kinase in the B cell receptor signaling. J Exp Med *179*, 1725-1729.

Law, D. A., Chan, V. W., Datta, S. K., and DeFranco, A. L. (1993). B-cell antigen receptor motifs have redundant signalling capabilities and bind the tyrosine kinases PTK72, Lyn and Fyn. Curr Biol *3*, 645-657.

Levkowitz, G., Waterman, H., Ettenberg, S. A., Katz, M., Tsygankov, A. Y., Alroy, I., Lavi, S., Iwai, K., Reiss, Y., Ciechanover, A., et al. (1999). Ubiquitin ligase activity and tyrosine phosphorylation underlie suppression of growth factor signaling by c-Cbl/Sli-1. Mol Cell *4*, 1029-1040.

Levkowitz, G., Waterman, H., Zamir, E., Kam, Z., Oved, S., Langdon, W. Y., Beguinot, L., Geiger, B., and Yarden, Y. (1998). c-Cbl/Sli-1 regulates endocytic sorting and ubiquitination of the epidermal growth factor receptor. Genes Dev *12*, 3663-3674.

Li, Q., and Verma, I. M. (2002). NF-κB regulation in the immune system. Nat Rev Immunol *2*, 725-734.

Li, T., Tsukada, S., Satterthwaite, A., Havlik, M. H., Park, H., Takatsu, K., and Witte, O. N. (1995). Activation of Bruton's tyrosine kinase (BTK) by a point mutation in its pleckstrin homology (PH) domain. Immunity *2*, 451-460.

Li, W., Llopis, J., Whitney, M., Zlokarnik, G., and Tsien, R. Y. (1998). Cell-permeant caged InsP3 ester shows that Ca^{2+} spike frequency can optimize gene expression. Nature *392*, 936-941.

Lin, X., and Wang, D. (2004). The roles of CARMA1, Bcl10, and MALT1 in antigen receptor signaling. Semin Immunol *16*, 429-435.

Liou, J., Kiefer, F., Dang, A., Hashimoto, A., Cobb, M. H., Kurosaki, T., and Weiss, A. (2000). HPK1 is activated by lymphocyte antigen receptors and negatively regulates AP-1. Immunity *12*, 399-408.

Liou, J., Kim, M. L., Heo, W. D., Jones, J. T., Myers, J. W., Ferrell, J. E., Jr., and Meyer, T. (2005). STIM is a Ca^{2+} sensor essential for Ca^{2+}-store-depletion-triggered Ca^{2+} influx. Curr Biol *15*, 1235-1241.

Liu, Q., Oliveira-Dos-Santos, A. J., Mariathasan, S., Bouchard, D., Jones, J., Sarao, R., Kozieradzki, I., Ohashi, P. S., Penninger, J. M., and Dumont, D. J. (1998). The inositol polyphosphate 5-phosphatase ship is a crucial negative regulator of B cell antigen receptor signaling. J Exp Med *188*, 1333-1342.

Liu, Y. C. (2004). Ubiquitin ligases and the immune response. Annu Rev Immunol *22*, 81-127.

Lu, Q., Hope, L. W., Brasch, M., Reinhard, C., and Cohen, S. N. (2003). TSG101 interaction with HRS mediates endosomal trafficking and receptor down-regulation. Proc Natl Acad Sci U S A *100*, 7626-7631.

Macian, F., Garcia-Cozar, F., Im, S. H., Horton, H. F., Byrne, M. C., and Rao, A. (2002). Transcriptional mechanisms underlying lymphocyte tolerance. Cell *109*, 719-731.

Maeda, A., Kurosaki, M., Ono, M., Takai, T., and Kurosaki, T. (1998). Requirement of SH2-containing protein tyrosine phosphatases SHP-1 and SHP-2 for paired immunoglobulin-like receptor B (PIR-B)-mediated inhibitory signal. J Exp Med *187*, 1355-1360.

Maeda, A., Scharenberg, A. M., Tsukada, S., Bolen, J. B., Kinet, J. P., and Kurosaki, T. (1999). Paired immunoglobulin-like receptor B (PIR-B) inhibits BCR-induced activation of Syk and Btk by SHP-1. Oncogene *18*, 2291-2297.

Mahajan, S., Fargnoli, J., Burkhardt, A. L., Kut, S. A., Saouaf, S. J., and Bolen, J. B. (1995). Src family protein tyrosine kinases induce autoactivation of Bruton's tyrosine kinase. Mol Cell Biol *15*, 5304-5311.

Manetz, T. S., Gonzalez-Espinosa, C., Arudchandran, R., Xirasagar, S., Tybulewicz, V., and Rivera, J. (2001). Vav1 regulates phospholipase cγ activation and calcium responses in mast cells. Mol Cell Biol *21*, 3763-3774.

Marchant, J. S. (2005). Cellular signalling: STIMulating calcium entry. Curr Biol *15*, R493-495.

Miyakawa, T., Maeda, A., Yamazawa, T., Hirose, K., Kurosaki, T., and Iino, M. (1999). Encoding of Ca^{2+} signals by differential expression of IP3 receptor subtypes. EMBO J *18*, 1303-1308.

Miyakawa, T., Mizushima, A., Hirose, K., Yamazawa, T., Bezprozvanny, I., Kurosaki, T., and Iino, M. (2001). Ca(2+)-sensor region of IP(3) receptor controls intracellular Ca^{2+} signaling. EMBO J *20*, 1674-1680.

Montell, C. (2004). Exciting trips for TRPs. Nat Cell Biol *6*, 690-692.

Mori, Y., Wakamori, M., Miyakawa, T., Hermosura, M., Hara, Y., Nishida, M., Hirose, K., Mizushima, A., Kurosaki, M., Mori, E., et al. (2002). Transient receptor potential 1 regulates capacitative Ca^{2+} entry and Ca^{2+} release from endoplasmic reticulum in B lymphocytes. J Exp Med *195*, 673-681.

Morris, S. C., Moroldo, M., Giannini, E. H., Orekhova, T., and Finkelman, F. D. (2000). *In vivo* survival of autoreactive B cells: characterization of long-lived B cells. J Immunol *164*, 3035-3046.

Morrogh, L. M., Hinshelwood, S., Costello, P., Cory, G. O., and Kinnon, C. (1999). The SH3 domain of Bruton's tyrosine kinase displays altered ligand binding properties when auto-phosphorylated in vitro. Eur J Immunol *29*, 2269-2279.

Nagaoka, H., Takahashi, Y., Hayashi, R., Nakamura, T., Ishii, K., Matsuda, J., Ogura, A., Shirakata, Y., Karasuyama, H., Sudo, T., et al. (2000). Ras mediates effector pathways responsible for pre-B cell survival, which is essential for the developmental progression to the late pre-B cell stage. J Exp Med *192*, 171-182.

Nakamura, A., Kobayashi, E., and Takai, T. (2004). Exacerbated graft-versus-host disease in Pirb-/- mice. Nat Immunol *5*, 623-629.

Newton, C. L., Mignery, G. A., and Sudhof, T. C. (1994). Co-expression in vertebrate tissues and cell lines of multiple inositol 1,4,5-trisphosphate (InsP3) receptors with distinct affinities for InsP3. J Biol Chem *269*, 28613-28619.

Newton, K., and Dixit, V. M. (2003). Mice lacking the CARD of CARMA1 exhibit defective B lymphocyte development and impaired proliferation of their B and T lymphocytes. Curr Biol *13*, 1247-1251.

Nishida, M., Sugimoto, K., Hara, Y., Mori, E., Morii, T., Kurosaki, T., and Mori, Y. (2003). Amplification of receptor signalling by Ca^{2+} entry-mediated translocation and activation of PLC-γ2 in B lymphocytes. EMBO J *22*, 4677-4688.

Oh-hora, M., Johmura, S., Hashimoto, A., Hikida, M., and Kurosaki, T. (2003). Requirement for Ras guanine nucleotide releasing protein 3 in coupling phospholipase C-γ2 to Ras in B cell receptor signaling. J Exp Med *198*, 1841-1851.

Okada, H., Bolland, S., Hashimoto, A., Kurosaki, M., Kabuyama, Y., Iino, M., Ravetch, J. V., and Kurosaki, T. (1998). Role of the inositol phosphatase SHIP in B cell receptor-induced Ca^{2+} oscillatory response. J Immunol *161*, 5129-5132.

Okada, T., Maeda, A., Iwamatsu, A., Gotoh, K., and Kurosaki, T. (2000). BCAP: the tyrosine kinase substrate that connects B cell receptor to phosphoinositide 3-kinase activation. Immunity *13*, 817-827.

Okamura, H., Aramburu, J., Garcia-Rodriguez, C., Viola, J. P., Raghavan, A., Tahiliani, M., Zhang, X., Qin, J., Hogan, P. G., and Rao, A. (2000). Concerted dephosphorylation of the transcription factor NFAT1 induces a conformational switch that regulates transcriptional activity. Mol Cell *6*, 539-550.

Ono, M., Bolland, S., Tempst, P., and Ravetch, J. V. (1996). Role of the inositol phosphatase SHIP in negative regulation of the immune system by the receptor FcγRIIB. Nature *383*, 263-266.

Ono, M., Okada, H., Bolland, S., Yanagi, S., Kurosaki, T., and Ravetch, J. V. (1997). Deletion of SHIP or SHP-1 reveals two distinct pathways for inhibitory signaling. Cell *90*, 293-301.

Parekh, A. B., and Penner, R. (1997). Store depletion and calcium influx. Physiol Rev *77*, 901-930.

Pike, K. A., and Ratcliffe, M. J. (2005). Dual requirement for the Igα immunoreceptor tyrosine-based activation motif (ITAM) and a conserved non-Igα ITAM tyrosine in supporting Igα β-mediated B cell development. J Immunol *174*, 2012-2020.

Pleiman, C. M., Abrams, C., Gauen, L. T., Bedzyk, W., Jongstra, J., Shaw, A. S., and Cambier, J. C. (1994). Distinct p53/56lyn and p59fyn domains associate with nonphosphorylated and phosphorylated Igα. Proc Natl Acad Sci U S A *91*, 4268-4272.

Pomerantz, J. L., Denny, E. M., and Baltimore, D. (2002). CARD11 mediates factor-specific activation of NF-κB by the T cell receptor complex. EMBO J *21*, 5184-5194.

Putney, J. W., Jr. (2005). Capacitative calcium entry: sensing the calcium stores. J Cell Biol *169*, 381-382.

Rameh, L. E., Arvidsson, A., Carraway, K. L., 3rd, Couvillon, A. D., Rathbun, G., Crompton, A., VanRenterghem, B., Czech, M. P., Ravichandran, K. S., Burakoff, S. J., et al. (1997). A comparative analysis of the phosphoinositide binding specificity of pleckstrin homology domains. J Biol Chem *272*, 22059-22066.

Randriamampita, C., and Tsien, R. Y. (1993). Emptying of intracellular Ca^{2+} stores releases a novel small messenger that stimulates Ca^{2+} influx. Nature *364*, 809-814.

Randriamampita, C., and Tsien, R. Y. (1995). Degradation of a calcium influx factor (CIF) can be blocked by phosphatase inhibitors or chelation of Ca^{2+}. J Biol Chem *270*, 29-32.

Rao, A., Luo, C., and Hogan, P. G. (1997). Transcription factors of the NFAT family: regulation and function. Annu Rev Immunol *15*, 707-747.

Rawlings, D. J., Scharenberg, A. M., Park, H., Wahl, M. I., Lin, S., Kato, R. M., Fluckiger, A. C., Witte, O. N., and Kinet, J. P. (1996). Activation of BTK by a phosphorylation mechanism initiated by SRC family kinases. Science *271*, 822-825.

Reichlin, A., Gazumyan, A., Nagaoka, H., Kirsch, K. H., Kraus, M., Rajewsky, K., and Nussenzweig, M. C. (2004). A B cell receptor with two Igkα cytoplasmic domains supports development of mature but anergic B cells. J Exp Med *199*, 855-865.

Reichlin, A., Hu, Y., Meffre, E., Nagaoka, H., Gong, S., Kraus, M., Rajewsky, K., and Nussenzweig, M. C. (2001). B cell development is arrested at the immature B cell stage in mice carrying a mutation in the cytoplasmic domain of immunoglobulin β. J Exp Med *193*, 13-23.

Reth, M. (1989). Antigen receptor tail clue. Nature *338*, 383-384.

Reuther, G. W., and Der, C. J. (2000). The Ras branch of small GTPases: Ras family members don't fall far from the tree. Curr Opin Cell Biol *12*, 157-165.

Rolli, V., Gallwitz, M., Wossning, T., Flemming, A., Schamel, W. W., Zurn, C., and Reth, M. (2002). Amplification of B cell antigen receptor signaling by a Syk/ITAM positive feedback loop. Mol Cell *10*, 1057-1069.

Roos, J., DiGregorio, P. J., Yeromin, A. V., Ohlsen, K., Lioudyno, M., Zhang, S., Safrina, O., Kozak, J. A., Wagner, S. L., Cahalan, M. D., et al. (2005). STIM1, an essential and conserved component of store-operated Ca^{2+} channel function. J Cell Biol *169*, 435-445.

Ruefli-Brasse, A. A., French, D. M., and Dixit, V. M. (2003). Regulation of NF-κB-dependent lymphocyte activation and development by paracaspase. Science *302*, 1581-1584.

Ruland, J., Duncan, G. S., Wakeham, A., and Mak, T. W. (2003). Differential requirement for Malt1 in T and B cell antigen receptor signaling. Immunity *19*, 749-758.

Rzigalinski, B. A., Willoughby, K. A., Hoffman, S. W., Falck, J. R., and Ellis, E. F. (1999). Calcium influx factor, further evidence it is 5, 6-epoxyeicosatrienoic acid. J Biol Chem *274*, 175-182.

Saijo, K., Mecklenbrauker, I., Santana, A., Leitger, M., Schmedt, C., and Tarakhovsky, A. (2002). Protein kinase C b controls nuclear factor κB activation in B cells through selective regulation of the IκB kinase α. J Exp Med *195*, 1647-1652.

Saito, K., Tolias, K. F., Saci, A., Koon, H. B., Humphries, L. A., Scharenberg, A., Rawlings, D. J., Kinet, J. P., and Carpenter, C. L. (2003). BTK regulates PtdIns-4,5-P2 synthesis: importance for calcium signaling and PI3K activity. Immunity *19*, 669-678.

Salim, K., Bottomley, M. J., Querfurth, E., Zvelebil, M. J., Gout, I., Scaife, R., Margolis, R. L., Gigg, R., Smith, C. I., Driscoll, P. C., et al. (1996). Distinct specificity in the recognition of phosphoinositides by the pleckstrin homology domains of dynamin and Bruton's tyrosine kinase. EMBO J *15*, 6241-6250.

Sanchez, M., Misulovin, Z., Burkhardt, A. L., Mahajan, S., Costa, T., Franke, R., Bolen, J. B., and Nussenzweig, M. (1993). Signal transduction by immunoglobulin is mediated through Igα and Igβ. J Exp Med *178*, 1049-1055.

Saouaf, S. J., Mahajan, S., Rowley, R. B., Kut, S. A., Fargnoli, J., Burkhardt, A. L., Tsukada, S., Witte, O. N., and Bolen, J. B. (1994). Temporal differences in the activation of three classes of non-transmembrane protein tyrosine kinases following B-cell antigen receptor surface engagement. Proc Natl Acad Sci U S A *91*, 9524-9528.

Shibasaki, F., Price, E. R., Milan, D., and McKeon, F. (1996). Role of kinases and the phosphatase calcineurin in the nuclear shuttling of transcription factor NF-AT4. Nature *382*, 370-373.

Shinohara H, Yasuda T, Aiba Y, Sanjo H, Hamadate M, Watarai H, Sakurai H, and Kurosaki T (in press). PKC β regulates BCR-mediated IKK activation by facilitating the interaction between TAK1 and CARMA1. J Exp Med.

Siebenlist, U., Brown, K., and Claudio, E. (2005). Control of lymphocyte development by nuclear factor-κB. Nat Rev Immunol *5*, 435-445.

Siemasko, K., Skaggs, B. J., Kabak, S., Williamson, E., Brown, B. K., Song, W., and Clark, M. R. (2002). Receptor-facilitated antigen presentation requires the recruitment of B cell linker protein to Igα. J Immunol *168*, 2127-2138.

Simons, K., and Toomre, D. (2000). Lipid rafts and signal transduction. Nat Rev Mol Cell Biol *1*, 31-39.

Singh, D. K., Kumar, D., Siddiqui, Z., Basu, S. K., Kumar, V., and Rao, K. V. (2005). The strength of receptor signaling is centrally controlled through a cooperative loop between Ca^{2+} and an oxidant signal. Cell *121*, 281-293.

Smani, T., Zakharov, S. I., Csutora, P., Leno, E., Trepakova, E. S., and Bolotina, V. M. (2004). A novel mechanism for the store-operated calcium influx pathway. Nat Cell Biol *6*, 113-120.

Smani, T., Zakharov, S. I., Leno, E., Csutora, P., Trepakova, E. S., and Bolotina, V. M. (2003). Ca^{2+}-independent phospholipase A2 is a novel determinant of store-operated Ca^{2+} entry. J Biol Chem *278*, 11909-11915.

Sohn, H. W., Gu, H., and Pierce, S. K. (2003). Cbl-b negatively regulates B cell antigen receptor signaling in mature B cells through ubiquitination of the tyrosine kinase Syk. J Exp Med *197*, 1511-1524.

Su, T. T., Guo, B., Kawakami, Y., Sommer, K., Chae, K., Humphries, L. A., Kato, R. M., Kang, S., Patrone, L., Wall, R., et al. (2002). PKC-β controls IκB kinase lipid raft recruitment and activation in response to BCR signaling. Nat Immunol *3*, 780-786.

Sugawara, H., Kurosaki, M., Takata, M., and Kurosaki, T. (1997). Genetic evidence for involvement of type 1, type 2 and type 3 inositol 1,4,5-trisphosphate receptors in signal transduction through the B-cell antigen receptor. EMBO J *16*, 3078-3088.

Suzuki, A., Kaisho, T., Ohishi, M., Tsukio-Yamaguchi, M., Tsubata, T., Koni, P. A., Sasaki, T., Mak, T. W., and Nakano, T. (2003). Critical roles of Pten in B cell homeostasis and immunoglobulin class switch recombination. J Exp Med *197*, 657-667.

Suzuki, H., Terauchi, Y., Fujiwara, M., Aizawa, S., Yazaki, Y., Kadowaki, T., and Koyasu, S. (1999). Xid-like immunodeficiency in mice with disruption of the p85α subunit of phosphoinositide 3-kinase. Science *283*, 390-392.

Takai, T. (2005). Paired immunoglobulin-like receptors and their MHC class I recognition. Immunology *115*, 433-440.

Takata, M., Homma, Y., and Kurosaki, T. (1995). Requirement of phospholipase C-γ2 activation in surface immunoglobulin M-induced B cell apoptosis. J Exp Med *182*, 907-914.

Takata, M., and Kurosaki, T. (1996). A role for Bruton's tyrosine kinase in B cell antigen receptor-mediated activation of phospholipase C-γ2. J Exp Med *184*, 31-40.

Takata, M., Sabe, H., Hata, A., Inazu, T., Homma, Y., Nukada, T., Yamamura, H., and Kurosaki, T. (1994). Tyrosine kinases Lyn and Syk regulate B cell receptor-coupled Ca^{2+} mobilization through distinct pathways. EMBO J *13*, 1341-1349.

Takemura, H., and Ohshika, H. (1999). Capacitative Ca^{2+} entry involves Ca^{2+} influx factor in rat glioma C6 cells. Life Sci *64*, 1493-1500.

Tamir, I., Stolpa, J. C., Helgason, C. D., Nakamura, K., Bruhns, P., Daeron, M., and Cambier, J. C. (2000). The RasGAP-binding protein p62dok is a mediator of inhibitory FcγRIIB signals in B cells. Immunity *12*, 347-358.

Tarakhovsky, A. (1997). Xid and Xid-like immunodeficiencies from a signaling point of view. Curr Opin Immunol *9*, 319-323.

Tedder, T. F., Inaoki, M., and Sato, S. (1997). The CD19-CD21 complex regulates signal transduction thresholds governing humoral immunity and autoimmunity. Immunity *6*, 107-118.

Teixeira, C., Stang, S. L., Zheng, Y., Beswick, N. S., and Stone, J. C. (2003). Integration of DAG signaling systems mediated by PKC-dependent phosphorylation of RasGRP3. Blood *102*, 1414-1420.

Thien, C. B., and Langdon, W. Y. (2001). Cbl: many adaptations to regulate protein tyrosine kinases. Nat Rev Mol Cell Biol *2*, 294-307.

Thomas, D., and Hanley, M. R. (1995). Evaluation of calcium influx factors from stimulated Jurkat T-lymphocytes by microinjection into Xenopus oocytes. J Biol Chem *270*, 6429-6432.

Thome, M. (2004). CARMA1, BCL-10 and MALT1 in lymphocyte development and activation. Nat Rev Immunol *4*, 348-359.

Tomida, T., Hirose, K., Takizawa, A., Shibasaki, F., and Iino, M. (2003). NFAT functions as a working memory of Ca^{2+} signals in decoding Ca^{2+} oscillation. EMBO J *22*, 3825-3832.

Tonks, N. K. (2005). Redox redux: revisiting PTPs and the control of cell signaling. Cell *121*, 667-670.

Torres, R. M., Flaswinkel, H., Reth, M., and Rajewsky, K. (1996). Aberrant B cell development and immune response in mice with a compromised BCR complex. Science *272*, 1804-1808.

Trepakova, E. S., Csutora, P., Hunton, D. L., Marchase, R. B., Cohen, R. A., and Bolotina, V. M. (2000). Calcium influx factor directly activates store-operated cation channels in vascular smooth muscle cells. J Biol Chem *275*, 26158-26163.

Turner, M., and Billadeau, D. D. (2002). VAV proteins as signal integrators for multi-subunit immune-recognition receptors. Nat Rev Immunol *2*, 476-486.

Uren, A. G., O'Rourke, K., Aravind, L. A., Pisabarro, M. T., Seshagiri, S., Koonin, E. V., and Dixit, V. M. (2000). Identification of paracaspases and metacaspases: two ancient families of caspase-like proteins, one of which plays a key role in MALT lymphoma. Mol Cell *6*, 961-967.

van de Graaf, S. F., Hoenderop, J. G., Gkika, D., Lamers, D., Prenen, J., Rescher, U., Gerke, V., Staub, O., Nilius, B., and Bindels, R. J. (2003). Functional expression of the epithelial Ca^{2+} channels (TRPV5 and TRPV6) requires association of the S100A10-annexin 2 complex. EMBO J *22*, 1478-1487.

van Rossum, D. B., Patterson, R. L., Kiselyov, K., Boehning, D., Barrow, R. K., Gill, D. L., and Snyder, S. H. (2004). Agonist-induced Ca^{2+} entry determined by inositol 1,4,5-trisphosphate recognition. Proc Natl Acad Sci U S A *101*, 2323-2327.

Vennekens, R., Hoenderop, J. G., Prenen, J., Stuiver, M., Willems, P. H., Droogmans, G., Nilius, B., and Bindels, R. J. (2000). Permeation and gating properties of the novel epithelial Ca^{2+} channel. J Biol Chem *275*, 3963-3969.

Voets, T., Prenen, J., Fleig, A., Vennekens, R., Watanabe, H., Hoenderop, J. G., Bindels, R. J., Droogmans, G., Penner, R., and Nilius, B. (2001). CaT1 and the calcium release-activated calcium channel manifest distinct pore properties. J Biol Chem *276*, 47767-47770.

Wang, D., Feng, J., Wen, R., Marine, J. C., Sangster, M. Y., Parganas, E., Hoffmeyer, A., Jackson, C. W., Cleveland, J. L., Murray, P. J., and Ihle, J. N. (2000). Phospholipase C-γ2 is essential in the functions of B cell and several Fc receptors. Immunity *13*, 25-35.

Watanabe, D., Hashimoto, S., Ishiai, M., Matsushita, M., Baba, Y., Kishimoto, T., Kurosaki, T., and Tsukada, S. (2001). Four tyrosine residues in phospholipase C-γ2, identified as Btk-dependent phosphorylation sites, are required for B cell antigen receptor-coupled calcium signaling. J Biol Chem *276*, 38595-38601.

Weiss, A., and Littman, D. R. (1994). Signal transduction by lymphocyte antigen receptors. Cell *76*, 263-274.

Wojcikiewicz, R. J., Furuichi, T., Nakade, S., Mikoshiba, K., and Nahorski, S. R. (1994). Muscarinic receptor activation down-regulates the type I inositol 1,4,5-trisphosphate receptor by accelerating its degradation. J Biol Chem *269*, 7963-7969.

Xu, X. Z., and Sternberg, P. W. (2003). A C. elegans sperm TRP protein required for sperm-egg interactions during fertilization. Cell *114*, 285-297.

Xu, Y., Harder, K. W., Huntington, N. D., Hibbs, M. L., and Tarlinton, D. M. (2005). Lyn tyrosine kinase: accentuating the positive and the negative. Immunity *22*, 9-18.

Xue, L., Morris, S. W., Orihuela, C., Tuomanen, E., Cui, X., Wen, R., and Wang, D. (2003). Defective development and function of Bcl10-deficient follicular, marginal zone and B1 B cells. Nat Immunol *4*, 857-865.

Yamanashi, Y., Tamura, T., Kanamori, T., Yamane, H., Nariuchi, H., Yamamoto, T., and Baltimore, D. (2000). Role of the rasGAP-associated docking protein p62(dok) in negative regulation of B cell receptor-mediated signaling. Genes Dev *14*, 11-16.

Yanagi, S., Sugawara, H., Kurosaki, M., Sabe, H., Yamamura, H., and Kurosaki, T. (1996). CD45 modulates phosphorylation of both autophosphorylation and negative regulatory tyrosines of Lyn in B cells. J Biol Chem *271*, 30487-30492.

Yasuda, T., Maeda, A., Kurosaki, M., Tezuka, T., Hironaka, K., Yamamoto, T., and Kurosaki, T. (2000). Cbl suppresses B cell receptor-mediated phospholipase C (PLC)-γ2 activation by regulating B cell linker protein-PLC-γ2 binding. J Exp Med *191*, 641-650.

Yasuda, T., Tezuka, T., Maeda, A., Inazu, T., Yamanashi, Y., Gu, H., Kurosaki, T., and Yamamoto, T. (2002). Cbl-b positively regulates Btk-mediated activation of phospholipase C-γ2 in B cells. J Exp Med *196*, 51-63.

Yokouchi, M., Kondo, T., Houghton, A., Bartkiewicz, M., Horne, W. C., Zhang, H., Yoshimura, A., and Baron, R. (1999). Ligand-induced ubiquitination of the epidermal growth factor receptor involves the interaction of the c-Cbl RING finger and UbcH7. J Biol Chem *274*, 31707-31712.

Yuan, J. P., Kiselyov, K., Shin, D. M., Chen, J., Shcheynikov, N., Kang, S. H., Dehoff, M. H., Schwarz, M. K., Seeburg, P. H., Muallem, S., and Worley, P. F. (2003). Homer binds TRPC family channels and is required for gating of TRPC1 by IP3 receptors. Cell *114*, 777-789.

Zheng, Y., Liu, H., Coughlin, J., Zheng, J., Li, L., and Stone, J. C. (2005). Phosphorylation of RasGRP3 on threonine 133 provides a mechanistic link between PKC and Ras signaling systems in B cells. Blood *105*, 3648-3654.

Chapter 11

DT40 MUTANTS: A MODEL TO STUDY TRANSCRIPTIONAL REGULATION OF B CELL DEVELOPMENT AND FUNCTION

Jukka Alinikula, Olli Lassila and Kalle-Pekka Nera
Turku Graduate School of Biomedical Sciences, Department of Medical Microbiology, University of Turku, Kiinamyllynkatu 13, 20520 Turku, Finland

Abstract: A key issue in understanding the hematopoietic system and B cell biology is to define the function of transcription factors. B lymphocyte development and function is controlled by a hierarchy of transcription factors including PU.1, Ikaros, E2A, EBF, Pax5 and Aiolos. Mouse knockout models provide information about the developmental and physiological importance of the disrupted gene. However, an early block in the development or a lethal phenotype prevents the studies of the functional importance of the gene at the later developing system such as the immune system. The chicken B cell line DT40 is used to circumvent these problems. Studies with DT40 have revealed a role for Ikaros transcription factor in B cell receptor signaling and Aiolos has been shown to regulate immunoglobulin gene conversion and cell survival. On the other hand, findings with Pax5 deficient mutants support DT40 targeting system as a valid model for the plasma cell differentiation and demonstrate the genetic plasticity of the cell line. This system is an excellent model to study transcription factors in B cell specific functions, antibody production and B cell differentiation.

Key words: B cell; transcription factor; PU.1; Ikaros; Pax5; Aiolos; DT40; chicken; bursa of Fabricius.

1.   INTRODUCTION

The generation of an efficient immune system in vertebrates involves development of antibody producing B cells. One of the main issues in understanding the hematolymphoid system is to define the functions of

J.-M. Buerstedde and S. Takeda (eds.), Reviews and Protocols in DT40 Research, 189–205.
© 2006 *Springer.*

regulatory genes. Germline gene targeting in the mouse has been used extensively in the studies of transcriptional regulation of B lymphocyte differentiation and function (Busslinger, 2004; Matthias and Rolink, 2005). While mouse knockout models provide information about the developmental and physiological importance of the disrupted gene, a developmental block or a lethal phenotype prevents the functional studies of the B cell specific genes. Moreover, the germline gene disruptions may create partial developmental blocks that lead to selection for abnormal phenotypes. Due to these disadvantages, alternative methods are required for the functional studies of transcriptional regulators. Conditional gene inactivation at later stages of B cell development in mice is an achievable solution. However, so far only few B cell specific conditional knockouts have been reported (Busslinger, 2004; Matthias and Rolink, 2005). DT40 B cell line has been used to overcome the disadvantages of the mouse model with very promising results.

B cell development is guided by a hierarchical gene regulatory network involving various transcription factors such as PU.1, Ikaros, E2A, early B cell factor (EBF) and paired box protein 5 (Pax5) (Medina et al., 2004). As the gene disruption of these factors leads to a total developmental block at early stages in the B cell differentiation (Bain et al., 1994; Scott et al., 1994; Urbanek et al., 1994; Zhuang et al., 1994; Lin and Grosschedl, 1995; Wang et al., 1996), their importance for B cell function has been difficult to access. In this review, we show how DT40 knockouts of *Ikaros* family genes and *Pax5* have expanded our knowledge about the function of these factors.

2.   B CELL TRANSCRIPTION FACTORS

Hematopoietic stem cells give rise to all the blood cells. They produce progeny that progressively lose their self-renewal capacity and become restricted to a single lineage (Matthias and Rolink, 2005). During the early B cell development transcription factors like Ikaros, PU.1, Pax5 and Aiolos are expressed (Fig. 11-1). The expression is maintained from early progenitors throughout the B cell lineage (Busslinger, 2004).

2.1   Transcription factors in the early hematopoiesis

Initial stages in the differentiation of hematopoietic stem cells into B lymphocytes rely on the action of transcription factors PU.1 and Ikaros. *PU.1*$^{-/-}$ mice do not generate B cells, T cells, monocytes or granulocytes and die before or rapidly after birth (Scott et al., 1994; McKercher et al., 1996). PU.1 expression is not restricted to the B cell lineage. In fact, a high level of PU.1 protein promotes macrophage differentiation and prevents B cell development whereas low level of PU.1 expression induces B cell fate (DeKoter and Singh, 2000). *PU.1^{-/-}* DT40 cells grow more slowly than the wild type cells indicating a role for PU.1 in cell proliferation (Matsudo et al., 2003). However, *PU.1^{-/-}* DT40 cells had no marked change in the expression Igβ or Ig λ L chain mRNAs as implicated in studies with mice (Simon, 1998).

Ikaros is expressed as early as at the second day of chick embryonic development (ED2), many days before colonization of the lymphoid primordia, the thymus and the bursa of Fabricius (Liippo and Lassila, 1997). The expression increases during the lymphocyte differentiation, peaking in double positive thymocytes, but is reduced along the monocyte/macrophage and erythroid differentiation pathways in mice (Klug et al., 1998). Ikaros expression is maintained throughout the mammalian B cell development (Morgan et al., 1997).

Gene inactivation studies in mice have established an important function for Ikaros in the early lymphopoiesis. A complete developmental block in *Ikaros^{-/-}* mice is observed at the level of hematopoietic stem cell (Fig. 11-1). Ikaros null mutant mice lack all fetal B and T cells and all their earliest progenitors as well as NK cells (Wang et al., 1996). Also long-term repopulation activity of the hematopoietic stem cells is reduced (Nichogiannopoulou et al., 1999). T cells and their precursors, but no B cells are observed after birth, demonstrating that Ikaros has a fundamental role in the B lineage development (Wang et al., 1996). Because of the early block in the B cell differentiation, the actual function of Ikaros in B cells has been difficult to address in the mouse model. However, Ikaros deficient DT40 cells have revealed that Ikaros is needed for efficient B cell receptor (BCR) signaling (Nera et al., 2006a) providing the first molecular evidence how Ikaros functions in B cells. In other words, Ikaros regulates the function of B cells by enhancing the BCR signals. Given the decisive role of the BCR signal strength in B cell fate determination (Casola et al., 2004) and the fact, that all resting mature B cells need adequate basal BCR signals for survival (Kraus et al., 2004), the findings with Ikaros deficient DT40 cells provide an explanation for the absolute requirement for Ikaros in B cells.

2.2   Maintaining B cell identity

The differentiation of developing lymphoid precursors to committed B cells depends critically on E2A, EBF and Pax5. These three transcription factors specify the primary characteristics of the B cell lineage. In addition to Ikaros, Pax5 gives another example where developmental block in mice makes functional analyses at later stages very difficult.

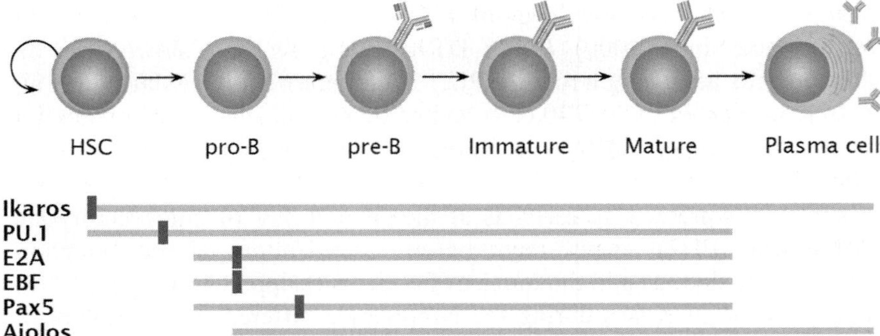

Figure 11-1. Expression of B cell transcription factors during the B cell development. The expression of the transcription factors (gray horizontal bar) is sustained after the observed block in the mice (dark vertical bar) suggesting a function for these molecules also at the later stages of development. DT40 system provides a tool to circumvent these problems and to study the transcriptional regulation in the B cells when a total block is observed in mice (e.g. *Ikaros$^{-/-}$*, *PU.1$^{-/-}$*, *E2A$^{-/-}$*, *EBF$^{-/-}$* and *Pax5$^{-/-}$*). DT40 gene targeting system is an easily manipulated model to study the function of the transcription factors even if knockout mice have no developmental block (e.g. *Aiolos$^{-/-}$*).

Pax5 is expressed from pro-B to mature B cells (Busslinger, 2004). Mice with a homozygous disruption in the *Pax5* gene have a complete developmental block before entering the pre-B cell stage (Fig. 11-1 and Table 11-1), prior to the V_H to $D_H J_H$ rearrangement (Urbanek et al., 1994). The findings that *Pax5$^{-/-}$* pre-B cells have a plasticity to generate functional lineages other than B cells (Nutt et al., 1999) indicate that Pax5 is the factor that commits the precursor cells to the B lineage (Busslinger, 2004). Furthermore, expression of *mb-1*, *CD19* and *BLNK* are activated and *MPO*, *M-CSFR*, *Notch1*, and *XBP1* are repressed by Pax5 (Calame et al., 2003; Busslinger, 2004), stressing the importance of Pax5 for the B cell phenotype.

To circumvent the early developmental block, conditional *Pax5* knockout mice have been generated (Horcher et al., 2001). In this model, a *Cre* recombinase is under the control of endogenous *CD19* locus and a functionally essential exon is flanked with *loxP* sites (Horcher et al., 2001). The basic finding with this model is that when Pax5 is inactivated, B cell development is reversed, since mature B cells are preferentially lost, several B cell specific genes are downregulated and lineage inappropriate genes upregulated (Horcher et al., 2001). The *Pax5$^{-/-}$* DT40 cells shows decreased expression of BCR signaling molecules (Nera et al., 2006b), which is in line with earlier reports (Busslinger, 2004). Importantly, plasma cell genes *Blimp-1* and *XBP-1* are significantly upregulated with concomitant drop in *Bcl-6* expression and increased secretion of immunoglobulin, suggesting that Pax5 sets a brake for plasma cell differentiation (Nera et al., 2006b).

Unraveling of the B cell regulatory network demonstrates the robustness of the DT40 system. Whereas the *Pax5* deficient mice have been instrumental in uncovering the requirement for Pax5 in B cell commitment, the DT40 model has revealed Pax5 as a key inhibitor of plasma cell differentiation (Table 11-1). Therefore, Pax5 constitutes an example in which the fate determination is coupled to inhibition of terminal differentiation and both are regulated by a single factor. *Pax5[-/-]* DT40 model has created new avenues for research concerning plasma cell differentiation.

Table 11-1. Comparison of the phenotypes observed in mouse and DT40 mutants of genes involved in the B cell development and function.

Gene	Mouse phenotype	DT40 phenotype
Ikaros	*Ikaros[-/-]*: reduced hematopoietic stem cell activity, lack of fetal lymphoid progenitors and postnatal B cells[1] *Ikaros[DN/DN]*: lack of DCs and NK-cells, lack of B and T cells and all precursors[2] *Ikaros[L/L]*: impaired B cell development[3]	Slowed growth, weakened BCR signaling with reduced PLCγ2 activation and intracellular calcium mobilization, increased Cbl phosphorylation[4]
PU.1	Block early in lymphoid an myeloid development[5]	Slowed growth[6]
Pax5	*Pax5[-/-]*: developmental block in pro-B to pre-B transition, developmental plasticity of B cell progenitors[7] *Pax5[F/-]*: reduced IgG secretion and loss of B cell specific gene expression[8]	Upregulation of plasma cell genes, immunoglobulin secretion and downregulation of genes encoding BCR signaling machinery[9]
Aiolos	Production of autoantibodies, increased germinal center formation, lack of marginal zone and peritoneal B cells, hyperresponsive to BCR signaling[10]	Gene conversion defect, sensitivity to apoptosis.[11]

[1]Wang et al., 1996; Nichogiannopoulou et al., 1999. [2]Georgopoulos et al., 1994. [3]Kirstetter et al., 2002. [4]Nera et al., 2006a. [5]Scott et al., 1994; McKercher et al., 1996. [6]Matsudo et al., 2003. [7]Urbanek et al., 1994; Nutt et al., 1999. [8]Horcher et al., 2001. [9]Nera et al., 2006b. [10]Morgan et al., 1997. [11]Narvi et al., 2006.

2.3 Survival of mature B cells

Aiolos is another transcription factor in the Ikaros family. Like Ikaros, it is expressed as several isoforms (Liippo et al., 1999; Liippo et al., 2001). Aiolos is exclusively expressed in lymphoid tissues (Wang et al., 1998; Liippo et al., 1999). The expression increases gradually during B cell maturation with highest levels observed in bursal cells at the same time as Ig V gene modification is started by gene conversion (Liippo et al., 1999). The transcripts are not detectable in early hematopoietic foci or before colonization of the primary lymphoid organs (Liippo et al., 1999).

Unlike *Ikaros*[-/-] mice, *Aiolos*[-/-] mice develop mature B cells (Wang et al., 1998), indicating that Aiolos does not have a decisive role in B cell development. However, a great proportion of the splenic *Aiolos*[-/-] B cells have an activated cell surface phenotype and the Aiolos deficient mice have increased number of germinal centers. They also have higher levels of serum immunoglobulin and develop a systemic lupus erythematosus type of autoimmune disease (Wang et al., 1998). These findings suggest an inhibitory role for Aiolos in mature B cells.

The gene conversion rate is dramatically reduced in *Aiolos*[-/-] DT40 cells (Narvi et al., 2006), suggesting that Aiolos acts to promote antibody diversification. Indeed, the expression of activation induced deaminase (AID), an enzyme required for somatic hypermutation, gene conversion and class switch recombination (Muramatsu et al., 2000; Revy et al., 2000; Arakawa et al., 2002), is downregulated (Narvi et al., 2006). Aiolos deficient DT40 cells have increased susceptibility to apoptosis induced by γ- or UV-irradiation, B cell receptor crosslinking or nutritional stress (Narvi et al., 2006). Sensitivity to BCR-induced apoptosis in *Aiolos*[-/-] DT40 is in line with the findings in the mouse system where a reduced threshold for proliferative response to BCR stimulation and elevated production of autoantibodies is observed (Wang et al., 1998).

Aiolos is suggested to form heteromeric complexes with Ikaros and Helios in B cells (Morgan et al., 1997; Hahm et al., 1998; Kelley et al., 1998; Wang et al., 1998). Ikaros complexes bind to their cognate sites in gene regulatory elements and recruit chromatin remodeling complexes to target genes (Georgopoulos, 2002). Inactivation of *Helios*, a third member of the family in DT40 cells (Kohonen et al., 2004), has an effect on BCR signaling in DT40 cells (K-P Nera, unpublished observations). As all of these three members of the *Ikaros* gene family are expressed in DT40 cells, and their gene targeting provides a phenotype, DT40 constitutes a valid model to study interactions of these nuclear proteins in the regulation of B cell function.

Table 11-2. DT40 mutants with disrupted or deleted genes related to transcriptional regulation.

Gene	Processes	Phenotype	References
c-myb	Development of the hematopoietic system	*Pdcd4* expression reduced	Appl and Klempnauer, 2002
Ikaros	Chromatin remodeling, development of lymphoid lineages, threshold for T cell activation	Reduced BCR signaling	Nera et al., 2006a
PU.1	Development of myeloid and lymphoid lineages	Slowed growth	Matsudo et al., 2003
Pax5	B cell commitment	Characteristics of plasma cells	Nera et al., 2006b
Aiolos	B cell function and homeostasis	Defective gene conversion and sensitivity to apoptosis	Narvi et al., 2006
*Notch-1**	Development of various tissues	Growth suppression and induction of apoptosis by constitutively active Notch	Morimura et al., 2000
B-Myb	Regulation of cell cycle	Sensitivity to DNA damage	Ahlbory et al., 2005
HDAC1	Histone deacetylase, chromatin remodeling	No clear phenotype	Takami et al., 1999
HDAC2	Histone deacetylase, chromatin remodeling	Increased levels of IgH and IgL mRNA, shift from membrane μ to secretory μ	Takami et al., 1999
HDAC3	Histone deacetylase, chromatin remodeling	Compromised cell viability	Takami and Nakayama, 2000
GCN5	Histone acetyl transferase, chromatin remodeling	Slowed growth, decreased acetylation of histones, altered expression of genes involved in apoptosis and cell cycle regulation	Kikuchi et al., 2005
PCAF	Histone acetyl transferase, chromatin remodeling	No clear phenotype	Kikuchi et al., 2005
YB-1	Y-box binding protein, translational regulation, mRNP binding	Slowed growth	Swamynathan et al., 2002; Matsumoto et al., 2005
TLP	Embryogenesis and spermiogenesis	Shortened G_2 phase and insufficient G_2 checkpoint arrest, changes in stress response and cell cycle regulatory genes	Shimada et al., 2003

* Ectopic expression of constitutively active Notch-1

3.   OTHER NUCLEAR REGULATORS

Ikaros transcription factors are thought to regulate transcription by recruiting multimeric histone deacetylase (HDAC) complexes to target genes (Georgopoulos, 2002). Many proteins that are involved in transcriptional regulation are implicated to play a role in the B cell development, including those that are involved in chromatin remodeling (Yu et al., 2000; Maes et al., 2001; Georgopoulos, 2002). Some of these have been disrupted in DT40 cells (Table 11-2). HDACs are enzymes that catalyze deacetylation of core histone proteins. A pattern of acetylation of histone tails provides a code in which the acetyl modifications alter the chromatin structure, which either permits or prevents the binding of transcription factors to their cognate sites in the DNA and the factors activate or repress transcription (Georgopoulos, 2002).

Disruption of *HDAC1* in a mouse leads to severe defects in the cell proliferation and to lethality in the embryonic period (Lagger et al., 2002). The phenotype may be caused by a deregulation of cell cycle inhibitors (Lagger et al., 2002). *HDAC1* inactivation in DT40 did not result in any major changes in the protein expression (Takami et al., 1999). In contrast, *HDAC2* inactivation in DT40 cells reveals a clear B cell phenotype. The cells have increased production of total IgM transcripts with increased secretory form of IgM heavy chain and decreased membrane bound form (Takami et al., 1999). HDAC3 on the other hand was shown to be essential for the viability of DT40 cells (Takami and Nakayama, 2000) indicating a more profound role. Complementation of *HDAC3*$^{-/-}$ DT40 cells with mutant HDAC3 constructs revealed that the N- and C-terminal regions, the nuclear export signal and the deacetylation activity were required to keep the cells alive (Takami and Nakayama, 2000). This is an example of how genetic complementation provides a tool to study detailed molecular mechanisms. The DT40 targeting system has proved to be useful also in establishing different functions for the HDAC1-3 proteins that are members of the same family and presumably have similar mechanisms in chromatin remodeling (Struhl, 1998).

Histone acetyl transferases (HATs) are a group of enzymes that add an acetyl modification to histones (Marmorstein, 2001). Two HAT genes, *GCN5* and *PCAF*, have been inactivated in the DT40 cell line (Kikuchi et al., 2005). Disruption of *GCN5* results in changes of acetylation pattern at various lysine residues of the histone proteins. The *GCN5* deficient cells also have a slowed growth rate with suppressed transition at the G1/S phase of the cell cycle and a concomitant change in the expression of cell cycle regulators, which might be a consequence of an altered chromatin structure (Kikuchi et al., 2005). In contrast to *GCN5*, disruption of *PCAF* has no effect on the growth rate reflecting the low expression level of the gene (Kikuchi et al., 2005).

The *c-myb* gene has a role in development of all the hematopoietic lineages controlling proliferation, differentiation and apoptosis of the stem and progenitor cells (Oh and Reddy, 1999). Traditional inactivation of *c-myb* in mice results in embryonic lethality (Mucenski et al., 1991), but conditional knockout mice with CD19 expression-coupled inactivation of *c-myb* have a partial block during the pro-B to pre-B cell transition (Thomas et al., 2005). The gene has been disrupted in DT40 cells (Appl and Klempnauer, 2002), but further studies are needed to find out whether c-Myb has a role in DT40 cells. However, transfection of *v-myb* into DT40 cells is known to increase the frequency of gene conversion (H. Arakawa and J.-M. Buerstedde, unpublished observations). The mechanism for this phenomenon is yet unknown.

Recent data suggests that HDACs and HATs regulate hematopoietic transcription factors by reversible acetylation (Huo and Zhang, 2005). As HDAC and HAT mutants are available, it will be interesting to find out how DT40 system provides new insights for the function of acetylation in regulation of B cell function.

4.   DT40 CELL LINE AND SYSTEMS BIOLOGY

Gene expression analysis with microarrays provides a robust tool to find putative target genes and changes in the functional gene regulation when transcription factors are studied by targeted gene inactivation. A targeted gene array experiment with bursa enriched genes was carried out to study the gene expression profiles in avian B cell development (Koskela et al., 2003). The gene expression profiles of prebursal, bursal, DT40, germinal center B cells and plasma cells from the harderian gland were compared. DT40 transcript profile was found to be similar to that of the bursal B cells (Koskela et al., 2003).

In another experiment, gene expression profiling of *Pax5*$^{-/-}$ DT40 cells displayed expression pattern with reduced expression of B cell genes (Nera et al., 2006b). In this array analysis, the gene expression profiles of the *Pax5*$^{-/-}$ and wild type DT40 cells were compared by using a novel BursaEST array (ArrayExpress accession: A-MEXP-155), which contained about 15 000 probes in duplicate from bursal B cells. Array analysis (ArrayExpress accession: E-MEXP-270) revealed downregulation of genes involved in the BCR signaling pathway, upregulation of genes involved in protein secretion and stress response in the absence of Pax5. Results were further confirmed by quantitative real-time PCR (Nera et al, 2006b), and they provided information that eventually led to the functional studies of plasmacytic phenotype of the *Pax5*$^{-/-}$ DT40 cells. Moreover, a comparison of the gene expression profiles of bursal B cells and plasma cells from harderian gland show several similarities with the expression profiles between *Pax5-/-* and wild type DT40 cells (P. Kohonen, personal communication), thus constituting further evidence that Pax5 deficient cells are expressing a plasma cell gene expression program. These studies show that DT40 system is useful in the studies of transcriptional regulation of B cell function also using systemic approaches.

Chicken genomic information is now available as chicken genome is sequenced at approximately 7-fold coverage (Hillier et al., 2004). Useful information on annotations from bursal sequencing efforts and comparisons of bursal and DT40 gene expression are freely accessible over the internet (http://pheasant.gsf.de/DEPARTMENT/dt40.html, see also Caldwell et al., this issue, Chapter 3). Also commercial gene expression arrays are available. These recent advances increase the suitability of the DT40 cell line for transcriptional studies and broaden the applicability of DT40 system as a model for functional B cell studies.

Table 11-3. Comparison of the gene targeting systems between DT40 and mouse as models for B cell physiology.

	DT40	Mouse
Advantages	Target gene function can be studied despite of the developmental block	Phenotype at the level of the whole organism
	Easy and fast disruption of practically any gene including conditional and drug-induced gene inactivation	Access to early B cell development Possibility of conditional gene inactivation *in vivo*
	Easy reconstitution of knockout cells with desired constructs e.g. splice variants	
	Mutants may be viable even if mouse mutants were lethal	
	Manipulability of growth conditions	
	Cells are easily available in any needed quantity for biochemical studies	
	Low cost maintenance	
Disadvantages	General drawbacks of an immortalized cell line	Block in B cell development prevents the functional studies
	Chicken specific reagents (monoclonal antibodies, cytokines, etc.) scarcely available	Selection of non-physiological subpopulations Expensive and labor intensive Ethical points

5.   BENEFITS OF THE DT40 SYSTEM IN STUDIES OF B CELL FUNCTION

As a model for processes involved in B cell function and the B cell physiology DT40 targeting system brings new possibilities and some restrictions compared to the *in vivo* models (Table 11-3). If the manipulation of the gene of interest in mice is lethal or a block in the B cell development is observed, it is possible to reveal the function of the gene in the DT40 system.

With the DT40 system genetic complementation is relatively easy. Restoring the functional gene in the null background verifies that the gene in question is responsible for the phenotype. This is rarely accomplished in mouse models. The null background can also be complemented with mutants of the gene of interest to dissect the molecular characteristics, as was shown with the *HDAC3*$^{-/-}$ DT40 cells (Takami and Nakayama, 2000). Additionally, individual isoforms of a factor that are produced from alternatively spliced transcripts can be expressed individually to study the roles of splice variants (Narvi et al., 2006), such as the isoforms of the Ikaros family members Ikaros and Aiolos (Liippo et al., 2000; Liippo et al., 2001). An elegant way to disrupt genes in DT40 is to use targeting constructs that have *loxP* sites flanking the selection marker cassette (Arakawa and Buerstedde, this issue, Chapter 1). This way the marker can be excised after a successful targeting event and used subsequently for complementation or for further genetic modifications in the cell line (Arakawa et al., 2001).

The most obvious benefit of the DT40 cell line over other B cell lines is the ease of accomplishing a true knockout. The other cell lines can be subjected to knockdown experiments using RNA interference (RNAi). However, small interfering (si) RNAs decrease the expression at best to a low level (Mittal, 2004), whereas with conventional disruption methods in the DT40 cell line the expression of a functional protein is totally lost. The residual gene expression can in fact be a real problem when phenotypes are modest. RNAi experiments in B cells do not seem to work in an efficient manner and it would be interesting to find out the underlying molecular basis.

6.   FROM DT40 CELLS TO BURSA OF FABRICIUS

The chicken has been the model system in B cell immunology and developmental biology for decades due to the easy manipulability of its embryonic development. In addition, chickens have an organ specialized for B cell development, the bursa of Fabricius. Moreover, the chicken bursa-derived

B cell line DT40 gives an opportunity to study the molecular mechanisms of central aspects of B cell immunology (Fig. 11-2).

Bursa, an organ for avian B cell development, was originally described in 1621 by Hieronymus Fabricius. It wasn't until 1956 that the important function of bursa in antibody production was discovered, when the bursa was surgically removed from hatching birds resulting in a loss of specific antibody production (Glick et al., 1956).

Bursa of Fabricius is a pouch-like organ connecting to the intestine or cloaca via bursal duct (Fig. 11-2). A developing bursal structure is observed on ED4 as a small epithelial bud in the cloaca. Definitive hematopoiesis was demonstrated to take place in intraembryonic regions using chick-chick embryo chimeras (Lassila et al., 1978). The primary source of the hematopoietic progeny is the aorta-associated splancnopleura. Later on, between ED6 and ED8, the para-aortic mesenchyme is formed ventral to the aorta providing a hematopoietic compartment, from where the progenitors migrate to colonize the bursa between ED8 and ED14 (Lassila et al., 1979). Avian B cell progenitors start to rearrange their immunoglobulin V_H and V_L genes prebursally in the ED7 para-aortic mesenchyme, indicating that the progenitors are committed to the B cell lineage before the bursal colonization (Mansikka et al., 1990).

The surface immunoglobulin positive cells undergo a rapid burst of proliferation in the bursal follicles from the initial 2-5 prebursal "stem" cells in each follicle (Pink et al., 1985) to about 150 000 B lymphocytes at the age of 2 months in each of the about 10 000 follicles. In contrast to the mammalian species, avian species use upstream V pseudogenes as templates in the process of somatic gene conversion to diversify their Ig genes (Reynaud et al., 1987). Further diversity is added by somatic hypermutation in a similar way as in mammals (Arakawa et al., 1996). Just before hatching, bursal B cells start to emigrate and enter the peripheral lymphoid organs, where they can form germinal centers (Veistinen and Lassila, 2005).

As studies in the chicken have revealed the mechanisms of antibody production and B lymphocyte development, DT40 studies give a molecular insight into the B cell physiology (Fig. 11-2). Pax5, Blimp-1 and XBP1 are proteins involved in the regulation of antibody secretion (Calame et al., 2003; Nera et al., 2006b), AID and Aiolos are examples of proteins regulating gene conversion (Arakawa et al., 2002; Narvi et al., 2006) and Lyn, Syk, Btk and PLCγ2 operate in BCR signal transduction (Kurosaki, 2002). Most of these proteins have been studied in the DT40 *in vitro* system with targeted gene inactivation (Fig. 11-2). The functions of these proteins are crucial for B cells to be able to develop a sufficient immunoglobulin repertoire and to differentiate into antibody secreting plasma cells in response to pathogen challenge. Therefore the nanoscale phenomena that are studied *in vitro* highlight the importance of the single genes in constituting systems level phenomena such as antibody production and B cell development. Hence, the DT40 cell line is a pertinent model to study B cell biology using the reverse genetic approach (Fig. 11-2).

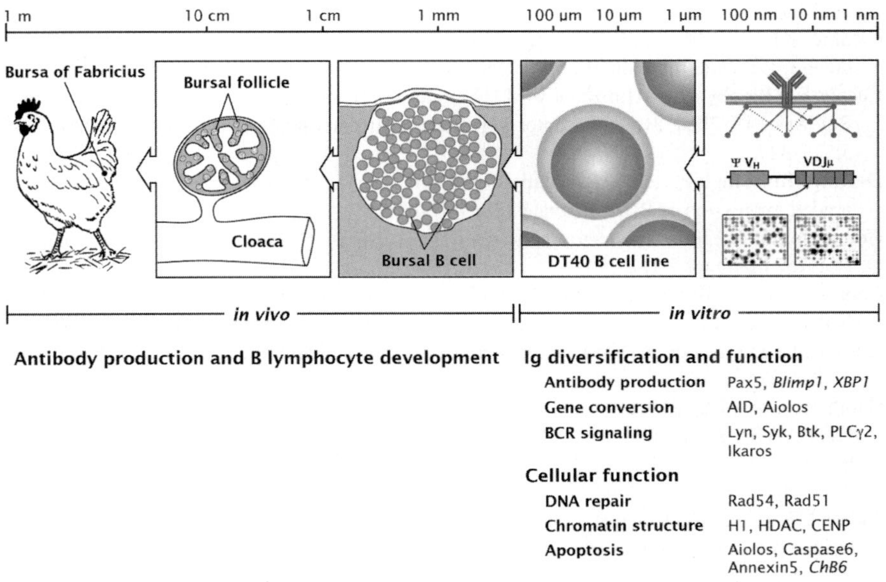

Figure 11-2. DT40 reverse genetics and the B cell immune system. When going from visual and microscopic *in vivo* studies to nanoscale *in vitro* experiments DT40 cells play an important role. Historically antibody production and B cell development are studied *in vivo*. With DT40 the same phenomena can be studied in molecular detail *in vitro*. Therefore the studies will reveal important aspects of B cell function. Some of the genes involved in the biological processes are given. The genes in *italics* have not been disrupted in the DT40 cell line.

REFERENCES

Ahlbory D, Appl H, Lang D, Klempnauer KH. 2005. Disruption of B-myb in DT40 cells reveals novel function for B-Myb in the response to DNA-damage. Oncogene.

Appl H, Klempnauer KH. 2002. Targeted disruption of c-myb in the chicken pre B-cell line DT40. Oncogene 21:3076-3081.

Arakawa H, Furusawa S, Ekino S, Yamagishi H. 1996. Immunoglobulin gene hyper-conversion ongoing in chicken splenic germinal centers. EMBO J 15:2540-2546.

Arakawa H, Hauschild J, Buerstedde JM. 2002. Requirement of the activation-induced deaminase (AID) gene for immunoglobulin gene conversion. Science 295:1301-1306.

Arakawa H, Lodygin D, Buerstedde JM. 2001. Mutant loxP vectors for selectable marker recycle and conditional knock-outs. BMC Biotechnol 1:7.

Bain G, Maandag EC, Izon DJ, Amsen D, Kruisbeek AM, Weintraub BC, Krop I, Schlissel MS, Feeney AJ, van Roon M, et al., 1994. E2A proteins are required for proper B cell development and initiation of immunoglobulin gene rearrangements. Cell 79:885-892.

Busslinger M. 2004. Transcriptional control of early B cell development. Annu Rev Immunol 22:55-79.

Calame KL, Lin KI, Tunyaplin C. 2003. Regulatory mechanisms that determine the development and function of plasma cells. Annu Rev Immunol 21:205-230.

Casola S, Otipoby KL, Alimzhanov M, Humme S, Uyttersprot N, Kutok JL, Carroll MC, Rajewsky K. 2004. B cell receptor signal strength determines B cell fate. Nat Immunol 5:317-327.

DeKoter RP, Singh H. 2000. Regulation of B lymphocyte and macrophage development by graded expression of PU.1. Science 288:1439-1441.

Georgopoulos K. 2002. Haematopoietic cell-fate decisions, chromatin regulation and ikaros. Nat Rev Immunol 2:162-174.

Georgopoulos K, Bigby M, Wang JH, Molnar A, Wu P, Winandy S, Sharpe A. 1994. The Ikaros gene is required for the development of all lymphoid lineages. Cell 79:143-156.

Glick B, Chang T, Jaap R. 1956. The bursa of Fabricius and antibody production. Poultry Sci 35:224-226.

Hahm K, Cobb BS, McCarty AS, Brown KE, Klug CA, Lee R, Akashi K, Weissman IL, Fisher AG, Smale ST. 1998. Helios, a T cell-restricted Ikaros family member that quantitatively associates with Ikaros at centromeric heterochromatin. Genes Dev 12: 782-796.

Hillier LW et al., 2004. Sequence and comparative analysis of the chicken genome provide unique perspectives on vertebrate evolution. Nature 432:695-716.

Horcher M, Souabni A, Busslinger M. 2001. Pax5/BSAP maintains the identity of B cells in late B lymphopoiesis. Immunity 14:779-790.

Huo X, Zhang J. 2005. Important roles of reversible acetylation in the function of hematopoietic transcription factors. J Cell Mol Med 9:103-112.

Kelley CM, Ikeda T, Koipally J, Avitahl N, Wu L, Georgopoulos K, Morgan BA. 1998. Helios, a novel dimerization partner of Ikaros expressed in the earliest hematopoietic progenitors. Curr Biol 8:508-515.

Kikuchi H, Takami Y, Nakayama T. 2005. GCN5: a supervisor in all-inclusive control of vertebrate cell cycle progression through transcription regulation of various cell cycle-related genes. Gene 347:83-97.

Kirstetter P, Thomas M, Dierich A, Kastner P, Chan S. 2002. Ikaros is critical for B cell differentiation and function. Eur J Immunol 32:720-730.

Klug CA, Morrison SJ, Masek M, Hahm K, Smale ST, Weissman IL. 1998. Hematopoietic stem cells and lymphoid progenitors express different Ikaros isoforms, and Ikaros is localized to heterochromatin in immature lymphocytes. Proc Natl Acad Sci U S A 95: 657-662.

Kohonen P, Nera KP, Lassila O. 2004. Avian Helios and evolution of the Ikaros family. Scand J Immunol 60:100-107.

Koskela K, Kohonen P, Nieminen P, Buerstedde JM, Lassila O. 2003. Insight into lymphoid development by gene expression profiling of avian B cells. Immunogenetics 55:412-422.

Kraus M, Alimzhanov MB, Rajewsky N, Rajewsky K. 2004. Survival of resting mature B lymphocytes depends on BCR signaling via the Ig alpha/beta heterodimer. Cell 117: 787-800.

Kurosaki T. 2002. Regulation of B-cell signal transduction by adaptor proteins. Nat Rev Immunol 2:354-363.

Lagger G, O'Carroll D, Rembold M, Khier H, Tischler J, Weitzer G, Schuettengruber B, Hauser C, Brunmeir R, Jenuwein T, Seiser C. 2002. Essential function of histone deacetylase 1 in proliferation control and CDK inhibitor repression. Embo J 21:2672-2681.

Lassila O, Eskola J, Toivanen P. 1979. Prebursal stem cells in the intraembryonic mesenchyme of the chick embryo at 7 days of incubation. J Immunol 123:2091-2094.

Lassila O, Eskola J, Toivanen P, Martin C, Dieterlen-Lievre F. 1978. The origin of lymphoid stem cells studied in chick yold sac-embryo chimaeras. Nature 272:353-354.

Liippo J, Lassila O. 1997. Avian Ikaros gene is expressed early in embryogenesis. Eur J Immunol 27:1853-1857.

Liippo J, Mansikka A, Lassila O. 1999. The evolutionarily conserved avian Aiolos gene encodes alternative isoforms. Eur J Immunol 29:2651-2657.

Liippo J, Nera KP, Kohonen P, Lampisuo M, Koskela K, Nieminen P, Lassila O. 2000. The Ikaros family and the development of early intraembryonic hematopoietic stem cells. Curr Top Microbiol Immunol 251:51-58.

Liippo J, Nera KP, Veistinen E, Lahdesmaki A, Postila V, Kimby E, Riikonen P, Hammarstrom L, Pelkonen J, Lassila O. 2001. Both normal and leukemic B lymphocytes express multiple isoforms of the human Aiolos gene. Eur J Immunol 31:3469-3474.

Lin H, Grosschedl R. 1995. Failure of B-cell differentiation in mice lacking the transcription factor EBF. Nature 376:263-267.

Maes J, O'Neill LP, Cavelier P, Turner BM, Rougeon F, Goodhardt M. 2001. Chromatin remodeling at the Ig loci prior to V(D)J recombination. J Immunol 167:866-874.

Mansikka A, Sandberg M, Lassila O, Toivanen P. 1990. Rearrangement of immunoglobulin light chain genes in the chicken occurs prior to colonization of the embryonic bursa of Fabricius. Proc Natl Acad Sci U S A 87:9416-9420.

Marmorstein R. 2001. Structure of histone acetyltransferases. J Mol Biol 311:433-444.

Matsudo H, Otsuka A, Ozawa Y, Ono M. 2003. Disruption of the PU.1 gene in chicken B lymphoma DT40 cells and its effect on reported target gene expression. Gene 322: 169-174.

Matsumoto K, Tanaka KJ, Tsujimoto M. 2005. An acidic protein, YBAP1, mediates the release of YB-1 from mRNA and relieves the translational repression activity of YB-1. Mol Cell Biol 25:1779-1792.

Matthias P, Rolink AG. 2005. Transcriptional networks in developing and mature B cells. Nat Rev Immunol 5:497-508.

McKercher SR, Torbett BE, Anderson KL, Henkel GW, Vestal DJ, Baribault H, Klemsz M, Feeney AJ, Wu GE, Paige CJ, Maki RA. 1996. Targeted disruption of the PU.1 gene results in multiple hematopoietic abnormalities. EMBO J 15:5647-5658.

Mittal V. 2004. Improving the efficiency of RNA interference in mammals. Nat Rev Genet 5:355-365.

Morgan B, Sun L, Avitahl N, Andrikopoulos K, Ikeda T, Gonzales E, Wu P, Neben S, Georgopoulos K. 1997. Aiolos, a lymphoid restricted transcription factor that interacts with Ikaros to regulate lymphocyte differentiation. Embo J 16:2004-2013.

Morimura T, Goitsuka R, Zhang Y, Saito I, Reth M, Kitamura D. 2000. Cell cycle arrest and apoptosis induced by Notch1 in B cells. J Biol Chem 275:36523-36531.

Mucenski ML, McLain K, Kier AB, Swerdlow SH, Schreiner CM, Miller TA, Pietryga DW, Scott WJ, Jr., Potter SS. 1991. A functional c-myb gene is required for normal murine fetal hepatic hematopoiesis. Cell 65:677-689.

Muramatsu M, Kinoshita K, Fagarasan S, Yamada S, Shinkai Y, Honjo T. 2000. Class switch recombination and hypermutation require activation-induced cytidine deaminase (AID), a potential RNA editing enzyme. Cell 102:553-563.

Narvi E, Nera K-P, Terho P, Granberg J, Lassila O. 2006. Aiolos controls gene conversion and cell death in DT40 B cells. Submitted.

Nera K-P, Alinikula J, Terho P, Narvi E, Törnquist K, Kurosaki T, Buerstedde J-M, Lassila O. 2006a. Ikaros has a crucial role in regulation of B cell receptor signaling. Eur J Immunol 36:516-525.

Nera K-P, Kohonen P, Narvi E, Peippo A, Mustonen L, Terho P, Koskela K, Buerstedde J-M, Lassila O. 2006b. Loss of Pax5 promotes plasma cell differentiation. Immunity 24:283-293.

Nichogiannopoulou A, Trevisan M, Neben S, Friedrich C, Georgopoulos K. 1999. Defects in hemopoietic stem cell activity in Ikaros mutant mice. J Exp Med 190:1201-1214.

Nutt SL, Heavey B, Rolink AG, Busslinger M. 1999. Commitment to the B-lymphoid lineage depends on the transcription factor Pax5. Nature 401:556-562.

Oh IH, Reddy EP. 1999. The myb gene family in cell growth, differentiation and apoptosis. Oncogene 18:3017-3033.

Pink JR, Vainio O, Rijnbeek AM. 1985. Clones of B lymphocytes in individual follicles of the bursa of Fabricius. Eur J Immunol 15:83-87.

Revy P et al., 2000. Activation-induced cytidine deaminase (AID) deficiency causes the autosomal recessive form of the Hyper-IgM syndrome (HIGM2). Cell 102:565-575.

Reynaud CA, Anquez V, Grimal H, Weill JC. 1987. A hyperconversion mechanism generates the chicken light chain preimmune repertoire. Cell 48:379-388.

Scott EW, Simon MC, Anastasi J, Singh H. 1994. Requirement of transcription factor PU.1 in the development of multiple hematopoietic lineages. Science 265:1573-1577.

Shimada M, Nakadai T, Tamura TA. 2003. TATA-binding protein-like protein (TLP/TRF2/TLF) negatively regulates cell cycle progression and is required for the stress-mediated G(2) checkpoint. Mol Cell Biol 23:4107-4120.

Simon MC. 1998. PU.1 and hematopoiesis: lessons learned from gene targeting experiments. Semin Immunol 10:111-118.

Struhl K. 1998. Histone acetylation and transcriptional regulatory mechanisms. Genes Dev 12:599-606.

Swamynathan SK, Varma BR, Weber KT, Guntaka RV. 2002. Targeted disruption of one allele of the Y-box protein gene, Chk-YB-1b, in DT40 cells results in major defects in cell cycle. Biochem Biophys Res Commun 296:451-457.

Takami Y, Kikuchi H, Nakayama T. 1999. Chicken histone deacetylase-2 controls the amount of the IgM H-chain at the steps of both transcription of its gene and alternative processing of its pre-mRNA in the DT40 cell line. J Biol Chem 274:23977-23990.

Takami Y, Nakayama T. 2000. N-terminal region, C-terminal region, nuclear export signal, and deacetylation activity of histone deacetylase-3 are essential for the viability of the DT40 chicken B cell line. J Biol Chem 275:16191-16201.

Thomas MD, Kremer CS, Ravichandran KS, Rajewsky K, Bender TP. 2005. c-Myb Is Critical for B Cell Development and Maintenance of Follicular B Cells. Immunity 23:275-286.

Urbanek P, Wang ZQ, Fetka I, Wagner EF, Busslinger M. 1994. Complete block of early B cell differentiation and altered patterning of the posterior midbrain in mice lacking Pax5/BSAP. Cell 79:901-912.

Wang JH, Avitahl N, Cariappa A, Friedrich C, Ikeda T, Renold A, Andrikopoulos K, Liang L, Pillai S, Morgan BA, Georgopoulos K. 1998. Aiolos regulates B cell activation and maturation to effector state. Immunity 9:543-553.

Wang JH, Nichogiannopoulou A, Wu L, Sun L, Sharpe AH, Bigby M, Georgopoulos K. 1996. Selective defects in the development of the fetal and adult lymphoid system in mice with an Ikaros null mutation. Immunity 5:537-549.

Veistinen E, Lassila O. 2005. Bursa of Fabricius. In: Encyclopedia of Life Sciences. John Wiley & Sons, Ltd: Chichester http://www.els.net/ [doi: 10.1038/npg.els.0003974].

Yu J, Angelin-Duclos C, Greenwood J, Liao J, Calame K. 2000. Transcriptional repression by blimp-1 (PRDI-BF1) involves recruitment of histone deacetylase. Mol Cell Biol 20:2592-2603.

Zhuang Y, Soriano P, Weintraub H. 1994. The helix-loop-helix gene E2A is required for B cell formation. Cell 79:875-884.

Chapter 12

TRANSCRIPTION AND RNA PROCESSING FACTORS PLAY COMPLEX ROLES IN DT40 CELLS

Stephanie Bush and James L. Manley
Department of Biological Sciences, Columbia University, New York, New York 10027

Abstract: Functional analysis of factors involved in transcription and RNA processing is difficult *in vivo* because of the importance of these processes to maintain cell viability. The ability to perform conditional knockouts and to generate stable transfectants in DT40 cells has facilitated the study of general transcription factors such as TATA binding proteins, essential splicing factors such as ASF/SF2 and crucial polyadenylation factors such as CstF64, as well as a number of more specific regulators of gene expression. The complexity of transcriptional regulation and the coupled mechanisms of RNA processing lead to the delineation of sometimes surprising roles for these factors in DT40 cells. Specifically, many of the components of the RNA transcriptional/processing machinery function to regulate cell proliferation in some manner, causing cell cycle abnormalities or cell death when cellular expression of any one of these components are altered. Additionally, only one of the factors described affected RNA synthesis in a global manner.

Key words: Transcription, RNA processing, splicing, polyadenylation, general transcription factors, gene-specific activators.

1. INTRODUCTION

Genetic regulation in a cell takes place at many levels; transcription, mRNA processing, RNA export from the nucleus, rate of RNA degradation,

J.-M. Buerstedde and S. Takeda (eds.), Reviews and Protocols in DT40 Research, 207–223.

rate of translation; as well as post-translational mechanisms that regulate the activity, localization or stability of a particular gene product. Ostensibly, the most effective method for a cell to regulate expression is at the level of transcription initiation, the interaction of *cis*-acting components and *trans*-acting proteins that provides for assembly of the components of the transcriptional machinery on a specific gene promoter. Eukaryotic transcription of genes that encode proteins begins with the assembly of a preinitiation complex (PIC) on the promoter of the gene to be transcribed (Hochheimer and Tjian, 2003; Muller and Tora, 2004; Roeder, 2005). The PIC contains RNA polymerase II (RNAPII), and several general transcription factors (GTFs), TFIIA, TFIIB, TFIID, TFIIE, TFIIF and TFIIH. GTFs were originally defined as factors required in addition to RNAPII for accurate transcription initiation *in vitro* (Conaway and Conaway, 1997; Reinberg et al., 1998). However, gene-specific *trans*-acting factors also bind to core promoter elements, modulating the rate of transcription initiation (Engel et al., 1992; Roeder, 1998). These specific transcription factors can bind to the promoter or, unlike GTFs, bind sequences some distance from the promoter. While seemingly clear cut, the regulation of transcription of nearly every gene, whether expressed only during a short time during development or produced constantly during the lifetime of a cell, is complex and the mechanisms governing initiation are just beginning to be understood.

Another level of gene regulation is RNA processing, where intronic sequences are excised from the nascent RNA and both the 5' and 3' ends of the RNA are modified. Splicing is dependent, like transcription, on both *cis*-acting factors and *trans*-acting factors (Beyersmann, 2000; Zheng, 2004). Furthermore, alternative splicing vastly increases the repertoire of any given genetic element (for review see Wang and Manley, 1997; Keegan et al., 2001). However, alternative splicing is not the only system involving RNA processing that contributes to genetic diversity. Polyadenylation, which involves endonucleolytic cleavage and poly(A) synthesis at the 3' end of mRNA precursors, is also frequently regulated by the use of alternative poly(A) sites (Colgan and Manley, 1997; Zhao et al., 1999). Current models show that RNA processing and transcription mechanisms are coupled (Hirose and Manley, 2000; Maniatis and Reed, 2002; Manley, 2002), fundamentally dependent upon each other to produce functional mRNA, further compounding the complexity of gene regulation and highlighting the necessity of *in vivo* studies.

Comprehension of the factors and mechanisms involved in regulating gene expression *in vivo* has been considerably enhanced by the use of DT40 cells. The ability of researchers to reduce or increase expression of specific factors required in transcription and RNA processing, mutate those same elements or to eliminate expression completely is a direct consequence of the high rate of targeted integration exhibited by the DT40 cell line (Buerstedde and Takeda, 1991). Disruption of one or both alleles of a component of the RNA machinery in DT40 by replacing part of the genetic sequence with antibiotic resistance has been extremely useful in analyzing the function of those components *in vivo* (Buerstedde and Takeda, 1991; Sonoda et al., 2001). Exogenous cDNA under a conditional promoter, most often controlled by tetracycline (tet), can be inserted to allow study of factors that are essential to the cell (Wang et al., 1996). Reporter vectors containing genetic elements of interest, such as promoters (Chen and Manley, 2003a) and plasmids containing genetic elements that signal for further processing, such as splicing (Liu et al., 2003) can be inserted into DT40 cells to investigate the function of components of the RNA processing machinery. The use of the above procedures with DT40 cells has alleviated many of the problems inherent in studying RNA processing, and begun to facilitate sophisticated genetic analysis of these critical factors.

2. TRANSCRIPTION

DT40 cells have been used to study both GTFs and gene-specific transcription factors. We describe below examples of each of these.

2.1 Reduced TBP expression affects transcription of only a subset of genes in DT40 cells

The most widely studied general transcription factor in DT40 cells is TATA-binding protein (TBP). The central component of TFIID, TBP was once recognized as the universal transcription factor and is essential for transcription of genes transcribed by all three RNA polymerases. TBP consists of two domains; an amino-terminus that is species-specific and a conserved carboxy-terminal domain. The conserved core domain of TBP directly contacts the TATA box of promoters and nucleates the assembly of other factors required for initiation of transcription (for review see Hernandez, 1993). TBP expression has been shown to be enhanced in some human tumors. Furthermore, the same study demonstrated that over-expression of TBP mutants deficient in DNA binding did not cause transformation, whereas overexpression of unmutated TBP caused tumors to form in athymic mice (Johnson et al., 2003).

To analyze the effects of TBP levels in DT40 cells, a line of cells heterozygous for the *tbp* gene (TBP-het) was generated (Um et al., 2001). Wild-type levels of TBP expression depend on the presence of both alleles, as Western blots show reduced levels of TBP protein in the heterozygous cell lines. Surprisingly, overall basal transcription was relatively unaffected and TBP-het cells are viable. However, phenotypic irregularities demonstrated the importance of constant levels of TBP during the cell cycle. TBP-het are significantly larger than wild-type DT40 cells and doubling time for the heterozygotes is about 24 hours, more than twice that of wild-type DT40 cells. Treatment with agents that cause cell cycle arrest and FACS analysis established that lower levels of TBP corresponded to an accumulation of cells in the G2/M phase of the cell cycle. The mitotic delay phenotype was attributed to a significant reduction in the expression of the M-phase inducer phosphatase cdc25B (Um et al., 2001). The absence of any drop in overall RNA synthesis in TBP-het cells was perhaps more surprising than the fact that TBP levels are implicated in regulation of cell cycle controlled genes. However, the recent isolation of TBP homologues (Berk, 2000; Bartfai et al., 2004; Jallow et al., 2004) has provided clues that transcriptional regulation is considerably more complex and interesting than provided for by a 'universal transcription factor.'

2.2 TLP regulates a subset of genes involved in proliferation of DT40 cells

One of the several TBP homologues that have been recently identified, TBP-like protein (TLP, TRF-2, TLF) has been shown to be homologous to the conserved C-terminal domain of TBP (Shimada et al., 1999; Berk, 2000). Further analysis of TBP-het cells revealed that TBP-het cells express RNA encoding the TBP homologue TBP-like protein (TLP) at more than twice the level of wild-type cells (Bush and Manley, unpublished data). Although it is appealing to speculate that TLP is up-regulated to compensate for lower levels of TBP, it seems more likely that TLP is expressed at higher levels because of the mitotic delay exhibited by the TBP-het cells, as TLP has been implicated in cell cycle control. Disruption of both alleles of the *tlp* gene in DT40 cells resulted in a cell cycle phenotype that directly contrasts with the TBP-het phenotype (Shimada et al., 2003). Instead of a longer G2 phase leading to mitotic delay, as in TBP-het cells (Um et al., 2001), DT40-TLP cells grew faster than wild-type cells due to a shorter G2 phase. Apoptosis in response to stress in cells lacking TLP was also reduced, implying a role for TLP in the G2 checkpoint mechanism (Shimada et al., 2003).

For some time it was speculated that TLP regulated transcription by sequestering essential factors, preventing their utilization in initiation complexes containing TBP. Indeed, TLP sequences necessary for binding TFIIA and TFIIB are identical to those found in TBP (Dantonel et al., 1999; Berk, 2000), and TLP does in fact repress transcription from the adenovirus major late promoter which contains a TATA box (Ohbayashi et al., 2003). However, *in vitro* analysis demonstrated that artificial recruitment of TLP to a promoter by the addition of a Gal-4 DNA binding domain supported transcription (Ohbayashi et al., 2001) and transcription from a terminal deoxynucleotidyl transferase (TdT) promoter lacking a TATA box was stimulated by TLP in transient transfection assays (Ohbayashi et al., 2003). RNA interference studies established that TLP is essential for embryogenesis in *C. elegans* and *X. laevis* (Dantonel et al., 2000; Kaltenbach et al., 2000; Veenstra et al., 2000). While mice lacking TLP were viable, the males are sterile due to defects in spermiogenesis (Martianov et al., 2002). The increase in TBP levels that correlate with increased TLP levels during spermatogenesis (Sugiura et al., 2003) provides further proof that although structurally similar, TBP and TLP have functionally divergent roles in transcriptional regulation. Taken together, the requirement for TLP during embryogenesis in *C. elegans* and *X. laevis* (Dantonel et al., 2000; Kaltenbach et al., 2000; Veenstra et al., 2000), regulation of spermatogenesis in mice by TLP (Martianov et al., 2002), and involvement in cell cycle checkpoints of TLP in DT40 cells, all seem to dictate a role for TLP in regulating transcription of a subset of genes required for cell proliferation in a species-specific manner.

2.3   TAF9 is essential for viability in DT40 cells

TBP-associated factors (TAFs), which together with TBP form the TFIID, are intricately involved in the process of promoter recognition. Activated transcription initiation *in vitro* is stimulated by their presence in TFIID, whereas TBP alone supports only basal transcription in *in vitro* assays (Burley and Roeder, 1996). About 15 conserved TAFs have been identified, of which 13 are essential in yeast (Li et al., 2002; Shen et al., 2003). Identification of variable combinations of TAFs and multiple homologues of TBP have redefined the concept of transcription initiation, auguring a complexity of regulation in metazoans that is being slowly unraveled (for review see Albright and Tjian, 2000; Veenstra and Wolffe, 2001; Hochheimer and Tjian, 2003).

Initial studies of one TAF, TAF9 (initially TAF (II) 31; for revised TAF nomenclature see Tora, 2002), in DT40 began with the construction of a conditional TAF9 knockout cell line (DT40-TAF9). Depletion of TAF9 by addition of tetracycline caused depletion of not only the tet regulated TAF9, but several other TAFs, although TBP levels were not significantly effected (Chen and Manley, 2000). TAF9 is essential for cell viability, as are most TAFs (for review see Lee and Young, 1998) and DT40-TAF9 cells stopped growing forty-eight hours after the addition of tet (Chen and Manley, 2000). TAF9 depleted cells displayed no definitive cell cycle abnormalities until cell death began, and cells began to accumulate in sub-G0/G1 (Chen and Manley, 2000). Unexpectedly, global regulation of transcription was not significantly affected by the absence of TAF9, even though TFIID complexes at least partially dissociated (Chen and Manley, 2003b). Attempts at rescue of the TAF9 phenotype with a C-terminal truncated version of

TAF9 restored the integrity of TFIID complexes, although expression of a TAF9 protein lacking the characteristic histone fold motif, characteristic of many TAFs, did not rescue cells depleted of TAF9 (Chen and Manley, 2003b). These studies extended previous work in yeast and provided the first evidence that many TAFs play a gene-specific but not general role in RNA polymerase II transcription.

Chen and Manley used the DT40-TAF9 cells to further our knowledge of the function of core promoter elements and TAF participation in vertebrate transcription regulation (Chen and Manley, 2003a). To investigate the roles of core promoter elements in transcriptional activation, they examined expression and factor occupancy on representative promoters. Characterized core elements, including TATA box-flanking regions and the downstream promoter element, were found to play significant roles in determining promoter strength, response to activators, and factor occupancy and recruitment. By adding tet to the medium, the dependence for TAF9 was found to be highly promoter specific, and TAF9 dependence and promoter occupancy were not always correlated (Chen and Manley, 2003a). These experiments illustrate both the important roles of core promoter elements and the diversity that characterizes transcriptional activation mechanisms in vertebrates. The systematic methods used in this series of experiments could be readily adapted to investigate the role of other general factors and corresponding *cis*-regulatory sequences involved in gene expression.

2.4   C-Myb exhibits functional redundancy in DT40 cells

The c-Myb proto-oncogene encodes a transcription factor thought to be essential for immature, proliferating hematopoietic cells (for review see Weston, 1998, Ness, 2003). Mice lacking a functional *c-myb* gene die during embryogenesis. (Mucenski et al., 1991) C-Myb has been definitively shown to be essential for the development of T cells from murine ES cells; however, recent evidence has shown that c-Myb is crucial for the development of all hematopoietic cells, including B cells (Reddy, 1999; Smith and Sigvardsson, 2004; Ezoe et al., 2004). Disruption of *c-myb* in DT40 was expected to have profound effects on the cell cycle, since previous studies indicated that c-Myb was involved in transcription of c-Myc, another transcription factor necessary for cell cycle progression (Evans et al., 1990; Nakagoshi et al., 1992). Paradoxically, c-Myb-null DT40 cells were not only viable, but displayed only minor variations in the cell cycle. FACs analysis indicated a slight rise in populations of cells in G-1, but no detectable difference in doubling time between c-Myb-null cells and wild-type DT40 was found (Appl and Klempnauer, 2002). Expression of the anti-apoptotic *bcl-2* gene has been similarly attributed to c-Myb in T-cells (Badiani et al., 1996). However, c-Myb-null DT40 cells produced similar amounts of bcl-2 mRNA as did wild-type cells and apoptotic time courses between the two cell lines were similar (Appl and Klempnauer, 2002). One gene product found to be down-regulated in DT40 cells lacking c-Myb was Pdcd4, a gene not fully characterized but implicated in inhibition of neoplastic transformation (Cmarik et al., 1999). The lack of cell cycle abnormalities in cells in c-Myb-null cells implies a functional redundancy of c-Myb in DT40 cells.

2.5 An Ets family member plays a limited role in development of DT40 cells

The Ets family of transcription factors regulates gene expression during the development of hematopoietic cells and, furthermore, these transcription factors play an important role in tumor formation and metastasis (for review see Sharrocks, et al., 2001). The characteristic Ets DNA binding domain is found in PU.1, which is expressed in immature B cells. Differential expression of PU.1 has been implicated in cell fate, with the lowest concentrations corresponding to B-cell differentiation (Dahl and Simon, 2003). Bi-allelic disruption of the gene encoding PU.1 in DT40 cells created a line of cells that produce a mutant protein lacking the conserved Ets DNA binding domain (Matsudo et al., 2003). The mutant cell line had a slightly prolonged doubling time, 1.3 times that of wild-type cells (Matsudo et al., 2003), demonstrating the expected effect of PU.1 on proliferation of B cells. However, Ig β and Ig λ, putative targets of the transcription factor, proved to be expressed at the same levels in wild-type and mutant cells, although reporter assays had implicated PU.1 binding sites in regulation of Ig λ in a murine cell line (Eisenbeis et al., 1993). PU.1 mutants were viable, indicating that PU.1 is not essential. The increased doubling time of the PU.1 mutants implies some function for PU.1 in regulation of cell cycle related genes, but these studies suggest that PU > 1 plays an unexpectedly minor role in DT40 cells.

2.6 Heat shock transcription factors have multiple roles in DT40 cells

Heat shock transcription factors (HSFs) undergo activating conformational changes when cells are exposed to elevated temperatures. They then activate transcription of heat shock protein encoding genes, and these proteins protect the cell and cause temporary cell cycle arrest (for review see Kuhl and Rensing, 2000). Disruption of both *hsf3* alleles in DT40 cells not only negatively affected the thermotolerance of the cells as expected, but HSF3-null cells also display a reduced growth rate in the absence of cellular stress (Tanabe et al., 1998). When *hsf1* was also disrupted in the HSF3-null cells, mRNA encoding the constitutively expressed heat shock protein Hsp90ά was reduced (Nakai and Ishikawa, 2001). Hsp90ά has been shown to interact with proteins expressed in a cell cycle-dependent manner in the absence of heat shock conditions (Aligue et al., 1994; Stepanova et al., 1997). Furthermore, cell cycle transition is blocked in wild-type DT40 cells after heat shock, a phenomenon lost in cells lacking HSF1 and HSF3 (Nakai and Ishikawa, 2001). C-Myb has been shown to directly interact with HSF3 during the cell cycle, thus providing further evidence linking HSF3 to cellular proliferation (Kanei-Ishii et al., 1997). Additional study is indicated to discover the mechanism whereby HSFs cooperate with c-Myb to regulate cell proliferation, and why only HSF, not c-Myb, knockout DT40 cells, show signs of cell cycle irregularities. But these studies have highlighted the importance of HSFs in multiple cell functions.

2.7   GAS41 broadly affects transcription in DT40 cells

V-Myb, the virally-encoded truncated version of c-Myb, regulates transcription of the glioma-amplified sequence, GAS41 (Braas et al., 2004). Lysozyme, a previously identified target of Myb regulation (Introna et al., 1990), is not expressed in DT40 cells. However, immediately downstream of lysozyme is a CpG island (Phi-van and Stratling, 1999) that when mapped led to the discovery of a novel gene encoding a putative transcription factor, GAS41. Targeted disruption of one allele of *gas41* reduces expression by nearly half, but attempts to eliminate expression completely were unsuccessful until a GAS41 cDNA driven by a tet-induced promoter was introduced. Depletion of GAS41 in cells grown in the presence of [3H]uridine demonstrated that cells gradually stopped synthesizing RNA in the absence of GAS41. Although this may be an indirect effect, restoring GAS41 expression before cell death began completely rescued RNA synthesis within 18 hours (Zimmermann et al., 2002). It is rare that knockout of a single transcription factor can lead to such global effects on RNA synthesis and cell proliferation. Analysis of the mechanism by which GAS41 depletion inhibits RNA synthesis may provide new insights into the complexes and mechanisms involved in basal transcription.

3.   RNA PROCESSING

3.1   ASF/SF2 affects multiple processes in DT40 cells

SR proteins are among the most prominent of protein factors required for splicing of pre-mRNA. SR proteins contain one or two RNP-type RNA binding domains and a carboxy-terminal domain that is rich in serine and arginine residues (RS domain). Required for both constitutive and alternative splicing, SR proteins came under intense scrutiny and new roles for these splicing factors in mRNA metabolism have been elicited (Manley and Tacke, 1996; Graveley, 2000; Huang and Steitz, 2005; Sanford et al., 2005). The prototypical SR protein, ASF/SF2 (Alternative Splicing Factor/Splicing Factor 2) was originally identified as an SR protein necessary to re-establish splicing activity to S100 (cytosolic) extracts *in vitro* and as a factor influencing the choice of alternative 5' splice sites (Ge and Manley, 1990; Krainer et al., 1990). The first DT40 line with a conditional knockout was established with exogenous ASF/SF2 under control of a tet-repressible promoter (DT40-ASF). Cells depleted of ASF/SF2 ceased growing and began to die between 48 and 60 hours after addition of tet (Wang et al., 1996). Analysis of these cells upon ASF/SF2 depletion indicated that the rate of mRNA processing, based on size analysis of pulse-labeled RNA, was slightly reduced, and alternative splicing of reporter RNAs was altered. However, general splicing was not significantly affected, indicating that ASF/SF2 function is largely redundant (Wang et al., 1996), although two other SR proteins SC35 and SRp40 failed to rescue the lethal phenotype of ASF/SF2 depletion (Wang et al., 1996). DT40-ASF cells depleted of ASF/SF2 expressing chimeric proteins consisting of the RNA binding domains of ASF/SF2 fused with the RS domains of other SR proteins grew at the same rate as wild-type cells (Wang et al., 1998), indicating that RS domains are largely interchangeable.

Analysis of DT40-ASF cells also revealed a role for ASF/SF2 in modulating mRNA stability. A screen of DT40-ASF cells to detect differential gene expression upon depletion of tet revealed that a single mRNA encoding a protein affecting protein kinase C activity, PKCI-r, was upregulated (Lemaire et al., 2002). Transcription and splicing patterns of the PKCI-r pre-mRNA were unchanged and the rate at which the second intron was spliced was actually lower in cells lacking ASF/SF2 than in wild-type cells. Inhibition of RNA synthesis revealed that the stability of the mature mRNA was enhanced in by ASF/SF2 depletion (Lemaire et al., 2002). Sequence analysis revealed a purine-rich sequence in the PKCI-r 3' UTR similar to the consensus binding site of ASF/SF2 (Tacke and Manley, 1995) and this sequence was shown to be important for ASF/SF2 mediated regulation. These studies indicate that a protein initially characterized in splicing of mRNA precursors can function in mRNA turnover.

ASF/SF2 depleted cells undergo cell cycle arrest and apoptosis (Li et al., 2005). The cells accumulate in G2 phase and exhibit several characteristics of apoptotic cell death (Li et al., 2005). Cleavage of DNA during apoptisis produces large DNA fragments of about 0.4-1 Mbp, 50 kbp and smaller fragments, produced by internucleosomal cleavage, that result in a characteristic DNA ladder (Samejima and Earnshaw, 2005). Analysis of apoptotic chromosomal DNA in DT40-ASF cells showed that the larger fragments were present, but the DNA laddering produced by internucleosomal cleavage was absent. Altered alternative splicing of the ICAD pre-mRNA, which produces an inhibitor of caspase-activated DNase, responsible for intranucleosomal cleavage, was found to be the cause of the lack of internucleosomal cleavage (Li et al., 2005).

An unexpected finding was observed during the course of the above experiments. Growth of DT40-ASF cells in the presence of tet for extended times resulted in a number of tet-resistant colonies in which ASF/SF2 had escaped tetracycline regulation. Upon further investigation, it was found that depletion of ASF/SF2 caused double stranded breaks (DSB) in chromosomal DNA, leading to the appearance of high molecular weight fragments and DNA rearrangements (Li and Manley, 2005). Recent experiments describing the coupling of RNA processing and transcription require that splicing factors associate with the transcriptional machinery (Hirose and Manley, 2000; Maniatis and Reed, 2002; Bentley, 2005). The lack of ASF/SF2 disrupts this association, leaving parts of the nascent RNA transcript uncoated and therefore able to anneal to template DNA, causing R-loops, in which the template strand is paired with the recently transcribed RNA and the nontemplate strand forms a "loop". In at least one case, the R-loop was located at the site where the DSB occurred, and stable RNAse H overexpression rescued both hypermutation and DSB phenotypes, providing evidence that R-loops indeed caused the DNA damage (Li and Manley, 2005). ASF/SF2 thus seems to play a role in packaging of nascent mRNAs and this is important for the maintenance of genome stability. Depletion of ASF/SF2 thus has the ability to affect cells in an extremely broad manner, altering RNA processing rates and mRNA stability, affecting genome stability, arresting cells in G2 phase, and causing cells to undergo apoptosis, as well as affecting splice site selection in alternative splicing.

3.2   CstF-64 affects alternative polyadenylation and the cell cycle in DT40 cells

The addition of the 3' poly(A) tail to mRNA has been linked to mRNA export, stability and translational efficiency. Polyadenylation involves an endonucleolytic cleavage of the nascent RNA followed by poly(A) synthesis (Colgan and Manley, 1997; Zhao et al., 1999). Multiple protein factors are required for cleavage and polyadenylation. One of these factors, cleavage stimulation factor (CstF), is a hetrotrimeric complex consisting of a 77 kDa subunit, a 64kDa subunit and a 50kDa subunit, all which have been shown to be essential in *in vitro* polyadenylation assays (Takagaki and Manley, 1994).

In B-cells differentiation is accompanied by a switch in the polyadenylation site of the immunoglobulin M heavy chain (IgM), from mRNA that specifies membrane bound IGM (μm) to mRNA that specifies secreted IgM (μs) (Peterson and Perry, 1986; Peterson et al., 1991). Levels of CstF64 are significantly reduced in mouse primary B-cells and higher in differentiated plasma cells, suggesting that this might affect the ratio of μm to μs (Takagaki et al., 1996). Indeed, overexpression of CstF64, the subunit that directly contacts RNA during polyadenylation, significantly increased the concentration of the entire CstF complex and, more importantly, increased levels of μs mRNA and decreased μm (Takagaki et al., 1996). To extend these results, a conditional knockout of CstF64 in DT40 cells (DT40-CstF64) was established and revealed a decrease in μs and increase in μm mRNA upon CstF64 depletion. This confirms that the level of CstF64 was responsible for poly(A) site selection of the IgM mRNA. In addition, analysis of these cells revealed that depletion of CstF64 caused cell cycle arrest and apoptotic-like cell death (Takagaki and Manley, 1998). A model, supported also by *in vitro* RNA binding and processing data, was proposed that elevated levels of CstF favor use of the upstream, but weaker, μs poly(A) site. A similar model was subsequently put forward to explain changes in polyadenylation of NF-ATc, a transcription factor expressed in T-cells (Chuvpilo et al., 1999) These findings illustrate how altering the concentration of a general, essential polyadenylation factor can contribute to gene regulation.

3.3 SAM68 is involved in all aspects of RNA processing

Signal transducers and activators of RNA (STAR) proteins contain a single KH domain and are involved in regulation of RNA metabolism (Vernet and Artzt, 1997; Lukong and Richard, 2003). Sam68, a prototypical STAR protein, has been implicated in regulation of nearly every step of RNA processing; influencing transcription by binding the transcriptional cofactor CBP (Hong et al., 2002), participating in splice site selection (Hartmann et al., 1999; Matter et al., 2002), and stimulating polyadenylation (McLaren et al., 2004), as well as documented roles in the cell cycle and apoptosis (Li et al., 2002; Taylor et al., 2004). Disruption of both alleles of SAM68 in DT40 produced a line of cells that do not express SAM68 (DT40-SAM68). Cells lacking SAM68 were viable, but had doubling times about half again as long as wild type DT40 cells due to a retardation of G2/M phase progression. However, cdc2 kinase activity, the key indicator of G2/M progression, did not differ significantly between wild-type and DT40-SAM 68 cell lines (Li et al., 2002). This data implies a mechanism other than the cyclin B/cdc2 kinase that is affected during cell cycle progression in DT40-SAM68 cells. Further studies in NIH 3T3 cell lines have implicated overexpression of SAM68 in G1 arrest and apoptosis (Taylor et al., 2004). The precise function of SAM68 in the cell remains elusive. However, the multiple RNA processes apparently affected by SAM68 suggest that further analysis of the knockout cells will be informative.

CONCLUSIONS AND FUTURE DIRECTIONS

The establishment of the DT40 cell line has been useful in the study of transcription and RNA processing factors *in vivo*. It is difficult to overlook the fact that disruption of all but one of the factors cited above affected cell proliferation in some manner, often drastically. Surprisingly, altered proliferation rates, cell cycle abnormalities and apoptosis are the most frequent phenotypes associated with altering factors that regulate RNA production, indicating the paramount importance and complexity of regulating genes involved in cell proliferation. The possibility exists that functionally redundant transcriptional and processing mechanisms are not affected by the altered expression of one factor in the vast array of proteins necessary for gene expression. Equally likely is that changes in levels of mature mRNA from genes related to less crucial processes are obscured by the overwhelming phenotypes present when the cell cycle is affected. It is difficult to analyze accurately the effects on RNA transcription or processing from cells undergoing cell cycle arrest or apoptosis, as these processes have inherent effects on gene expression (e.g. Burns and Gould, 1999; Blencowe, 2003). Global RNA production is seldom affected by a change in expression of any one factor involved in the production of mature mRNA, although cell death or cell cycle arrest may obscure some effects. Continued use of the DT40 system should facilitate the unraveling of the multiple functions of proteins involved in the complex process of gene expression.

REFERENCES

Aligue, R., Akhavan-Niak, H., and Russell, P. (1994). A role for Hsp90 in cell cycle control: Wee1 tyrosine kinase activity requires interaction with Hsp90. Embo J *13*, 6099-6106.

Appl, H., and Klempnauer, K. H. (2002). Targeted disruption of c-myb in the chicken pre B-cell line DT40. Oncogene *21*, 3076-3081.

Badiani, P. A., Kioussis, D., Swirsky, D. M., Lampert, I. A., and Weston, K. (1996). T-cell lymphomas in v-Myb transgenic mice. Oncogene *13*, 2205-2212.

Bartfai, R., Balduf, C., Hilton, T., Rathmann, Y., Hadzhiev, Y., Tora, L., Orban, L., and Muller, F. (2004). TBP2, a vertebrate-specific member of the TBP family, is required in embryonic development of zebrafish. Curr Biol *14*, 593-598.

Bentley, D. L. (2005). Rules of engagement: co-transcriptional recruitment of pre-mRNA processing factors. Curr Opin Cell Biol *17*, 251-256.

Berk, A. J. (2000). TBP-like factors come into focus. Cell *103*, 5-8.

Beyersmann, D. (2000). Regulation of mammalian gene expression. Exs *89*, 11-28.

Blencowe, B. J. (2003). Splicing regulation: the cell cycle connection. Curr Biol *13*, R149-151.

Braas, D., Gundelach, H., and Klempnauer, K. H. (2004). The glioma-amplified sequence 41 gene (GAS41) is a direct Myb target gene. Nucleic Acids Res *32*, 4750-4757.

Buerstedde, J. M., and Takeda, S. (1991). Increased ratio of targeted to random integration after transfection of chicken B cell lines. Cell *67*, 179-188.

Burley, S. K., and Roeder, R. G. (1996). Biochemistry and structural biology of transcription factor IID (TFIID). Annu Rev Biochem *65*, 769-799.

Burns, C. G., and Gould, K. L. (1999). Connections between pre-mRNA processing and regulation of the eukaryotic cell cycle. Front Horm Res *25*, 59-82.

Chen, Z., and Manley, J. L. (2000). Robust mRNA transcription in chicken DT40 cells depleted of TAF(II)31 suggests both functional degeneracy and evolutionary divergence. Mol Cell Biol *20*, 5064-5076.

Chen, Z., and Manley, J. L. (2003a). Core promoter elements and TAFs contribute to the diversity of transcriptional activation in vertebrates. Mol Cell Biol *23*, 7350-7362.

Chen, Z., and Manley, J. L. (2003b). *In vivo* functional analysis of the histone 3-like TAF9 and a TAF9-related factor, TAF9L. J Biol Chem *278*, 35172-35183.

Chuvpilo, S., Zimmer, M., Kerstan, A., Glockner, J., Avots, A., Escher, C., Fischer, C., Inashkina, I., Jankevics, E., Berberich-Siebelt, F., et al. (1999). Alternative polyadenylation events contribute to the induction of NF-ATc in effector T cells. Immunity *10*, 261-269.

Cmarik, J. L., Min, H., Hegamyer, G., Zhan, S., Kulesz-Martin, M., Yoshinaga, H., Matsuhashi, S., and Colburn, N. H. (1999). Differentially expressed protein Pdcd4 inhibits tumor promoter-induced neoplastic transformation. Proc Natl Acad Sci U S A *96*, 14037-14042.

Colgan, D. F., and Manley, J. L. (1997). Mechanism and regulation of mRNA polyadenylation. Genes Dev *11*, 2755-2766.

Conaway, R. C., and Conaway, J. W. (1997). General transcription factors for RNA polymerase II. Prog Nucleic Acid Res Mol Biol *56*, 327-346.

Dahl, R., and Simon, M. C. (2003). The importance of PU.1 concentration in hematopoietic lineage commitment and maturation. Blood Cells Mol Dis *31*, 229-233.

Dantonel, J. C., Quintin, S., Lakatos, L., Labouesse, M., and Tora, L. (2000). TBP-like factor is required for embryonic RNA polymerase II transcription in C. elegans. Mol Cell *6*, 715-722.

Dantonel, J. C., Wurtz, J. M., Poch, O., Moras, D., and Tora, L. (1999). The TBP-like factor: an alternative transcription factor in metazoa? Trends Biochem Sci *24*, 335-339.

Eisenbeis, C. F., Singh, H., and Storb, U. (1993). PU.1 is a component of a multiprotein complex which binds an essential site in the murine immunoglobulin lambda 2-4 enhancer. Mol Cell Biol *13*, 6452-6461.

Engel, J. D., Beug, H., LaVail, J. H., Zenke, M. W., Mayo, K., Leonard, M. W., Foley, K. P., Yang, Z., Kornhauser, J. M., Ko, L. J., and et al. (1992). cis and trans regulation of tissue-specific transcription. J Cell Sci Suppl *16*, 21-31.

Evans, J. L., Moore, T. L., Kuehl, W. M., Bender, T., and Ting, J. P. (1990). Functional analysis of c-Myb protein in T-lymphocytic cell lines shows that it trans-activates the c-myc promoter. Mol Cell Biol *10*, 5747-5752.

Ezoe, S., Matsumura, I., Satoh, Y., Tanaka, H., and Kanakura, Y. (2004). Cell cycle regulation in hematopoietic stem/progenitor cells. Cell Cycle *3*, 314-318.

Ge, H., and Manley, J. L. (1990). A protein factor, ASF, controls cell-specific alternative splicing of SV40 early pre-mRNA in vitro. Cell *62*, 25-34.

Graveley, B. R. (2000). Sorting out the complexity of SR protein functions. Rna *6*, 1197-1211.

Hartmann, A. M., Nayler, O., Schwaiger, F. W., Obermeier, A., and Stamm, S. (1999). The interaction and colocalization of Sam68 with the splicing-associated factor YT521-B in nuclear dots is regulated by the Src family kinase p59(fyn). Mol Biol Cell *10*, 3909-3926.

Hernandez, N. (1993). TBP, a universal eukaryotic transcription factor? Genes Dev *7*, 1291-1308.

Hirose, Y., and Manley, J. L. (2000). RNA polymerase II and the integration of nuclear events. Genes Dev *14*, 1415-1429.

Hochheimer, A., and Tjian, R. (2003). Diversified transcription initiation complexes expand promoter selectivity and tissue-specific gene expression. Genes Dev *17*, 1309-1320.

Hong, W., Resnick, R. J., Rakowski, C., Shalloway, D., Taylor, S. J., and Blobel, G. A. (2002). Physical and functional interaction between the transcriptional cofactor CBP and the KH domain protein Sam68. Mol Cancer Res *1*, 48-55.

Huang, Y., and Steitz, J. A. (2005). SRprises along a messenger's journey. Mol Cell *17*, 613-615.

Introna, M., Golay, J., Frampton, J., Nakano, T., Ness, S. A., and Graf, T. (1990). Mutations in v-myb alter the differentiation of myelomonocytic cells transformed by the oncogene. Cell *63*, 1289-1297.

Jallow, Z., Jacobi, U. G., Weeks, D. L., Dawid, I. B., and Veenstra, G. J. (2004). Specialized and redundant roles of TBP and a vertebrate-specific TBP paralog in embryonic gene regulation in Xenopus. Proc Natl Acad Sci U S A *101*, 13525-13530.

Johnson, S. A., Dubeau, L., Kawalek, M., Dervan, A., Schonthal, A. H., Dang, C. V., and Johnson, D. L. (2003). Increased expression of TATA-binding protein, the central transcription factor, can contribute to oncogenesis. Mol Cell Biol *23*, 3043-3051.

Kaltenbach, L., Horner, M. A., Rothman, J. H., and Mango, S. E. (2000). The TBP-like factor CeTLF is required to activate RNA polymerase II transcription during C. elegans embryogenesis. Mol Cell *6*, 705-713.

Kanei-Ishii, C., Tanikawa, J., Nakai, A., Morimoto, R. I., and Ishii, S. (1997). Activation of heat shock transcription factor 3 by c-Myb in the absence of cellular stress. Science *277*, 246-248.

Keegan, L. P., Gallo, A., and O'Connell, M.A. (2001). The many roles of an RNA editor. Nat Rev Genet *2*, 869-878.

Krainer, A. R., Conway, G. C., and Kozak, D. (1990). The essential pre-mRNA splicing factor SF2 influences 5' splice site selection by activating proximal sites. Cell *62*, 35-42.

Kuhl, N. M., and Rensing, L. (2000). Heat shock effects on cell cycle progression. Cell Mol Life Sci *57*, 450-463.

Lemaire, R., Prasad, J., Kashima, T., Gustafson, J., Manley, J. L., and Lafyatis, R. (2002). Stability of a PKCI-1-related mRNA is controlled by the splicing factor ASF/SF2: a novel function for SR proteins. Genes Dev *16*, 594-607.

Li, Q. H., Haga, I., Shimizu, T., Itoh, M., Kurosaki, T., and Fujisawa, J. (2002). Retardation of the G2-M phase progression on gene disruption of RNA binding protein Sam68 in the DT40 cell line. FEBS Lett *525*, 145-150.

Li, X., and Manley, J. L. (2005). Inactivation of the SR protein splicing factor ASF/SF2 results in genomic instability. Cell *122*, 365-378.

Li, X., Wang, J., and Manley, J. L. (2005). Loss of splicing factor ASF/SF2 induces G2 cell cycle arrest and apoptosis, but inhibits internucleosomal DNA fragmentation. Genes and and Dev, in press.

Li, X. Y., Bhaumik, S. R., Zhu, X., Li, L., Shen, W. C., Dixit, B. L., and Green, M. R. (2002). Selective recruitment of TAFs by yeast upstream activating sequences. Implications for eukaryotic promoter structure. Curr Biol *12*, 1240-1244.

Liu, X., Mayeda, A., Tao, M., and Zheng, Z. M. (2003). Exonic splicing enhancer-dependent selection of the bovine papillomavirus type 1 nucleotide 3225 3' splice site can be rescued in a cell lacking splicing factor ASF/SF2 through activation of the phosphatidylinositol 3-kinase/Akt pathway. J Virol *77*, 2105-2115.

Lukong, K. E., and Richard, S. (2003). Sam68, the KH domain-containing superSTAR. Biochim Biophys Acta *1653*, 73-86.

Maniatis, T., and Reed, R. (2002). An extensive network of coupling among gene expression machines. Nature *416*, 499-506.

Manley, J. L. (2002). Nuclear coupling: RNA processing reaches back to transcription. Nat Struct Biol *9*, 790-791.

Manley, J. L., and Tacke, R. (1996). SR proteins and splicing control. Genes Dev *10*, 1569-1579.

Martianov, I., Brancorsini, S., Gansmuller, A., Parvinen, M., Davidson, I., and Sassone-Corsi, P. (2002). Distinct functions of TBP and TLF/TRF2 during spermatogenesis: requirement of TLF for heterochromatic chromocenter formation in haploid round spermatids. Development *129*, 945-955.

Matsudo, H., Otsuka, A., Ozawa, Y., and Ono, M. (2003). Disruption of the PU.1 gene in chicken B lymphoma DT40 cells and its effect on reported target gene expression. Gene *322*, 169-174.

Matter, N., Herrlich, P., and Konig, H. (2002). Signal-dependent regulation of splicing via phosphorylation of Sam68. Nature *420*, 691-695.

McLaren, M., Asai, K., and Cochrane, A. (2004). A novel function for Sam68: enhancement of HIV-1 RNA 3' end processing. Rna *10*, 1119-1129.

Mucenski, M. L., McLain, K., Kier, A. B., Swerdlow, S. H., Schreiner, C. M., Miller, T. A., Pietryga, D. W., Scott, W. J., Jr., and Potter, S. S. (1991). A functional c-myb gene is required for normal murine fetal hepatic hematopoiesis. Cell *65*, 677-689.

Muller, F., and Tora, L. (2004). The multicoloured world of promoter recognition complexes. Embo J *23*, 2-8.

Nakagoshi, H., Kanei-Ishii, C., Sawazaki, T., Mizuguchi, G., and Ishii, S. (1992). Transcriptional activation of the c-myc gene by the c-myb and B-myb gene products. Oncogene *7*, 1233-1240.

Nakai, A., and Ishikawa, T. (2001). Cell cycle transition under stress conditions controlled by vertebrate heat shock factors. Embo J *20*, 2885-2895.

Ness, S. A. (2003). Myb protein specificity: evidence of a context-specific transcription factor code. Blood Cells Mol Dis *31*, 192-200.

Ohbayashi, T., Shimada, M., Nakadai, T., and Tamura, T. A. (2001). TBP-like protein (TLP/TLF/TRF2) artificially recruited to a promoter stimulates basal transcription *in vivo*. Biochem Biophys Res Commun *285*, 616-622.

Ohbayashi, T., Shimada, M., Nakadai, T., Wada, T., Handa, H., and Tamura, T. (2003). Vertebrate TBP-like protein (TLP/TRF2/TLF) stimulates TATA-less terminal deoxynucleotidyl transferase promoters in a transient reporter assay, and TFIIA-binding capacity of TLP is required for this function. Nucleic Acids Res *31*, 2127-2133.

Peterson, M. L., Gimmi, E. R., and Perry, R. P. (1991). The developmentally regulated shift from membrane to secreted mu mRNA production is accompanied by an increase in cleavage-polyadenylation efficiency but no measurable change in splicing efficiency. Mol Cell Biol *11*, 2324-2327.

Peterson, M. L., and Perry, R. P. (1986). Regulated production of mu m and mu s mRNA requires linkage of the poly(A) addition sites and is dependent on the length of the mu s-mu m intron. Proc Natl Acad Sci U S A *83*, 8883-8887.

Phi-van, L., and Stratling, W. H. (1999). An origin of bidirectional DNA replication is located within a CpG island at the 3" end of the chicken lysozyme gene. Nucleic Acids Res *27*, 3009-3017.

Reddy, T. R., Xu, W., Mau, J. K., Goodwin, C.D., Suhasini, M., Tang, H., Frimpong, K., Rose, D. W., and Wong-Staal, F. (1999). Inhibition of HIV replication by dominant negative mutants of Sam68, a functional homolog of HIV-1. Rev Nat Med *5*, 635-642.

Reinberg, D., Orphanides, G., Ebright, R., Akoulitchev, S., Carcamo, J., Cho, H., Cortes, P., Drapkin, R., Flores, O., Ha, I., et al. (1998). The RNA polymerase II general transcription factors: past, present, and future. Cold Spring Harb Symp Quant Biol *63*, 83-103.

Roeder, R. G. (1998). Role of general and gene-specific cofactors in the regulation of eukaryotic transcription. Cold Spring Harb Symp Quant Biol *63*, 201-218.

Roeder, R. G. (2005). Transcriptional regulation and the role of diverse coactivators in animal cells. FEBS Lett *579*, 909-915.

Samejima, K., and Earnshaw, W. C. (2005). Trashing the genome: the role of nucleases during apoptosis. Nat Rev Mol Cell Biol.

Sanford, J. R., Ellis, J., and Caceres, J. F. (2005). Multiple roles of arginine/serine-rich splicing factors in RNA processing. Biochem Soc Trans *33*, 443-446.

Sharrocks, A. D. (2000). Introduction: the regulation of eukaryotic transcription factor function. Cell Mol Life Sci *57*, 1147-1148.

Shen, W. C., Bhaumik, S. R., Causton, H. C., Simon, I., Zhu, X., Jennings, E. G., Wang, T. H., Young, R. A., and Green, M. R. (2003). Systematic analysis of essential yeast TAFs in genome-wide transcription and preinitiation complex assembly. Embo J *22*, 3395-3402.

Shimada, M., Nakadai, T., and Tamura, T. A. (2003). TATA-binding protein-like protein (TLP/TRF2/TLF) negatively regulates cell cycle progression and is required for the stress-mediated G(2) checkpoint. Mol Cell Biol *23*, 4107-4120.

Shimada, M., Ohbayashi, T., Ishida, M., Nakadai, T., Makino, Y., Aoki, T., Kawata, T., Suzuki, T., Matsuda, Y., and Tamura, T. (1999). Analysis of the chicken TBP-like protein(tlp) gene: evidence for a striking conservation of vertebrate TLPs and for a close relationship between vertebrate tbp and tlp genes. Nucleic Acids Res *27*, 3146-3152.

Smith, E., and Sigvardsson, M. (2004). The roles of transcription factors in B lymphocyte commitment, development, and transformation. J Leukoc Biol *75*, 973-981.

Sonoda, E., Morrison, C., Yamashita, Y. M., Takata, M., and Takeda, S. (2001). Reverse genetic studies of homologous DNA recombination using the chicken B-lymphocyte line, DT40. Philos Trans R Soc Lond B Biol Sci *356*, 111-117.

Stepanova, L., Leng, X., and Harper, J. W. (1997). Analysis of mammalian Cdc37, a protein kinase targeting subunit of heat shock protein 90. Methods Enzymol *283*, 220-229.

Sugiura, S., Kashiwabara, S., Iwase, S., and Baba, T. (2003). Expression of a testis-specific form of TBP-related factor 2 (TRF2) mRNA during mouse spermatogenesis. J Reprod Dev *49*, 107-111.

Tacke, R., and Manley, J. L. (1995). The human splicing factors ASF/SF2 and SC35 possess distinct, functionally significant RNA binding specificities. Embo J *14*, 3540-3551.

Takagaki, Y., and Manley, J. L. (1994). A polyadenylation factor subunit is the human homologue of the Drosophila suppressor of forked protein. Nature *372*, 471-474.

Takagaki, Y., Seipelt, R. L., Peterson, M. L., and Manley, J. L. (1996). The polyadenylation factor CstF-64 regulates alternative processing of IgM heavy chain pre-mRNA during B cell differentiation. Cell *87*, 941-952.

Takagaki, Y., and Manley, J. L. (1998). Levels of polyadenylation factor CstF-64 control IgM heavy chain mRNA accumulation and other events associated with B cell differentiation. Mol Cell *2*, 761-771.

Tanabe, M., Kawazoe, Y., Takeda, S., Morimoto, R. I., Nagata, K., and Nakai, A. (1998). Disruption of the HSF3 gene results in the severe reduction of heat shock gene expression and loss of thermotolerance. Embo J *17*, 1750-1758.

Taylor, S. J., Resnick, R. J., and Shalloway, D. (2004). Sam68 exerts separable effects on cell cycle progression and apoptosis. BMC Cell Biol *5*, 5.

Um, M., Yamauchi, J., Kato, S., and Manley, J. L. (2001). Heterozygous disruption of the TATA-binding protein gene in DT40 cells causes reduced cdc25B phosphatase expression and delayed mitosis. Mol Cell Biol *21*, 2435-2448.

Veenstra, G. J., Weeks, D. L., and Wolffe, A. P. (2000). Distinct roles for TBP and TBP-like factor in early embryonic gene transcription in Xenopus. Science *290*, 2312-2315.

Vernet, C., and Artzt, K. (1997). STAR, a gene family involved in signal transduction and activation of RNA. Trends Genet *13*, 479-484.

Wang, J., and Manley, J. L. (1997). Regulation of pre-mRNA splicing in metazoa. Curr Opin Genet Dev *7*, 205-211.

Wang, J., Takagaki, Y., and Manley, J. L. (1996). Targeted disruption of an essential vertebrate gene: ASF/SF2 is required for cell viability. Genes Dev *10*, 2588-2599.

Wang, J., Xiao, S. H., and Manley, J. L. (1998b). Genetic analysis of the SR protein ASF/SF2: interchangeability of RS domains and negative control of splicing. Genes Dev *12*, 2222-2233.

Warren, A. J. (2002). Eukaryotic transcription factors. Curr Opin Struct Biol *12*, 107-114.

Weston, K. M. (1998). The myb genes. Semin Cancer Biol *1*, 371-382.

Zhao, J., Hyman, L., and Moore, C. (1999). Formation of mRNA 3' ends in eukaryotes: mechanism, regulation, and interrelationships with other steps in mRNA synthesis. Microbiol Mol Biol Rev *63*, 405-445.

Zheng, Z. M. (2004). Regulation of alternative RNA splicing by exon definition and exon sequences in viral and mammalian gene expression. J Biomed Sci *11*, 278-294.

Zimmermann, K., Ahrens, K., Matthes, S., Buerstedde, J. M., Stratling, W. H., and Phi-van, L. (2002). Targeted disruption of the GAS41 gene encoding a putative transcription factor indicates that GAS41 is essential for cell viability. J Biol Chem *277*, 18626-18631.

PARTICIPATION OF HISTONES, HISTONE MODIFYING ENZYMES AND HISTONE CHAPERONES IN VERTEBRATE CELL FUNCTIONS

Hidehiko Kikuchi[1,2], Hirak Kumar Barman[2,3], Masami Nakayama[2], Yasunari Takami[2] and Tatsuo Nakayama[1,2]

[1]*Department of Life Science, Frontier Science Research Center,* [2]*Section of Biochemistry and Molecular Biology, Department of Medical Sciences, Miyazaki Medical College, University of Miyazaki, 5200, Kihara, Kiyotake, Miyazaki 889-1692, Japan;* [3]*Indian Council of Agricultural Research, New Delhi, India.*

Abstract: Alterations in the chromatin structure are essential for easy accesses to chromosomal DNA. Such architectural alterations can be achieved by four means: (i) variants of histone subtypes, (ii) chromatin remodeling, (iii) post-translational modification, and (iv) chromatin assembly. This chapter discusses mainly on the first, third and fourth mechanisms, and especially on the acetylation of core histones, one of the third mechanisms. Taking the advantage of the gene targeting technique, we systematically established numerous mutant DT40 cell lines, each lacking particular gene, of interest such that encoding histones, histone deacetylases (HDACs), acetyltransferases (HATs) and chaperones, etc. Every subtype member of the histone gene family is capable of compensating the loss of others to maintain the mRNA level of each histone subtype, and most of histone variants are involved positively or negatively in the transcription regulation of particular genes. Regarding HDACs, HDAC-2 controls the amount of the IgM H-chain at the steps of both transcription and alternative pre-mRNA processing, and HDAC-3 is indispensable for cell viability. Concerning HATs, GCN5 has tremendous impact on growth kinetics by preferentially acting as a supervisor in the normal cell cycle progression. The distinct participatory roles of the N-terminal and C-terminal halves of HIRA, one of histone chaperones, in both cell growth and transcription regulations of cell cycle-related genes, have also been highlighted. Therefore, the gene targeting technique in the DT40 cell line can be used as a powerful tool for the functional analysis of histones, histone modifying enzymes and histone chaperones relevant to chromatin biology.

J.-M. Buerstedde and S. Takeda (eds.), Reviews and Protocols in DT40 Research, 225–243.

Key words: Histone; histone acetyltransferase (HAT); histone deacetylase (HDAC); histone chaperone; DT40

1.   INTRODUCTION

In eukaryotes, genetic information is present in a compact structure called chromatin, which is not only responsible for packaging genomic DNA into nucleus efficiently but also provides the place for various DNA catalyzed reactions, such as replication, recombination, repair and transcription events. The fundamental repeating unit of chromatin is the nucleosome, which consists of ~146 base pairs of DNA wrapped around a histone octamer, comprising two molecules each of core histones H2A, H2B, H3 and H4. In higher eukaryotes, a linker histone H1 (or H5) remains associated with linker DNA (0-80 base pairs) between nucleosomes forming a higher order organization. A large number of each histone subtype essential for maintenance of chromatin structure must be rapidly accumulated in nucleus prior to the division of cells. For this purpose, the genes encoding all histone subtypes are present in multiple copies in higher eukaryotes, ranging from several dozen to hundreds, whereas yeast has only two genes for each of the core histones. With the assistance of a number of non-histone proteins, including high-mobility group (HMG) proteins, genomic DNA folds around nucleosomes to form 10 nm fibers that again fold helically into 30 nm chromatin fibers. These further form loops observed in the prophase chromosome axis that coils to form the fully condensed metaphase chromosome.

Epigenetic mechanisms can define alterations in cellular phenotypes without altering genotypes (Biel et al., 2005). According to this theory, the epigenetic control of transcriptional activation or silencing is mostly influenced by the intricately and timely modifications of chromatin-bound histones mediated by acetylation, methylation, phosphorylation and ubiquitination so called known as histone code. One particular interest among these delicate modifications is the post-translational and reversible core histone acetylation catalyzed by histone acetyltransferases (HATs) and deacetylation catalyzed by histone deacetylases (HDACs) as epigenetic phenomena that play critical roles in the modulation of chromatin topology and thereby regulation of gene expressions. However, the detailed understanding with regard to physiological roles played by individual HATs and HDACs is still not known completely in vertebrate cells.

On the other hand, histone chaperones are the group of proteins that aid in the dynamics of chromatin organization during various cellular processes. According to the recent studies, *de novo* nucleosome assembly is mediated by H3/H4 chaperones, i.e. chromatin assembly factor-1 (CAF-1), consisting of p150, p60 and p48 subunits, that loads H3/H4 onto replicated DNA; and anti silencing function 1 (ASF1) facilitates this process (Loyola and Almouzni, 2004). In addition, CAF-1 is involved in replication-dependent nucleosome assembly with the major histone H3 (H3.1), and HIRA, a homolog of *Saccharomyces cerevisiae* transcriptional co-repressors Hir1p and Hir2p, facilitates replication-independent nucleosome assembly with another histone H3 variant (H3.3), while ASF1 could act as a histone donor and maintenance of cytosolic histones H3 and H4 pool (Tagami et al., 2004). However, physiological significances of these are also poorly understood, especially in higher eukaryotes.

To better understand participatory roles of these proteins concerning histone metabolisms, modifications, nucleosome assembly and transcription regulations, we generated several homozygous DT40 mutant cell lines each devoid of histones, HATs, HDACs and histone chaperones, using gene targeting techniques (Buerstedde and Takeda, 1991). In this chapter, we will focus on physiological roles being played by some of histone variants, histone-modifying enzymes and histone chaperones by analyzing these mutant DT40 cells.

2. HISTONE GENES AND VARIANTS

To ensure a steady supply of a large amount of every histone subtype, histone-encoding genes are present in multiple copies in higher eukaryotes. In chickens, 39 of 44 histone genes: six H1, nine H2A, eight H2B, eight H3 and eight H4, are located within a major histone gene cluster of ~110 kb, while others are in 4 different regions (Takami et al., 1996). On the other hand, the structures of histone proteins are highly conserved from lower to higher eukaryotes. Almost all the histone proteins identified in divergent species have several variants with specific amino acid substitution(s) instead of a unique amino acid sequence. This led to suppose that each variants of histone subtypes must be involved in particular type of cellular processes besides the maintenance of the chromatin structure, for which a unique amino acid sequence should be sufficient for each histone subtype. For example, a set of histones, consisting of six H1, at least three H2A, four H2B and two H3 variants, and an H4 protein, exist in chickens (Takami et al., 1996). This also supported the proposition that histone H4 is one of the most extremely evolutionarily conserved protein, with little amino acid sequence difference even between plants and mammals. These observations suggested that each variant of histone subtypes has been evolved to conduct specific purpose with regard to regulation of gene expressions, in addition to their vital role in the chromatin organization. In this regard, ample evidences have been accumulated in *Saccharomyces cerevisiae, Drosophila melanogaster, Xenopus* and *Tetrahymena thermophila* (Nakayama and Takami, 2001).

Using the gene targeting technique, we generated various homozygous DT40 mutants devoid of particular histone gene(s). Interestingly, deletion of one allele of the 110 kb gene cluster in DT40 cells did not affect in their proliferative rate (Takami and Nakayama, 1997). One allele of this major histone gene cluster was shown to be sufficient for normal cell function as the steady state levels of total mRNAs for respective proteins of H1, H2A, H2B, H3 and H4 in the heterozygous mutant remained constant as those of wild-type cells. When two alleles of both H3-IV and H3-V accounting ~24% of mRNA levels from all the 11 H3 genes, and those of H2B-V encoding a particular H2B variant sharing ~10% of mRNA levels from all the eight H2B genes, respectively, were disrupted, the mRNA levels of remaining respective genes were increased so as to compensate the loss of them (Takami et al., 1995a; Takami et al., 1995b). Hence, the growth rate was not deviated in these mutant cells. Similar genotypic and phenotypic findings were recorded in DT40 mutants lacking each of H1 variants, i.e. 02H1, 10H1, 03H1, 01H1, H1L and H1R (Seguchi et al., 1995; Takami et al., 2000). In support of these findings, even knocking out of two alleles that contained approximately half set of histone genes, i.e. 21 histone gene segment of 57 kb out of 42 genes encoding variants of all histone subtypes, maintained constant expression levels of corresponding histone subtype genes (Takami et al., 1997). Interestingly, however, 2D-PAGE analyses revealed the quantitative appearance or disappearance of some cellular proteins even though overall protein patterns remained unchanged in these histone variant-deficient mutant clones, when compared to wild-type DT40 cells. Thus, our systematic analyses on histone genes using the gene targeting technique lead at least three important results as follows. Firstly, all of the histone gene family has the inherent ability to compensate for their deletion and to maintain the amount of each histone subtype. Secondly, only one copy of the H1 genes (12 copies) is enough for cell viability. Thirdly, most of the H1 and core histone variants should participate positively or negatively in transcription regulation of particular genes, probably through alterations in the chromatin structure. All our results concerning histone genes were briefly summarized in Table 1.

3. HISTONE-MODIFYING ENZYMES

3.1 Histone deacetylases (HDACs)

The catalytic activity of HDACs is to deacetylate by removing acetyl groups from Lys residues of acetylated histones, particularly from the N-terminal tail of all core histones, which extend outside the nucleosome core

Table 13-1. Histone genes disrupted in the DT40 cell line.

Genes		Function	Reference
01H1	H1 variant	chromatin organization transcription regulation	Seguchi et al., 1995
02H1	H1 variant	chromatin organization transcription regulation	Takami et al., 2000
.10H1	H1 variant	chromatin organization transcription regulation	Takami et al., 2000
03H1	H1 variant	chromatin organization transcription regulation	Takami et al., 2000
H1L	H1 variant	chromatin organization transcription regulation	Takami et al., 2000
H1R	H1 variant	chromatin organization transcription regulation	Takami et al., 2000
H2B-V	H2B variant	chromatin organization transcription regulation	Takami et al., 1995b
H3-IV/V	H3 variants	chromatin organization	Takami et al., 1995a
Half set of histone genes (57kb deletion)		chromatin organization	Takami et al., 1997
Allele of major histone gene cluster (110kb deletion)		chromatin organization	Takami and Nakayama 1997

particle. Histone deacetylation is associated with transcriptoin repression of various genes, because the removal of acetyl groups by HDACs imposes positive charge in the histones allowing their tighter interactions with DNA and thereby limiting accessibility of DNA for trans-acting factors (Biel et al., 2005; Mai et al., 2005). HDACs have been grouped into at least three classes. Class I HDACs are originally derived from the HDAC RPD3 in yeast. In chickens, HDAC-1, 2, 3 and 8 are known as the member of class I HDACs, because of their strong structural resemblance, nuclear localization and expressions in most cells. On the other hand, Class II HDACs (HDAC-4, 5, 6, 7 and 9 in chickens) are homologous to yeast HDAC HDA1, show cell type-specific expression patterns, and are ubiquitously present in both nucleus and cytoplasm. Last, class III HDACs (SIRT1~7) have similar structure derived from silent information regulator (Sir) 2 of yeast and utilize nicotinamide adenine dinucleotide (NAD) as a cofactor for their catalytic activity.

Recently, HDAC inhibitors, including trichostatin A (TSA), attract attention as anticancer agents (Mai et al., 2005). Inhibition of HDAC activity maintains the acetylated state causing hyperacetylation of histone proteins. As a result of which gene regulations will definitely be altered leading to hindrances in cell proliferation and differentiation, and finally eventual induction of apoptotic cell death. Indeed, several HDAC inhibitors could effectively induce apoptotic cell death in many types of tumor cells by arresting at the G1 or G2/M phase of the cell cycle. This raised the expectation to identify and develop inhibitors as anticancer agents specifically targeting HDAC isozymes. To aim that the studies concerning individual functions of HDACs have become more and more important. In order to shed light on physiological functions of individual HDACs, we generated several independent clones of homozygous HDAC-deficient DT40 mutant cell lines.

3.1.1   HDAC-1 and 2

We cloned cDNAs encoding two chicken HDACs, HDAC-1 and 2, comprising of 480 and 488 amino acids and exhibiting 93.8% and 97.1% homology to human HDAC-1 and 2, respectively. The C-terminal regions of both HDAC-1 and 2 contain the consensus sequence for the Rb-binding motif, but considerable difference of approximately 50 amino acids was detected between the C-terminal regions of these two enzymes. Like in other organisms, HDAC-1 and 2 are preferentially localized in nuclei as detected by the immunofluorescence imazing using antibodies raised against HDAC-1 and 2, respectively.

Next, we generated two homozygous DT40 mutants, ΔHDAC-1 and ΔHDAC-2, devoid of two alleles of either HDAC-1 and -2, respectively. In order to distinguish the individual roles being played by HDAC-1 and HDAC-2, we compared total cellular proteins from ΔHDAC-1 and ΔHDAC-2 with those from the wild-type cell line (Takami et al., 1999). The protein patterns on 2D-PAGE for ΔHDAC-2 were obviously distinct from those for the wild-type cell line, while changes were insignificant for ΔHDAC-1. Interestingly, the transcriptional up-regulation of IgM H- and L-chain genes was detected resulting in the accumulated amounts of IgM H- and L-chains in ΔHDAC-2. Therefore, we concluded that HDAC-1 and 2 must have distinct biological roles. The differential participatory roles of HDAC-1 and 2 in the transcriptional regulation of IgM H and L-chain genes should be due to the structural difference in their C-terminal amino acid sequences having considerably low homology (~50%) as compared to their N-terminal ends exhibiting extensive homology (~94%).

Furthermore, the secreted form (μs) of IgM H-chain increased inΔHDAC-2, but the membrane-bound form (μm) of it decreased. As a matter of fact, the alternative processing from μm to μs mRNA of IgM H-chain pre-mRNA preferentially occurred in the mutant, resulting in increased amounts of μs mRNA without affecting the stabilities of these two types of mRNAs. Interestingly, TSA treated DT40 cells also showed the similar effect of increasing amounts of IgM H-chain mRNAs and the switch from μm to μs mRNA. Based on these results, a model regarding the participatory roles of HDAC-2 in the DT40 cell line has been proposed (Figure 1). HDAC-2 mediated transcription regulation probably occurs in association with the hypothetical signal(s) that reduces HDAC-2 activity considerably surrounding the IgM H-chain gene, resulting in increased transcription of this target gene. Simultaneous regulation of the gene(s) encoding putative switch-related factor(s) should promote the switch from μm to μs mRNA. Thus, HDAC-2 maintains the μs form of IgM H-chain by dual controlling at the steps of both the transcription and alternative pre-mRNA processing via acetylation mediated modulation of chromatin structure. Moreover, HDAC-2 is also involved in control of the transcription of other genes encoding a set of B cell-specific proteins (unpublished data).

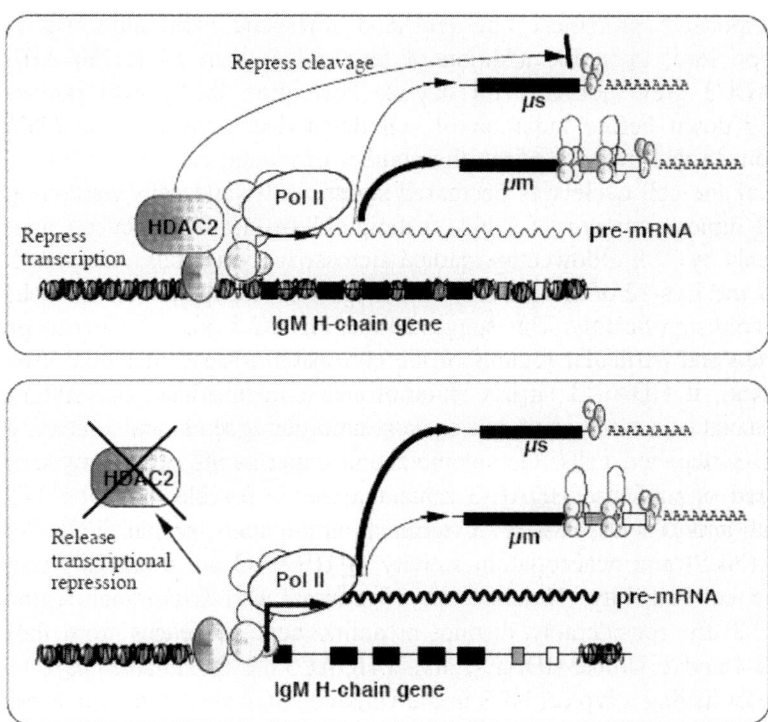

Figure 13-1. A model for a role of HDAC-2 in the control of the amount of the IgM H-chain (see plate 32).

3.1.2 HDAC-3

We cloned both cDNA and genomic DNA for HDAC-3 that differed remarkably in size from both HDAC-1 and 2 (Takami and Nakayama, 2000).

The HDAC-3 protein encoded by the cloned cDNA comprises 428 amino acids and exhibits approximately 97% identity to human and mouse HDAC-3s. Immunofluorescence assay confirmed its presence in both nuclei and cytoplasms. The TSA treatment could also abolish HDAC activity in the HDAC-3 immunoprecipitate.

Using the conventional gene targeting technique, we could easily generate a heterozygous HDAC-3-deficient mutant but failed to do so for a homozygous one after several attempts, indicating that HDAC-3 might be essential for the survival of DT40 cells. Accordingly, we obtained a conditional homozygous HDAC-3-deficient mutant, ΔHDAC-3/FHDAC3, carrying a HDAC-3 transgene, encoding FLAG-tagged HDAC-3 driven by a tetracycline (tet)-responsive promoter. The HDAC-3 transgene expression was below detection level upon the addition of tet for less than 24 h, but ΔHDAC-3/FHADC3 grew normally till day 2. Thereafter, the growth rate started slowing down before initiation of cell death due to loss of the FHDAC3 function. In the presence of tet, the number of mutant cells in the S or G2/M phase of the cell cycle was decreased significantly, and cells containing less than a diploid amount of DNA (subdiploid or sub-G1 fraction) appeared obviously by 72 h followed by marked increase in dying cells. Acetylated state (Lys-8 and Lys-12 of histone H4) for the bulk chromosomal core histones was not altered significantly. This suggested that HDAC-3 possibly deacetylate the restricted and particular regions of the chromatin, instead of global effect on chromatin, if HDAC-3 targets chromosomal core histones. In addition, the expressions of a number of cell cycle-related genes remained constant in the HDAC-3 depleted cells. Complementation experiments, involving series of truncated or missense HDAC-3 mutant proteins, revealed that the 1-23 N-terminal amino acids, 389-417 C-terminal amino acids, proper nuclear export signal (NES) and deacetylation activity of HDAC-3 are essentially required for the cell viability. Notably, the N-terminal and C-terminal regions of HDAC-3 are considerably distinct in amino acid sequences from those of HDAC-1 and 2. Unlike HDAC-1 and 2, HDAC3 has the notable characteristic feature of having a typical NES in the variable N-terminal end, which helps to export from nucleus to cytoplasm. HDAC-3 possibly performs the scavenging function in the cytoplasm by deacetylating abnormally or unnecessarily acetylated histone proteins before their import into nucleus for nucleosome assembly and thereby contributes in the proliferation of DT40 cells.

3.1.3   Other HDACs

Concerning chicken SIRT1 and 2 belong to class III HDACs, both SIRT1 and 2-deficient DT40 cells had mild growth defects (Matsushita et al., 2005). Moreover, SIRT1 and 2-deficient DT40 cells showed moderate and mild sensitivity to cisplatin, a DNA damaging agent, respectively, while the SIRT1 knockout clone only conferred sensitivity to ionizing radiation. These findings, together with others, indicated that SIRT1 and 2 regulate stress-induced cell death pathways. Since EST database for SIRT3~7 are now available, their biological significance are expected to be investigated in DT40 cells.

Besides the above, we have systematically generated several mutant DT40 cells, lacking of genes encoding remaining HDACs, using the gene targeting technique. All our data concerning HDACs were briefly summarized in Table 2. Individual deficiency of each of HDAC-4, 7 and 8 exhibited insignificant deviations in growth and proliferative properties. We are now in the process of investigating about particular functions of these enzymes.

Table 13-2. HDAC genes disrupted in the DT40 cell line.

	Genes	Cell growth	Function	Reference
Class I	HDAC1	no effect	in analysis	Takami et al., 1999
	HDAC2	no effect	IgM expression	Takami et al., 1999
	HDAC3	lethal	cell viability	Takami and Nakayama, 2000
	HDAC8	no effect	in analysis	in preparation
Class II	HDAC4	no effect	in analysis	in preparation
	HDAC7	no effect	in analysis	in preparation
	HDAC4/7	no effect	in analysis	in preparation
Class III	SIRT1	delayed	stress-tolerance	Matsushita et al., 2005
	SIRT2	delayed	stress-tolerance	Matsushita et al., 2005

3.2   Histone acetyltransferases (HATs)

Opposing to deacetylation reaction, HATs catalyze the transfer of acetyl moiety from acetyl coenzyme A (acetyl-CoA) onto acceptor residues (i.e. the ε-amino groups of conserved Lys residues) of core histones. Acetylation neutralizes positive charge and by doing so enhances hydrophobicity of histones resulting in the reduced affinity of acetylated N-tails of histones to DNA (Roth et al., 2001; Hasan and Hottiger, 2002). The attenuated histone-DNA interactions facilitate melted and permissive chromatin structure for transcriptional activation. Histone tails are exposed outside of chromatin structure and modified marks provide an effective signaling stage that may mediate interactions with various effector proteins (histone-modifying enzymes, chromatin remodeling factors and histone chaperones).

HATs were grouped into two types (A and B) based on their intracellular distribution and substrate specificity. The A-type HATs (GCN5, PCAF, MORF, MOZ, TIP60, and so on) are nuclear enzymes catalyzing histone acetylation related to transcription. On the other hand, the cytoplasmic B-type HATs (HAT1 and HAT2) can catalyze histone acetylation for newly synthesized histones in the cytoplasm. However, almost all true connections between their physiological roles and other HAT activities are still open to question.

3.2.1   GCN5 and PCAF

Both GCN5 and PCAF are members of the A-type HAT as well as GNAT (GCN5-related N-acetyltransferase) superfamily, which have several highly conserved acetylation-related motifs. GCN5 and PCAF independently show tissue (or cell type) specific expression characteristics, and therefore each is expected to play the distinct role, and may be at a particular time.

To investigate physiological roles of GCN5 and PCAF, we first cloned cDNAs encoding the chicken GCN5 and PCAF from DT40 cells, respectively. Subsequently, we generated two DT40 mutants, ΔGCN5 and ΔPCAF, lacking GCN5 and PCAF, respectively (Kikuchi et al., 2005). The electrophoretic patterns on 2D-PAGE of proteins from these mutants were virtually the same as in the case of DT40 cells, indicating that the deficiency of either GCN5 or PCAF has no significant effects on most major cellular proteins. While the PCAF-deficiency exhibited no effect on growth kinetics, the GCN5-deficiency caused delayed growth rate, reduced number of cells in S phase and suppressed G1/S phase transition. The systematic semi-quantitative RT-PCR analyses for ΔGCN5 showed the decrease in mRNA levels of E2F-1, E2F-3, E2F-4, E2F-6, DP-2, cyclin A, cyclin D3, PCNA and cdc25B, while the increase in those of c-myc, cyclin D2 and cyclin G1. Further, the mRNA level of p107 was decreased but that of p27 was increased, while steady-state mRNA levels were maintained for other major factors, such as p53, p130, Rb, some cyclins and cdks. These findings highlighted the fact that GCN5 is involved in both positive and negative transcription regulations of many genes, especially required for G1/S phase transition. Moreover, in ΔGCN5 cells the mRNA level of bcl-xL, an apoptosis-related gene, was reduced significantly, while that of bcl-2, another apoptosis-related gene, was drastically increased.

Based on the results obtained for ΔGCN5 and ΔPCAF, a model was proposed illustrating the possible contribution of GCN5 in the cell cycle progression and/or cell proliferation of vertebrate cells (Figure 2). The impact of GCN5 preferentially lies in the transcriptional regulations of a number of genes involved in the two biologically important processes such as G1/S phase transition of the cell cycle and apoptotic process. Further studies, including chromatin immunoprecipitation assays for these numerous altered genes, are needed to explore how exactly GCN5 participates in all-inclusive control of the cell cycle progression of vertebrate cells.

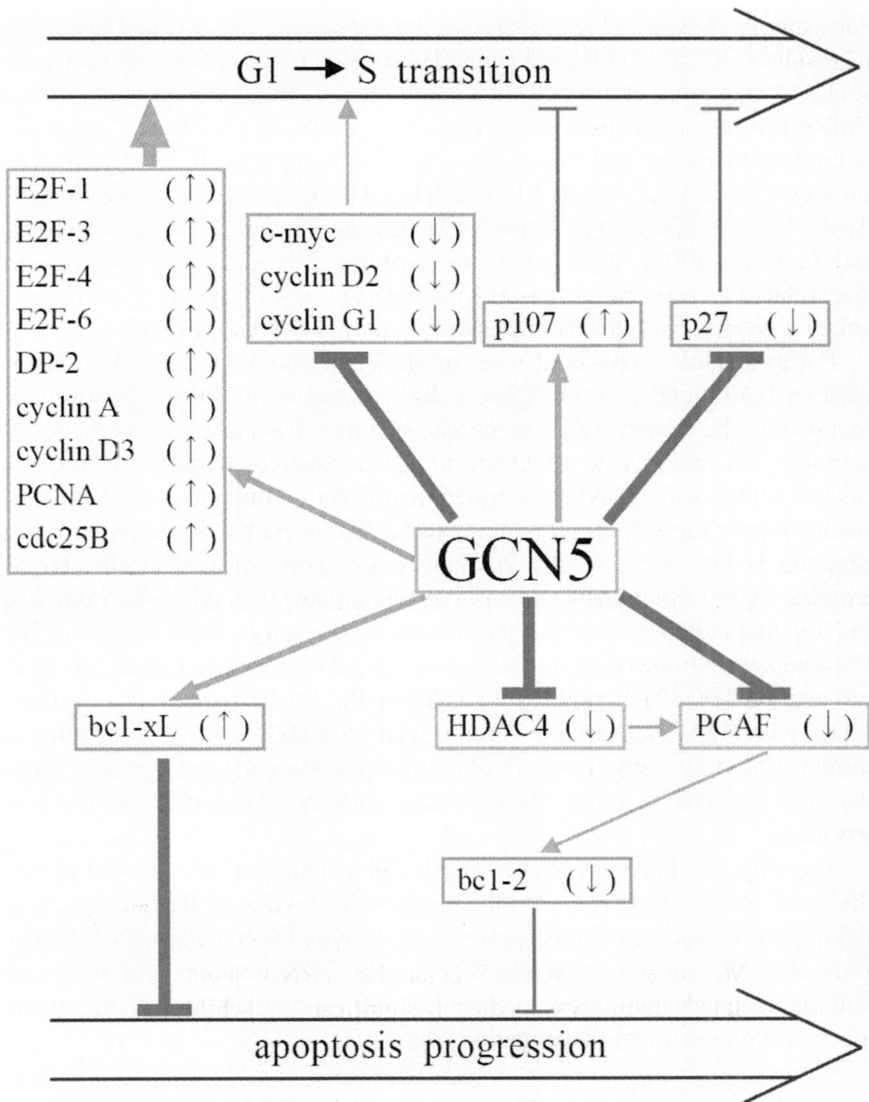

Figure 13-2. A model for a role of GCN5 as a supervisor in all-inclusive control of cell cycle progression of vertebrate cells (see plate 33).

3.2.2   HAT1

The newly synthesized histones in the cytoplasm are suitably modified as known as post-translational modifications prior to their migration into nucleus for nucleosome assembly. One of such important modifications is the site-specific acetylation of nascent histones H3 and H4 by type B HATs. For example, HAT1, which is structurally conserved, diacetylates Lys-5 and 12 residues (K5, K12) of free histone H4 in wide ranges of organisms (Sobel et al., 1995; Chang et al., 1997; Verreault et al., 1998; Ahmad et al., 2001). Unlike H4, an acetylation pattern of newly synthesized H3 in the cytoplasm is random in nature and varies from species to species. Recently, in yeast another type B HAT, termed HATB3.1, a GCN5-containing complex with ADA3 but not ADA2, was identified as nascent histone H3 specific (Sklenar and Parthun, 2004). The impact of evolutionarily conseved and distinct diacetylated isoform of histone H4 mediated by HAT1 seems to be directly linked with *de novo* nucleosome assembly during DNA replication.

For systematic analysis of roles of HAT1 in vertebrates, we first cloned chicken HAT1 cDNA from DT40 cells (Ahmad et al., 2001). The protein encoded by the cloned cDNA comprises 408 amino acids and exhibits 80.4% homology in amino acid sequence to the human counterpart. A series of deleted mutational analyses revealed that HAT1 contains a characteristic leucine zipper motif (380-408 amino acids) that is essentially required for its interaction with the seventh WD dipeptide motif of CAF-1p48. HAT1 remains tightly bound with p46 polypeptide (p46), a CAF-1p48 homolog, and together is believed to take part in diacetylating Lys-5 and Lys-12 of H4. The chicken p46 encoded by the cloned cDNA comprises 424 amino acids and exhibits 90.3% homology to CAF-1p48. While human p46 remains bound with H2A, chicken p46 is associated with H2B instead, in addition to common binding partners of HAT1 and H4 (Ahmad et al., 2000). Thus, HAT1 is expected to be involved in many aspects of chromatin modulating processes.

Recently, we have established the DT40 mutant cell line devoid of two alleles of HAT1 (in preparation). The growth kinetics of the mutant clone lacking HAT1 showed almost similar to wild-type DT40 cells. Physiological roles of HAT1 in the dynamics of cytosolic H4-containing complexes as well as its participation in nuclear events via modulation of chromatin configuration are under investigation.

Table 13-3. HAT genes disrupted in the DT40 cell line.

	Genes	Cell growth	Function	Reference
GNAT family	GCN5	delayed	cell cycle regulation	Kikuchi et al., 2005
	PCAF	no effect	apoptosis (bcl-2 gene expression)	Kikuchi et al., 2005
	HAT1	delayed	*de novo* nucleosome assembly	in submission
MYST family	MOZ	no effect	in analysis	in preparation
	MORF	no effect	in analysis	in preparation
	TIP60	lethal	in analysis	in preparation

3.2.3 Other HATs

Besides the above, to study the biological relevance of other HATs, we have systematically generated several DT40 mutant cells, lacking of genes encoding remaining HATs. All our data concerning HATs were briefly summarized in Table 3. Individual deficiency of each of other HATs exhibited insignificant deviations in growth and proliferative properties. We are now in the process of investigating about their particular functions in detail.

4. HISTONE CHAPERONES

During and after S phase in the cell cycle, post-replicative nascent DNA chains must be assembled into a copy of the parental chromatin structure to maintain proper genome function (McNairn and Gilbert, 2003). Through the passage of the replication fork, parental nucleosomes are transiently disassembled and transferred to the newly synthesized DNA (parental nucleosome segregation), and the nascent histone H3-H4 tetramers are also deposited onto newly replicated DNA behind the replication fork (*de novo* nucleosome assembly), followed by loading of H2A-H2B dimers to complete nucleosome formation (Gruss et al., 1993).

CAF-1, a well-conserved protein complex of three polypeptides of p150, p60, and p48, was originally purified from human cell nuclear extract as an activity to promote nucleosome assembly of replicating DNA in the Simian virus 40 (SV40) replication system (Smith and Stillman, 1989). The smallest subunit of CAF-1, p48 (also known as RbAp48), is an element of multiple complexes involved in different aspects of histone metabolism (Roth and

Allis, 1996; Verreault et al., 1996). The other two larger subunits, p150 and p60, remain associated with each other and interact preferentially with histones H3 and H4 for *de novo* deposition and with proliferating cell nuclear antigen (PCNA) to be recruited into in replication foci (Kaufman et al., 1995; Shibahara and Stillman, 1999). Other histone binding proteins, ASF1 and HIRA, also interact with histones H3-H4, and may act in this pathway (Lorain et al., 1998; Tyler et al., 1999).

As the above, these histone chaperones (CAF-1, ASF1 and HIRA) play important roles in various DNA-utilizing processes in eukayotes. However, the individual biological significance of each of these has been unclear in vertebrate cells. To examine their roles, we generated DT40 knockout cells devoid of each of CAF-1p150, CAF-1p60, CAF-1p48, ASF1 and HIRA.

4.1   HIRA

The mammalian HIRA family (HIR is an acronym for histone regulator) was named for its homology to yeast Hir1p and Hir2p, and cloned from the DiGeorge syndrome (DGS) critical region. In recent years, participatory roles of the mammalian HIRA family in the chromatin assembly, cell cycle progression and transcription regulation of histone genes have gradually been elucidated. We cloned full-length cDNA encoding chicken HIRA (Ahmad et al., 2004; Ahmad et al., 2005). WD dipeptide motifs in the N-terminal half of HIRA and an LXXLL motif in the C-terminal half are biologically significant as binding platforms for its *in vitro* and *in vivo* interactions with CAF-1p48 and HDAC-2, respectively. Both the WD dipeptide motifs and LXXLL motif are individually, essentially required for the global transcription regulations *in vivo*, and hence the two halves should be differentially involved in numerous DNA-utilizing processes, probably through distinct participations in transcription regulations.

Very recently, we generated the homozygous HIRA-deficient DT40 mutant, ΔHIRA (Ahmad et al., 2005). The HIRA deficiency led to the delayed cell growth as well as the opposing influences on transcription regulation of cell cycle-related genes, i.e. repressions for p18, cdc25B and bcl-2, and activations for p19, cyclin A and core histones. Most interestingly, the N-terminal half of HIRA, but not the C-terminal half, was identified as vital for growth of DT40 cells. The detailed analysis of transcriptional regulation of chicken p18 gene, which was repressed in ΔHIRA, revealed that the N-terminal half of HIRA interacts with its putative promoter sequence GCGGGCGC at positions -1157 to -1150 to up-regulate the p18 gene. Accordingly, a model has been proposed depicting that distinguished roles of the N-terminal and C-terminal halves of HIRA in both cell growth and transcription regulations of cell cycle-related genes (Figure 3). The N-terminal and C-terminal halves of HIRA are dually and oppositely involved in transcription regulation of cell cycle-related and apoptosis-related genes as well as core histone genes. The regulatory role of the N-terminal half is mediated by its binding capability with various putative promoters such as for p18 gene, probably including cdc25B and bcl-2 genes, directly or indirectly through formation of complex(es) with or without CAF-1p48 and/or other proteins. As a result of which a set of cell cycle-related and apoptosis-related genes is being expressed. On the other hand, the C-terminal half mediated regulation possibly takes place by binding to the putative promoters of a different set of genes, i.e. p19 and cyclin A genes, like those of histones H2A, H2B, H3 and H4 genes, directly or indirectly through formation of complex(es) with or without HDAC-2 (and/or HDAC-1) and/or other proteins, resulting in down-regulations of genes encoding these cell cycle-related factors and core histones.

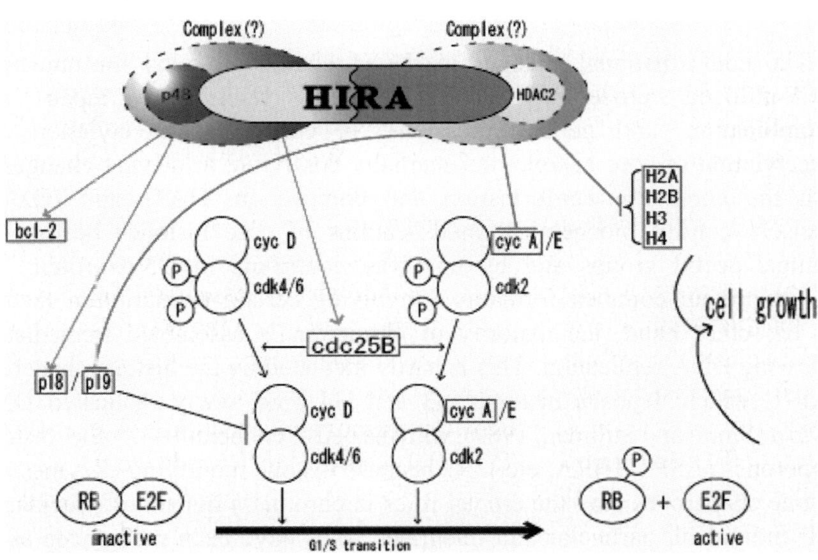

Figure 13-3. A model for different roles of N-terminal and C-terminal helves of HIRA in control of vertebrate cell growth (see plate 34).

4.2 Other histone chaperones

Concerning other histone chaperones, knockout DT40 cells, devoid of CAF-1p150, CAF-1p60, CAF-1p48 and ASF1 are available in our hand (in preparation). The results obtained from these mutants were briefly summarized in Table 4.

Table 13-4. Histone chaperone genes disrupted in the DT40 cell line.

Genes	Cell growth	Function	Reference
CAF-1p150	lethal	replication coupled chromatin assembly	in submission
CAF-1p60	lethal	replication coupled chromatin assembly	in submission
CAF-1p48	lethal	replication coupled chromatin assembly maintenance of heterochromatin domain	in submission
ASF1	lethal	replication coupled chromatin assembly	in submission
HIRA	delayed	replication coupled chromatin assembly transcription of histone genes cell cycle regulation	Ahmad et al., 2005

CONCLUSIONS

The conformational changes in the chromatin are vital for numerous DNA-utilizing processes, including DNA replication, repair and recombination, and gene expressions, in eukaryotes. Acetylation and deacetylation of core histones substantially contribute to driving changes in both the chromatin conformation and compaction. HATs and HDACs precisely control biochemical modifications of core histones bearing N-terminal acetyl groups and by this way contribute in DNA-protein and protein-protein complex formations involving certain transcription factors. On the other hand, the majority of chromatin is assembled immediately following DNA replication. This is partly mediated by the histone chaperone CAF-1, which deposits histones H3 and H4 onto newly replicated DNA *in vitro* (Smith and Stillman, 1989), with the active cooperation of other histone chaperones (ASF1, HIRA, etc.). As because histone modifying enzymes and histone chaperones play the crucial roles in chromatin dynamics, elucidating their individual, particular and integrated roles have been recognized as an active research area.

The utilization of the gene targeting technique in the DT40 cell line has already gained momentum as a powerful tool for analysis of physiological roles of associated molecules linked to chromatin dynamics. Using this technique, we have provided some evidences of distinguished individual and integrated roles, maintaining the intracellular levels of histones and changing the acetylated state of core histones that is related to transcription activity, played by each member as well as in combination with other member of HAT and HDAC families. For example, HDAC-2 takes care of IgM synthesis and alternative pre-mRNA processing, HDAC-3 is linked DT40 cell viability, and GCN5 preferentially acts as a supervisor in the normal cell cycle progression having comprehensive control over expressions of many cell cycle-related and apoptosis-related genes. Unlike histone variants, any member of HAT or HDAC family is not capable of delivering compensatory function for the remaining member, possibly because of their distinct substrate specificities, distinguished binding partners and targeted chromatin localizations. However, it remains a challenge to explore very exact mechanisms of their interactive roles, being the hallmark of complex chromatin-based processes, required for specific gene expressions, and the growth and survival of vertebrate cells. Our laboratory is actively engaged in search of further insights regarding the specific roles of several HATs, HDACs and histone chaperones in mechanisms of DNA replication, recombination and repair, and transcriptions via formation and modulation of the chromatin structure.

REFERENCES

Ahmad A, Kikuchi H, Takami Y, Nakayama T. 2005. Different roles of N-terminal and C-terminal halves of HIRA in transcription regulation of cell cycle-related genes that contribute to control of vertebrate cell growth. J Biol Chem 280:32090-32100.

Ahmad A, Nagamatsu N, Kouriki H, Takami Y, Nakayama T. 2001. Leucine zipper motif of chicken histone acetyltransferase-1 is essential for *in vivo* and *in vitro* interactions with the p48 subunit of chicken chromatin assembly factor-1. Nucleic Acids Res 29:629-637.

Ahmad A, Takami Y, Nakayama T. 2000. Distinct regions of the chicken p46 polypeptide are required for its *in vitro* interaction with histones H2B and H4 and histone acetyltransferase-1. Biochem Biophys Res Commun 279:95-102.

Ahmad A, Takami Y, Nakayama T. 2004. WD dipeptide motifs and LXXLL motif of chicken HIRA are essential for interactions with the p48 subunit of chromatin assembly factor-1 and histone deacetylase-2 *in vitro* and *in vivo*. Gene 342:125-136.

Biel M, Wascholowski V, Giannis A. 2005. Epigenetics–an epicenter of gene regulation: histones and histone-modifying enzymes. Angew Chem Int Ed Engl 44:3186-3216.

Buerstedde JM, Takeda S. 1991. Increased ratio of targeted to random integration after transfection of chicken B cell lines. Cell 67:179-188.

Chang L, Loranger SS, Mizzen C, Ernst SG, Allis CD, Annunziato AT. 1997. Histones in transit: cytosolic histone complexes and diacetylation of H4 during nucleosome assembly in human cells. Biochemistry 36:469-480.

Gruss C, Wu J, Koller T, Sogo JM. 1993. Disruption of the nucleosomes at the replication fork. Embo J 12:4533-4545.

Hasan S, Hottiger MO. 2002. Histone acetyl transferases: a role in DNA repair and DNA replication. J Mol Med 80:463-474.

Kaufman PD, Kobayashi R, Kessler N, Stillman B. 1995. The p150 and p60 subunits of chromatin assembly factor I: a molecular link between newly synthesized histones and DNA replication. Cell 81:1105-1114.

Kikuchi H, Takami Y, Nakayama T. 2005. GCN5: a supervisor in all-inclusive control of vertebrate cell cycle progression through transcription regulation of various cell cycle-related genes. Gene 347:83-97.

Lorain S, Quivy JP, Monier-Gavelle F, Scamps C, Lecluse Y, Almouzni G, Lipinski M. 1998. Core histones and HIRIP3, a novel histone-binding protein, directly interact with WD repeat protein HIRA. Mol Cell Biol 18:5546-5556.

Loyola A, Almouzni G. 2004. Histone chaperones, a supporting role in the limelight. Biochim Biophys Acta 1677:3-11.

Mai A, Massa S, Rotili D, Cerbara I, Valente S, Pezzi R, Simeoni S, Ragno R. 2005. Histone deacetylation in epigenetics: an attractive target for anticancer therapy. Med Res Rev 25:261-309.

Matsushita N, Takami Y, Kimura M, Tachiiri S, Ishiai M, Nakayama T, Takata M. 2005. Role of NAD-dependent deacetylases SIRT1 and SIRT2 in radiation and cisplatin-induced cell death in vertebrate cells. Genes Cells 10:321-332.

McNairn AJ, Gilbert DM. 2003. Epigenomic replication: linking epigenetics to DNA replication. Bioessays 25:647-656.

Nakayama T, Takami Y. 2001. Participation of histones and histone-modifying enzymes in cell functions through alterations in chromatin structure. J Biochem (Tokyo) 129:491-499.

Roth SY, Allis CD. 1996. The subunit-exchange model of histone acetylation. Trends Cell Biol 6:371-375.

Roth SY, Denu JM, Allis CD. 2001. Histone acetyltransferases. Annu Rev Biochem 70: 81-120.

Seguchi K, Takami Y, Nakayama T. 1995. Targeted disruption of 01H1 encoding a particular H1 histone variant causes changes in protein patterns in the DT40 chicken B cell line. J Mol Biol 254:869-880.

Shibahara K, Stillman B. 1999. Replication-dependent marking of DNA by PCNA facilitates CAF-1-coupled inheritance of chromatin. Cell 96:575-585.

Sklenar AR, Parthun MR. 2004. Characterization of yeast histone H3-specific type B histone acetyltransferases identifies an ADA2-independent Gcn5p activity. BMC Biochem 5:11.

Smith S, Stillman B. 1989. Purification and characterization of CAF-I, a human cell factor required for chromatin assembly during DNA replication in vitro. Cell 58:15-25.

Sobel RE, Cook RG, Perry CA, Annunziato AT, Allis CD. 1995. Conservation of deposition-related acetylation sites in newly synthesized histones H3 and H4. Proc Natl Acad Sci U S A 92:1237-1241.

Tagami H, Ray-Gallet D, Almouzni G, Nakatani Y. 2004. Histone H3.1 and H3.3 complexes mediate nucleosome assembly pathways dependent or independent of DNA synthesis. Cell 116:51-61.

Takami Y, Higashio M, Fukuoka T, Takechi S, Nakayama T. 1996. Organization of the chicken histone genes in a major gene cluster and generation of an almost complete set of the core histone protein sequences. DNA Res 3:95-99.

Takami Y, Kikuchi H, Nakayama T. 1999. Chicken histone deacetylase-2 controls the amount of the IgM H-chain at the steps of both transcription of its gene and alternative processing of its pre-mRNA in the DT40 cell line. J Biol Chem 274:23977-23990.

Takami Y, Nakayama T. 1997. One allele of the major histone gene cluster is enough for cell proliferation of the DT40 chicken B cell line. Biochim Biophys Acta 1354:105-115.

Takami Y, Nakayama T. 2000. N-terminal region, C-terminal region, nuclear export signal, and deacetylation activity of histone deacetylase-3 are essential for the viability of the DT40 chicken B cell line. J Biol Chem 275:16191-16201.

Takami Y, Nishi R, Nakayama T. 2000. Histone H1 variants play individual roles in transcription regulation in the DT40 chicken B cell line. Biochem Biophys Res Commun 268:501-508.

Takami Y, Takeda S, Nakayama T. 1995a. Targeted disruption of an H3-IV/H3-V gene pair causes increased expression of the remaining H3 genes in the chicken DT40 cell line. J Mol Biol 250:420-433.

Takami Y, Takeda S, Nakayama T. 1995b. Targeted disruption of H2B-V encoding a particular H2B histone variant causes changes in protein patterns on two-dimensional polyacrylamide gel electrophoresis in the DT40 chicken B cell line. J Biol Chem 270:30664-30670.

Takami Y, Takeda S, Nakayama T. 1997. An approximately half set of histone genes is enough for cell proliferation and a lack of several histone variants causes protein pattern changes in the DT40 chicken B cell line. J Mol Biol 265:394-408.

Tyler JK, Adams CR, Chen SR, Kobayashi R, Kamakaka RT, Kadonaga JT. 1999. The RCAF complex mediates chromatin assembly during DNA replication and repair. Nature 402:555-560.

Verreault A, Kaufman PD, Kobayashi R, Stillman B. 1996. Nucleosome assembly by a complex of CAF-1 and acetylated histones H3/H4. Cell 87:95-104.

Verreault A, Kaufman PD, Kobayashi R, Stillman B. 1998. Nucleosomal DNA regulates the core-histone-binding subunit of the human Hat1 acetyltransferase. Curr Biol 8:96-108.

Chapter 14

ANALYSIS OF GENE EXPRESSION, COPY NUMBER AND PALINDROME FORMATION WITH A DT40 ENRICHED cDNA MICROARRAY

Paul E. Neiman[1], Joan Burnside[2], Katrina Elsaesser[1], Harry Hwang[1] Bruce E. Clurman[1], Robert Kimmel[1] and Jeff Delrow[1]

[1]Fred Hutchinson Cancer Research Center, Seattle Washington 98109, USA; [2]Department of Animal and Food Sciences, Unversity of Deleware, Newark, Deleware19717, USA

Abstract: DT40 presents a unique opportunity to exploit newly available tools for chicken genomic analysis. A 13K chicken cDNA microarray representing 11447 non-overlapping ESTs has been developed. This array detects expression of 7086 DT40 genes of which 644 are over-expressed 3-fold or greater and 1585 are under-expressed 3-fold or greater relative to normal post-hatch bursal cell populations. Changes in RNA expression due to single gene alterations can be detected by expression profiling. For example, by this method, over expression of the oncogenic micro RNA *bic* up-regulates expression of VBP, a known regulator of Avian Leukosis Virus LTR- driven transcription with very little additional expression change, A degree of cytogenetic abnormality and instability of DT40 cells has been observed, which is characterized at the fine structure level using microarray-based comparative genome hybridization (array-CGH). The relationship between gene copy number and RNA expression levels can be assessed in the same tissue samples using the same microarray. A newly introduced technique for genome-wide analysis of palindrome formation (GAPF) detects long   inverted repeats, or palindromes, which are early events in gene amplification and possibly other DNA structural change. Since both array CGH-detected copy number changes and GAPF-detected palindromes are abundant in DT40, these techniques, coupled with targeted gene deletion and replacement, may provide a powerful tool for analysis of genomic instability and its underlying genetic mechanisms.

Key words: Microarray, expression profiling, genetic instability, Myc, bic.

J.-M. Buerstedde and S. Takeda (eds.), Reviews and Protocols in DT40 Research, 245–256.
© 2006 *Springer.*

1.   INTRODUCTION

Exploitation of the unique opportunities presented by experimental chicken models in general and the properties of DT40 cells in particular, has been limited by the relative lack of genomic resources in comparison with other experimental systems. This barrier is rapidly being overcome. One tool we have developed, and made available to the research community, is a chicken cDNA microarray (Burnside et al., 2005). In this chapter we describe this array and its application to analysis of gene expression and genetic stability of DT40 cells. Three types of microarray-based analyses have been carried out. The first is a comparison with normal bursal cells of RNA levels for most of the genes expressed in DT40. We illustrate the power of this approach to detect changes in expression induced by a single gene. The second is comparative genomic hybridization (array CGH) to determine changes in gene copy number (Pollack et al., 1999) in DT40 cells. This technique was also exploited to analyze genomic instability in DT 40 cell cultures. Finally, using a bursal lymphoma like the one from which DT40 was derived, we illustrate a new technique, GAPF, for detecting genome-wide formation of palindromes (Tanaka et al., 2005), which are structural changes shown to precede gene amplification (Tanaka et al., 2002). This approach if carried out in DT40 has promise for probing more deeply into genes controlling stability in general.

1.1   Expression Array Analysis

The ability to introduce specific gene changes in DT40 cells, when combined with microarray technology, provides an opportunity to probe the effects of such mutations on genome-wide RNA expression. The currently available chicken cDNA microarray is based on an earlier version composed of DT40 and other chicken immune ESTs. The approximately 3500 features on this precursor array was used successfully to analyze transcriptional change during early bursal development and Myc-oncogene induced bursal neoplasia (Neiman et al., 2001) as well as such change associated with Avian Leukosis Virus (ALV)-induced mutation of c- *myb* (Neiman et al., 2001). DT40 RNA expression profiles were similar to those of *in-vivo* Myc-induced bursal lymphomas.

Principally by selecting a large number of non-overlapping ESTs from the Biotechnology and Biological Sciences Research Council (BBSRC) chicken cDNA project (Boardman et al., 2002) this initial array was expanded to include 13007 features representing 11447 non-overlapping ESTs. On this array 7433 features give useful signals above background when hybridized to bursal RNA (Burnside et al., 2005). This number would be expected to represent the large majority of all bursal-expressed genes. An example of the fluorescent scan image of an expression analysis in which DT40 was compared to normal 2 week post-hatching bursa from the inbred white leghorn line from which DT40 was derived (Baba and Humphries, 1985) is depicted in Figure 1A. ESTs relatively over expressed in DT40 are represented by red spots, those over expressed in normal bursa relative to DT40 by green spots and genes expressed at about the same level in both cell types are yellow. In this experiment 7086 ESTs gave useful signals from DT40, RNA from 644 genes was at least 3-fold more abundant in DT40 than normal bursa and from 1585 genes was at least 3-fold less abundant in DT40 than in normal bursa. The details, including gene identifications, are provided as supplemental tables in a Microsoft Excel spreadsheet format at: http://www.fhcrc.org/science/basic/neiman/

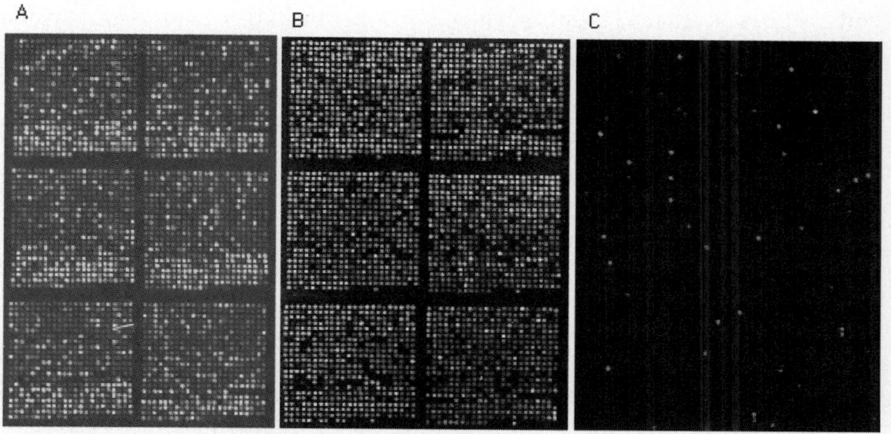

Figure 14-1. Microarray fluorescence scans images of DT40 or bursal lymphoma (Cy-5 dye label, red) vs. normal chicken bursa (Cy-3 dye label, green) experiments. Six of 32 feature blocks are shown (A): DT40 expression array, original DT40 and immune system ESTs are in the lower seven rows of each block, (B): DT40 arrayCGH, (C): DT40 GAPF (see plate 35).

To test the sensitivity of the array to the effects of single gene changes in DT40, we examined the effects of over expression of a small non-coding RNA called *bic*. This putative microRNA was discovered to be activated by ALV insertional mutagenesis in a significant portion of bursal lymphomas, and thought to represent a cooperating oncogene in progression of Myc-induced neoplasms (Clurman and Hayward, 1989). Recently *bic* over expression has been implicated in the pathogenesis of several human neoplasms (van den Berg et al., 2003; Metzler et al., 2004; Eis et al., 2005). We thought it would be useful to see if we could identify *bic* target genes in

DT40 using our expression array. The experiment shown in Fig. 1A and preceding analyses (Neiman et al., 2001) indicated that *bic* is expressed at very low levels, if at all, in our DT40 cells. Therefore, retroviral vector-mediated overexpression of *bic* was carried out in DT40 cells and microarray analysis was carried out by comparison of DT40 cells infected with a control empty vector.

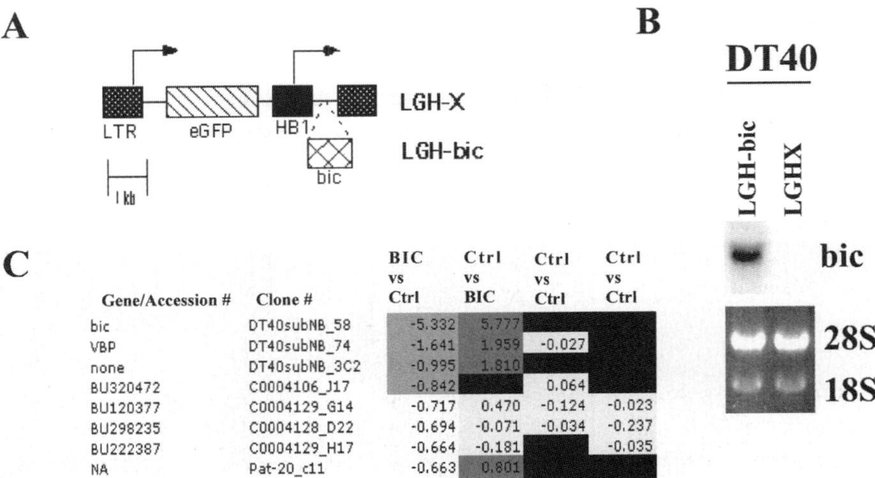

Gene/Accession #	Clone #	BIC vs Ctrl	Ctrl vs BIC	Ctrl vs Ctrl	Ctrl vs Ctrl
bic	DT40subNB_58	-5.332	5.777		
VBP	DT40subNB_74	-1.641	1.959	-0.027	
none	DT40subNB_3C2	-0.995	1.810		
BU320472	C0004106_J17	-0.842		0.064	
BU120377	C0004129_G14	-0.717	0.470	-0.124	-0.023
BU298235	C0004128_D22	-0.694	-0.071	-0.034	-0.237
BU222387	C0004129_H17	-0.664	-0.181		-0.035
NA	Pat-20_c11	-0.663	0.801		

Figure 14-2. (A): Schematic of LGH-bic and LGH-X control vectors. LTR = Moloney Leukemia Virus long terminal repeat. eGFP = enhanced green fluorescent protein. HB1 = promoter-enhancer from the HB1 Avian Myelocytomatosis Virus (Enrietto et al., 1983). (B): Northern blot analysis of DT-40 cell lines infected with LGH-bic and LGH-X virus. 28S and 18S rRNA demonstrates equal loading, (C): Data analysis showing upregulated genes in bic expressing DT-40 cells. Log2 fluorescence.

Diagrams of the LGH-*bic* and LGH-X control vectors are shown in figure 2A. Overexpression of *bic* mRNA was confirmed by Northern analysis as shown in Figure 2B. RNA was isolated from LGH-*bic* and LGH-X expressing DT40 cell line, amplified, and used to synthesize cDNA, which was then labeled with Cy3 and Cy5 dyes. Labeled cDNA was combined and then hybridized to our DT40 specific expression array. Analysis of the results showed that in replicate experiments, which included dye switching in labeling of control and *bic* expressing samples, there was overexpression of a small set of genes in addition to *bic*. Most prominently vitellogenin binding protein (VBP) was over expressed 3 to 4 fold in replicate experiments with dye swaps as, to a lesser extent, were other genes of unknown function (Fig. 2C). Genes found to be consistently down regulated were not detected in these studies. The presence of consistently upregulated genes raises the possibility that *bic* may regulate their expression, through siRNA/microRNA mechanisms, by repression of a negative regulator for these genes. Intriguingly, a regulatory role for VBP in ALV-LTR driven transcription has been previously reported suggesting a link in this bursal derived tumor system (Curristin et al., 1997).

1.2 Gene Copy Number Change

The genetic stability of DT40 cells may be important for interpretation of many experiments. Recent cytogenetic analysis of DT40 macro chromosomes indicates that modal chromosome numbers vary among DT40 cells with different passage histories, and that DT40 cultures are cytogenically heterogeneous producing condition-dependent emergence of non-modal karyotypes (Chang and Delany, 2004). Array CGH can reveal fine structural detail of gene copy number change (Pollack et al., 1999). Using this approach we measured copy number change relative to a reference normal chicken DNA sample of all of the features on the cDNA microarray. Depicted in Figure 1B, loci with increased copy number relative to normal chicken DNA are represented by red spots and those deleted in DT40 by green spots. Unchanged loci are yellow.

Simple inspection of the figure indicates that copy number change occurs at many loci in DT40 cells. The release of a draft chicken genome sequence (Hillier et al., 2004) and related resources allowed us to map these changes on chicken chromosomes.

Figure 4A is a box plot that summarizes the Log_2 fluorescent signal ratios (red vs. green, or Cy-5 dye label DT40 vs. Cy 3 dye label normal chicken) relative to normal chicken (see Fig 1) for specific DT40 chromosomes. These data are to be compared with ratios for all autosomes, right hand column, which give the expected mean Log_2 value of 0 (Log_2 of 0 = 1). DT40 has a female karyotype (ZW), and the reference normal DNA in this experiment was from a male (ZZ). The median fluorescence ratio of signals from Z chromosome loci, second column from right, was, as expected, about half that of autosomal loci Similarly, on the basis of fluorescent signal ratios compared with other autosomes, micro chromosome 24 is clearly trisomomic. Monosomy of chromosome 4 and trisomy of chromosome 2, previously reported by cytogenetic analysis of other DT40 cultures (Chang and Delany, 2004) is not

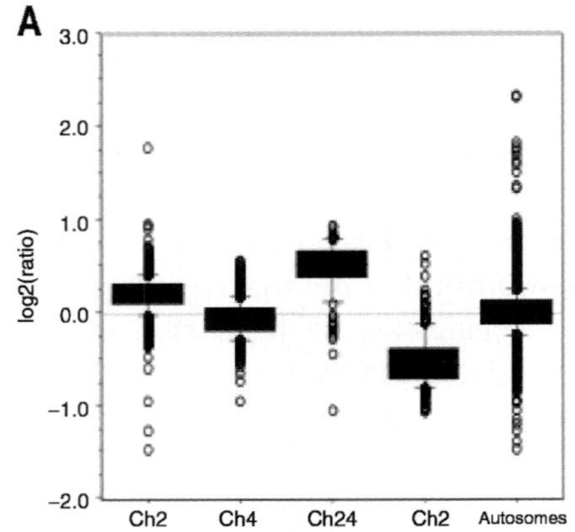

Figure 14-3. Mapping gene copy number on DT40 Chromosomes. (A) Box plots of Log2 fluorescence signal ratios for DT40 chromosomes 2,4,24 and Z relative to other autosomes. Boxes show inter quartile range, horizontal lines in the boxes are the median values, brackets are 10 - 90 percentile range and open circles are outliers. (B) Map of gene amplifications and deletions on ~190 MB chromosome 1 for three DT40 clones. Solid diamonds along line show position of array features. Verticals above line in light, medium and dark gray map positions of amplified loci for DT40 clones 3, 8 and 9 respectively, verticals below line are positions of deleted loci.

as complete (left hand columns) as the changes in Chromosomes Z and 24 perhaps due to heterogeneity in the cultures examined. Mapping of other chromosomes shows widespread copy number change across the whole genome as illustrated by mapping of amplifications and deletions on chromosome 1 (Figure 4B).

DT40 cells in this laboratory were obtained shortly after their establishment (Baba and Humphries, 1985) and cultured for many years. We reasoned that if they were genetically unstable (as assayed by array CGH), examination of different clones from a current parental culture would reveal the degree of such instability.

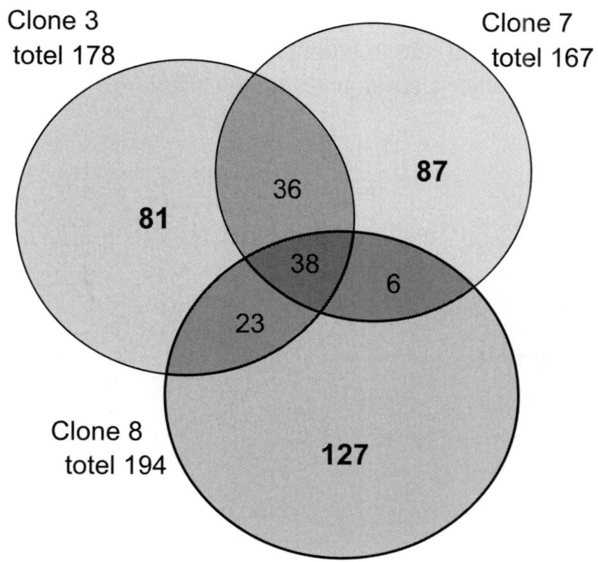

Figure 14-4. Venn diagram of numbers of loci showing copy number change shared between, or unique to, each of three DT40 fresh agar colony clones.

We therefore grew a large number of soft agar DT40 clones from the culture used in Figure 1, randomly selected 12 of these for DNA extraction (see protocol 8) and carried out array CGH in triplicate experiments with three of these, clones 3,7 and 8, and the same reference normal bursal DNA. To be scored as indicating a copy number change an array feature had to demonstrate signal fluorescence ratios of >1.4 or < −1.4 in all three replicates, and be outside the noise range for each feature as determined by triplicate array CGH experiments with the reference normal DNA vs. other normal bursal DNA samples. Figure 5 shows a Venn diagram that enumerates loci changed in common among the three clones and those with changes unique to each clone. These data indicate that 45 - 64% of loci showing copy number change in this assay are unique to each clone. Therefore, despite some large common chromosomal aberrations, extensive gene copy number change occurs at many loci in DT40 cultures over time.

DT40 cells are derived from a Myc-oncogene-induced bursal lymphoma (Baba and Humphries, 1985). Based on the expression microarray data illustrated in Figure 1A c-*myc* RNA is over expressed 5-fold in these cells relative to normal bursa; and, in addition, c-*myc* shows a net one gene copy increase by array CGH (nine experiments with DNA preparations from three DT40 clones). Thus the total level of c-myc mRNA in DT40 may be the sum sum of ALV LTR-driven transcription and gene copy number increase.

Displaying fluorescence signal ratios for expression and copy number in the same chart can reveal the relationship between gene copy number and expression for the other altered genes on the array.

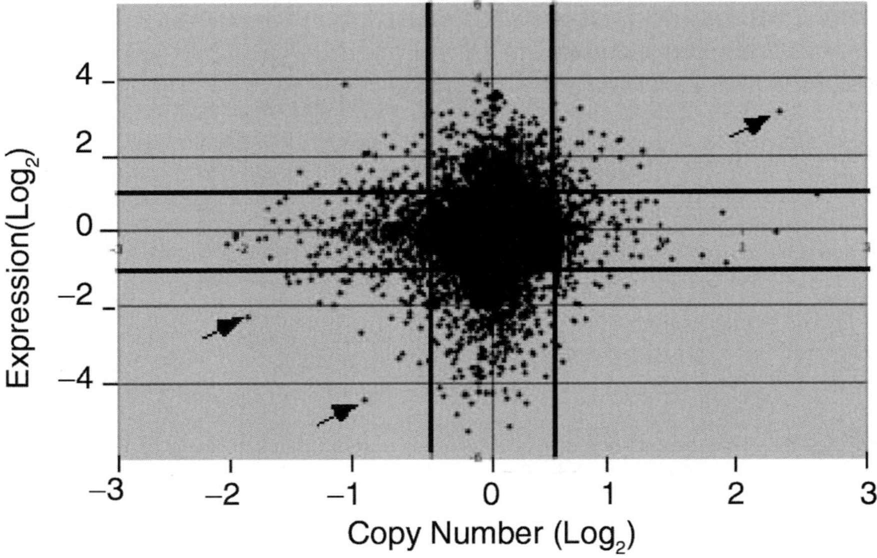

Figure 14-5. Relationship of gene copy number to expression in DT40. X-axis, log2 fluorescence signal ratio by array CGH, and Y-axis, log2 fluorescence ratio for expression in DT40 relative to normal bursa. Bold vertical and horizontal lines enclose arbitrary noise ranges for each parameter. Arrows are explained in text.

Figure 6 depicts such a plot for the data from experiments illustrated in figures 1A and B. Genes appearing in the upper right box are candidates for over-expression, due at least in part, to gene amplification. The arrow in this box indicates cytochrome C, which, with respect to normal bursa, is amplified 4 to 5-fold and over-expressed 9-fold in DT40. Genes appearing in the lower left box are candidates for decreased expression due to deletions. The upper of the two arrows with respect to the Y-axis indicates an inhibitor of apoptosis (IAP) associated factor, VIAF, thought to modulate caspase activation in programmed cell death (Wilkinson et al., 2004). The lower arrow indicates a gene for a hypothetical protein of unknown function. The data suggest that gene copy number instability may affect gene expression in DT40.

1.3   Palindrome Formation

The formation of long inverted repeats or palindromes has been implicated as a conserved early, rate limiting step in gene amplification (Yasuda and Yao, 1991; Butler et al., 1996; Tanaka et al., 2002), found in specific tumor associated gene amplifications (Ford and Fried, 1986) and recently, using the GAPF approach, as a widespread manifestation of genetic instability in human neoplasms (Tanaka et al., 2005). Given the extent and variation of gene copy number change in DT40, widespread palindrome formation would be expected in these cells. We have adapted the GAPF technique for use with this chicken cDNA microarray as described in Protocol # 8 and illustrated in Figure 1C. In this approach palindromes are isolated by virtue of their ability to "snap-back" to form hairpin duplex structures in single-stranded DNA preparation. These snap-back duplexes are then isolated by selective digestion of single-stranded DNA, further digested with restriction enzymes, subjected to ligation-mediated PCR using Cy-dye labeled primers and hybridized to the microarray. Figure 1C shows red (Cy-5-labeled) spots representing palindromes in DNA from DT40 no green (Cy-3 labeled) or yellow spots as would occur if these palindromes were present only in normal DNA, or both normal and lymphoma DNA respectively. Thus GAPF-detected long inverted repeats are, like array CGH detected gene copy number change, widely prevalent in these Myc-induced bursal lymphoma cells.

1.4   Conclusions and Opportunities

The current chicken 13K microarray is an effective tool for all of the types of analyses illustrated in this review, and is readily available to academic investigators (www.fhcrc.org/science/shared-resources/genomics/ dna- array/spotted-arrays/. The technology, however, is rapidly evolving. For RNA expression analysis, Affymetrics has announced the release of a whole chicken genome chip that includes some important pathogens as well (www.affymetrix.com/products/arrays/specific/chicken.affx.). While, at this writing, not yet available for chicken, glass slide 70mer oligonucleotide arrays can improve the quality of data obtained in array CGH experiments, and when combined with single nucleotide polymorphisms, become a powerful tool for detecting allelic copy number alterations genome wide (Bignell et al., 2004; Zhao et al., 2004). Nevertheless the present technology is both useful and relatively economical.

Application of this technology extends preceding cytogenetic findings (Chang and Delany, 2004) and demonstrates gene deletions and amplifications on all DT 40 chromosomes. Furthermore these alterations are unstable and change over time, although the time frame and influence of specific conditions for genetic alteration in DT40 remain to be elucidated. These facts need to be taken into account in planning and interpreting experiments, such as targeted gene deletion or replacement, in DT40.

Finally, since genetic instability itself is an important phenomenon, for example in cancer and stem cell biology, DT40 cells and their special properties may present an opportunity for dissecting underlying mechanisms. The instability described here may result directly from Myc-oncogene deregulation, indirectly from some other feature of the neoplastic phenotype, the bursal stem cell origins of these cells or some combination of these factors. The ability of DT40 cells to undergo immunoglobulin gene diversification and to incorporate exogenous DNA efficiently into chromosomal DNA by homologous recombination is reviewed in other chapters. It may be revealing to determine if the genes which have been shown to influence these processes by targeted deletion in DT40 cells also play a role in the pathological manifestations of copy number instability: palindrome formation, amplification and/or deletion, which can be detected experimentally, on a genome-wide basis, by microarray-based techniques.

ACKNOWLEDGEMENTS

Work described here uniquely was supported by NIH grants R01 CA20068 and CA 200725 to PEN and R01 CA 102742 to BEC. The authors thank Amalia Icreversi, Lillian Ambroggio and Gil Loring for technical assistance and Alanna Ruddell for helpful comments.

REFERENCES

Baba TW, Humphries EH. 1985. Cell lines derived from avian lymphomas exhibit two distinct phenotypes. Virology 144:139-51.

Bignell GR, Huang J, Greshock J, Watt S, Butler A, West S, Grigorova M, Jones KW, Wei W, Stratton MR, Futreal PA, Weber B, Shapero MH, Wooster R. 2004. High-resolution analysis of DNA copy number using oligonucleotide microarrays. Genome Res 14: 287-295.

Boardman E, Sanz-Ezquerro J, Overton IM, Burt DW, Bosch E, Fong WT, Tickle C, Brown WRA, Wilson SA, Hubbard SJ. 2002. A comprehensive collection of chicken cDNAs. Current Biology 12:1965-1969.

Burnside J, Neiman P, Tang J, Basom R, Talbot R, Aronszajn M, Burt D, Delrow J. 2005. Development of a cDNA array for chicken gene expression analysis. BMC Genomics 6:13.

Butler DK, Yasuda LE, Yao MC. 1996. Induction of large DNA palindrome formation in yeast: implications for gene amplification and genome stability in eukaryotes. Cell 87:1115-1122.

Chang H, Delany ME. 2004. Karyotype stability of the DT40 chicken B cell line: macrochromosome variation and cytogenetic mosaicism. Chromosome Res 12:299-307.

Clurman BE, Hayward WS. 1989. Multiple proto-oncogene activations in avian leukosis virus-induced lymphomas: Evidence for stage specific events. Molecular and Cellular Biology 9:2657-2664.

Curristin SM, Bird KJ, Tubbs RJ, Ruddell A. 1997. VBP and RelA regulate avian leukosis virus long terminal repeat-enhanced transcription in B cells. J Virol 71:5972-5981.

Eis PS, Tam W, Sun L, Chadburn A, Li Z, Gomez MF, Lund E, Dahlberg JE. 2005. Accumulation of miR-155 and BIC RNA in human B cell lymphomas. Proc Natl Acad Sci U S A 102:3627-3632.

Enrietto PJ, Payne LN, Hayman MJ. 1983. A recovered avian myelocytomatosis virus that induces lymphomas in chickens: pathologic properties and their molecular basis. Cell 35:369-379.

Ford M, Fried M. 1986. Large inverted duplications are associated with gene amplification. Cell 45:425-430.

Hillier LW, Miller W, Birney E, Warren W, Hardison RC, Ponting CP, Bork P, Burt DW, Groenen MA, Delany ME, Dodgson JB, Chinwalla AT, Cliften PF, Clifton SW, Delehaunty KD, Fronick C, Fulton RS, Graves TA, Kremitzki C, Layman D, Magrini V, McPherson JD, Miner TL, Minx P, Nash WE, Nhan MN, Nelson JO, Oddy LG, Pohl CS, Randall-Maher J, Smith SM, Wallis JW, Yang SP, Romanov MN, Rondelli CM, Paton B, Smith J, Morrice D, Daniels L, Tempest HG, Robertson L, Masabanda JS, Griffin DK, Vignal A, Fillon V, Jacobbson L, Kerje S, Andersson L, Crooijmans RP, Aerts J, van der Poel JJ, Ellegren H, Caldwell RB, Hubbard SJ, Grafham DV, Kierzek AM, McLaren SR, Overton IM, Arakawa H, Beattie KJ, Bezzubov Y, Boardman PE, Bonfield JK, Croning MD, Davies RM, Francis MD, Humphray SJ, Scott CE, Taylor RG, Tickle C, Brown WR, Rogers J, Buerstedde JM, Wilson SA, Stubbs L, Ovcharenko I, Gordon L, Lucas S, Miller MM, Inoko H, Shiina T, Kaufman J, Salomonsen J, Skjoedt K, Wong GK, Wang J, Liu B, Yu J, Yang H, Nefedov M, Koriabine M, Dejong PJ, Goodstadt L, Webber C, Dickens NJ, Letunic I, Suyama M, Torrents D, von Mering C, Zdobnov EM, Makova K, Nekrutenko A, Elnitski L, Eswara P, King DC, Yang S, Tyekucheva S, Radakrishnan A, Harris RS, Chiaromonte F, Taylor J, He J, Rijnkels M, Griffiths-Jones S, Ureta-Vidal A, Hoffman MM, Severin J, Searle SM, Law AS, Speed D, Waddington D, Cheng Z, Tuzun E, Eichler E, Bao Z, Flicek P, Shteynberg DD, Brent MR, Bye JM, Huckle EJ, Chatterji S, Dewey C, Pachter L, Kouranov A, Mourelatos Z, Hatzigeorgiou AG, Paterson AH, Ivarie R, Brandstrom M, Axelsson E, Backstrom N, Berlin S, Webster MT, Pourquie O, Reymond A, Ucla C, Antonarakis SE, Long M, Emerson JJ, Betran E, Dupanloup I, Kaessmann H, Hinrichs AS, Bejerano G, Furey TS, Harte RA, Raney B, Siepel A, Kent WJ, Haussler D, Eyras E, Castelo R, Abril JF, Castellano S, Camara F, Parra G, Guigo R, Bourque G, Tesler G, Pevzner PA, Smit A, Fulton LA, Mardis ER, Wilson RK. 2004. Sequence and comparative analysis of the chicken genome provide unique perspectives on vertebrate evolution. Nature 432:695-716.

Metzler M, Wilda M, Busch K, Viehmann S, Borkhardt A. 2004. High expression of precursor microRNA-155/BIC RNA in children with Burkitt lymphoma. Genes Chromosomes Cancer 39:167-169.

Neiman PE, Ruddell A, Jasoni C, Loring G, Thomas SJ, Brandvold KA, Lee R, Burnside J, Delrow J. 2001. Analysis of gene expression during myc oncogene-induced lymphomagenesis in the bursa of Fabricius. Proceedings of the National Academy of Sciences of the United States of America 98:6378-6383.

Pollack JR, Perou CM, Alizadeh AA, Eisen MB, Pergamenschikov A, Williams CF, Jeffrey
    SS, Botstein D, Brown PO. 1999. Genome-wide analysis of DNA copy-number changes
    using cDNA microarrays. Nat Genet 23:41-46.
Tanaka H, Bergstrom DA, Yao MC, Tapscott SJ. 2005. Widespread and nonrandom
    distribution of DNA palindromes in cancer cells provides a structural platform for
    subsequent gene amplification. Nat Genet 37:320-327.
Tanaka H, Tapscott SJ, Trask BJ, Yao MC. 2002. Short inverted repeats initiate gene
    amplification through the formation of a large DNA palindrome in mammalian cells. Proc
    Natl Acad Sci U S A 99:8772-8777.
van den Berg A, Kroesen BJ, Koolstra K, de Jong D, Briggs J, Blokzijl T, Jacobs S, Kluiver J,
    Diepstra A, Maggio E, Poppema S. 2003. High expression of B-cell receptor inducible
    gene BIC in all subtypes of Hodgkins lymphoma. Genes Chromosomes & Cancer 37:
    20-28.
Wilkinson JC, Richter BW, Wilkinson AS, Burstein E, Rumble JM, Balliu B, Duckett CS.
    2004. VIAF, a conserved inhibitor of apoptosis (IAP)-interacting factor that modulates
    caspase activation. J Biol Chem 279:51091-51099.
Yasuda LF, Yao MC. 1991. Short inverted repeats at a free end signal large palindromic DNA
    formation in Tetrahymena. Cell 67:505-516.
Zhao X, Li C, Paez JG, Chin K, Janne PA, Chen TH, Girard L, Minna J, Christiani D, Leo C,
    Gray JW, Sellers WR, Meyerson M. 2004. An integrated view of copy number and allelic
    alterations in the cancer genome using single nucleotide polymorphism arrays. Cancer Res
    64:3060-3071.

Chapter 15

CALCIUM SIGNALING, ION CHANNELS AND MORE

THE DT40 SYSTEM AS A MODEL OF VERTEBRATE ION HOMEOSTASIS AND CELL PHYSIOLOGY

Anne-Laure Perraud[1], Carsten Schmitz[1] and Andrew M. Scharenberg[2]
1 Dept. of Immunology, University of Colorado and National Jewish Medical Research Hospital, Denver, CO; 2 Depts. of Pediatrics and Immunology, University of Washington, Seattle, WA

Abstract: The DT40 B-lymphocyte cell line is a chicken bursal lymphocyte tumor cell line which grows rapidly, expresses a variety of types of constitutive and signal dependent ion transport systems., and supports the efficient use of stable and conditional genetic manipulations. Below, we review the use of DT40 cells in dissecting molecular mechanisms involved in Ca^{2+}, Mg^{2+}, and Zn^{2+} transport physiology. These studies highlight the flexibility and advantages the DT40 environment offers to investigators interested in the study of basic vertebrate ion transport physiology.

Key words: Ion channel, DT40, patch clamp, magnesium, calcium, zinc, divalent cation, physiology.

1. INTRODUCTION

1.1 The DT40 "channelome"

1.2 Genetic analysis of divalent cation physiology in DT40 cells

Divalent cations are ubiquitously involved in the cell physiology of living organisms, and in vertebrates play a multitude of roles ranging from

J.-M. Buerstedde and S. Takeda (eds.), Reviews and Protocols in DT40 Research, 257–270.

promoting protein folding and maintenance of protein tertiary structures, cofactor to cellular nucleotides, participation in catalysis, and signal transduction. Because of their importance to such diverse processes, the regulatory mechanisms of the major divalent cations Ca^{2+}, Mg^{2+}, and Zn^{2+} have been the subject of investigation for many years. While the use of DT40 cells to study cation channel physiology has been slow to catch on, the past several years have seen an acceleration of their use in investigation of vertebrate ion homeostasis (fig. 1 and Table 1). Below, the application of genetically manipulated DT40 cells in the study of the physiology of the three major cellular divalent cations is reviewed, with an emphasis on how data from DT40 cells has made unique contributions to our understanding of the cellular roles of specific proteins in divalent cation physiology.

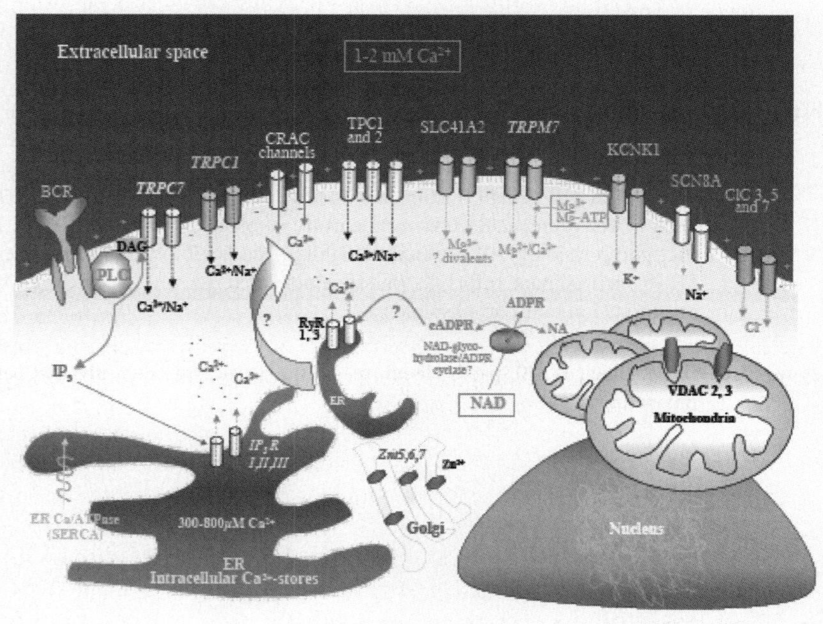

Figure 15-1. The known DT40 channelome: The genes encoding channels and transporters in bold italic have been deleted in the DT40 system, and the phenotype published (see plate 36).

1.2.1 Issues in unraveling regulation of ion homeostasis

Not surprisingly, regulation of divalent cation uptake and intracellular homeostasis is complex, particularly in higher organisms such as vertebrates, and involves a wide range of channels, transporters, and pumps specific for each type of cation. The complexity of higher organisms coupled with the fundamental roles which divalent cations play in many cellular processes has rendered the study of divalent cation transport (in the context of a living organism) using modern genetic techniques problematic due to the existence of multiple channel/transporter homologues, and the capacity of related channels to compensate through up or down regulation of their function. For example, animal models in which specific channels or transporters have been deleted have been useful in understanding the contribution of a protein to the development of the organism or to pathophysiologic processes manifesting at the whole organism level, but have been less useful in providing insights into molecular mechanisms which regulate particular channels/transporters, or to intracellular biochemical processes directly influenced by their function. These problems are further exacerbated by the costs of animal care and effort/time required to execute complex genetic manipulations at the whole animal level.

Because of the issues inherent in the application of genetic techniques at the whole animal level, a great deal of our present understanding of divalent cation channel physiology is derived from the use of pharmacologic inhibitors or inferred from the behavior of ion channels or transporters analyzed after heterologous expression in whole cell patch clamp experiments. While these techniques are indispensable for generating biophysical data on the function of ion channels and transporters, methods which allow the correlation of pharmacological effects and *in-vitro* data with the responses of intact cells are important for understanding how particular molecular functions of a channels and transporting proteins fit within the multitude of processes ongoing in living cells. It is here that the DT40 system stands out, as it offers physiologists and biophysicists a unique, inexpensive, and relatively simple system in which to combine physiological studies with the power of molecular genetics:

1) DT40 cells are derived from a vertebrate organism, and therefore findings in this system are relevant to a wide variety of higher organisms, including humans.
2) DT40 cells provide a reasonable model of the behavior of a continuously growing hematopoietic cell.
3) DT40 cells are large enough to be suitable for various forms of imaging, and to allow the application of patch clamp and other electrophysiologic methods without the need for isolation/dissection.
4) DT40 cells support the facile use of stable and conditional genetic modifications for manipulation of genes, either to produce conditional knockouts or for the purpose of more subtle manipulations to confer inhibitor activity to desired target ion channels.

1.2.2 Ca^{2+} physiology in DT40 cells

Tomohiro Kurosaki's group's investigation of the biochemical mechanisms activated by engagement of the B-cell antigen receptor was the first application of DT40 cell genetics to the study of signal transduction. Dr. Kurosaki's body of work using the DT40 system is a noteworthy example of what can be accomplished using the DT40 system to systematically apply physiology, biochemistry and genetics to unravel a biological puzzle. As an early elevation of cytosolic Ca^{2+} is an important component of the signal produced by BCR engagement, this work is also an important contribution to the present knowledge of vertebrate divalent cation physiology.

Dr. Kurosaki's group's first major contribution in this area was the demonstration that IP3 receptors are absolutely required for both BCR-induced calcium release and calcium entry, but are dispensable for the activation of store operated calcium entry. This observation firmly established that BCR-mediated IP3 production has a crucial role in the calcium entry required for generation of a sustained calcium signal (Sugawara et al., 1997). The technical achievement of producing this cell line was also significant, as it required deletion of both alleles of each of three IP3 receptor genes using six different resistance markers to generate IP3R-triple knockout cells - the first time that any cell type had been so extensively genetically manipulated. These cells have now been utilized by laboratory's around the world to study the role which IP3-receptors have in a variety of cell physiological or receptor-mediated signaling processes, such as the IP3 receptor-independent regulation of TRPC channels and the role of ryanodine receptors in activation of store operated CRAC channels (Broad et al., 2001; Bultynck et al., 2004; Cui et al., 2004; Guillemette et al., 2005; Kiselyov et al., 2001; Laude et al., 2005; Ma et al., 2001; Ma et al., 2002; Morita et al., 2004; Prakriya and Lewis, 2001; Vazquez et al., 2001; Vazquez et al., 2002; Vazquez et al., 2003; Venkatachalam et al., 2001; Wedel et al., 2003; Yogo et al., 2004).

A second important area of study undertaken by Dr. Kurosaki's laboratory in conjunction with the laboratory of Dr. Yasuo Mori has been a genetic analysis of the role of TRPC ion channels in BCR-activated and store-operated calcium signals in vertebrate cells. Prior to work in the DT40 system, multiple studies on TRPC channel gating mechanisms had provided a diverse set of conflicting data on TRPC channel function. However, by demonstrating that knockout of the TRPC1 gene in DT40 cells left store operated calcium entry intact, but produced alterations in both BCR-induced calcium release and calcium entry, this work provided one of the first pieces of evidence placing TRP channel function at the level of IP3 production, as opposed to direct participation in store operated calcium entry (Mori et al., 2002). Subsequent analysis of a TRPC7-knockout DT40 line by the Mori and Putney groups has also implicated TRPC7 channels in receptor-operated, but not store operated cation entry (Lievremont et al., 2005). These data have provided important supportive evidence that TRPC1 and TRPC7 channels probably do not underly CRAC channel activity, although subtle alterations in the measurement of CRAC currents in TRPC1 knockout cells leave open the possibility of some linkage between TRPC proteins and CRAC channels.

A final area in which DT40 cell genetics have made an important contribution to understanding the physiology of calcium signaling is in defining the role of TRPM7, a ubiquitous ion channel which underlies MagNuM/MIC currents present in many types of cells (Nadler et al., 2001). As TRPM7 is a member of the TRP superfamily, it had been assumed that it was primarily permeating Na^+ and Ca^{2+}, A collaboration between our group and Dr. Kurosaki's group led the creation of a conditional knockout of TRPM7. Analysis of this cell line demonstrated that TRPM7 was required for cell growth, and to a later observation that TRPM7-deficient cells could be grown if sufficient extracellular Mg^{2+} were provided (discussed in more detail below (Nadler et al., 2001; Schmitz et al., 2004)). The capacity to grow TRPM7-deficient cells has allowed us to study their Ca^{2+} signaling physiology, and we have observed no apparent differences between these cells and WT DT40 cells, providing evidence against a major role for TRPM7 in receptor-mediated or store-operated calcium signaling (C. Schmitz, A-L Perraud, and A.M. Scharenberg, unpublished data).

1.2.3 Mg^{2+} physiology in DT40 cells

As mentioned above, the generation of a knockout of TRPM7 in DT40 cells led to the observation that TRPM7-deficient DT40 cells are not able to grow in regular media, providing the first demonstration that an ion channel has a crucial role in the regulation of cell growth. Subsequently, a homologue of TRPM7, TRPM6, was shown to have a role in the regulation of organismal Mg^{2+} homeostasis, and this observation led to our own exploration of whether TRPM7 might play an analogous role in Mg^{2+} homeostasis at the cellular level (Schmitz et al., 2004). Our studies have shown that TRPM7-deficient DT40 cells can be grown indefinitely if provided with supplemental Mg^{2+} in their media. We have built on this observation by utilizing the capacity of DT40 cells to allow conditional expression of mutant forms of TRPM7 using a Tet-suppressor based conditional expression approach (Schmitz et al., 2004). These studies have shown that the capacity of TRPM7 channels to complement cell growth correlates with their net capacity to transport Mg^{2+} across the plasma membrane (i.e. the number of channels and/or their sensitivity to suppression by free Mg^{2+} determines their capacity to support cell prolifertion), but is independent of their intrinsic protein kinase activity. Overall these studies support a key role for TRPM7 in the regulation of DT40 cell Mg^{2+}

homeostasis. We have subsequently utilized our TRPM7-deficient cells to show that SLC41 proteins, which have distant homology to a class of prokaryotic Mg^{2+} transporters, are able to also complement the growth defect of TRPM7-deficient cells (J. Sahni and A. M. Scharenberg, in preparation). This observation further supports the involvement of TRPM7 in Mg^{2+} uptake, and also provides the first evidence that SLC41 proteins have a role in vertebrate cellular Mg^{2+} uptake.

My laboratory is presently utilizing the DT40 system to further analyze mechanisms which regulate Mg^{2+} homeostasis by attempting to create both stable and conditional knockouts of SLC41 proteins. Furthermore, we are exploiting another advantage of DT40 cells, which is the capacity to control their extracellular ionic environment to better understand how transport of a particular ion is regulated. As an example, we have determined that DT40 cells are able to grow at a normal rate when extracellular Mg^{2+} is lowered to 100 μM, or when 95 mM KCl is substituted for the normal 95 mM NaCl in their extracellular media (J. Sahni and A. M. Scharenberg, in preparation). The latter substitution largely collapses the membrane potential, enhancing DT40 cells' dependence on chemical gradients to drive uptake of individual nutrients and ions. However, under the collapsed Vmem conditions, DT40 cells grow at a normal rate if Mg^{2+} is at a physiologic level of 0.5 mM, but their rate of growth slows in proportion to the extracellular Mg^{2+} concentration if the extracellular Mg^{2+} concentration is dropped below 0.4-0.5 mM free Mg^{2+}, in the range of what free $[Mg^{2+}]$ is thought to be inside the cell. This observation strongly suggests that the majority of DT40 Mg^{2+} uptake occurs via a Vmem dependent mechanism, consistent with a major contribution from an ion channel transport mechanism such as TRPM7.

1.2.4 Zn^{2+} physiology in DT40 cells

Most recently, Suzuki, Kambe and colleagues have been exploring the use of DT40 cells in analysis of Zn^{2+} transport and homeostasis in vertebrate cells (Suzuki et al., 2005a; Suzuki et al., 2005b). They have utilized DT40 genetics to evaluate the role of the ZnT5, ZnT6, and ZnT7 zinc transporters by producing ZnT5, ZnT6, and ZnT7 single, double, and triple knockout cell lines (Suzuki et al., 2005a; Suzuki et al., 2005b). Using these cell lines, they have performed analyses of vertebrate Zn^{2+} regulatory mechanisms which demonstrate that ZnT5 and ZnT7 transporters have overlapping functions in loading zinc into the Golgi lumen for assembly with enzymes such as alklaline phosphatase apoenzymes, which must bind Zn^{2+} prior to maturing into their active forms. They have also utilized re-expression of ZnT5, ZnT6, and ZnT7 proteins to show that ZnT5 and ZnT6 work in a similar pathway and hetero-oligomerize, and that each protein may complement the function of the other, while ZnT7 works in a different pathway as a homo-oligomer. In combination, these data provide clear evidence on the function of ZnT5, ZnT6, and ZnT7 in vertebrate Zn^{2+} physiology at the cellular level. They also illustrate the utility which DT40 cells have as a model for the study of the individual contributions which various Zn^{2+} transporters make to maintaining Zn^{2+} homeostasis in vertebrate cells.

1.3 Summary

Involvement of ion transport proteins in diverse aspects of organism physiology in many cases renders the analysis of their regulation and role in cell physiology through the creation of organism level-knockouts either prohibitively expensive or non-feasible due to issues of profoundly altered organism development or complete non-viability. However, in the DT40 system, genetic manipulations have been achieved for up to three distinct homologues (six different alleles) of a given protein family, and have allowed a number of detailed studies to be performed on cells with well defined defects in specific cation transport pathways. In the future, the use of floxed selection markers that can be recycled by excision through the Cre-recombinase should allow the deletion of virtually any desired number of genes of interest in the same DT40 cell line. Furthermore, the stable phenotype of genetically manipulated DT40 cells coupled with their capacity to be transfected with mutant forms of deficient proteins allows "structure/function" analyses to be performed with a functional readout - an important issue for proteins such as transporters and channels where in-vitro functional assays may not be available. Overall, the DT40 system offers a well defined vertebrate system where modern genetic manipulations can be combined with electrophysiological and cell biological methods to gain novel insights into the cell physiology of ion transport.

Table 15-1. The DT40 "Channelome": Overview of the expression pattern of ion channels and transporters in the DT40 cell line.

Name	Major features	DT40 Expression, ref	DT40 KO phenotype, ref
CATIONIC CHANNELS			
TRPC1	Regulated by PLC, store-depletion? Non-selective, Ca^{2+}-permeable	**YES** MORI ET AL (2002)	Both Ca^{2+} release-activated Ca^{2+} currents and IP_3-mediated Ca^{2+} release from endoplasmic reticulum (ER) are reduced, resulting in decreased B cell antigen receptor-mediated Ca^{2+} oscillations and NF-AT activation (Mori et al., 2002).
TRPC2	Regulated by PLC, DAG, store-depletion? Non-selective, Ca^{2+}-permeable	No representation in the DT40/Bursa EST database Note:dkfz426_12a16r 1.blsp and dkfz426_14n8r1.blge do not appear to be real matches to TRPC2 sequences	N/A
TRPC3-TRPC6	Non-selective, Ca^{2+}-permeable	No representation in the DT40/Bursa EST database	N/A
TRPC7	Regulated by PLC, DAG, store-depletion? Non-selective, Ca^{2+}-permeable	**YES** Lievremont et al (2005)	No diminution in whole-cell I_{CRAC} current, increased size of the Ca^{2+}-stores. Non store-operated cation entry in response to the activation of the BCR, PAR2, intracellular dialysis with GTPγS, or application of DAG is absent (Lievremont et al., 2005)
TRPV channels		No representation in the DT40/Bursa EST	N/A

		database	
TRPM1	?	**Yes** (Kurosaki, Scharenberg, Perraud, unpublished)	N/A
TRPM2	C-terminal Nudix ADP-ribose hydrolase Gated by: ADP-ribose, oxidants Non-selective, Ca^{2+}-permeable	**No** (Scharenberg, Perraud, unpublished) Note: riken1_9j1r1 corresponds to the chicken TRPM8 homologue, and not to TRPM2 as suggested.	N/A
TRPM3	activated by sphingolipids? Ca^{2+}-permeable	No representation in the DT40/Bursa EST database	N/A
TRPM4	Gated by Ca^{2+}, voltage-sensitive Monovalent specific	**No** (Scharenberg, Perraud, unpublished)	N/A
TRPM5	Gated by Ca^{2+}, voltage-sensitive Monovalent specific	No representation in the DT40/Bursa EST database	N/A
TRPM6	C-terminal α-kinase ? Mg^{2+}	**No** (Scharenberg, Perraud, unpublished)	N/A
TRPM7	C-terminal α-kinase inhibited by intracellular Mg^{2+}/Mg.nucleotides Divalent cation selective (Mg^{2+}, Ca^{2+}...)	**YES** Nadler et al. (2001), Schmitz et al. (2003)	Homozygous deletion lethal unless medium supplemented with mM Mg^{2+}. In the absence of supplementary Mg^{2+}, total cellular Mg^{2+} drops, and the cells go into growth arrest and ultimately die (Schmitz et al., 2003).
TRPM8	Cold, menthol, icilin Non-selective, Ca^{2+}-permeable	**No** (Scharenberg, Perraud, unpublished) Note: riken1_9j1r1 corresponds to TRPM8, but we have not been able to confirm DT40 expression of cTRPM8 by RT-PCR	N/A
TPC1 and TPC2	Two-pore channels 1 and 2 Ca^{2+}/Na^+ permeable	**Yes** (Ping Liu and A.M. Scharenberg, unpublished)	N/A

Table 15-1 (Cont)

Table 15-1 (Cont)

IP$_3$-receptors type I, II, III	Intracellular (ER Ca^{2+}-store membrane) IP$_3$-gated Ca^{2+} channels	**YES** Sugawara et al. (1997)	DT40s in which a single type of IP$_3$R has been deleted still mobilize Ca^{2+} in response to BCR stimulation, whereas this Ca^{2+} mobilization is abrogated in B cells lacking all three types of IP3R (Sugawara et al., 1997). IP3R-1 is highly sensitive to ATP and mediates less regular Ca^{2+} oscillations. IP$_3$R-3 is the least sensitive to IP$_3$ and Ca^{2+}, and tends to generate monophasic Ca^{2+} transients (Miyakawa et al., 1999).
Ryanodine receptors, type 1 and 3	Intracellular (ER Ca^{2+}-store membrane) Ca^{2+}-channels	**Yes** (no Type 2) Kiselyov et al. (2001)	N/A
KCNK1 (TWIK-1)	Two-pore domain K$^+$ channel	Represented in the DT40/Bursa EST database: riken1_6b3r1; riken1_6b3	N/A
Kv channels and other K$^+$-channels		No convincing matches could be identified in the DT40/Bursa EST database	
SCN8A	Na$^+$ channel, voltage gated, type VIII, alpha	Represented in the DT40/Bursa EST database: dkfz426_25g18r1	N/A

ZnT5, ZnT6, ZnT7	Zinc-transporter, mainly in the Golgi and vesicular subcompartment	Suzuki et al. (2005)	ZnT5$^{-/-}$/ZnT6$^{-/-}$/Znt7-/- look healthy and total cellular Zn^{2+} like in WT, confirming that these transporters supply Zn^{2+} into the lumens of the secretory compartments. Activity og Golgi enzymes requiring Zn^{2+} severely reduced. Znt5 and ZnT6 associate and are in the same pathway, Znt7 works alone.
SLC41A2	Mg^{2+} transporter, possible transporter of other divalent cations	**Yes**, J. Sahni and A. M. Scharenberg, unpublished	N/A
ANIONIC CHANNELS			
ClC-3	Voltage-gated chloride channel	Represented in the DT40/Bursa EST database: riken1_4d21r1	N/A
ClC-5	Voltage-gated chloride channel Cl$^-$/H$^+$ exchangers	Represented in the DT40/Bursa EST database: dkfz426_4i15r1	N/A
ClC-7	endosomal/lysosomal chloride channel	riken1_7i18r1	N/A
CFTR	ABC (ATP-binding cassette)-transporter; cAMP dependent chloride channel	No representation in the DT40/Bursa EST database dkfz426_8l16r1 and riken1_17b24r1.blsp (corresponds tp the multidrug transporter ABCC4) do not appear to be a real match to CFTR sequences.	N/A

Table 15-1 (Cont)

Table 15-1 (Cont)

CLIC2	Intracellular chloride channel? Inhibits RyR, might limit Ca^{2+} release from internal stores	Represented in the DT40/Bursa EST database: riken1_10k5r1	N/A
VDAC 2	Predominantly expressed in the outer mitochondrial membrane; almost freely permeable to low molecular-weight molecules, slight preference for anions over cations	Represented in the DT40/Bursa EST database: dkfz426_15a24r1	N/A
VDAC3	Same as above	Represented in the DT40/Bursa EST database: dkfz426_1j21r1	N/A

Note: Only one EST clone ID is indicated per gene, although there are for all the listed genes multiple entries. Overlapping clones can be easily identified in the database.

Kiselyov K, Shin DM, Shcheynikov N, Kurosaki T, Muallem S (2001): Regulation of Ca^{2+}-release-activated Ca^{2+} current (Icrac) by ryanodine receptors in inositol 1,4,5-trisphosphate-receptor-deficient DT40 cells. Biochem J; 360(Pt 1):17-22.

Lievremont J-P, Numaga T, Vazquez G, Lemonnier L, Hara Y, Mori E, Trebak M, Moss SE, Bird GS, Mori Y, Putney JW Jr. (2005): The Role of Canonical Transient Receptor Potential 7 in B-cell Receptor-activated Channels. J. Biol. Chem. 280: 35346-35351.

Miyakawa T, Maeda A, Yamazawa T, Hirose K, Kurosaki T, Iino M (1999): Encoding of Ca^{2+} signals by differential expression of IP3 receptor subtypes. EMBO J; 18(5):1303-8.

Mori Y, Wakamori M, Miyakawa T, Hermosura M, Hara Y, Nishida M, Hirose K, Mizushima A, Kurosaki M, Mori E, Gotoh K, Okada T, Fleig A, Penner R, Iino M, Kurosaki T (2002): Transient receptor potential 1 regulates capacitative Ca^{2+} entry and Ca^{2+} release from endoplasmic reticulum in B lymphocytes. J Exp Med; 195(6):673-81.

Nadler MJ, Hermosura MC, Inabe K, Perraud AL, Zhu Q, Stokes AJ, Kurosaki T, Kinet JP, Penner R, Scharenberg AM, Fleig A (2001): LTRPC7 is a Mg.ATP-regulated divalent cation channel required for cell viability. Nature; 411(6837):590-5.

Schmitz C, Perraud AL, Johnson CO, Inabe K, Smith MK, Penner R, Kurosaki T, Fleig A, Scharenberg AM (2003): Regulation of vertebrate cellular Mg^{2+} homeostasis by TRPM7.Cell;114(2):191-200.

Sugawara H, Kurosaki M, Takata M, Kurosaki T (1997): Genetic evidence for involvement of type 1, type 2 and type 3 inositol 1,4,5-trisphosphate receptors in signal transduction through the B-cell antigen receptor. EMBO J; 16(11):3078-88.

Suzuki T, Ishihara K, Migaki H, Ishihara K, Nagao M, Yamaguchi-Iwai Y, Kambe T (2005): Two different zinc transport complexes of cation diffusion facilitator proteins localized in the secretory pathway operate to activate alkaline phosphatases in vertebrate cells. J Biol Chem; 280(35):30956-62.

REFERENCES

Broad L. M., Braun F. J., Lievremont J. P., Bird G. S., Kurosaki T. and Putney J. W., Jr. 2001. Role of the phospholipase C-inositol 1,4,5-trisphosphate pathway in calcium release-activated calcium current and capacitative calcium entry. *J Biol Chem* **276**, 15945-52.

Bultynck G., Szlufcik K., Kasri N. N., Assefa Z., Callewaert G., Missiaen L., Parys J. B. and De Smedt H. 2004. Thimerosal stimulates Ca^{2+} flux through inositol 1,4,5-trisphosphate receptor type 1, but not type 3, via modulation of an isoform-specific Ca2+-dependent intramolecular interaction. *Biochem J* **381**, 87-96.

Cui J., Matkovich S. J., deSouza N., Li S., Rosemblit N. and Marks A. R. 2004. Regulation of the type 1 inositol 1,4,5-trisphosphate receptor by phosphorylation at tyrosine 353. *J Biol Chem* **279**, 16311-6.

Guillemette J., Caron A. Z., Regimbald-Dumas Y., Arguin G., Mignery G. A., Boulay G. and Guillemette G. 2005. Expression of a truncated form of inositol 1,4,5-trisphosphate receptor type III in the cytosol of DT40 triple inositol 1,4,5-trisphosphate receptor-knockout cells. *Cell Calcium* **37**, 97-104.

Kiselyov K., Shin D. M., Shcheynikov N., Kurosaki T. and Muallem S. 2001. Regulation of Ca^{2+}-release-activated Ca^{2+} current (Icrac) by ryanodine receptors in inositol 1,4,5-trisphosphate-receptor-deficient DT40 cells. *Biochem J* **360**, 17-22.

Laude A. J., Tovey S. C., Dedos S. G., Potter B. V., Lummis S. C. and Taylor C. W. 2005. Rapid functional assays of recombinant IP3 receptors. *Cell Calcium* **38**, 45-51.

Lievremont J. P., Numaga T., Vazquez G., Lemonnier L., Hara Y., Mori E., Trebak M., Moss S. E., Bird G. S., Mori Y. and Putney J. W., Jr. 2005. The Role of Canonical Transient Receptor Potential 7 in B-cell Receptor-activated Channels. *J Biol Chem* **280**, 35346-51.

Ma H. T., Venkatachalam K., Li H. S., Montell C., Kurosaki T., Patterson R. L. and Gill D. L. 2001. Assessment of the role of the inositol 1,4,5-trisphosphate receptor in the activation of transient receptor potential channels and store-operated Ca^{2+} entry channels. *J Biol Chem* **276**, 18888-96.

Ma H. T., Venkatachalam K., Parys J. B. and Gill D. L. 2002. Modification of store-operated channel coupling and inositol trisphosphate receptor function by 2-aminoethoxydiphenyl borate in DT40 lymphocytes. *J Biol Chem* **277**, 6915-22.

Mori Y., Wakamori M., Miyakawa T., Hermosura M., Hara Y., Nishida M., Hirose K., Mizushima A., Kurosaki M., Mori E., Gotoh K., Okada T., Fleig A., Penner R., Iino M. and Kurosaki T. 2002. Transient receptor potential 1 regulates capacitative Ca^{2+} entry and Ca^{2+} release from endoplasmic reticulum in B lymphocytes. *J Exp Med* **195**, 673-81.

Morita T., Tanimura A., Nezu A., Kurosaki T. and Tojyo Y. 2004. Functional analysis of the green fluorescent protein-tagged inositol 1,4,5-trisphosphate receptor type 3 in Ca^{2+} release and entry in DT40 B lymphocytes. *Biochem J* **382**, 793-801.

Nadler M. J., Hermosura M. C., Inabe K., Perraud A. L., Zhu Q., Stokes A. J., Kurosaki T., Kinet J. P., Penner R., Scharenberg A. M. and Fleig A. 2001. LTRPC7 is a Mg.ATP-regulated divalent cation channel required for cell viability. *Nature* **411**, 590-5.

Prakriya M. and Lewis R. S. 2001. Potentiation and inhibition of Ca^{2+} release-activated Ca^{2+} channels by 2-aminoethyldiphenyl borate (2-APB) occurs independently of IP(3) receptors. *J Physiol* **536**, 3-19.

Schmitz C., Perraud A. L., Fleig A. and Scharenberg A. M. 2004. Dual-Function Ion Channel/Protein Kinases: Novel Components of Vertebrate Magnesium Regulatory Mechanisms. *Pediatr Res*.

Sugawara H., Kurosaki M., Takata M. and Kurosaki T. 1997. Genetic evidence for involvement of type 1, type 2 and type 3 inositol 1,4,5-trisphosphate receptors in signal transduction through the B-cell antigen receptor. *Embo J* **16**, 3078-88.

Suzuki T., Ishihara K., Migaki H., Matsuura W., Kohda A., Okumura K., Nagao M., Yamaguchi-Iwai Y. and Kambe T. 2005a. Zinc transporters, ZnT5 and ZnT7, are required for the activation of alkaline phosphatases, zinc-requiring enzymes that are glycosylphosphatidylinositol-anchored to the cytoplasmic membrane. *J Biol Chem* **280**, 637-43.

Suzuki T., Ishihara K., Migaki H., Nagao M., Yamaguchi-Iwai Y. and Kambe T. 2005b. Two different zinc transport complexes of cation diffusion facilitator proteins localized in the secretory pathway operate to activate alkaline phosphatases in vertebrate cells. *J Biol Chem* **280**, 30956-62.

Vazquez G., Lievremont J. P., St J. B. G. and Putney J. W., Jr. 2001. Human Trp3 forms both inositol trisphosphate receptor-dependent and receptor-independent store-operated cation channels in DT40 avian B lymphocytes. *Proc Natl Acad Sci U S A* **98**, 11777-82.

Vazquez G., Wedel B. J., Bird G. S., Joseph S. K. and Putney J. W. 2002. An inositol 1,4,5-trisphosphate receptor-dependent cation entry pathway in DT40 B lymphocytes. *Embo J* **21**, 4531-8.

Vazquez G., Wedel B. J., Trebak M., St John Bird G. and Putney J. W., Jr. 2003. Expression level of the canonical transient receptor potential 3 (TRPC3) channel determines its mechanism of activation. *J Biol Chem* **278**, 21649-54.

Venkatachalam K., Ma H. T., Ford D. L. and Gill D. L. 2001. Expression of functional receptor-coupled TRPC3 channels in DT40 triple receptor InsP3 knockout cells. *J Biol Chem* **276**, 33980-5.

Wedel B. J., Vazquez G., McKay R. R., St J. B. G. and Putney J. W., Jr. 2003. A calmodulin/inositol 1,4,5-trisphosphate (IP3) receptor-binding region targets TRPC3 to the plasma membrane in a calmodulin/IP3 receptor-independent process. *J Biol Chem* **278**, 25758-65.

Yogo T., Kikuchi K., Inoue T., Hirose K., Iino M. and Nagano T. 2004. Modification of intracellular Ca^{2+} dynamics by laser inactivation of inositol 1,4,5-trisphosphate receptor using membrane-permeant probes. *Chem Biol* **11**, 1053-8.

Chapter 16

ANALYSIS OF DNA REPLICATION DAMAGE BYPASS AND ITS ROLE IN IMMUNOGLOBULIN REPERTOIRE DEVELOPMENT

Julian E. Sale, Anna-Laura Ross and Laura J. Simpson
M.R.C. Laboratory of Molecular Biology, Hills Road, Cambridge, CB2 2QH

Abstract: Cells possess very effective mechanisms for repairing DNA damage. However, in some circumstances repair cannot be carried out before passage of a replication fork, leading to polymerase stalling and an unreplicated gap. Left unrepaired, such gaps will lead to misegregation of genetic information or apoptosis. Cells can ensure that replication is completed by bypassing the damaged bases in the DNA template either directly by translesion synthesis or indirectly by employing an alternative undamaged template. This cellular activity, originally termed post-replication repair is now often referred to as replication damage bypass. Extensively studied in bacteria and yeast, replication damage bypass is now receiving much attention in higher eukaryotes because of its close link to mutagenesis and genetic instability. Work in DT40 has given many insights into the roles of and relationships between the genes involved in DNA damage tolerance and recombination, not least because the cell line can tolerate disruption of many genes that result in early embryonic lethality in mice. In this review we examine current thinking about vertebrate replication damage bypass in the context of studies in bacteria and yeast. We focus particularly on the contribution made by DT40 to these studies and discuss how immunoglobulin diversification in this cell line can contribute to our understanding of replication damage bypass in vertebrates.

Key words: DNA damage, replication, translesion synthesis, homologous recombination, specialised polymerases, DT40, immunoglobulin, gene conversion, somatic hypermutation.

J.-M. Buerstedde and S. Takeda (eds.), Reviews and Protocols in DT40 Research, 271–294.
© 2006 *Springer.*

1.   INTRODUCTION & GENERAL PRINCIPLES

All cells have a series of intricate pathways for correcting the DNA damage that continuously threatens the integrity of their genetic code. This threat largely arises from the ability of many lesions to block advancing DNA polymerases during replication. While DNA repair is able to fix the majority of lesions, some inevitably slip through the net and are met by the replication machinery. Thus, cells have evolved a second set of mechanisms that allows replication to be completed in the face of a damaged DNA template. A very simplistic, but nonetheless useful, view is that cells have two choices when faced with this problem: either to change polymerases, replacing the stalled replicative polymerase with one that is able to synthesise across the lesion (translesion synthesis) or to change template, employing a homologous stretch of DNA that is not damaged (homologous recombination) [Figure 1]. Translesion synthesis employs specialised polymerases that have active sites which are able to accommodate damaged or distorted templates or primer termini (Goodman, 2002; Prakash et al., 2005). The downside of this approach is an increased risk of mutation, due to both the non- or mis-instructional nature of the template coupled to an often increased intrinsic error rate of these polymerases. Recombinational modes of bypass are generally accurate since the donor template is usually an exact copy of the damaged template. However, recombination too has its risks, principally in loss of heterozygosity.

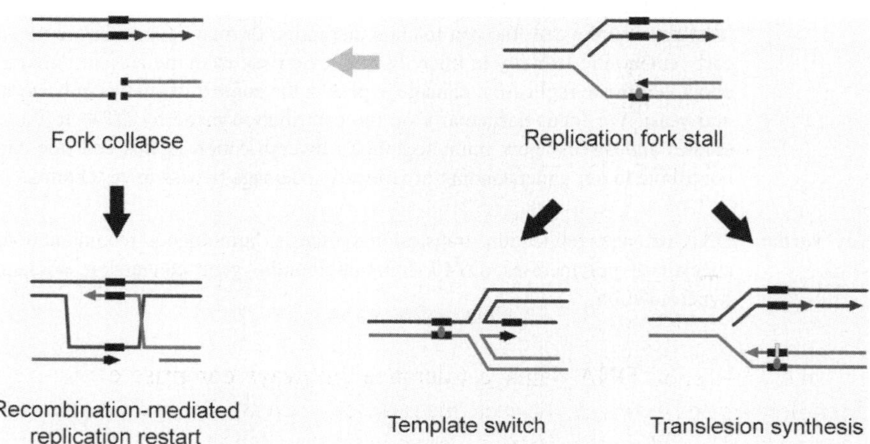

Figure 16-1. A simplified scheme of DNA damage bypass during replication (see plate 37).

The nomenclature of these replication damage tolerance and repair pathways is somewhat confusing and, to an extent, reflects uncertainty about their exact mode of action. Diverse evidence from different systems points to a role for these pathways both in filling in gaps left at the end of replication, and in the bypass of lesions while preserving replication fork integrity [Figure 2]. The truth is, of course, likely to be that both are used and that the numerous overlapping possibilities for bypass of DNA damage underscores the importance to cells of completing replication. It may also reflect the differing circumstances under which damage is encountered, for instance on a leading- versus lagging-strand template.

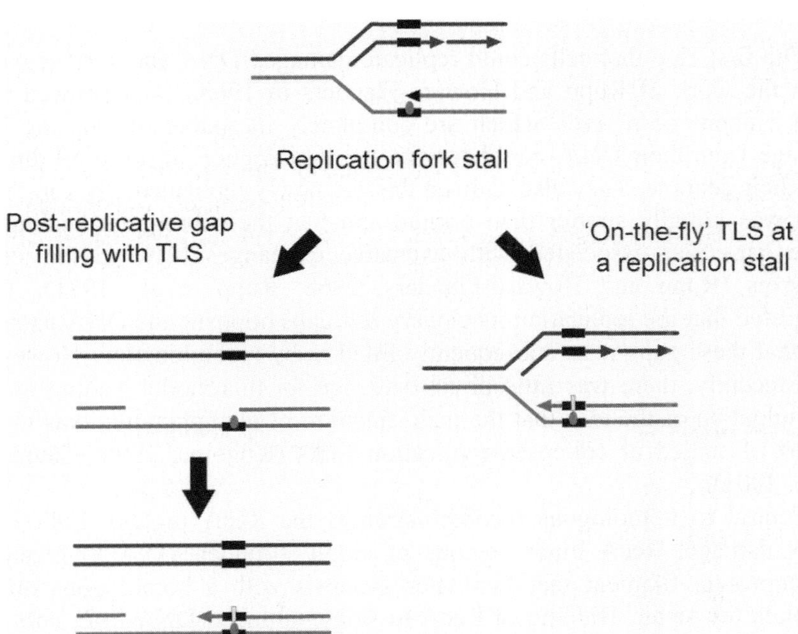

Figure 16-2. The timing of TLS use - rescue of stalled replication vs. post-replicative gap filling (see plate 38).

The replication DNA damage tolerance pathways comprise elements of homologous recombination and mechanisms known as post-replication repair (PRR). Although the term post-replication repair is widely used, it does not strictly describe a repair pathway (the lesion is not actually removed) and, since some components may well work during, rather than after, replication we prefer the more recently coined term replication damage bypass (RDB). Here we use RDB to encompass all pathways responsible for

DNA damage tolerance during S phase, including both translesion synthesis and recombinational based mechanisms, such as template switching.

In this review, we will examine our current understanding of replication damage bypass pathways in higher eukaryotes and their relationship to homologous recombination in the cellular tolerance of DNA damage. We will highlight the contribution made by studying these processes in DT40 and examine what can be learned about these pathways from the constitutively diversifying immunoglobulin loci in this cell line.

2. THE PARADIGM OF REPLICATION DAMAGE TOLERANCE: POST-REPLICATION REPAIR IN *E. COLI*

The first clue that cells could replicate damaged DNA and survive came from the work of Rupp and Howard-Flanders in 1968. They showed that *uvrA* mutants of *E. coli*, which are completely incapable of excising UV damage from their DNA, could survive the introduction of up to 50 dimers into their genome. They also showed that the newly replicated DNA in these cells was initially shorter than normal and that the subsequent healing of these gaps was associated with extensive exchanges between the sister duplexes (Rupp and Howard-Flanders, 1968; Rupp et al., 1971). This suggested that the replication machinery left gaps opposite the DNA damage and that these gaps were subsequently filled in by recombination. However, until recently, there was little direct evidence for this model leading to the promulgation of the idea that the main function of recombination was in the restart of stalled or collapsed replication forks (Kogoma, 1996; Courcelle et al., 2001).

Central to homologous recombination is the RecA protein. Following DNA damage, RecA binds regions of single stranded DNA to create a nucleoprotein filament that facilitates synapsis with a homologous donor template for repair. Binding of RecA to single-stranded DNA in *E. coli* has an additional effect, the cleavage of the LexA transcriptional repressor. This leads to the elevated expression of about 40 genes involved in responding to DNA damage – the SOS response. Among the genes induced in the SOS response are the subunits of the translesion polymerase, DNA polymerase V, UmuC and UmuD. Following further activation, this enzyme is able to mediate filling of postreplicative gaps by direct DNA synthesis over the lesion (Reuven et al., 1999; Tang et al., 1999). However, in the SOS-induced state, where both recombinational repair and translesion synthesis can operate, recombination repair is the dominant mode of gap filling by a ratio of six to one. Only when there is no available homologous template for repair does translesion synthesis dominate (Berdichevsky et al., 2002). Thus, for gap filling in *E. coli*, translesion synthesis appears to be largely a backup mechanism used *in extremis* either when recombinational repair cannot cope or, perhaps, when there are types of damage that cannot be repaired by recombination.

3.   REPLICATION DAMAGE BYPASS PATHWAYS IN BUDDING YEAST

Much of the current effort in understanding replication damage bypass in higher eukaryotes has been guided by the genetic and biochemical framework developed in the budding yeast, *Saccharomyces cerevisiae* and while significant differences in some aspects of replication damage bypass are emerging between yeast and vertebrates, an understanding of these pathways in yeast has proved a good starting point.

Genes involved in DNA repair in yeast fall into three major epistasis groups. The RAD3 group encompasses genes involved in excision repair, the RAD52 group genes involved in homologous recombination and the RAD6 group, the most heterogenous of the three, genes involved in replication damage bypass.

The core members of the RAD6 group, RAD6 itself and RAD18, are epistatic to all other members of the group with respect to DNA damage senstivity (reviewed in Broomfield et al., 2001). RAD6 is an E2 ubiquitin conjugating enzyme (Jentsch et al., 1987) that is involved in a number of cellular processes. For its role in replication damage bypass it pairs with the RING domain E3 ubiquitin ligase, RAD18 (Bailly et al., 1994; Bailly et al., 1997). *rad6* and *rad18* mutants are both defective in the ability to restore high molecular weight DNA following replication over a damage DNA template (Prakash, 1981), a classical phenotype of post-replication repair mutants. They are also defective in UV induced mutagenesis. These two phenotypes are a consequence of the loss of two sub-pathways within the RAD6 group. The members of the first subpathway, *UBC13, MMS2* and *RAD5* also encode genes involved in protein ubiquitination. Mutants in these genes are defective in restoration of high molecular weight DNA after DNA synthesis on UV-damaged templates and they exhibit elevated spontaneous mutation rates. They have therefore been assigned to an error-free bypass pathway. Genetic evidence also implicates the replicative DNA polymerase δ, PCNA and RAD52 in this process (Prakash, 1981; Torres-Ramos et al., 1996; Torres-Ramos et al., 1997) and although it has been proposed to reflects a form of copy-choice or template switch mechanism (Higgins et al., 1976), no direct evidence for this exists.

That *rad6* and *rad18* mutants are defective in UV induced mutagenesis (Lawrence and Christensen, 1976; Cassier-Chauvat and Fabre, 1991) is due to their genetic interaction with three other genes *REV1*, *REV3* and *REV7* which form a second RAD6-dependent subpathway (Lawrence and Hinkle, 1996). *REV3* and *REV7* encode the translesion polymerase pol ζ (Nelson et al., 1996) while *REV1* encodes a polymerase-like enzyme capable only of *in vitro* deoxycytidyl transfer across a limited number of lesions (Lawrence and Hinkle, 1996). REV1 nevertheless plays an important role in mutagenesis induced by a much wider range of DNA damaging agents, including UV light (reviewed in Lawrence and Maher, 2001). DNA polymerase ζ mutants are able to reconstitute high molecular weight DNA after DNA synthesis on a UV damaged template suggesting that pol ζ-dependent translesion synthesis is not required for post-replicative gap filling in yeast (Prakash, 1981; Torres-Ramos et al., 2002).

A further translesion polymerase, pol η, encoded by the *RAD30* gene, has been identified through its homology to *umuC* (McDonald et al., 1997). Epistasis analysis revealed RAD30 to be in the RAD6 group but in a separate pathway to both RAD5 and REV1/3/7. *rad30* mutants exhibit a much smaller defect in post-replicative gap filling as assessed by alkaline sucrose gradient centrifugation than *rad6* or *rad18* (Torres-Ramos et al., 2002) and do not have a defect in UV induced reversion (McDonald et al., 1997). This led to the proposal that RAD30 forms a second error-free branch of the RAD6 group. For UV damage, this is certainly the case as DNA polymerase η is able to replicate over T-T cyclobutane pyrimidine dimers and 6-4 photoproducts accurately (Yu et al., 2001). However, it tends to misincorporate opposite other forms of damage, such as O^6-methylguanine (Haracska et al., 2000).

Much sense of the rather confusing genetic data on the yeast RAD6 group was made by the observation that PCNA is a target for the ubiquitin ligase activity of RAD6/RAD18 (Hoege et al., 2002; Watanabe et al., 2004). PCNA is the homotrimeric sliding clamp that coordinates the replicative polymerases and various repair factors at the replication fork. RAD6/RAD18 monoubiquitinate PCNA at lysine 164 in a DNA damage dependent manner and this modification is required for translesion synthesis. This single ubiquitin can itself be further ubiquitinated to form a chain, a reaction requiring UBC13/MMS2/RAD5. However, the chain formed is not of the canonical form in which each ubiquitin is linked to its neighbour by lysine 48, but is instead linked via lysine 63. This lysine 63-linked chain does not target PCNA for degradation by the proteosome. Although the exact function of these non-canonical ubiquitin chains in error-free bypass is unclear, it has been suggested that they may somehow facilitate interaction between site of replication stalling and the newly synthesised template on the sister chromatid (Hoege et al. 2002).

In addition to these two post-translational modifications with ubiquitin, PCNA can also be modified by SUMO (Small Ubiquitin-like MOdifier).

SUMOylation appears to compete for ubiquitination at the same site (lysine 164), but can also occur to a lesser extent at a nearby lysine at residue 127. SUMOylated PCNA has been shown to preferentially bind the anti-recombinogenic DNA helicase SRS2 providing biochemical support for the genetic evidence that the function of PCNA SUMOylation is to prevent homologous recombination (Pfander et al., 2005).

4. HOMOLOGUES OF YEAST RAD6 GROUP PROTEINS IN HIGHER EUKARYOTES

Many of the genes involved in the DNA damage response have homologues in vertebrates and members of the RAD6 group are no exception (Table 1). As is frequently the case, the vertebrate array of homologues is complicated by gene duplications and this is evident in the case of RAD6 itself, MMS2 and RAD30. In addition, the range of translesion polymerases has expanded from just two (plus REV1) in budding yeast to probably over ten in mammals.

There is extra potential complexity in examples where additional functions have been ascribed to the vertebrate homologues of some members of the yeast RAD6 group. Thus, UBC13 in mammals is also involved in regulating signalling in the NF-κB pathway, a system that does not exist in yeast (Deng et al., 2000).

There are also apparent gaps in the vertebrate protein repertoire when compared with yeast. No functional homologue of RAD5 has yet been identified in vertebrates. The closest homologue, hSMARCA3, whose C terminus contains a similar arrangement to RAD5 of SNF2, RING and helicase domains, does not interact with hRAD18 (ALR & JES, unpublished) while in yeast, RAD18 and RAD5 interact tightly (Ulrich and Jentsch, 2000).

In addition, no clear vertebrate homologue of the antirecombinogenic DNA helicase SRS2 has yet been identified. SRS2, which is able to displace RAD51 from DNA (Krejci et al., 2003) is recruited to replication forks by SUMOylation of PCNA (Papouli et al., 2005; Pfander et al., 2005). This is a constitutive post-translational modification in yeast (Hoege et al., 2002; Stelter and Ulrich, 2003; Haracska et al., 2004) but has not yet been seen in mammalian cells (Hoege et al., 2002; Kannouche et al., 2004). Thus, it seems likely that, while some aspects of replication damage bypass are conserved between yeast and vertebrates, there remain important differences.

Table 16-1. Vertebrate homologues of the budding yeast RAD6 epistasis group.

Yeast	Chicken	Mammals	chicken:human sequence identity
RAD6	RAD6A	RAD6A	99%
	RAD6B	RAD6B	100%
RAD18	RAD18	RAD18	47%
UBC13	UBC13	UBC13	100%
MMS2	UBE2V1	UBE2V1	98%*
	UBE2V2	UBE2V2	95%
RAD5	not identified	not identified	–
PCNA K164	conserved	conserved	94%†
PCNA K127	not conserved; nearby SUMOylation consensus at K138	not conserved; nearby SUMOylation consensus at K138	–
SRS2	not identified	not identified	–
REV1	REV1	REV1	76%
REV3§	REV3	REV3	61%
REV7	REV7	REV7	96%
RAD30	POLη	POLη	52%
		POLι	
	POLκ	POLκ	57%

*UBE2V1 has a number of exon 1 splice variants. This alignment is for the core UBE region (exon 2 to the end). For discussion see Simpson & Sale, 2005. †This alignment is of PCNA itself. §REV3/REV7 form the two subunits of POLζ, REV3 being catalytic. In addition there are a number of other specialised polymerases in vertebrates including, POLλ, POLμ, POLθ and POLσ, which we will not cover in this review.

Chicken RAD6 group genes and TLS polymerases are highly homologous to those in mouse and human with amino acid identities between human and chicken ranging from 100% for RAD6B and 94% for PCNA to 47% for RAD18 and 51% for DNA polymerase η (Table 1).

It is, however, also important to note that there are potentially important differences between chicken and mammals as well. For example a search of the completed chicken genome reveals only a single RAD30 homologue, DNA polymerase η. Further, chickens, like all vertebrates except placental mammals, possess a homologue of photolyase, an enzyme capable of direct reversal of some UV-induced DNA damage (Sancar, 2003). Whether this enzyme makes any contribution to DNA repair in DT40, or is indeed active in chickens, is not known.

5. PHENOTYPES OF MAMMALIAN AND DT40 CELLS DEFICIENT IN DNA DAMAGE TOLERANCE PATHWAYS

Mutants of vertebrate homologues of the yeast RAD6 epistasis group exhibit essential phenotypes that largely mirror those of their yeast counterparts. However, the emerging view, to which studies in DT40 have contributed significantly, is that there are important differences in the relative contributions of individual genes to overall DNA damage tolerance. Although this is perhaps not surprising given the additional complexity of replicating a vertebrate genome, it emphasises the importance of genetic studies in vertebrates as well as in yeast. In this section we look at the analysis of DNA damage tolerance pathways in single mutants and then examine how the ability to readily combine mutations in DT40 allows the relationships of genes to be established.

5.1 The spectrum of damage sensitivity in RDB mutants

To date, analysis of the DNA damage sensitivity of mammalian and DT40 RDB mutants has revealed two patterns. Mutants such as *rad18*, *rev1*, and *rev3* are sensitive to a wide range of agents (Tateishi et al., 2000; Yamashita et al., 2002; Simpson and Sale, 2003; Sonoda et al., 2003; Tateishi et al., 2003), suggesting that these genes play multiple or regulatory roles in DNA damage tolerance. On the other hand, mutants in genes such as *polκ* are sensitive only to a limited range of insults (Okada et al., 2002), suggesting that such enzymes play a role in the bypass of specific lesions.

5.2 When is bypass used in the cell cycle?

Analysis of the damage sensitivity of several DT40 RDB mutants at different stages of the cell cycle reveals, as might be expected, that the cells are most sensitive to UV irradiation after the onset of DNA replication (Okada et al., 2002; Yamashita et al., 2002; Sonoda et al., 2003). Further, *rad18* mutants of both DT40 and mouse ES cells, as well as human cell lines expressing a dominant negative form of RAD18, exhibit defective restoration of high molecular weight DNA following replication after DNA damage (Tateishi et al., 2000; Yamashita et al., 2002; Tateishi et al., 2003). In contrast, the *rev3* mutant DT40 does not exhibit this phenotype (Sonoda et al., 2003). This suggests that differential involvement of members of the RAD6 group in post-replicative gap filling. Despite these phenotypic similarities between yeast and vertebrate mutants, yeast *rev3* exhibits a relatively mild phenotype in terms of survival while *rev3* mutation in mice results in inviability, even with a concurrent *p53* mutation (Bemark et al., 2000; Wittschieben et al., 2000; Kajiwara et al., 2001; O-Wang et al., 2002; Van Sloun et al., 2002). Nevertheless, viable, but damage sensitive, *rev3/p53* null murine fibroblasts have been established demonstrating that *rev3* is not absolutely essential for mammalian cell viability (Zander and Bemark, 2004).

5.3   Elevated use of homologous recombination in RDB mutants

DT40 with disabled replication damage bypass pathways frequently exhibit elevated sister chromatid exchange (SCE). SCEs reflect crossing over arising during homologous recombination and an increased level may reflect increased use of recombination pathways or a biasing of the outcome of recombination towards crossing over. Increased SCE are most strikingly observed in the *rad18* and *rev3* mutants (Yamashita et al., 2002; Sonoda et al., 2003) and the suggestion that these result from channeling of lesions into homologous recombination is consistent with double mutations of *rad18* or *rev3* with *rad54* resulting in cell death (Yamashita et al., 2002). However, there are some paradoxes yet to be addressed. Both *rad18* and, particularly, *rev3* mutants exhibit some features of disabled recombination with decreased frequencies of targeted integration and increased sensitivity to ionising radiation after the completion of replication (Yamashita et al., 2002; Sonoda et al., 2003). Thus, the relationship between recombination and translesion synthesis is likely not to be one of simple alternatives. Rather the two processes are to some degree interdependent.

To date, little is known about the role in vertebrates of the homologues of the RAD5/UBC13/MMS2 error-free pathway of damage tolerance in yeast. As discussed above, at least one of the key players, RAD5, has not been unambigously identified and a DT40 mutant of UBE2V2, the vertebrate homologue of MMS2, does not exhibit a DNA damage response phenotype (Simpson and Sale, 2005). While this latter observation may be the result of functional redundancy with another closely related gene, UBE2V1, it remains possible that the modality of DNA damage tolerance mediated by this ubiquitin ligase system in yeast is not functional in vertebrates, having been supplanted by other mechanisms. Indeed, one such more recently evolved pathway, defined by the Fanconi Anaemia group of genes, has mutants which have a number of features suggestive of involvement in a form of recombinational damage bypass (Hirano et al., 2005; Yamamoto et al., 2005).

5.4   Epistasis analysis using DT40

A particularly important feature of DT40 is the ability to readily combine knockouts and thus to examine genetic interactions (epistasis analysis). As so powerfully demonstrated in yeast and bacteria, this form of genetic analysis can provide vital clues to the order and dependencies of genes within a pathway and therefore provides a roadmap to understanding the functioning of the pathway.

A simple example of epistasis analysis in DT40 shows that, as is the case in yeast, disabling nucleotide excision repair puts more strain on the replication damage tolerance pathways: a double mutation of the nucleotide excision repair factor XPA with DNA polymerase κ results in the double mutant being more sensitive to UV light than either single mutant (Okada et al., 2002).

More revealingly, epistasis analysis of RAD18 with other translesion polymerases in DT40 reveals potentially important differences to yeast. While RAD18 is required for recruitment of TLS polymerases in yeast, this does not appear to be the case for TLS in chicken. While both *polκ* and *rad18* exhibit similar sensitivities to UV irradiation, a double mutant is considerably more sensitive than either single mutant suggesting that the recruitment of polκ to sites of UV damage is not dependent on RAD18 (Okada et al., 2002). A similar picture is seen in the interaction between RAD18 and REV1, with a *rad18/rev1* double mutant being more sensitve than either single mutant (Ross et al., 2005). Despite the demonstration that RAD18 is likely to be responsible for ubiquitination of PCNA in mammals as well as yeast (Kannouche et al., 2004), these genetic results suggest that a substantial amount of translesion synthesis in vertebrates is independent of RAD18.

6.   IMMUNOGLOBULIN DIVERSIFICATION IN DT40 AS A MODEL FOR DNA DAMAGE BYPASS

6.1   Immunoglobulin diversification in chicken

At this point it is necessary to introduce another important feature of DT40, its ability to diversify its immunoglobulin genes constitutively in culture (Buerstedde et al., 1990; Kim et al., 1990). Here we will review the aspects of this process that are relevant to the analysis of mechanisms of DNA damage tolerance and examine the ways in which immunoglobulin diversification in DT40 provides a unique opportunity for studying the site specific bypass of a defined DNA lesion.

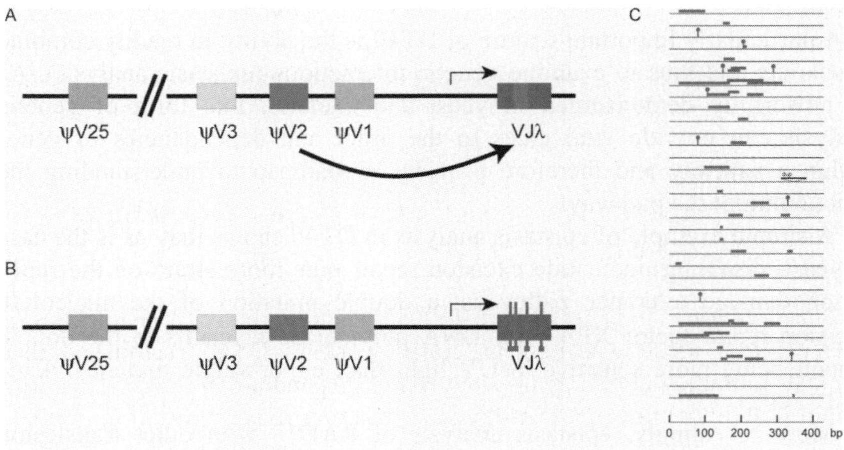

Figure 16-3. Gene conversion and point mutation in the Ig light chain locus of DT40. A. Gene conversion results in the introduction of blocks of sequence change into the rearranged V gene, templated from one of 25 upstream pseudogene donors in the case of the light chain illustrated here. B. Gene conversion is accompanied by point mutations which cannot be explained by templation from the pseudogenes. C. Light chain sequences from DT40 showing a typical patchwork of gene conversion and point mutation. Each grey line represents a single Vλ sequence. Gene conversion tracts are shown as blue lines and point mutations as red lollipops (see plate 39).

In contrast to mice and humans, chickens generate very little of their immunoglobulin repertoire through V(D)J recombination. Instead they employ a V(D)J step from a very limited repertoire of gene segments mainly to generate single functional heavy and light immunoglobulin chains (Reynaud et al., 1985). Much of the pre-immune immunoglobulin diversity in chicken is generated within the Bursa of Fabricius by gene conversion (Reynaud et al., 1987; Reynaud et al., 1989). Gene conversion involves the non-reciprocal transfer of sequence from one of an array of V pseudogenes into the rearranged and expressed V gene [Figure 3]. Accompanying this gene conversion there are also point mutations, which are not derived by templated gene conversion. The balance of gene conversion and point mutation shifts during B cell development and, during the antigen-dependent stages of the immune response, immunoglobulin diversification in chicken comes closer to resembling the somatic hypermutation seen in mouse and human (Arakawa et al., 1998). DT40, being derived from a bursal lymphoma exhibits predominantly gene conversion with, on average, about one point mutation for every three or four gene conversions. There is no evidence that the point mutations occur in association with detectable gene conversion tracts. However, the targeting of the somatic mutation in DT40 differs from that seen in chicken in that the mutations occur predominantly at G/C basepairs rather than being spread more evenly between G/C and A/T. This is reminiscent of the difference in mutation spectrum between hypermutating B cell lines and that seen *in vivo* in mouse and in human (Denépoux et al., 1997; Sale and Neuberger, 1998).

6.2 The mechanisms of initiation of immunoglobulin diversification in DT40

Until recently however, the mechanism by which immunoglobulin diversification arose was unclear. Early work in DT40 established that efficient immunoglobulin gene conversion depended on RAD54, a key protein in the homologous recombination reaction (Bezzubova et al., 1997). The first clue that gene conversions and point mutations shared a common initiating event came from the observation that deletion of the RAD51 paralogues, XRCC2 or XRCC3 resulted in the inhibition of gene conversion and a swing to point mutation (Sale et al., 2001). This observation suggested that inhibition of homologous recombination left an initiating lesion available for bypass by translesion synthesis. The nature of the common initiating event became clear following the demonstration that Activation Induced Deaminase (AID), a gene shown to be essential for immunoglobulin hyper-mutation and class switch recombination in mouse and human (Muramatsu et al., 2000; Revy et al., 2000), could act directly on DNA (Petersen-Mahrt et al., 2002) (Figure 4). AID is a cytidine deaminase whose target was initially thought to be RNA based on its homology with the known RNA editing enzyme APOBEC1 (Muramatsu et al., 1999). However, subsequent work demonstrated that AID can deaminate C to U in DNA and that much of the point mutation at G/C is a direct result of its enzymatic activity (Petersen-Mahrt et al., 2002; Bransteitter et al., 2003; Chaudhuri et al., 2003; Ramiro et al., 2003). U is not normally found in DNA and is readily removed by uracil DNA glycosylase (UNG) to leave an abasic site (Krokan et al., 2002). The uracils generated in the immunoglobulin loci are no exception. Inhibition of UNG in hypermutating *xrcc2* DT40 by expression of the bacteriophage protein Ugi resulted in a marked shift from transversions at G/C to transitions (Di Noia and Neuberger, 2002). This not only provided evidence for dC to dU deamination *in vivo* but suggested that mutations at G/C are the result of DNA synthesis across a non-instructional abasic site. Further support for this model came from the demonstration that UNG-deficient mice and humans also exhibit a shift from transversions to transitions at G/C as well as a defect in class switch recombination (Rada et al., 2002; Imai et al., 2003).

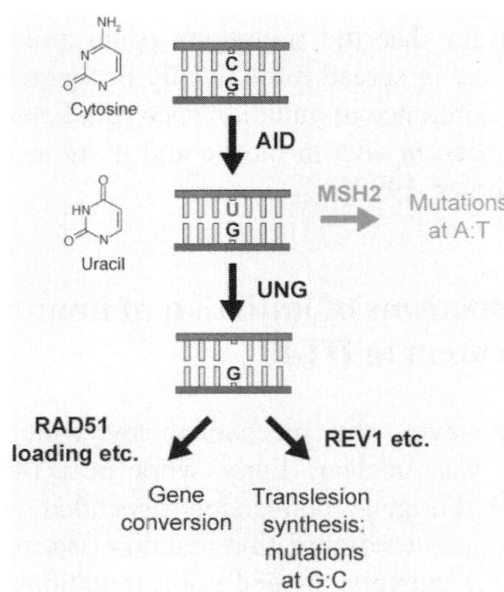

Figure 16-4. The deamination model of Ig diversification applied to DT40. While in mouse and human mutations at A:T, arising in consequence of MSH2 recognition of U:G mismatches, make a substantial contribution, mutation in DT40 is largely the result of TLS across abasic sites formed at C. Gene conversion is also predominantly initiated from abasic sites. (Adapted from Petersen-Mahrt et al., 2002) (see plate 40).

AID is also absolutely required for both the point mutation and gene conversions seen during immunoglobulin diversification in DT40, confirming their common origin (Arakawa et al., 2002; Harris et al., 2002). That gene conversion is also initiated predominantly from an abasic site rather than a U/G mismatch was suggested by a reduction in gene conversion in DT40 cells expressing Ugi (Di Noia and Neuberger, 2004) or carrying an *ung* disruption (Jean-Marie Buerstedde, personal communication).

Thus, the diversifying immunoglobulin loci of DT40 provide a unique model system in which to study the mechanisms of processing a defined DNA lesion. So what has been learned so far about the mechansism by which immunoglobulin gene abasic sites are processed and to what extent can immunoglobulin diversification in DT40 act as a model of DNA damage tolerance?

6.3 The genetic requirements for gene conversion in the Ig loci of DT40

It is clear from work in DT40 that immunoglobulin gene conversion requires genes involved in the early stages of homologous recombination. To facilitate the invasion of a homologous donor template, a free 3' DNA end from the recipient site becomes coated in RAD51 (Wyman et al., 2004).

Deletion of RAD51 itself in DT40 results in death due to chromosome fragmentation within a couple of cell cycles (Sonoda et al., 1998) and therefore analysis of Ig diversification in this mutant is not possible. However, the efficient loading of RAD51 onto single stranded DNA to form the nucleoprotein filament requires a number of so-called 'mediator' proteins. These include the RAD51 paralogues (XRCC2, XRCC3, RAD51B, C and D) and BRCA2. Disruption of any of these genes in DT40 results in failure to form subnuclear RAD51 foci in response to ionising radiation and a defect in homologous recombination, suggesting that these mutants cannot effectively load RAD51 onto single stranded DNA (Takata et al., 2000; Takata et al., 2001; Hatanaka et al., 2004). Their disruption also has a similar effect on Ig diversification, namely a swing in the pattern from predominantly gene conversion to predominantly point mutation (Sale et al., 2001; Hatanaka et al., 2004). Further, the same swing in the pattern of diversification is seen if the donor pseudogenes are removed (Arakawa et al., 2004). Thus, initiating abasic sites appear to be frequently left intact if recombination cannot be efficiently initiated.

Following formation of the RAD51 nucleoprotein filament, strand invasion of the donor template is facilitated by RAD54, a double-stranded DNA dependent ATPase, closely related to the SWI/SNF family of chromatin remodelling factors (reviewed in Tan et al., 2003). Its main role is thought to be in assisting joint molecule formation by disrupting base pairing in the template molecule. Disruption of RAD54 in DT40 has a subtly different effect on Ig diversification than the RAD51 paralogues in that it reduces, but does not abolish, gene conversion (Bezzubova et al., 1997) but no marked increase in point mutation. This may reflect the role of RAD54 further downstream in the recombination reaction or the presence of a second RAD54-like protein, RAD54B (Tanaka et al., 2000).

Immunoglobulin gene conversion in chicken B lymphocytes is a somewhat ususual form of homologous recombination in that the donor template, the V pseudogene array, is frequently markedly heterologous to the rearranged V gene. Indeed, the ability to introduce such heterologous sequence into the rearranged V gene is key to its ability to generate an immunoglobulin repertoire. However, the polarity of gene conversion is such that the conversion tract is always initiated in a region of homology (McCormack and Thompson, 1990). Indeed, the fact that the number of point mutations in hyperpointmutating variants appears to exceed the number of observed gene conversions plus point mutations in wild-type cells suggests that normally many gene conversions are 'silent'. In other words they result in restoration of the original sequence. Further, the observation that deletion of the light chain V pseudogene array results in hypermutation similar to that seen in the RAD51 paralogue mutants (Arakawa et al., 2004) suggests that both the 'silent' as well as the observable conversions are templated from the pseudogenes. This suggests that during Ig conversion, the donors are looped back to be brought into proximity to the expressed V gene in a manner similar to that proposed for the mating type switching loci of both budding and fission yeast (Haber, 1998; Egel, 2005) (Figure 5). Of note, the gene conversion mechanism appears to favour donors in the pseudogene array that are in the antisense orientation to the expressed V gene (McCormack and Thompson, 1990). A recent and interesting observation is that the transcriptionally active and gene converting light chain allele in DT40 is hyperacetylated at histone H4 and that the rate of gene conversion could be increased following treatment of DT40 with the histone deacetylase inhibitor trichostatin A (Seo et al., 2005). This shows that gene conversion is regulated by histone acetylation, although possibly simply through regulating the levels of transcription.

Another important feature of immunoglobulin gene conversion that provides clues to its mechanism is that the donor pseudogenes do not become reciprocally diversified. Thus, while early genes involved in homologous recombination are required, it is not likely that a Holliday junction is formed since crossover between the rearranged V gene and the pseudogene donors is not observed. An attractive model, which bears strong similarities to that proposed by Thompson and colleagues in 1990 (Carlson et al., 1990), which explains many of these features is synthesis-dependent strand annealing (SDSA) (reviewed in Paques and Haber, 1999). In the classical form of SDSA initiated by a double-strand DNA break, the invading strand primes new DNA synthesis in D-loop, but instead of this resulting in the formation of a Holliday junction, the nascent DNA is released to form a long 3' overhang that can now reanneal with the other end of the break (Figure 5).

It is not yet clear precisely which polymerases are involved in the DNA synthesis step of Ig gene conversion. However, a recent study has shown that the frequency of gene conversion is decreased in *rev1*, but not *rev3* or *rev7* DT40 cells. Additionally, the flap endonuclease FEN-1 is required for immunoglobulin gene conversion involving heterologous sequence (Kikuchi et al., 2005). FEN-1 is a structure-specific nuclease that is able to cleave the 5' flaps of branched DNA structures and *fen-1* DT40 exhibit decreased Ig gene conversion and show, in the conversions observed, a marked preference for use of the most homologous donor.

6.4   Translesion synthesis in the immunoglobulin loci of DT40

The pattern of point mutation in both wild type and hypermutating variants of DT40 is highly biased to G/C base pairs and likely reflects almost exclusively the bypass of abasic sites formed following C deamination on either strand. It therefore provides a unique model for studying the processing of a defined and common DNA lesion, the abasic site, in a chromosomal context.

Confirmation that mutations generated at G/C in the immunoglobulin loci do require functions associated with translesion synthesis came from the analysis of the DT40 *rev1* mutant (Simpson and Sale, 2003). In this mutant, the number of point mutations is reduced some six fold. The role of REV1 in vertebrate translesion synthesis has received much attention in the past few years. Although a member of the Y family of DNA polymerases, *in vitro* it is only capable of deoxycytidyl transferase activity (Lin et al., 1999; Zhang et al., 2002). However, several lines of evidence, including a requirement for its function in tolerance of lesions which are not substrates for its dC transferase activity, have suggested a second role in translesion synthesis. A clue to this came from the finding that human and mouse REV1 is able to bind several of the other known translesion polymerases through its extreme C terminus (Murakumo et al., 2001; Guo et al., 2003; Ohashi et al., 2004; Tissier et al., 2004). Further, a region immediately N terminal of this region mediates an interaction with PCNA (Ross et al., 2005). Together these data suggest that REV1 plays a role in coordination or recruitment of the other translesion polymerases. For tolerance of DNA damage, the C terminus of REV1 is crucial while the transferase activity is dispensable (Ross et al., 2005). However, the catalytic activity of REV1 does appear to be required for insertion of C opposite abasic site in the immunoglobulin loci (Ross & Sale, submitted). The identity of the other polymerases playing a part in the bypass of immunoglobulin abasic sites is not yet established, although DNA polymerases ζ and κ do not appear to play a role (Sonoda et al., 2003; Shunichi Takeda, personal communication).

While in yeast, recruitment of the translesion polymerases critically depends on the ubiquitin ligation to lysine 164 of PCNA by RAD6/RAD18, the situation appears more complex in DT40. Not only is RAD18 not apparently epistatic to REV1 for DNA damage tolerance (Ross et al., 2005) but it also does not appear to have the same effect on non-templated mutation as REV1. Point mutation in *rad18* mutants of DT40 is much less affected than *rev1* mutants (Simpson and Sale, 2005). Clearly the link between the generation of abasic sites by the combined action of AID/UNG and point mutation is one that requires further investigation. A central role for PCNA is likely given the recruitment of UNG to PCNA during replication (Otterlei et al., 1999) and the interaction of REV1 with PCNA (Ross et al., 2005), although much more work is needed to dissect the fine choreography of these interactions.

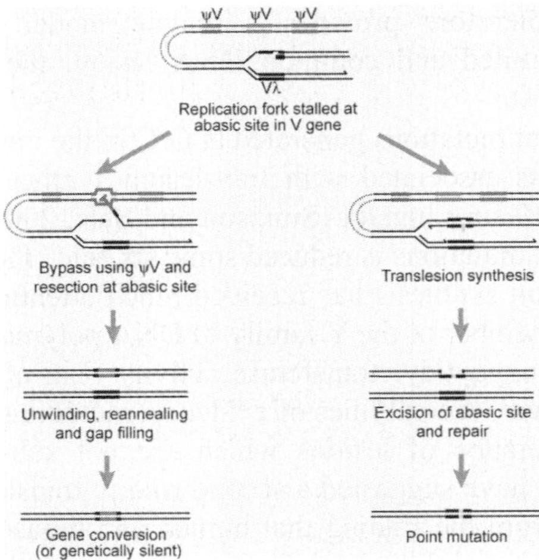

Figure 16-5. A model for gene conversion and point mutation initiated by abasic sites
(see plate 41).

7.   DIVERSIFICATION IN THE IMMUNOGLOBULIN LOCI AS A MODEL FOR REPLICATION DNA DAMAGE BYPASS IN VERTEBRATES

So two competing mechanisms operate following formation of abasic sites in the immunoglobulin loci of DT40, translesion synthesis and gene conversion. DT40 provides a unique system in vertebrate biology in which the processing of a defined lesion can be followed through to the generation of mutations. So, to what extent do they reflect mechanisms of DNA damage bypass? For the non-templated mutations, the bulk of evidence points to replication damage bypass. The point mutations seen in DT40 are most readily explained by translesion synthesis over an abasic site, these sites being repaired efficiently at a later stage. However, whether the recombination mechanism responsible for immunoglobulin gene conversion is triggered directly by the abasic site or by a nick resulting from a repair intermediate is not yet clear. Indeed, the exact events, illustrated in Figure 5, between the formation of the abasic site and the generation of a gene conversion tract are still largely a matter of speculation. Further, although immunoglobulin gene conversion perhaps represents a rather atypical recombination scenario, with high levels of damage concentrated in a small area and the immediate and close availability of homeologous donors, we nonetheless believe that it is likely that the recombination mechanisms employed reflect those that are used in lesion bypass in general. There is still much to learn about both the mechanisms of DNA damage tolerance in humans and much new is likely to come from this remarkable chicken cell line.

REFERENCES

Arakawa H, Hauschild J, Buerstedde JM. 2002. Requirement of the activation-induced deaminase (AID) gene for immunoglobulin gene conversion. Science 295:1301-1306.

Arakawa H, Kuma K, Yasuda M, Furusawa S, Ekino S, Yamagishi H. 1998. Oligoclonal development of B cells bearing discrete Ig chains in chicken single germinal centers. J Immunol 160:4232-4241.

Arakawa H, Saribasak H, Buerstedde JM. 2004. Activation-induced cytidine deaminase initiates immunoglobulin gene conversion and hypermutation by a common intermediate. PLoS Biol 2:E179.

Bailly V, Lamb J, Sung P, Prakash S, Prakash L. 1994. Specific complex formation between yeast RAD6 and RAD18 proteins: a potential mechanism for targeting RAD6 ubiquitin-conjugating activity to DNA damage sites. Genes Dev 8:811-820.

Bailly V, Lauder S, Prakash S, Prakash L. 1997. Yeast DNA repair proteins Rad6 and Rad18 form a heterodimer that has ubiquitin conjugating, DNA binding, and ATP hydrolytic activities. J Biol Chem 272:23360-23365.

Bemark M, Khamlichi AA, Davies SL, Neuberger MS. 2000. Disruption of mouse polymerase ζ (Rev3) leads to embryonic lethality and impairs blastocyst development in vitro. Curr Biol 10:1213-1216.

Berdichevsky A, Izhar L, Livneh Z. 2002. Error-free recombinational repair predominates over mutagenic translesion replication in E. coli. Mol Cell 10:917-924.

Bezzubova O, Silbergleit A, Yamaguchi-Iwai Y, Takeda S, Buerstedde JM. 1997. Reduced X-ray resistance and homologous recombination frequencies in a RAD54-/- mutant of the chicken DT40 cell line. Cell 89:185-193.

Bransteitter R, Pham P, Scharff MD, Goodman MF. 2003. Activation-induced cytidine deaminase deaminates deoxycytidine on single-stranded DNA but requires the action of RNase. Proc Natl Acad Sci U S A 100:4102-4107.

Broomfield S, Hryciw T, Xiao W. 2001. DNA postreplication repair and mutagenesis in Saccharomyces cerevisiae. Mutat Res 248:167-184.

Buerstedde JM, Reynaud CA, Humphries EH, Olson W, Ewert DL, Weill JC. 1990. Light chain gene conversion continues at high rate in an ALV-induced cell line. Embo J 9: 921-927.

Carlson LM, McCormack WT, Postema CE, Humphries EH, Thompson CB. 1990. Templated insertions in the rearranged chicken IgL V gene segment arise by intrachromosomal gene conversion. Genes Dev 4:536-547.

Cassier-Chauvat C, Fabre F. 1991. A similar defect in UV-induced mutagenesis conferred by the rad6 and rad18 mutations of Saccharomyces cerevisiae. Mutat Res 254:247-253.

Chaudhuri J, Tian M, Khuong C, Chua K, Pinaud E, Alt FW. 2003. Transcription-targeted DNA deamination by the AID antibody diversification enzyme. Nature 422:726-730.

Courcelle J, Ganesan AK, Hanawalt PC. 2001. Therefore, what are recombination proteins there for? Bioessays 23:463-470.

Denépoux S, Razanajaona D, Blanchard D, Meffre G, Capra JD, Banchereau J, Lebecque S 1997. Induction of somatic mutation in a human B cell line in vitro. Immunity 6:35-46.

Deng L, Wang C, Spencer E, Yang L, Braun A, You J, Slaughter C, Pickart C, Chen ZJ. 2000. Activation of the IkappaB kinase complex by TRAF6 requires a dimeric ubiquitin-conjugating enzyme complex and a unique polyubiquitin chain. Cell 103:351-361.

Di Noia J, Neuberger MS. 2002. Altering the pathway of immunoglobulin hypermutation by inhibiting uracil-DNA glycosylase. Nature 419:43-48.

Di Noia JM, Neuberger MS. 2004. Immunoglobulin gene conversion in chicken DT40 cells largely proceeds through an abasic site intermediate generated by excision of the uracil produced by AID-mediated deoxycytidine deamination. Eur J Immunol 34:504-508.

Egel R. 2005. Fission yeast mating-type switching: programmed damage and repair. DNA Repair (Amst) 4:525-536.

Goodman MF. 2002. Error-prone repair DNA polymerases in prokaryotes and eukaryotes. Annu Rev Biochem 71:17-50.

Guo C, Fischhaber PL, Luk-Paszyc MJ, Masuda Y, Zhou J, Kamiya K, Kisker C, Friedberg EC. 2003. Mouse Rev1 protein interacts with multiple DNA polymerases involved in translesion DNA synthesis. Embo J 22:6621-6630.

Haber JE. 1998. Mating-type gene switching in Saccharomyces cerevisiae. Annu Rev Genet 32:561-599.

Haracska L, Prakash S, Prakash L. 2000. Replication past O(6)-methylguanine by yeast and human DNA polymerase η. Mol Cell Biol 20:8001-8007.

Haracska L, Torres-Ramos CA, Johnson RE, Prakash S, Prakash L. 2004. Opposing effects of ubiquitin conjugation and SUMO modification of PCNA on replicational bypass of DNA lesions in Saccharomyces cerevisiae. Mol Cell Biol 24:4267-4274.

Harris RS, Sale JE, Petersen-Mahrt SK, Neuberger MS. 2002. AID is essential for immunoglobulin V gene conversion in a cultured B cell line. Curr Biol 12:435-438.

Hatanaka A, Yamazoe M, Sale JE, Takata M, Yamamoto K, Kitao H, Sonoda E, Kikuchi K, Yonetani Y, Takeda S. 2005. Similar effects of Brca2 truncation and Rad51 paralog deficiency on immunoglobulin V gene diversification in DT40 cells support an early role for Rad51 paralogs in homologous recombination. Mol Cell Biol 25:1124-1134.

Higgins NP, Kato K, Strauss B. 1976. A model for replication repair in mammalian cells. J Mol Biol 101:417-425.

Hirano S, Yamamoto K, Ishiai M, Yamazoe M, Seki M, Matsushita N, Ohzeki M, Yamashita YM, Arakawa H, Buerstedde JM, Enomoto T, Takeda S, Thompson LH, Takata M. 2005. Functional relationships of FANCC to homologous recombination, translesion synthesis, and BLM. Embo J 24:418-427.

Hoege C, Pfander B, Moldovan GL, Pyrowolakis G, Jentsch S. 2002. RAD6-dependent DNA repair is linked to modification of PCNA by ubiquitin and SUMO. Nature 419:135-141.

Imai K, Slupphaug G, Lee WI, Revy P, Nonoyama S, Catalan N, Yel L, Forveille M, Kavli B, Krokan HE, Ochs HD, Fischer A, Durandy A. 2003. Human uracil-DNA glycosylase deficiency associated with profoundly impaired immunoglobulin class-switch recombination. Nat Immunol 4:1023-1028.

Jentsch S, McGrath JP, Varshavsky A. 1987. The yeast DNA repair gene RAD6 encodes a ubiquitin-conjugating enzyme. Nature 329:131-134.

Kajiwara K, J OW, Sakurai T, Yamashita S, Tanaka M, Sato M, Tagawa M, Sugaya E, Nakamura K, Nakao K, Katsuki M, Kimura M. 2001. Sez4 gene encoding an elongation subunit of DNA polymerase ζ is required for normal embryogenesis. Genes Cells 6: 99-106.

Kannouche PL, Wing J, Lehmann AR. 2004. Interaction of human DNA polymerase η with monoubiquitinated PCNA: a possible mechanism for the polymerase switch in response to DNA damage. Mol Cell 14:491-500.

Kikuchi K, Taniguchi Y, Hatanaka A, Sonoda E, Hochegger H, Adachi N, Matsuzaki Y, Koyama H, van Gent DC, Jasin M, Takeda S. 2005. Fen-1 facilitates homologous recombination by removing divergent sequences at DNA break ends. Mol Cell Biol 25:6948-6955.

Kim S, Humphries EH, Tjoelker L, Carlson L, Thompson CB. 1990. Ongoing diversification of the rearranged immunoglobulin light-chain gene in a bursal lymphoma cell line. Mol Cell Biol 10:3224-3231.

Kogoma T. 1996. Recombination by replication. Cell 85:625-627.

Krejci L, Van Komen S, Li Y, Villemain J, Reddy MS, Klein H, Ellenberger T, Sung P. 2003. DNA helicase Srs2 disrupts the Rad51 presynaptic filament. Nature 423:305-309.

Krokan HE, Drablos F, Slupphaug G. 2002. Uracil in DNA—occurrence, consequences and repair. Oncogene 21:8935-8948.

Lawrence CW, Christensen R. 1976. UV mutagenesis in radiation-sensitive strains of yeast. Genetics 82:207-232.

Lawrence CW, Hinkle DC. 1996. DNA polymerase zeta and the control of DNA damage induced mutagenesis in eukaryotes. Cancer Surv 28:21-31.

Lawrence CW, Maher VM. 2001. Eukaryotic mutagenesis and translesion replication dependent on DNA polymerase ζ and Rev1 protein. Biochem Soc Trans 29:187-191.

Lin W, Xin H, Zhang Y, Wu X, Yuan F, Wang Z. 1999. The human REV1 gene codes for a DNA template-dependent dCMP transferase. Nucleic Acids Res 27:4468-4475.

McCormack WT, Thompson CB. 1990. Chicken IgL variable region gene conversions display pseudogene donor preference and 5' to 3' polarity. Genes Dev 4:548-558.

McDonald JP, Levine AS, Woodgate R. 1997. The Saccharomyces cerevisiae RAD30 gene, a homologue of Escherichia coli dinB and umuC, is DNA damage inducible and functions in a novel error-free postreplication repair mechanism. Genetics 147:1557-1568.

Murakumo Y, Ogura Y, Ishii H, Numata S, Ichihara M, Croce CM, Fishel R, Takahashi M. 2001. Interactions in the error-prone postreplication repair proteins hREV1, hREV3, and hREV7. J Biol Chem 276:35644-35651.

Muramatsu M, Kinoshita K, Fagarasan S, Yamada S, Shinkai Y, Honjo T. 2000. Class switch recombination and hypermutation require activation-induced cytidine deaminase (AID), a potential RNA editing enzyme. Cell 102:553-563.

Muramatsu M, Sankaranand VS, Anant S, Sugai M, Kinoshita K, Davidson NO, Honjo T. 1999. Specific expression of activation-induced cytidine deaminase (AID), a novel member of the RNA-editing deaminase family in germinal center B cells. J Biol Chem 274:18470-18476.

Nelson JR, Lawrence CW, Hinkle DC. 1996. Thymine-thymine dimer bypass by yeast DNA polymerase zeta. Science 272:1646-1649.

O-Wang J, Kajiwara K, Kawamura K, Kimura M, Miyagishima H, Koseki H, Tagawa M. 2002. An essential role for REV3 in mammalian cell survival: absence of REV3 induces p53-independent embryonic death. Biochem Biophys Res Commun 293:1132-1137.

Ohashi E, Murakumo Y, Kanjo N, Akagi J, Masutani C, Hanaoka F, Ohmori H. 2004. Interaction of hRev1 with three human Y-family DNA polymerases. Genes Cells 9:523-531.

Okada T, Sonoda E, Yamashita YM, Koyoshi S, Tateishi S, Yamaizumi M, Takata M, Ogawa O, Takeda S. 2002. Involvement of vertebrate polkappa in Rad18-independent postreplication repair of UV damage. J Biol Chem 277:48690-48695.

Otterlei M, Warbrick E, Nagelhus TA, Haug T, Slupphaug G, Akbari M, Aas PA, Steinsbekk K, Bakke O, Krokan HE. 1999. Post-replicative base excision repair in replication foci. Embo J 18:3834-3844.

Papouli E, Chen S, Davies AA, Huttner D, Krejci L, Sung P, Ulrich HD. 2005. Crosstalk between SUMO and ubiquitin on PCNA is mediated by recruitment of the helicase Srs2p. Mol Cell 19:123-133.

Paques F, Haber JE. 1999. Multiple pathways of recombination induced by double-strand breaks in Saccharomyces cerevisiae. Microbiol Mol Biol Rev 63:349-404.

Petersen-Mahrt SK, Harris RS, Neuberger MS. 2002. AID mutates E. coli suggesting a DNA deamination mechanism for antibody diversification. Nature 418:99-103.

Pfander B, Moldovan GL, Sacher M, Hoege C, Jentsch S. 2005. SUMO-modified PCNA recruits Srs2 to prevent recombination during S phase. Nature 436:428-433.

Prakash L. 1981. Characterization of postreplication repair in Saccharomyces cerevisiae and effects of rad6, rad18, rev3 and rad52 mutations. Mol Gen Genet 184:471-478.

Prakash S, Johnson RE, Prakash L. 2005. EUKARYOTIC TRANSLESION SYNTHESIS DNA POLYMERASES: Specificity of Structure and Function. Annu Rev Biochem 74:317-353.

Rada C, Williams GT, Nilsen H, Barnes DE, Lindahl T, Neuberger MS. 2002. Immunoglobulin isotype switching is inhibited and somatic hypermutation perturbed in UNG-deficient mice. Curr Biol 12:1748-1755.

Ramiro AR, Stavropoulos P, Jankovic M, Nussenzweig MC. 2003. Transcription enhances AID-mediated cytidine deamination by exposing single-stranded DNA on the nontemplate strand. Nat Immunol 4:452-456.

Reuven NB, Arad G, Maor-Shoshani A, Livneh Z. 1999. The mutagenesis protein UmuC is a DNA polymerase activated by UmuD', RecA, and SSB and is specialized for translesion replication. J Biol Chem 274:31763-31766.

Revy P, Muto T, Levy Y, Geissmann F, Plebani A, Sanal O, Catalan N, Forveille M, Dufourcq-Labelouse R, Gennery A, Tezcan I, Ersoy F, Kayserili H, Ugazio AG, Brousse N, Muramatsu M, Notarangelo LD, Kinoshita K, Honjo T, Fischer A, Durandy A. 2000. Activation-induced cytidine deaminase (AID) deficiency causes the autosomal recessive form of the Hyper-IgM syndrome (HIGM2). Cell 102:565-575.

Reynaud CA, Anquez V, Dahan A, Weill JC. 1985. A single rearrangement event generates most of the chicken immunoglobulin light chain diversity. Cell 40:283-291.

Reynaud CA, Anquez V, Grimal H, Weill JC. 1987. A hyperconversion mechanism generates the chicken light chain preimmune repertoire. Cell 48:379-388.

Reynaud CA, Dahan A, Anquez V, Weill JC. 1989. Somatic hyperconversion diversifies the single Vh gene of the chicken with a high incidence in the D region. Cell 59:171-183.

Ross AL, Simpson LJ, Sale JE. 2005. Vertebrate DNA damage tolerance requires the C-terminus but not BRCT or transferase domains of REV1. Nucleic Acids Res 33:1280-1289.

Rupp WD, Howard-Flanders P. 1968. Discontinuities in the DNA synthesized in an excision-defective strain of Escherichia coli following ultraviolet irradiation. J Mol Biol 31:291-304.

Rupp WD, Wilde CE, 3rd, Reno DL, Howard-Flanders P. 1971. Exchanges between DNA strands in ultraviolet-irradiated Escherichia coli. J Mol Biol 61:25-44.

Sale JE, Calandrini DM, Takata M, Takeda S, Neuberger MS. 2001. Ablation of XRCC2/3 transforms immunoglobulin V gene conversion into somatic hypermutation. Nature 412:921-926.

Sale JE, Neuberger MS. 1998. TdT-accessible breaks are scattered over the immunoglobulin V domain in a constitutively hypermutating B cell line. Immunity 9:859-869.

Sancar A. 2003. Structure and function of DNA photolyase and cryptochrome blue-light photoreceptors. Chem Rev 103:2203-2237.

Seo H, Masuoka M, Murofushi H, Takeda S, Shibata T, Ohta K. 2005. Rapid generation of specific antibodies by enhanced homologous recombination. Nat Biotechnol 23:731-735.

Simpson LJ, Sale JE. 2003. Rev1 is essential for DNA damage tolerance and non-templated immunoglobulin gene mutation in a vertebrate cell line. Embo J 22:1654-1664.

Simpson LJ, Sale JE. 2005. UBE2V2 (MMS2) is not required for effective immunoglobulin gene conversion or DNA damage tolerance in DT40. DNA Repair (Amst) 4:503-510.

Sonoda E, Sasaki MS, Buerstedde JM, Bezzubova O, Shinohara A, Ogawa H, Takata M, Yamaguchi-Iwai Y, Takeda S. 1998. Rad51-deficient vertebrate cells accumulate chromosomal breaks prior to cell death. Embo J 17:598-608.

Sonoda E, Okada T, Zhao GY, Tateishi S, Araki K, Yamaizumi M, Yagi T, Verkaik NS, van Gent DC, Takata M, Takeda S. 2003. Multiple roles of Rev3, the catalytic subunit of polζ in maintaining genome stability in vertebrates. Embo J 22:3188-3197.

Stelter P, Ulrich HD. 2003. Control of spontaneous and damage-induced mutagenesis by SUMO and ubiquitin conjugation. Nature 425:188-191.

Takata M, Sasaki MS, Sonoda E, Fukushima T, Morrison C, Albala JS, Swagemakers SM, Kanaar R, Thompson LH, Takeda S. 2000. The Rad51 paralog Rad51B promotes homologous recombinational repair. Mol Cell Biol 20:6476-6482.

Takata M, Sasaki MS, Tachiiri S, Fukushima T, Sonoda E, Schild D, Thompson LH, Takeda S. 2001. Chromosome instability and defective recombinational repair in knockout mutants of the five Rad51 paralogs. Mol Cell Biol 21:2858-2866.

Tan TL, Kanaar R, Wyman C. 2003. Rad54, a Jack of all trades in homologous recombination. DNA Repair (Amst) 2:787-794.

Tanaka K, Hiramoto T, Fukuda T, Miyagawa K. 2000. A novel human rad54 homologue, Rad54B, associates with Rad51. J Biol Chem 275:26316-26321.

Tang M, Shen X, Frank EG, O'Donnell M, Woodgate R, Goodman MF. 1999. UmuD'(2)C is an error-prone DNA polymerase, Escherichia coli pol V. Proc Natl Acad Sci U S A 96:8919-8924.

Tateishi S, Niwa H, Miyazaki J, Fujimoto S, Inoue H, Yamaizumi M. 2003. Enhanced genomic instability and defective postreplication repair in RAD18 knockout mouse embryonic stem cells. Mol Cell Biol 23:474-481.

Tateishi S, Sakuraba Y, Masuyama S, Inoue H, Yamaizumi M. 2000. Dysfunction of human Rad18 results in defective postreplication repair and hypersensitivity to multiple mutagens. Proc Natl Acad Sci U S A 97:7927-7932.

Tissier A, Kannouche P, Reck MP, Lehmann AR, Fuchs RP, Cordonnier A. 2004. Co-localization in replication foci and interaction of human Y-family members, DNA polymerase poleta and REV1 protein. DNA Repair (Amst) 3:1503-1514.

Torres-Ramos CA, Yoder BL, Burgers PM, Prakash S, Prakash L. 1996. Requirement of proliferating cell nuclear antigen in RAD6-dependent postreplicational DNA repair. Proc Natl Acad Sci U S A 93:9676-9681.

Torres-Ramos CA, Prakash S, Prakash L. 1997. Requirement of yeast DNA polymerase delta in post-replicational repair of UV-damaged DNA. J Biol Chem 272:25445-25448.

Torres-Ramos CA, Prakash S, Prakash L. 2002. Requirement of RAD5 and MMS2 for postreplication repair of UV-damaged DNA in Saccharomyces cerevisiae. Mol Cell Biol 22:2419-2426.

Ulrich HD, Jentsch S. 2000. Two RING finger proteins mediate cooperation between ubiquitin-conjugating enzymes in DNA repair. Embo J 19:3388-3397.

Van Sloun PP, Varlet I, Sonneveld E, Boei JJ, Romeijn RJ, Eeken JC, De Wind N. 2002. Involvement of mouse Rev3 in tolerance of endogenous and exogenous DNA damage. Mol Cell Biol 22:2159-2169.

Watanabe K, Tateishi S, Kawasuji M, Tsurimoto T, Inoue H, Yamaizumi M. 2004. Rad18 guides poleta to replication stalling sites through physical interaction and PCNA monoubiquitination. Embo J 23:3886-3896.

Wittschieben J, Shivji MK, Lalani E, Jacobs MA, Marini F, Gearhart PJ, Rosewell I, Stamp G, Wood RD. 2000. Disruption of the developmentally regulated Rev3l gene causes embryonic lethality. Curr Biol 10:1217-1220.

Wyman C, Ristic D, Kanaar R. 2004. Homologous recombination-mediated double-strand break repair. DNA Repair (Amst) 3:827-833.

Yamamoto K, Hirano S, Ishiai M, Morishima K, Kitao H, Namikoshi K, Kimura M, Matsushita N, Arakawa H, Buerstedde JM, Komatsu K, Thompson LH, Takata M. 2005. Fanconi anemia protein FANCD2 promotes immunoglobulin gene conversion and DNA repair through a mechanism related to homologous recombination. Mol Cell Biol 25: 34-43.

Yamashita YM, Okada T, Matsusaka T, Sonoda E, Zhao GY, Araki K, Tateishi S, Yamaizumi M, Takeda S. 2002. RAD18 and RAD54 cooperatively contribute to maintenance of genomic stability in vertebrate cells. Embo J 21:5558-5566.

Yu SL, Johnson RE, Prakash S, Prakash L. 2001. Requirement of DNA polymerase eta for error-free bypass of UV-induced CC and TC photoproducts. Mol Cell Biol 21:185-188.

Zander L, Bemark M. 2004. Immortalized mouse cell lines that lack a functional Rev3 gene are hypersensitive to UV irradiation and cisplatin treatment. DNA Repair (Amst) 3: 743-752.

Zhang Y, Wu X, Rechkoblit O, Geacintov NE, Taylor JS, Wang Z. 2002. Response of human REV1 to different DNA damage: preferential dCMP insertion opposite the lesion. Nucleic Acids Res 30:1630-1638.

Chapter 17

THE FANCONI ANEMIA PATHWAY PROMOTES HOMOLOGOUS RECOMBINATION REPAIR IN DT40 CELL LINE

Minoru Takata, Kazuhiko Yamamoto, Nobuko Matsushita, Hiroyuki Kitao, Seiki Hirano and Masamichi Ishiai
Department of Immunology and Molecular Genetics, Kawasaki Medical School, Kurashiki, Okayama, Japan 701-0192

Abstract: Fanconi anemia (FA) is a rare hereditary disorder characterized by bone marrow failure, compromised genome stability, and increased incidence of cancer. FA is caused by abnormalities that occur in components of the FA core complex, a key factor FancD2, breast cancer susceptibility protein BRCA2/FancD1, or BRIP1/FancJ. These proteins are proposed to function in a common biochemical process (FA pathway), however, its precise role is still unclear. In this chapter, we will summarize our genetic analysis on the FA pathway using DT40 cells line. Our data revealed that (1) FA pathway promotes DNA repair mediated by homologous recombination, and likely regulates translesion synthesis, thereby protecting cells against stalled replication forks; (2) BLM helicase can be regarded as an effector molecule of the FA pathway, since its subnuclear localization is regulated by FA pathway; (3) the FA core complex has multiple roles in the activation, relocalization, and DNA repair function of FANCD2.

Key words: Fanconi anemia, monoubiquitination, homologous recombination, translesion synthesis, gene conversion, DT40.

1. INTRODUCTION

Fanconi anemia (FA) is a rare hereditary disorder characterized by bone marrow failure, compromised genome stability and increased incidence of cancer (Sasaki and Tonomura, 1973; D'Andrea and Grompe, 2003). Cells

J.-M. Buerstedde and S. Takeda (eds.), Reviews and Protocols in DT40 Research, 295–311.

derived from FA patients display increased levels of chromosome breakage, particularly following exposure to drugs that induce DNA interstrand cross-links (ICLs) such as mitomycin C (MMC) (Sasaki and Tonomura, 1973). This property has been used as a diagnostic hallmark for FA, and also had led to the proposal by Dr Masao Sasaki that the basic cellular defects in FA lies in a subpathway of DNA repair that is specific for ICLs (Sasaki, 1975). However, until very recently, this proposal has not been substantiated, and confounded by another proposal that there is a defect in oxygen radical metabolism in FA cells (Ahmad et al., 2002).

FA has been attributed to multiple genetic abnormalities. So far eleven causative genes (FancA/B/C/D1/D2/E/F/G/J/L/M) have been cloned out of twelve complementation groups (D'Andrea and Grompe, 2003; Meetei et al., 2003a; Levitus et al., 2004; Meetei et al., 2004; Bridge et al., 2005; Levitus et al., 2005; Levran et al., 2005; Meetei et al., 2005; Mosedale et al., 2005), and products from eight of them (FancA/B/C/E/F/G/L/M) assemble to form the nuclear FA core complex (D'Andrea and Grompe, 2003). FancD2 appears to be a key factor, since it is converted from S-form to L-form by monoubiquitination on specific residue (Lys 561 in human protein), depending on the presence or integrity of the FA core complex (Garcia-Higuera et al., 2001) during S phase or following DNA double strand breaks (DSBs) or ICLs. The FA core complex most likely functions as multi-subunit E3 ubiquitin ligase for FancD2. Upon induction of monoubiquitination, FancD2 accumulates in subnuclear foci containing BRCA1 or Rad51 (Garcia-Higuera et al., 2001; Taniguchi et al., 2002). These observations gave a hint that FA pathway participates homologous recombination (HR), since Rad51 is a central player in DNA repair mediated by HR (West, 2003), and BRCA1 is one of the critical factors involved in HR (Venkitaraman, 2002). Importantly, FancD1 is found to be identical to BRCA2 (Howlett et al., 2002), which is a major regulator for Rad51 (Venkitaraman, 2002). This discovery highlighted the connection between the FA phenotype and HR. Furthermore, FA proteins are reported to interact with HR proteins in various combinations (Nakanishi et al., 2002; Hussain et al., 2003; Stewart et al., 2003).

Besides FancL (contains E3 ubiquitin ligase domain) (Meetei et al., 2003a) and FancM (contains helicase and nuclease motifs) (Meetei et al., 2005), none of the core complex components has shared motifs that indicate obvious protein function. Interestingly, any known mutation in components of the core complex appears to disrupt integrity of the complex, leading to the loss of E3 ligase function. Thus FA cells lacking any component of the core complex display defective FancD2 monoubiquitination (D'Andrea and Grompe, 2003), while in *BRCA2/FANCD1-* or *BRIP1/FANCJ*-deficient cells FancD2 is monoubiquitinated normally (Levitus et al., 2004; Wang et al., 2004; Bridge et al., 2005). Thus BRCA2 and BRIP1 likely function downstream of, or in parallel to, the FA core complex-FancD2 pathway.

It is well accepted that replication forks often stall at various DNA lesions including endogenous base damage (Cox et al., 2000; Cox, 2001). At least two basic processes, translesion synthesis (TLS) and homologous recombination (HR), function to ensure that replication can be resumed at stalled or broken forks in prokayotes and lower eukaryotes. In vertebtrates, the FA pathway probably facilitates the processing and restarting of the stalled or collapsed forks by the evolutionarily conserved TLS and HR enzymes (Thompson et al., 2005). It should be noted that Hef protein in Archea, a prototype of eukaryotic FancM, binds and processes forked DNA structure (Komori et al., 2002), which could represent stalled replication forks. The ICL sensitivity characteristic of FA cells might be related to ICL's obvious ability to prevent fork progression. Single-strand breaks or blocking lesions are converted to DSBs or gaps, respectively, during passage of replication forks. These chromatid discontinuities may be repaired by HR occurring between daughter chromatids (Sonoda et al., 1999). Alternatively, blocked replication forks are restarted by switching to specialized TLS polymerases that override blocking lesions (Kannouche et al., 2004; Watanabe et al., 2004). If such mechanisms are compromised, chromosomal and genetic instability would occur, leading to cancer predisposition as in patients with FA or BRCA1/2-mutated familial breast cancer.

While we were studying HR mechanisms in DT40, it became apparent that many HR mutants, particularly cells lacking Rad51 paralogs, display ICL sensitivity similarly to human FA cells (Takata et al., 2000; Takata et al., 2001). Therefore we planned to examine activity of homologous recombination by making DT40 FA mutants. In particular, FancD2 has been the focus of our effort. We now know that Rad51 paralogs and FA proteins participate HR in a quite distinct manner, as described below. However, precise function of the FA pathway still remains enigmatic.

1.1   DT40 is appropriate for studying FA pathway

The core complex components such as *FANCA/C/G* genes are not found in non-vertebrate species ranging from *E. coli* to *C. elegans*. Interestingly, among so far identified FA genes, *FANCM* is probably the most ancient, and a homologous gene exists even in *S. cerevisiae* (called Mph1) or *Archaea* (called Hef). *BRCA2/FANCD1* also seems to have quite ancient origin, and is found in fungus *Ustilago maydis* (Kojic et al., 2002), but not in *S. cerevisiae*. (Table 1). The widely held view is that the complete set of FA genes are only found in vertebrates. Thus, DT40 should provide an excellent system to study molecular mechanisms of the FA pathway including the core complex function.

Table 17-1. Evolutionary conservation of the FA genes.

	FANCM	*BRCA2 /FANCD1*	*FANCD2*	*FANCL*	*FANCA/C/G*
E.coli	–	–	–	–	–
S.cerevisiae	+	–	–	–	–
Archea	+	–	–	–	–
U.maydis	?	+	?	?	?
C.elegans	–	?	+	–	+
Vertebrates	+	+	+	+	+

We have obtained chicken *FANCG, FANCC*, and *FANCD2* cDNAs by library screening or RT-PCR, based on the sequence information in the chicken EST database (Yamamoto et al., 2003; Hirano et al., 2005; Yamamoto et al., 2005). The percentage amino acid sequence identity of full-length chicken FancG, FancC, and FancD2 with that of human counterpart is 39%, 49% or 57%, respectively. Thus levels of conservation seem rather low, although the overall similarities across species are highly significant by statistical analysis. Interestingly, HR genes in human and chicken in general exhibit higher levels of conservation compared to that of FA genes. For example, for Rad51 or Xrcc3 (one of Rad51 paralogs), the percent identity between chicken and human is 95% or 61%, respectively (Bezzubova et al., 1993; Takata et al., 2001).

Even though relatively low levels of sequence conservation, we are convinced that these putative counterparts are real orthologs of human FA genes from the following reasons. First, there are obvious functional similarity between human FA genes and the putative chicken counterpart as revealed by our genetic analysis (as described below). Second, anti-chicken FancD2 antibody (a kind gift from Prof Kenshi Komatsu, University of Kyoto) detected DNA damage-induced conversion of S-form to L-form of chicken FancD2 (Yamamoto et al., 2005), as in human protein. Indeed, the chicken FancD2 contains Lys residue at the conserved position (K563 in chicken protein). The monoubiquitination could not be observed in the absence of the core complex components, or on transfected mutant chicken FancD2 protein in which K563 was changed to Arg. These data demonstrate authenticity of these chicken candidates and conservation of the basic mechanisms in the chicken FA pathway.

Interestingly, single transfection with the targeting vector abrogated the wild-type band in Southern blot analysis in *FANCG* (Yamamoto et al., 2003) and *FANCC* (Hirano et al., 2005) targeting, indicating that there is only one *FANCG* or *FANCC* allele in DT40 cells. We later exploited this in making double knockouts. Human chromosome 9, which carries both

FANCG and *FANCC* genes, is obviously diploid, while the corresponding chicken counterpart is the sex chromosome Z and is hemizygous in DT40 (Sonoda et al., 1998). Extensive synteny has been reported between chicken Z and human chromosome 9 (Nanda et al., 1999). In addition, we found a neighboring gene in the 5' region of the chicken *FANCG* locus that has significant homology to the phosphatidylinositol glycan class o (*PIGO*) gene that lies next to *FANCG* in the human genome sequence as well.

We disrupted *FANCG*, *FANCC*, and *FANCD2* genes in DT40 cells, and examined the basic cellular phenotype (Yamamoto et al., 2003; Hirano et al., 2005; Yamamoto et al., 2005). These mutant cell lines generally displayed slightly reduced rate of growth, and drastically increased levels of chromosome aberrations following MMC exposure, which is a diagnostic feature of clinical FA. In addition, these cell lines were extremely sensitive to MMC or cisplatin treatment, while sensitivity to X-ray or UV was only mild (*fancc* and *fancd2*) or absent (*fancg*). This sensitivity pattern is consistent with the sensitivity specifically hightened to ICL in human FA cells. Collectively, these cells seemed to resemble FA cell line derived from human patients, and the data support the idea that DT40 provide an excellent opportunity for genetic study of the FA pathway.

Of note, *FANCC* or *FANCD2*-deficient cells (Hirano et al., 2005; Yamamoto et al., 2005) displayed similarly high levels of ICL sensitivity, while ICL sensitivity in *FANCG* mutant (Yamamoto et al., 2003) was much milder. We therefore recreated *fancg* cells with a targeting vector which was designed to delete larger portion of the *FANCG* locus. Again, the mutant cells were only mildly sensitive to ICL agents (unpublished). Right now we cannot provide a definite explanation for the difference between *fancd2* or *fancc* and *fancg*, however, it is possible that there is some residual E3 ligase activity of the core complex in *fancg* cells. As a result, a tiny amount of FancD2-L protein may be produced, which provides some protection against ICL damage. However, we could not detect any monoubiquitinated FancD2-L form in these cells.

1.2   Role of FA genes in homologous recombination

The role of BRCA2/FANCD1 in HR is now well established (Venkitaraman, 2002). To elucidate the roles of other FA genes in HR, we planned to exploit several HR assays that are available in DT40 cells. First, we measured ratio of random to targeted integration events using previously constructed gene targeting vectors. Prior studies established that mechanisms of targeted integration of homologous DNA require HR factors including Rad51, Rad52, Rad54, Rad51 paralogs, and Nbs1. As shown in Figure 1 (modified results from (Yamamoto et al., 2005)), *fancd2* cells displayed quite drastic reduction in gene targeting efficiency in all loci tested compared to wild type cells, and the defects were reversed by re-expression of chicken FancD2 at least to some extent. We also observed a decrease in gene targeting efficiency in *fancg* (Yamamoto et al., 2003) as well as in *fancc* cells (Hirano et al., 2005). The decrease in the former cells was mild, but the defects seemed convincing when compared with the data using *FANCG*-complemented cells (Yamamoto et al., 2003).

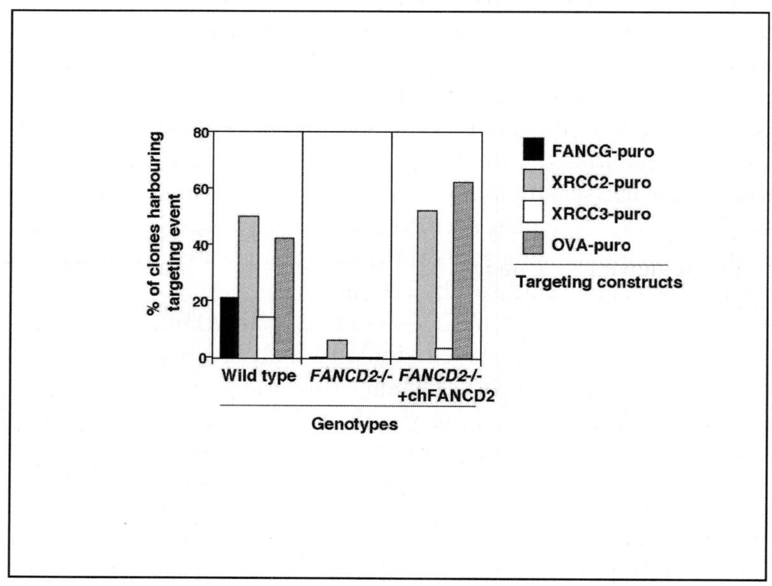

Figure 17-1. Targeted integration frequency is reduced in *fancd2* cells. Wild type, *fancd2*, and *FANCD2*-complemented *fancd2* cells were transfectd with indicated gene targeting vectors, and selected in the presence of puromycin. The clones were expanded and examined for targeting events by genomic southern blot analysis. The vertial axis represents % of clones with targeted events among clones tested.

Next, we disrupted the *FANCG* or *FANCD2* loci in wild type cells carrying the *SCneo* recombination substrate at *OVALBUMIN* locus (Yamamoto et al., 2003; Yamamoto et al., 2005). In this assay, transient expression of I-*Sce*I was used to induce DSB in a non-functional *neo* segment of *SCneo*. Then G418 selection was applied for ~2 weeks until the colonies develop. Cells should become G418-resistant only if the DSB is repaired correctly by HR using the upstream *neo* fragment as template. Significantly, the number of G418-resistant colonies was reduced ~40-fold or ~9-fold in *fancd2* and *fancg* cells, respectively, compared to wild type control cells. This assay depends on colony forming ability of clones after acquisition of G418 resistance, and DT40 FA mutants forms colonies less efficiently than wild type cells. Thus actual reduction in HR repair should be less drastic. For example, the mean plating efficiency of G418-resistant wild type (three clones) and *fancd2* (three clones) were 52% and 34%, respectively. Correcting this difference, the HR deficiency caused by *FANCD2* disruption was calculated to be ~26 fold. In addition to this quantitative difference, we detected aberrant repair products in a small number of G418-resistant *fancd2* clones by Southern blot analysis, raising the possibility that HR repair in the absence of *FANCD2* was qualitatively compromised (Yamamoto et al., 2005).

One may argue that these assays are artificial and might not reflect real cellular activities of HR. To evaluate the FA pathway in a more physiological HR, we measured immunoglobulin (Ig) gene conversion rate in *fancd2* cells (Yamamoto et al., 2005). DT40 cells continue to diversify the Ig V gene through gene conversion using upstream pseudogene segments as a donor template, which recapitulates B cells in the Bursa of Fabricius. Ig gene conversion depends on the activation-induced deaminase (AID) and the HR machinery (reviewed in Chapter 2). DT40Cre1 was a sIgM-negative sub-line, in which Ig conversion reverts the cells to sIgM-positive by correcting a frame shift mutation present in the Vλlight chain locus. Subclones of *fancd2* mutant and wild type DT40Cre1 were expanded for 4 weeks, and assayed for sIgM expression by flow cytometry or the changes in Vλ gene were directly determined by PCR and sequencing. Strikingly, gene conversion events in *fancd2* cells was reduced ~3.5-fold at protein level (sIgM re-expression) or ~6.3-fold at DNA level compared to heterozygous *fancd2* mutant cells. In addition, non-templated single nucleotide substitutions was reduced ~3.6-fold. Although the substitution events occur less frequently and therefore it is more difficult to measure accurately, the reduction was statistically significant. These single nucleotide changes are recently reported to depend on *Rev1* gene (Simpson and Sale, 2003), which is involved in translesion synthesis (TLS), and are equivalent to somatic hypermutation (SHM) in germinal center B cells.

Taken together, we conclude that *FANCG* and *FANCD2* are required for efficient HR-mediated repair of at least certain types of DSBs. *FANCD2* is also crucial for Ig gene conversion and possibly SHM, in which the initiating DNA lesion may be single nucleotide gap, not DSB, created by series of enzymatic reactions (reviewed in Chapters 2 and 16). Thus we propose that the FA pathway promote HR, and probably TLS as well. In support of the latter possibility, another group recently showed that *FANCC* deficiency is epistatic with *rev1* or *rev3* deficiency in terms of ICL sensitivity (Niedzwiedz et al., 2004). Thus ICL repair probably requires the complex and cooperative networking of HR factors, TLS polymerases and the FA pathway.

Of note, Rad51 focus formation induced by DNA damage was normal in *fancg*, *fancc*, and *fancd2* mutants (Yamamoto et al., 2003; Hirano et al., 2005; Yamamoto et al., 2005). Rad51 focus may reflect Rad51's polymer formation on single-stranded DNA, which is processed DNA end by poorly understood mechanisms, and depends on several proteins including the Rad51 paralogs and BRCA2/FancD1 (West, 2003). Thus the core complex-FancD2 pathway (excluding BRCA2/FancD1) functions either independently of Rad51 or in later steps after the focus formation. Consistently, we found that ectopic Rad51 overexpression, which may bypass requirements of the co-factors for Rad51, rescued cisplatin sensitivity in Rad51 paralog mutants but not in *fancg* or *fancd2* cells (Yamamoto et al., 2003; Yamamoto et al., 2005).

However, not all of the HR-related processes were similarly affected in FA mutants. In particular, the frequency of spontaneous sister chromatid exchanges (SCEs) was actually rather elevated in *fancc* and *fancd2* cells (Hirano et al., 2005; Yamamoto et al., 2005). Our original *fancg* cells displayed normal SCE levels (Yamamoto et al., 2003), but *fancg* cells made by another group have elevated spontaneous SCEs (Niedzwiedz et al., 2004). Thus it now appears that elevated spontaneous SCE levels are characteristics of DT40 FA mutants. SCE is a result of crossing-over event by which two Holliday junctions are resolved during HR at stalled replication forks (Sonoda et al., 1999). However, most mitotic HR events including ones that occur at stalled forks actually proceed without forming Holliday junctions (this pathway is called synthesis-dependent strand annealing, SDSA) (Paques and Haber, 1999). Taken together, we suggest a role of FancD2 might be restricted to a certain subpathway of HR such as SDSA. Adding even more complexity, the SCE levels following MMC treatment were found not to be elevated in *fancd2* or *fancc* cells (Hirano et al., 2005; Yamamoto et al., 2005), contrary to wild type cells. This finding probably indicates defective ICL processing in the absence of the core complex-FancD2 pathway.

1.3　Analysis of FA pathway using double knock-outs: relationship between FA pathway and BLM helicase

The presence of only single allele of *FANCG* and *FANCC* in DT40 cells provides an ideal opportunity to make double knockouts in various genetic backgrounds. To investigate the mechanism of SCE elevation in *fancc* cells, and to better define the FA pathway relative to other DNA repair pathways, we performed genetic analysis by disrupting *FANCC* in cells that are deficient in HR (*xrcc3*), TLS (*rad18*), and BLM helicase (*blm*) (Hirano et al., 2005).

In keeping with the above conclusion that the core complex-FancD2 pathway participates in HR, *xrcc3/fancc* mutant cells displayed identical levels of cisplatin sensitivity compared to *fancc* cells, indicating they function in the common pathway (epistasis). However, the levels of spontaneous SCE in *xrcc3/fancc* were decreased to that of *xrcc3* single mutant. Thus, SCE events that occur in *fancc* cells during replication are dependent on Xrcc3 (Hirano et al., 2005).

Since TLS mediates lesion bypass by a group of specialized polymerases, dysfunctional TLS might lead to more SCE if the stalled fork breaks and requires HR to restart. Several TLS-deficient vertebrate cells including Rad18-deficient cells (Yamashita et al., 2002) have been shown to have increased SCE (Okada et al., 2002; Sonoda et al., 2003), and only TLS mutants have the extreme levels of ICL sensitivity which is similar to cells lacking an FA gene. Thus there might be a link between the FA pathway and TLS defects. We tested this notion by making *rad18/fancc* double mutant, and found that the double mutants displayed even higher sensitivity to ICL, indicating their distinct functions (Hirano et al., 2005). Moreover, *fancc/rad18* cells had greater increased SCE than either *rad18* or *fancc* cells. Thus, we conclude that spontaneous SCEs in *fancc* cells are elevated by a mechanism independent of Rad18 function (Hirano et al., 2005). However, not all of the TLS events are dependent on Rad18 in vertebrates (Okada et al., 2002), and it is possible that the FA pathway functions with Rad18-independent branch of TLS subpathways, as suggested by the epistasis of *fancc* and *rev1/3* (Niedzwiedz et al., 2004).

There is another well-known condition called Bloom syndrome (BS) that displays highly elevated SCE levels (~10-fold) (Hickson, 2003). BS is a hereditary disorder caused by a defect in BLM helicase (reviewed by Enomoto in this book). We tested genetic relationship of *FANCC* and *BLM* by making double knockout, and unexpectedly found that the *blm* and *fancc/blm* mutants displayed about the same high levels of spontaneous SCE (Hirano et al., 2005). Furthermore, while MMS-treated *fancc/blm* cells survived more poorly than either single mutant (additive phenotype), survival following cisplatin or MMC treatment was similar between *fancc* and *fancc/blm* cells. These results suggest that FancC and BLM helicase act in a common pathway in ICL repair as well as in controlling SCE levels. However, in repairing DNA lesions created by MMS, they likely have more distinct functions (Hirano et al., 2005).

We hypothesized that increased SCE levels in *fancc* and *fancd2* cells is due to mildly compromised BLM function by the absence of the FA pathway. Since BLM is reported to relocalize to damage-induced foci, we visualized chicken BLM localization by transfecting GFP-chicken BLM expression construct into wild type, *fancc,* and *fancd2* cells. Formation of GFP-chBLM foci was clearly induced by MMC treatment in wild type cells, while in *fancc* or *fancd2* mutant cells the foci formation was largely suppressed (Hirano et al., 2005). We also found that MMC-induced GFP-BLM foci partially colocalized with FancD2 foci. These GFP-BLM data were reproduced in human FA cells lacking *FANCC* or *FANCD2* as well. Interestingly, we detected FancD2-L form protein in anti-GFP immuno-precipitates from wild type but not *fancc* cells expressing GFP-chBLM following MMC treatment (Hirano et al., 2005). These results suggest that monoubiquitinated FANCD2 physically interacts with BLM either directly or indirectly, and that MMC-induced nuclear relocalization of BLM is regulated by the core complex-FancD2 pathway.

A recent study reported physical interaction of BLM with the FA core complex (Meetei et al., 2003b). Thus it is possible that the core complex helps mobilize BLM to the sites of stalled/broken replication forks induced by DNA damage. Alternatively, the core complex may regulate BLM redistribution only through FancD2 activation. This is an interesting issue, as it is related to how the core complex is recruited to chromatin to participate DNA repair (see below). In any case, these data suggest that the FA pathway and BLM may function in a common pathway to repair ICLs and suppress SCE (Hirano et al., 2005).

We have also created DT40 double mutants that lack *FANCG* together with *Rad18*, *Ku70* (non-homologous end joining, NHEJ), or *XRCC3* (K. Yamamoto, unpublished). Analyses using these double mutants revealed that *fancg* and *xrcc3* are epistatic in terms of MMC sensitivity, but deletion of *FANCG* with *Rad18* or *Ku70* had an additive effect. These results are consistent with the data described above, and further indicated that the FA pathway promotes HR repair but not NHEJ or Rad18-dependent TLS.

1.4 FancD2-fusion proteins revealed monoubiquitination-independent function of the core complex

Following DNA damage, FancD2 is monoubiquitinated, most likely by E3 ligase subunit FancL contained in the core complex. Since FancD2-L form is mostly found in chromatin (Wang et al., 2004), and the mutant chicken FancD2 allele lacking monoubiquitination site (K563R) (made by "knock-in" gene targeting) is non-functional (our unpublished data), this modification appears to be a prerequisite for chromatin targeting, focus formation, and the subsequent DNA repair function of FancD2. It has been proposed, therefore, that the sole function of the core complex is to monoubiuqitinate FancD2, however, this has not been experimentally tested. Furthermore, the precise mechanism by which monoubiquitination affects FancD2 function was not clearly defined. Monoubiquitination could serve as a protein localization signal by interacting with proteins containing a number of ubiquitin binding motifs, and is emerging as a key regulator of various cellular pathways (Di Fiore et al., 2003) such as DNA repair, endocytic trafficking and nuclear export.

Recent studies have provided several examples in which physiological monoubiquitination was bypassed by fusion with ubiquitin (Haglund et al., 2003). To examine whether monoubiuqitination is sufficient for FancD2 function, we appended a single ubiquitin moiety to chicken FancD2 carrying a monoubiquitination site mutation (K563R) at its carboxyl-terminus (D2KR-Ub) (Matsushita et al., 2005). We found that D2KR-Ub protein expressed in *fancd2* background was constitutively present in the chromatin fraction, and reversed the extreme cisplatin sensitivity to near wild type levels. The D2KR-Ub protein could not form foci or did not further increase its amount in chromatin following MMC treatment. The chromatin targeting, hence DNA repair, was dependent on the presence of the core complex components FancC/G/L as well as the Ile 44 residue in the ubiquitin moiety. Interestingly, all ubiquitin-binding motifs found to date interact with ubiquitin through the critical residue Ile44 on hydrophobic surface of ubiquitin (Hicke et al., 2005). Thus it is highly likely that there is a FancD2-monubiquitin receptor that works in chromatin targeting of FancD2. The putative receptor may be a part of the FA core complex, or may be dependent on that complex for its function.

Then we tried to bypass the requirement of the core complex components in chromatin targeting of D2KR protein by fusing it with stable chromatin component, histone H2B (D2KR-H2B-GFP) (Matsushita et al., 2005). Remarkably, we could suppress cisplatin sensitivity by expressing D2KR-H2B-GFP in *fancd2* cells but not in *fancc*, *fancg*, and *fancl* cells. Taken together, these results clearly indicate monoubiquitination-independent roles of the core complex. It also suggests that monoubiquitination is dispensable for DNA repair once FancD2 is localized to chromatin and becomes available at DNA damage site. We thus conclude that the monoubiquitin moiety on FancD2 may function primarily as a targeting signal to chromatin.

Based on these results, we propose that the FA core complex works at three distinct levels to promote the activation and DNA repair function of FancD2 (Figure 2) (Matsushita et al., 2005). First, the FA core complex is essential for monoubiquitinating FancD2. Second, the core complex proteins are crucial for chromatin targeting of monoubiquitinated FancD2. Third, the core complex is required for the proper function of chromatin-bound FancD2 in DNA repair.

The model in Figure 2 is supported by several observations (Matsushita et al., 2005). First, we detected GFP-tagged core complex components in chromatin fraction, in keeping with the proposed role in DNA repair. Second, the complementation by D2KR-H2B expression was not seen in cells in which abundance of the core complex components was reduced in chromatin. For example, *FANCG*-disrupted cells were not complemented by D2KR-H2B, and the level of GFP-FancC in chromatin was drastically reduced in those cells. Third, the FA core complex appeared to relocalize to DNA damage sites, as we detected colocalization of GFP-FancC foci with FancD2 foci. Interestingly, MMC-induced foci formation of BLM (Hirano et al., 2005), a known component of the FA core complex (Meetei et al., 2003b), depends on FancC and FancD2 in both human and chicken cells (Hirano et al., 2005). It is therefore possible that accumulation of the core complex components at DNA damage sites may require chromatin-bound FancD2.

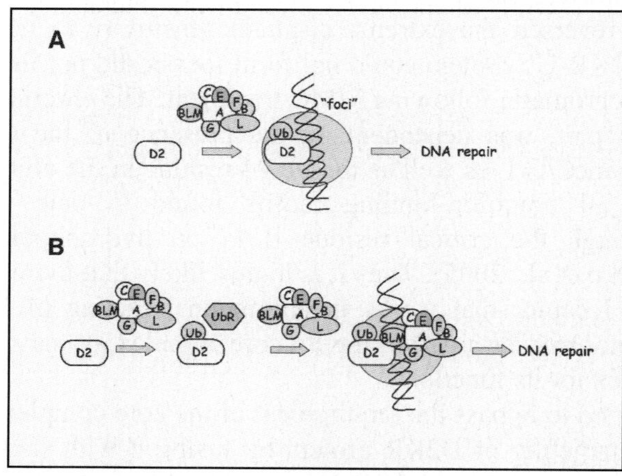

Figure 17-2. Models depicting roles of FANCD2 and the FA core complex. (A) A conventional model. The core complex only works in monoubiquitination of FancD2. (B) Our proposed model. The core complex functions in monoubiquitination, chromatin targeting of FancD2, and DNA repair in DNA-damage induced foci. UbR, putative ubiquitin receptor (see plate 42).

The FA core complex proteins might carry out these three roles indirectly - for example, through stabilization of the FA core complex or another component(s) of the core complex. It seems possible that the core complex has a specialized effector subunit(s) for DNA repair, as has E3 ligase subunit for monoubiquitination (Meetei et al., 2003a). Recently identified FancM (Meetei et al., 2005; Mosedale et al., 2005) and FancJ (Bridge et al., 2005; Levitus et al., 2005; Levran et al., 2005) proteins are of interest in this regard. FancM, one of the core complex components, might have a crucial role in DNA repair with its potential DNA-modifying activity, although its nuclease domain appears to be dispensable and the helicase domain does not have *in vitro* helicase activity (Meetei et al., 2005; Mosedale et al., 2005). In *fancj* cells the core complex is intact and FancD2 monoubiquitination occurs normally (Levitus et al., 2004; Bridge et al., 2005). It is likely, therefore, that the BRIP1/FancJ helicase is specifically required for the FancD2 chromatin targeting or the more downstream role of the core complex, and its function could in turn depend on other components of the core complex. Another possibility is that FancL may play a direct role by ubiquitinating additional substrates (other than FancD2) in chromatin. These possibilities are certainly not mutually exclusive.

1.5   The future perspective

There are a number of remaining questions. For example, we do not know how FancD2 is targeted to chromatin through the interaction with putative ubiquitin receptor. Although a recent study showed FancD2 can directly bind to naked DNA *in vitro* (Park et al., 2005), chromatin serves as an effective barrier, since FancD2 cannot access DNA without prior monoubiquitination. There should be a molecular mechanism, potentially analogous to the Rad51 loading by BRCA2 (Venkitaraman, 2002), in the FancD2 chromatin targeting. The ubiquitin receptor has to be identified to answer this question. Also, we would like to address what is an effector subunit(s) in the FA core complex. We may be able to identify this by targeting components of the core complex into chromatin one by one.

FancD2 might have an as yet unidentified enzymatic function. Alternatively, or in addition, it may provide a scaffold to recruit repair factors into vicinity of DNA damage sites. BLM might be relocalized to form foci by this mechanism. Whether the recruitment of the core complex needs chromatin-bound FancD2 is another important issue. The relationship of BRCA2/FancD1 with the core complex-FancD2 pathway also warrants more investigation.

What about the mechanisms for bone marrow failure? This is obviously beyond the scope of the research using DT40. Recent studies indicate genes involved in DNA damage response are required for maintenance of bone marrow stem cells (O'Driscoll et al., 2001; Bender et al., 2002; Ito et al., 2004), possibly protecting against oxidative damage (Ito et al., 2004). This line of investigation may lead to full understanding of the mechanism of bone marrow failure in FA, and the "oxygen radical hypothesis" may turn out to be partially correct.

The utility of DT40 cell system in studying human genetic disease is well exemplified in the recent discovery and functional characterization of FancJ and FancM genes. Since the chicken genome project has been finished, the gene targeting in DT40 is now much quicker and easier than before.

Application of proteomic approach such as pull-down of protein complex followed by mass spectrometric identification, combined with the reverse genetics, would expand potential in biological and medical sciences using DT40 system.

REFERENCES

Ahmad SI, Hanaoka F, Kirk SH. 2002. Molecular biology of Fanconi anaemia–an old problem, a new insight. Bioessays 24:439-448.

Bender CF, Sikes ML, Sullivan R, Huye LE, Le Beau MM, Roth DB, Mirzoeva OK, Oltz EM, Petrini JH. 2002. Cancer predisposition and hematopoietic failure in Rad50(S/S) mice. Genes Dev. 16:2237-2251.

Bezzubova O, Shinohara A, Mueller RG, Ogawa H, Buerstedde JM. 1993. A chicken RAD51 homologue is expressed at high levels in lymphoid and reproductive organs. Nucleic Acids Res. 21:1577-1580.

Bridge WL, Vandenberg CJ, Franklin RJ, Hiom K. 2005. The BRIP1 helicase functions independently of BRCA1 in the Fanconi anemia pathway for DNA crosslink repair. Nat. Genet. 37:953-957.

Cox MM. 2001. Recombinational DNA repair of damaged replication forks in Escherichia coli: questions. Annu. Rev. Genet. 35:53-82.

Cox MM, Goodman MF, Kreuzer KN, Sherratt DJ, Sandler SJ, Marians KJ. 2000. The importance of repairing stalled replication forks. Nature 404:37-41.

D'Andrea AD, Grompe M. 2003. The Fanconi anaemia/BRCA pathway. Nat. Rev. Cancer 3:23-34.

Di Fiore PP, Polo S, Hofmann K. 2003. When ubiquitin meets ubiquitin receptors: a signalling connection. Nat. Rev. Mol. Cell Biol. 4:491-497.

Garcia-Higuera I, Taniguchi T, Ganesan S, Meyn MS, Timmers C, Hejna J, Grompe M, D'Andrea AD. 2001. Interaction of the Fanconi anemia proteins and BRCA1 in a common pathway. Mol. Cell 7:249-262.

Haglund K, Di Fiore PP, Dikic I. 2003. Distinct monoubiquitin signals in receptor endocytosis. Trends Biochem. Sci. 28:598-603.

Hicke L, Schubert HL, Hill CP. 2005. Ubiquitin-binding domains. Nat. Rev. Mol. Cell Biol. 6:610-621.

Hickson ID. 2003. RecQ helicases: caretakers of the genome. Nat. Rev. Cancer 3:169-178.

Hirano S, Yamamoto K, Ishiai M, Yamazoe M, Seki M, Matsushita N, Ohzeki M, Yamashita YM, Arakawa H, Buerstedde JM, Enomoto T, Takeda S, Thompson LH, Takata M. 2005. Functional relationships of FANCC to homologous recombination, translesion synthesis, and BLM. EMBO J. 24:418-427.

Howlett NG, Taniguchi T, Olson S, Cox B, Waisfisz Q, De Die-Smulders C, Persky N, Grompe M, Joenje H, Pals G, Ikeda H, Fox EA, D'Andrea AD. 2002. Biallelic inactivation of BRCA2 in Fanconi anemia. Science 297:606-609.

Hussain S, Witt E, Huber PA, Medhurst AL, Ashworth A, Mathew CG. 2003. Direct interaction of the Fanconi anaemia protein FANCG with BRCA2/FANCD1. Hum. Mol. Genet. 12:2503-2510.

Ito K, Hirao A, Arai F, Matsuoka S, Takubo K, Hamaguchi I, Nomiyama K, Hosokawa K, Sakurada K, Nakagata N, Ikeda Y, Mak TW, Suda T. 2004. Regulation of oxidative stress by ATM is required for self-renewal of haematopoietic stem cells. Nature 431:997-1002.

Kannouche PL, Wing J, Lehmann AR. 2004. Interaction of human DNA polymerase eta with monoubiquitinated PCNA: a possible mechanism for the polymerase switch in response to DNA damage. Mol. Cell 14:491-500.

Kojic M, Kostrub CF, Buchman AR, Holloman WK. 2002. BRCA2 homolog required for proficiency in DNA repair, recombination, and genome stability in Ustilago maydis. Mol. Cell 10:683-691.

Komori K, Fujikane R, Shinagawa H, Ishino Y. 2002. Novel endonuclease in Archaea cleaving DNA with various branched structure. Genes Genet. Syst. 77:227-241.

Levitus M, Rooimans MA, Steltenpool J, Cool NF, Oostra AB, Mathew CG, Hoatlin ME, Waisfisz Q, Arwert F, De Winter JP, Joenje H. 2004. Heterogeneity in Fanconi anemia: evidence for two new genetic subtypes. Blood 103:2498-2503.

Levitus M, Waisfisz Q, Godthelp BC, Vries Y, Hussain S, Wiegant WW, Elghalbzouri-Maghrani E, Steltenpool J, Rooimans MA, Pals G, Arwert F, Mathew CG, Zdzienicka MZ, Hiom K, De Winter JP, Joenje H. 2005. The DNA helicase BRIP1 is defective in Fanconi anemia complementation group J. Nat. Genet. 37:934-935.

Levran O, Attwooll C, Henry RT, Milton KL, Neveling K, Rio P, Batish SD, Kalb R, Velleuer E, Barral S, Ott J, Petrini J, Schindler D, Hanenberg H, Auerbach AD. 2005. The BRCA1-interacting helicase BRIP1 is deficient in Fanconi anemia. Nat. Genet. 37: 931-933.

Matsushita N, Kitao H, Ishiai M, Nagashima N, Hirano S, Okawa K, Ohta T, Yu DS, McHugh PJ, Hickson ID, Venkitaraman AR, Kurumizaka H, Takata M. 2005. A FancD2-monoubiquitin fusion reveals hidden functions of Fanconi anemia core complex in DNA repair. Mol Cell 19:841-847.

Meetei AR, de Winter JP, Medhurst AL, Wallisch M, Waisfisz Q, van de Vrugt HJ, Oostra AB, Yan Z, Ling C, Bishop CE, Hoatlin ME, Joenje H, Wang W. 2003a. A novel ubiquitin ligase is deficient in Fanconi anemia. Nat. Genet. 35:165-170.

Meetei AR, Levitus M, Xue Y, Medhurst AL, Zwaan M, Ling C, Rooimans MA, Bier P, Hoatlin M, Pals G, de Winter JP, Wang W, Joenje H. 2004. X-linked inheritance of Fanconi anemia complementation group B. Nat. Genet. 36:1219-1224.

Meetei AR, Medhurst AL, Ling C, Xue Y, Singh TR, Bier P, Steltenpool J, Stone S, Dokal I, Mathew CG, Hoatlin M, Joenje H, de Winter JP, Wang W. 2005. A human ortholog of archaeal DNA repair protein Hef is defective in Fanconi anemia complementation group M. Nat. Genet. 37:958-963.

Meetei AR, Sechi S, Wallisch M, Yang D, Young MK, Joenje H, Hoatlin ME, Wang W. 2003b. A multiprotein nuclear complex connects Fanconi anemia and Bloom syndrome. Mol. Cell. Biol. 23:3417-3426.

Mosedale G, Niedzwiedz W, Alpi A, Perrina F, Pereira-Leal JB, Johnson M, Langevin F, Pace P, Patel KJ. 2005. The vertebrate Hef ortholog is a component of the Fanconi anemia tumor-suppressor pathway. Nat. Struct. Mol. Biol. 12:763-771.

Nakanishi K, Taniguchi T, Ranganathan V, New HV, Moreau LA, Stotsky M, Mathew CG, Kastan MB, Weaver DT, D'Andrea AD. 2002. Interaction of FANCD2 and NBS1 in the DNA damage response. Nat. Cell. Biol. 4:913-920.

Nanda I, Shan Z, Schartl M, Burt DW, Koehler M, Nothwang H, Grutzner F, Paton IR, Windsor D, Dunn I, Engel W, Staeheli P, Mizuno S, Haaf T, Schmid M. 1999. 300 million years of conserved synteny between chicken Z and human chromosome 9. Nat. Genet. 21:258-259.

Niedzwiedz W, Mosedale G, Johnson M, Ong CY, Pace P, Patel KJ. 2004. The Fanconi anaemia gene FANCC promotes homologous recombination and error-prone DNA repair. Mol. Cell 15:607-620.

O'Driscoll M, Cerosaletti KM, Girard PM, Dai Y, Stumm M, Kysela B, Hirsch B, Gennery A, Palmer SE, Seidel J, Gatti RA, Varon R, Oettinger MA, Neitzel H, Jeggo PA, Concannon P. 2001. DNA ligase IV mutations identified in patients exhibiting developmental delay and immunodeficiency. Mol. Cell 8:1175-1185.

Okada T, Sonoda E, Yamashita YM, Koyoshi S, Tateishi S, Yamaizumi M, Takata M, Ogawa O, Takeda S. 2002. Involvement of vertebrate polkappa in Rad18-independent postreplication repair of UV damage. J. Biol. Chem. 277:48690-48695.

Paques F, Haber JE. 1999. Multiple pathways of recombination induced by double-strand breaks in Saccharomyces cerevisiae. Microbiol. Mol. Biol. Rev. 63:349-404.

Park WH, Margossian S, Horwitz AA, Simons AM, D'Andrea AD, Parvin JD. 2005. Direct DNA binding activity of the fanconi anemia D2 protein. J. Biol. Chem. 280:23593-23598.

Sasaki MS. 1975. Is Fanconi's anaemia defective in a process essential to the repair of DNA cross links? Nature 257:501-503.

Sasaki MS, Tonomura A. 1973. A high susceptibility of Fanconi's anemia to chromosome breakage by DNA cross-linking agents. Cancer Res. 33:1829-1836.

Simpson LJ, Sale JE. 2003. Rev1 is essential for DNA damage tolerance and non-templated immunoglobulin gene mutation in a vertebrate cell line. EMBO J. 22:1654-1664.

Sonoda E, Okada T, Zhao GY, Tateishi S, Araki K, Yamaizumi M, Yagi T, Verkaik NS, van Gent DC, Takata M, Takeda S. 2003. Multiple roles of Rev3, the catalytic subunit of polzeta in maintaining genome stability in vertebrates. EMBO J. 22:3188-3197.

Sonoda E, Sasaki MS, Buerstedde JM, Bezzubova O, Shinohara A, Ogawa H, Takata M, Yamaguchi-Iwai Y, Takeda S. 1998. Rad51-deficient vertebrate cells accumulate chromosomal breaks prior to cell death. EMBO J. 17:598-608.

Sonoda E, Sasaki MS, Morrison C, Yamaguchi-Iwai Y, Takata M, Takeda S. 1999. Sister chromatid exchanges are mediated by homologous recombination in vertebrate cells. Mol. Cell. Biol. 19:5166-5169.

Stewart GS, Wang B, Bignell CR, Taylor AM, Elledge SJ. 2003. MDC1 is a mediator of the mammalian DNA damage checkpoint. Nature 421:961-966.

Takata M, Sasaki MS, Sonoda E, Fukushima T, Morrison C, Albala JS, Swagemakers SM, Kanaar R, Thompson LH, Takeda S. 2000. The Rad51 paralog Rad51B promotes homologous recombinational repair. Mol. Cell. Biol. 20:6476-6482.

Takata M, Sasaki MS, Tachiiri S, Fukushima T, Sonoda E, Schild D, Thompson LH, Takeda S. 2001. Chromosome instability and defective recombinational repair in knockout mutants of the five Rad51 paralogs. Mol. Cell. Biol. 21:2858-2866.

Taniguchi T, Garcia-Higuera I, Andreassen PR, Gregory RC, Grompe M, D'Andrea AD. 2002. S-phase-specific interaction of the Fanconi anemia protein, FANCD2, with BRCA1 and RAD51. Blood 100:2414-2420.

Thompson LH, Hinz JM, Yamada NA, Jones NJ. 2005. How Fanconi anemia proteins promote the four Rs: Replication, recombination, repair, and recovery. Environ. Mol. Mutagen. 45:128-142.

Venkitaraman AR. 2002. Cancer susceptibility and the functions of BRCA1 and BRCA2. Cell 108:171-182.

Wang X, Andreassen PR, D'Andrea AD. 2004. Functional interaction of monoubiquitinated FANCD2 and BRCA2/FANCD1 in chromatin. Mol. Cell. Biol. 24:5850-5862.

Watanabe K, Tateishi S, Kawasuji M, Tsurimoto T, Inoue H, Yamaizumi M. 2004. Rad18 guides pol h to replication stalling sites through physical interaction and PCNA monoubiquitination. EMBO J. 23:3886-3896.

West SC. 2003. Molecular views of recombination proteins and their control. Nat. Rev. Mol. Cell Biol. 4:435-445.

Yamamoto K, Hirano S, Ishiai M, Morishima K, Kitao H, Namikoshi K, Kimura M, Matsushita N, Arakawa H, Buerstedde JM, Komatsu K, Thompson LH, Takata M. 2005. Fanconi anemia protein FANCD2 promotes immunoglobulin gene conversion and DNA repair through a mechanism related to homologous recombination. Mol. Cell. Biol. 25: 34-43.

Yamamoto K, Ishiai M, Matsushita N, Arakawa H, Lamerdin JE, Buerstedde JM, Tanimoto M, Harada M, Thompson LH, Takata M. 2003. Fanconi anemia FANCG protein in mitigating radiation- and enzyme-induced DNA double-strand breaks by homologous recombination in vertebrate cells. Mol. Cell. Biol. 23:5421-5430.

Yamashita YM, Okada T, Matsusaka T, Sonoda E, Zhao GY, Araki K, Tateishi S, Yamaizumi M, Takeda S. 2002. RAD18 and RAD54 cooperatively contribute to maintenance of genomic stability in vertebrate cells. EMBO J. 21:5558-5566.

Chapter 18

PHENOTYPIC ANALYSIS OF CELLULAR RESPONSES TO DNA DAMAGE

Helfrid Hochegger and Shunichu Takeda
Dep. Of Radiation Genetics; Graduate School of Medicine; Kyoto University Sakyo-ku, Kyoto, Japan

Abstract: Since the establishment of the first mutant DT40 cell line, a large panel of DT40 mutants in DNA repair genes has been created. The systematic phenotypic analysis of these mutants has provided insight in the mechanisms of various DNA repair pathways in vertebrate. Especially, in the field of double strand break repair DT40 cells have contributed to the advance of knowledge. This review will give a brief overview of DNA double strand break repair, and will summarize phenotypes observed in DT40 mutants in these pathways, focusing on the methodological aspects of analysis.

Key words: Double strand break repair, homologous recombination, non-homologous endjoing, ionizing radiation, Rad51.

1.   INTRODUCTION

Cells are continuously challenged by DNA damage arising spontaneously during the cellular metabolism or deriving from external insults such as UV irradiation and environmental toxins. Maintaining the integrity of the genome is thus an essential task for cellular survival and a large variety of DNA repair pathways have evolved to deal with all forms of DNA damage (Wood et al., 2005). Genetic analysis of bacteria and yeast has helped us to investigate the contribution of a large variety of genes to different repair processes. Mammalian genetics have been slow to follow these pioneering studies, because many DNA repair factors proved to be essential for embryonic development. DT40 cells provide a unique opportunity to

J.-M. Buerstedde and S. Takeda (eds.), Reviews and Protocols in DT40 Research, 313–325.
© 2006 *Springer.*

perform a systematic genetic analysis of DNA repair in vertebrates. Several factors contribute to the advantage of DT40 cells for the analysis of DNA repair phenotypes. Firstly, the high gene targeting frequencies allow quick generation of single, double and even triple mutants with otherwise identical genetic background. Secondly, the overall stable genome in DT40 cells with very low levels of spontaneous chromosome aberrations allows direct visualization of the effects of DNA damage by scoring metaphase spreads. Thirdly, the absence of p53 reduces the level of rapid apoptosis induced by DNA damage and allows analysis of DNA repair defects in cells that would, under normal circumstances simply die. Lastly, the diversification of immunoglobulin light and heavy chain by gene conversion and somatic hyper mutation represents an ideal system to study homologous recombination and translesion synthesis pathways in close detail, and to analyze defects in these pathways by simple DNA sequencing (Arakawa and Buerstedde, this issue, chapter1).

This review will summarize recent progress in the analysis of DT40 mutants in DNA repair genes focusing on double strand break repair.

2.   PRINCIPLES OF DOUBLE STRAND BREAK REPAIR

Among the different types of damage that cells have to cope with double strand breaks (DSBs) certainly pose one of the most serious threats. In fact a single break of the double helix, can prove lethal to the cell, if left unrepaired. DSBs can be caused by exogenous agents such as ionizing radiation and by metabolically produced free radicals. They also occur spontaneously at stalled replication forks (Michel et al., 1997). The first stage of the response to DSB involves activation of Atm; phosphorylation of histone H2AX, binding of 53bp1 and the Mre11/Rad50/Nbs1 complex (Bradbury and Jackson, 2003). These DSB recognition factors are often involved in both stalling of the cell cycle by activating the DNA damage checkpoint, and initiating the repair of the lesions. There are two major pathways to deal with DSBs that have in principle been conserved throughout evolution. Firstly; homologous recombination (HR) uses the information of a homologous sequence, preferentially the sister chromatide, as a blueprint to accurately repair the break (Wyman et al., 2004). Secondly non-homologous endjoing (NHEJ) simply glues the broken ends together often resulting in loss of genetic information. Vertebrae HR is mediated by Rad51 and a variety of other proteins (as discussed below) and is mainly employed to repair DSB occurring during DNA replication and in G2 phase of the cell cycle when homologous sister chromatids are available. NHEJ on the other hand is initiated by the binding of the Ku hetero-dimer (consisting of Ku70 and Ku80) and DNA-PK to the DSB is the pre-dominant repair pathway during G1 phase of the cell cycle (Doherty and Jackson, 2001). The initial step after DSB formation decides which pathway will be chosen. Access of an exonuclease creates a 3' single strand tail that will be covered with Rad51 and initiate the subsequent HR reaction (Baumann and West, 1998). NHEJ, on the other, hand is initiated by the binding of the Ku hetero-dimer (consisting of Ku70 and Ku80) and Dna-Pk to the DSB (Downs and Jackson, 2004). Several studies have proposed a direct competition between NHEJ and HR (Frank-Vaillant and Marcand, 2002; Fukushima et al., 2001; Pierce et al., 2001), but the synergistic effect of a double deletion of NHEJ and HR genes (Takata et al., 1998), also suggests a considerable overlap between the two pathways. The precise regulation of the balance between these NHEJ and HR remains to be fully understood.

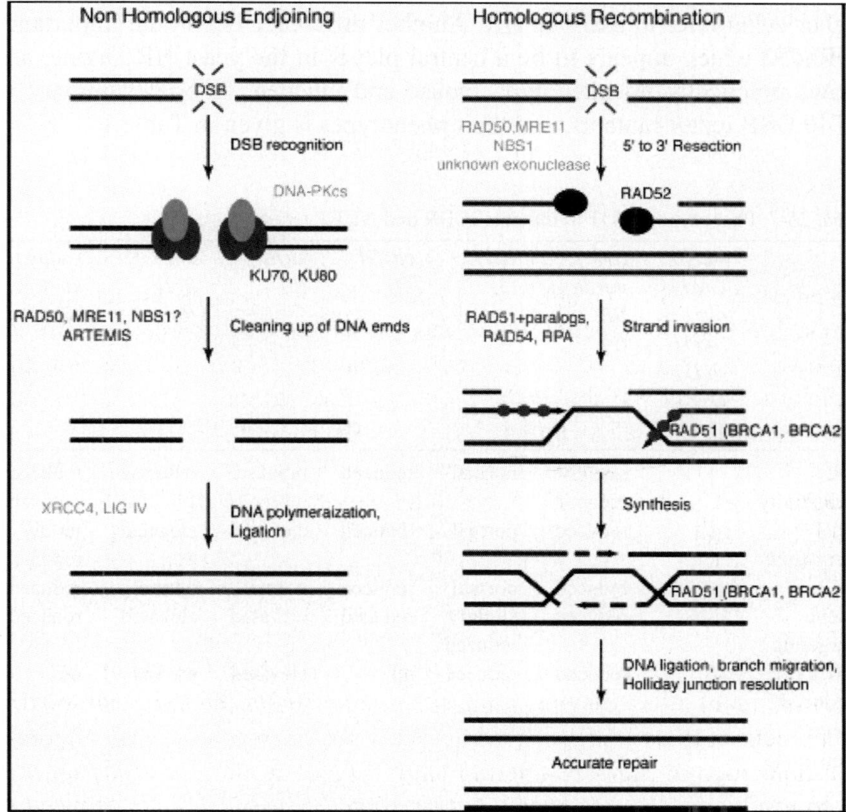

Figure 18-1. NHEJ and HR are the two major pathways of DSB repair. During NHEJ the two broken ends are ligated back together, often resulting in loss of information. HR uses a homologous sequence such as the sister chromatid to copy the sequence lost at the break and repair the lesion with accuracy (see plate 43).

3. GENETIC ANALYSIS OF DOUBLE STRAND BREAK REPAIR IN YEAST AND VERTEBRATES

The classic model system for the genetic analysis of DNA repair in eukaryotes is the budding yeast *Saccharomyces cervisiae* and we owe most of our insights into the molecular details of this field to the pioneering forward screens that have been accomplished in this organism. Double strand break repair has been extensively studied in budding yeast (Paques and Haber, 1999). Genetic analysis in mouse and DT40 cells, have confirmed the basic conservation of the HR and NHEJ reactions but also have pointed out important differences between fungi and vertebrae cells. In yeast cells, the HR pathway repairs most DSBs, while NHEJ appears to be employed much more frequently in vertebrates. Moreover besides the conserved players of the HR reaction such as Rad51, Rad52 and Rad54, several new factors such as the Rad51 paralogs and Brca2 have evolved in higher eukaryotes to assist in HR. Another difference lies in the importance of Rad52 which appears to be a central player in the yeast HR engine, and shows practically no phenotype mouse and chicken. A brief summary of DT40 DSB repair mutants, and their phenotypes is given in Table 1

Table 18-1. Phenotypes of DT40 mutants in HR and NHEJ genes.

	rad50, mre11 nbs1 rad51 rad52+ xrcc3 (1)	*rad51 - paralogs BRCA2 (2)*	*rad52 (3)*	*rad54 (4)*	*ku70 (5)*	*dna-PK ligIV (6)*	*rad54+ ku70 (7)*
IR sensitivity		slightly reduced	normal	reduced	biphasic	reduced	reduced
CPT sensitivity		reduced	normal	reduced	elevated	elevated	mildly reduced
SCE		reduced	normal	reduced	normal	normal	reduced
Gene targeting		reduced	slightly reduced	reduced	elevated	elevated	reduced
iSCE1 induced gene conversion		reduced	reduced	nd	elevated	normal	nd
Rad51 foci		reduced	normal	normal	normal	normal	normal
growth	lethal	reduced	normal	slightly reduced	normal	normal	reduced

References:
(1)   *rad50* (Nakahara et al., MS in preparation); *mre11* (Yamaguchi-Iwai et al., 1999); *nbs1* (Nakahara et al., MS in preparation); *rad51* (Sonoda et al., 1998); *rad52+paralog* (Fujimori et al., 2001)
(2)   *rad51* paralogs (Takata et al., 2001); *brca2* (Hatanaka et al., 2005)
(3)   *rad52* (Yamaguchi-Iwai et al., 1998)
(4)   *rad54 (Bezzubova* et al., 1997)
(5)   *ku70 (Takata* et al., *1998)*
(6)   *dna-pk (Fukushima* et al., *2001); ligIV* (Adachi et al., 2001; *Adachi* et al., *2004)*
(7)   *rad54+ ku70 (Takata* et al., *1998)*

In the following sections we will present the basic principles of the different assays, used to analyze DSB repair, and discuss the phenotypes of the various mutants presented in table 1; where particular phenotypes are discussed please refer to table 1 for references

4.   COLONY FORMATION ASSAYS

The most straightforward way to analyze the involvement of a particular gene in DSB repair is to analyze the mutants survival following increasing doses to ionizing irradiation (IR). Most DT40 mutants in the HR and NHEJ that our lab and others have prepared, display elevated IR-sensitivity (see table 1). Assaying an asynchronous population of cells in this way will, however, give you a mixed response to this treatment, if the mutant in question is more sensitive in one phase of the cell cycle than another. Thus, pre-synchronization by nocodazole or elutrition, and irradiation at different cell cycle phases can be very informative to evaluate the cell cycle specific involvement of a particular mutant in DSB repair. *ku70* cells, for example, are highly IR-sensitive in G1, while being IR resistant during G2. In an asynchronous culture this results in elevated sensitivity to low doses of IR, which kills the population of G1 cells. The remaining S and G2 cells are more resistant and change the slope of the curve. Conversely, HR mutants, such as *rad54*, are most sensitive during G2 (see Figure 2). Camptothecin, an inhibitor of topoismerase I, is a useful reagent to determine the involvement of a given mutant in the repair of replication associated DSBs. CPT blocks Topoisomerase I in a state where it is covalently linked to a single broken DNA strand. This DNA-protein fusion forms a unsurpassable obstacle to replicative polymerases, and cause replication for collapse often resulting in DSBs. DT40 mutants in HR genes, such as *XRCC3* and *BRCA2*, are sensitive to CPT. NHEJ mutants, such as *KU70, DNA-PK* and *LigIV* display an increase in CPT resistance. This result suggests that HR is predominantly involved in the repair of CPT induced DSBs while the NHEJ appears to interfere with this process.

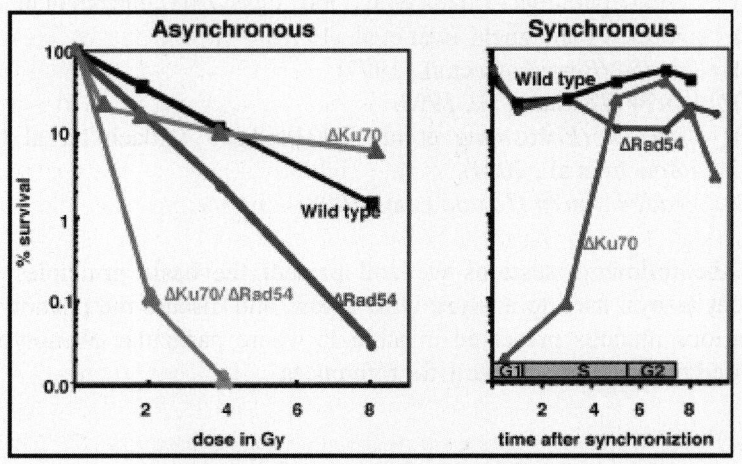

Figure 18-2. Cell cycle specific sensitivity of HR and NHEJ mutants. The left panel shows the sensitivity of *ku70* and *rad54* cells in asynchronous culture. The left panel represents IR-sensitivities of cell taken at different times after synchronization by elutriation. *Ku70* cells are highly sensitive in G1 but more resistant than WT in G2. *Rad54* cells, show the opposite pattern, being IR sensitive only in late S and G2-phase (see plate 44).

5. CHROMOSMAL ABERRATIONS

One advantage of working with vertebrate cells is the relative ease, with which chromosomes can be visualize. Counting gaps and breaks in mitotic spreads, one can easily score spontaneous or induced chromosome damage. In the case of IR, the timing of irradiation before entry into mitosis will reflect the cell cycle specific sensitivities of the analyzed mutants. Cells collected within 3 hours of mitotic arrest by colcemid after irradiation, will include all the cells in an asynchronous population that have been in G2 phase at time of IR exposure. Likewise, cells that have been irradiated and blocked to exit mitosis in a window of 3-6, or 6-9 hours after IR exposure, will cover the majority the S-phase or G1 population, respectively. Accordingly HR mutants like *rad54* and the *rad51* paralogs, display a strong increase in IR induced chromosome aberrations when irradiated in S and G2-phase, while NHEJ mutants show the strongest peak of DNA damage in G1 phase.

We previously observed that conditional *rad51* mutants die quickly after removal of the RAD51 transgene, displaying a massive increase in spontaneous chromosomal aberrations. Interestingly, the DSBs in these mutants span both chromatides (see Fig. 3). However, accidental damage from metabolic radicals is likely to affect only one chromatid and will appear as a single chromatid gap or break; the appearance of breaks that span the entire chromosome could point to problems occurring in connection with DNA replication. A collapsed and unresolved replication fork is likely to affect both sister chromatids simultaneously. We thus concluded that the essential role of Rad51 for survival is closely linked to the protection of DNA replication rather than with spontaneously occurring breaks in single chromatids. In this context, genetic data from mammals and DT40 concerning the importance of NHEJ and HR for survival should be viewed in a new light. If DSBs are induced exogenously, NHEJ plays a major role in repairing them; especially in cells that have extended G1 phases (Liang et al., 1998). However, such accidental breaks occur relatively rarely and indeed, mice lacking KU or other NHEJ genes are viable, as are DT40 *ku70* cells. HR, on the other hand, is an essential process and absolutely required for the survival of vertebrate cells. This is most likely due to the frequent occurrence of DSBs at stalled replication forks that have to be repaired by the HR pathway.

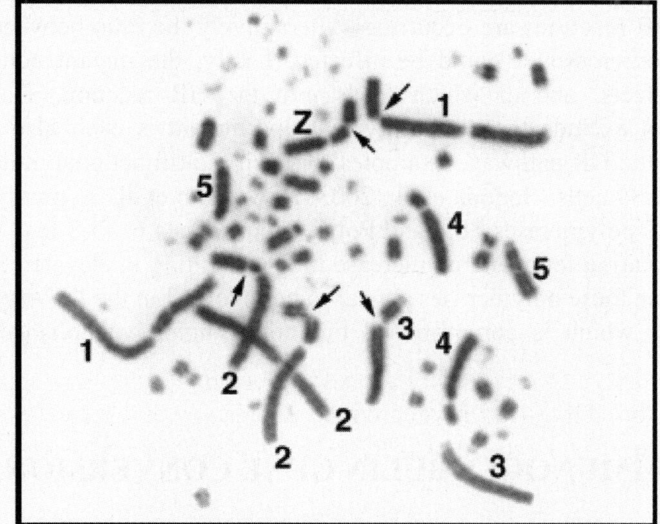

Figure 18-3. Chromosome aberrations in *rad51* cells after depletion of the RAD51 transgene. The cells die rapidly displaying a more then 10-fold increase in breaks and gaps that span all of the chromosome. This Picture was published by (Sonoda et al., 1998) (see plate 45).

6.   SISTER CHROMATID EXCHANGE (SCE)

Another way of looking at chromosomes in vertebrates is to score the exchange between the sister-chromatids (for a detailed description of the method see below in the method section) that is a consequence of gene-conversion associated with crossover between the homologous sisters (Latt, 1981). Such sister-chromatid exchange occurs only once or 2-3 per cell cycle in a healthy DT40 cell, whereas it is increased about fivefold after MMC exposure (Sonoda et al., 1999). Mutants in repair pathways, such as base excision repair (BER) or translesion synthesis (TLS) that lead to an increase in replication blocks and chromosome instability, generally cause an increase in SCE. This is a result of an increased usage of HR as a back up to the other repair pathways (Mizutani et al., 2004). Mutants in HR genes generally have very low levels of spontaneous and induced SCE. This method represents thus a reliable way to measure the capability of a mutant to undergo HR. It is, however, important to keep in mind that SCE only scores crossover events, whereas the HR reactions that are resolved without crossover do not result in visible exchanges between the sisters. In yeast, the majority of mitotic HR does not lead to cross over (Ira et al., 2003), whereas the precise ratio between the two is not known in vertebrates (Johnson and Jasin, 2000). SCE is not a reliable method to conclude on the actual number of HR events that are occurring in the cell. Caution is also recommended, when an increase in SCE is interpreted. Generally, increase in SCE means that more HR reactions are occurring. Alternatively the ratio between cross-over and non-crossover could be affected. Lastly, the mutant could have multiple defects, one of which feeds into the HR reaction. This does, however, not exclude the possibility that this mutant is itself also directly involved in the HR pathway. Examples for such multifunctional mutants are *rev3* and *rad30* cells (Sonoda et al., 2003; Kawamoto et al., in press), which lack the TLS polymerases Polξ and Polη. Their defects in TLS lead to more broken replication forks and an increase in HR resulting in elevated SCE. At the same time these polymerases appear to be involved in the DNA synthesis step of HR, which is compromised but not completely abolished in the mutants.

7.   IMMUNOGLOBULIN GENE CONVERSION

One of the major advantages of the DT40 system in analyzing defects in HR lays in the fact that chicken use gene conversion to diversify their immunoglobulin genes. This method is described in detail by Arakawa and Buerstedde (this issue, chapter1), and will not be further discussed here.

8.   GENE TARGETING

The high frequencies of gene targeting in DT40 cells are very useful to analyze the involvement of genes of interest in HR. The precise mechanism of this reaction is not known, but it certainly involves a specialized form of HR. Accordingly, all mutants in known HR genes that others and we have found, have reduced gene targeting rates (see Table 1). NHEJ mutants such as *ku70*, *dna-pk*, and *ligIV* are much less efficient in the random integration of the targeting constructs (Takeda lab, unpublished results). In these mutants the ratio of random, versus targeting integration is greatly shifted towards the latter. This means that a lot less clones have to be screened to identify a knock out cell. Transient knock down of these genes by siRNA could thus be useful to increase gene targeting rates in the case it is difficult to obtain a knock out cell line for a particular mutant. This strategy might also be useful in other cellular systems. Conversely other mutants show a strong increase in random integration resulting in a low ratio of gene targeting. This increase in random integration could be due to an increase in DNA breaks and nicks that are the consequence of defects in other repair pathways than HR. Interpretation of the results of gene targeting can thus be misleading and it is highly recommended to avoid quantification of gene targeting as the relative ratio between random and targeting events.

9.   FOCUS FORMATION OF PROTEINS INVOLVED IN DNA REPAIR

Many of the important repair factors accumulate sequentially at the site of DNA damage. This accumulation can be visualized as distinct foci that appear in timed intervals after DNA damage induction. Lisby et al., systematically analyzed this sequential appearance of various early and late repair and checkpoint factors fused to fluorescent proteins in living yeast cells. In DT40 such a GFP knock-in strategy has not yet been performed but foci can readily be observed in fixed specimen, if antibodies raised against the repair or checkpoint factor of interest, are available. Our laboratory regularly uses phospho-Histone H2AX, Rad51 and Rad41 antibodies to detect foci as markers for early, intermediate and late repair events. But with an increase in antibody availability and an increase in knock-in fusions of fluorescent proteins with repair factors, this method is likely to gain in precision and importance. Reduction in the formation of either of these foci points to a defect in damage recognition, HR execution or HR completion. Naturally defects in the earlier stages will also affect the later events as can be clearly seen in Lisby et al., s analysis. In DT40 we have identified various mutants that show defects in the formation of these repair foci (see Table1). Data obtained from the foci formation assay do not always correlate with data from other repair assays. *rad51* paralog mutants for example cannot form Rad51 foci after damaging the cells by various means. However, these mutants show only a very mild sensitivity to IR when compared to other mutants in HR genes. We concluded from these and other results that the paralogs are certainly involved albeit not essential in recruiting Rad51 to sites of damage. A sufficient amount of Rad51 seems to be loaded onto DNA to catalyze strand invasion and subsequent steps of HR in these mutants.

Hence, a more sensitive assay to measure the binding of DNA repair factors on DNA is missing in DT40 cells. One possibility would be to induce a large amount of damage in a defined area by laser and measure the recruitment of repair factors to these laser-exposed spots in the nucleus. Several laboratories have established this method in adherent cell lines (Kim et al., 2005; Lan et al., 2004), but it is difficult to perform a similar experiment in floating DT40 cells. The most sensitive assay to measure the recruitment of repair factors to DSBs, is CHIP analysis after induction of a break with a rare cutting endonuclease. This method based on the inducible expression of the HO endonuclease has been extensively used in yeast. However, a similar rapid and synchronous induction of endonucleases has not yet been achieved in DT40 cells.

10.   I-SCE1 INDUCED GENE CONVERSION

In the last decade Maria Jasin's laboratory developed a very elegant system to directly measure the efficacy of gene conversion in mouse embryonic stem cells (Liang and Jasin, 1999). The principle of this method is to align two mutant alleles of a suitable marker in tandem. The cleavage site of the rare cutting endonuclease I-Sce1 disrupts one of these mutant alleles, while the other serves as a donor in the HR reaction that restores the functional marker. Gene conversion can be induced by expression of I-Sce1 and scored by assaying for presence of the functional marker. In DT40 cells the Scneo construct, which uses neomycin as a marker, can be knocked in the OVALBUMIN locus, thus ensuring that a single copy of the construct will be inserted in the same genomic environment. After transient expression of I-Sce1 the various cell lines carrying the SCneo construct are grown in neomycin containing methylcellulose. Gene-conversion can be quantified by simply counting the neomycin resistant colonies. A similar assay can also be used to measure gene targeting at a defined DSB site. To this end the HR donor is not placed in tandem to the mutant neo allele containing the cleavage site, but co-transfected with the I-Sce1 endonuclease. Defects in this assay are a strong indication of compromised HR efficacy.

11.   OUTLOOK

Phenotypic analysis of DT40 DNA repair mutants has proven very useful to start the drawing of a genetic map of vertebrate DNA repair. The first steps in this direction, that involve the creation of various single, double and triple mutants of known repair factors and subject them to the various assays described above, is well underway but still far from completion. This undertaking has revealed a complex picture of interactions and overlaps between different repair pathways. There are, of course, possible pitfalls in generalizing the findings from one particular chicken cell line, but in principle the phenotypes of DT40 mutant correlate well with their mouse counterparts in the case they are available.

Future directions to make use of this panel of mutants will certainly involve more precise phenotypic and biochemical analysis. Once a system comparable to the above discussed HO-endonuclease based assay in yeast will be established, CHIP analysis and PCR based assays will help determine the precise sequence of events in vertebrate HR. We are also only starting to explore the close connection of HR and DNA replication. Inducing replication blocks at particular sites in the genome has yet to be achieved, but will be very useful to analyze how the switch from replication to repair is triggered and which genes this process involves. Analysis of replication forks by 2-dimensional gel electrophoresis has also not yet been established in DT40 cells; DNA combing methods, on the other hand are starting to be used to look closer at replication.

Finally we have started to use the panel of DT40 mutants to screen for sensitivity to carcinogenic and chemotherapeutic agents by clonogenic survival assays (Mizutani et al., 2004, Nojima et al., in press, Wu et al., in press). This approach can be very useful to clarify the mechanism by which a particular agent causes cancer. Identifying mutants that are particularly sensitive to chemotherapeutic drugs could also be helpful in increasing the efficacy of chemotherapy. The newly identified genes are ideal targets for the design of specific inhibitors. In this way phenotypic analysis of DT40 repair mutants could prove a potent tool to rationalize and improve drug development.

REFERENCES

Adachi, N., Ishino, T., Ishii, Y., Takeda, S. and Koyama, H. (2001) DNA ligase IV-deficient cells are more resistant to ionizing radiation in the absence of Ku70: Implications for DNA double-strand break repair. *Proc. Natl. Acad. Sci. U S A*, **98**, 12109-12113.

Adachi, N., So, S. and Koyama, H. (2004) Loss of nonhomologous end joining confers camptothecin resistance in DT40 cells. Implications for the repair of topoisomerase I-mediated DNA damage. *J. Biol. Chem.*, **279**, 37343-37348.

Baumann, P. and West, S.C. (1998) Role of the human RAD51 protein in homologous recombination and double-stranded-break repair. *Trends Biochem. Sci.*, **23**, 247-251.

Bezzubova, O., Silbergleit, A., Yamaguchi-Iwai, Y., Takeda, S. and Buerstedde, J.M. (1997) Reduced X-ray resistance and homologous recombination frequencies in a RAD54-/- mutant of the chicken DT40 cell line. *Cell*, **89**, 185-193.

Bradbury, J.M. and Jackson, S.P. (2003) The complex matter of DNA double-strand break detection. *Biochem. Soc. Trans.*, **31**, 40-44.

Doherty, A.J. and Jackson, S.P. (2001) DNA repair: how Ku makes ends meet. *Curr. Biol.*, **11**, R920-924.

Downs, J.A. and Jackson, S.P. (2004) A means to a DNA end: the many roles of Ku. *Nat. Rev. Mol. Cell Biol.*, **5**, 367-378.

Frank-Vaillant, M. and Marcand, S. (2002) Transient stability of DNA ends allows nonhomologous end joining to precede homologous recombination. *Mol. Cell*, **10**, 1189-1199.

Fujimori, A., Tachiiri, S., Sonoda, E., Thompson, L.H., Dhar, P.K., Hiraoka, M., Takeda, S., Zhang, Y., Reth, M. and Takata, M. (2001) Rad52 partially substitutes for the Rad51 paralog XRCC3 in maintaining chromosomal integrity in vertebrate cells. *Embo J.*, **20**, 5513-5520.

Fukushima, T., Takata, M., Morrison, C., Araki, R., Fujimori, A., Abe, M., Tatsumi, K., Jasin, M., Dhar, P.K., Sonoda, E., Chiba, T. and Takeda, S. (2001) Genetic analysis of the DNA-dependent protein kinase reveals an inhibitory role of Ku in late S-G2 phase DNA double-strand break repair. *J. Biol. Chem.*, **276**, 44413-44418.

Hatanaka, A., Yamazoe, M., Sale, J.E., Takata, M., Yamamoto, K., Kitao, H., Sonoda, E., Kikuchi, K., Yonetani, Y. and Takeda, S. (2005) Similar effects of Brca2 truncation and Rad51 paralog deficiency on immunoglobulin V gene diversification in DT40 cells support an early role for Rad51 paralogs in homologous recombination. *Mol. Cell Biol.*, **25**, 1124-1134.

Ira, G., Malkova, A., Liberi, G., Foiani, M. and Haber, J.E. (2003) Srs2 and Sgs1-Top3 suppress crossovers during double-strand break repair in yeast. *Cell*, **115**, 401-411.

Johnson, R.D. and Jasin, M. (2000) Sister chromatid gene conversion is a prominent double-strand break repair pathway in mammalian cells. *Embo J.*, **19**, 3398-3407.

Kim, J.S., Krasieva, T.B., Kurumizaka, H., Chen, D.J., Taylor, A.M. and Yokomori, K. (2005) Independent and sequential recruitment of NHEJ and HR factors to DNA damage sites in mammalian cells. *J. Cell Biol.*, **170**, 341-347.

Lan, L., Nakajima, S., Oohata, Y., Takao, M., Okano, S., Masutani, M., Wilson, S.H. and Yasui, A. (2004) In situ analysis of repair processes for oxidative DNA damage in mammalian cells. *Proc. Natl. Acad. Sci. U S A*, **101**, 13738-13743.

Latt, S.A. (1981) Sister chromatid exchange formation. *Annu. Rev. Genet.*, **15**, 11-55.

Liang, F., Han, M., Romanienko, P.J. and Jasin, M. (1998) Homology-directed repair is a major double-strand break repair pathway in mammalian cells. *Proc. Natl. Acad. Sci. U S A*, **95**, 5172-5177.

Liang, F. and Jasin, M. (1999) Extrachromosomal assay for DNA double-strand break repair. *Methods Mol. Biol.*, **113**, 487-497.

Michel, B., Ehrlich, S.D. and Uzest, M. (1997) DNA double-strand breaks caused by replication arrest. *Embo J.*, **16**, 430-438.

Mizutani, A., Okada, T., Shibutani, S., Sonoda, E., Hochegger, H., Nishigori, C., Miyachi, Y., Takeda, S. and Yamazoe, M. (2004) Extensive chromosomal breaks are induced by tamoxifen and estrogen in DNA repair-deficient cells. *Cancer Res.*, **64**, 3144-3147.

Paques, F. and Haber, J.E. (1999) Multiple pathways of recombination induced by double-strand breaks in *Saccharomyces cerevisiae*. *Microbiol/. Mol. Biol. Rev.*, **63**, 349-404.

Pierce, A.J., Hu, P., Han, M., Ellis, N. and Jasin, M. (2001) Ku DNA end-binding protein modulates homologous repair of double-strand breaks in mammalian cells. *Genes Dev.*, **15**, 3237-3242.

Sonoda, E., Okada, T., Zhao, G.Y., Tateishi, S., Araki, K., Yamaizumi, M., Yagi, T., Verkaik, N.S., van Gent, D.C., Takata, M. and Takeda, S. (2003) Multiple roles of Rev3, the catalytic subunit of polzeta in maintaining genome stability in vertebrates. *Embo J*, **22**, 3188-3197.

Sonoda, E., Sasaki, M.S., Buerstedde, J.M., Bezzubova, O., Shinohara, A., Ogawa, H., Takata, M., Yamaguchi-Iwai, Y. and Takeda, S. (1998) Rad51-deficient vertebrate cells accumulate chromosomal breaks prior to cell death. *Embo J.*, **17**, 598-608.

Sonoda, E., Sasaki, M.S., Morrison, C., Yamaguchi-Iwai, Y., Takata, M. and Takeda, S. (1999) Sister chromatid exchanges are mediated by homologous recombination in vertebrate cells. *Mol. Cell Biol.*, **19**, 5166-5169.

Takata, M., Sasaki, M.S., Sonoda, E., Morrison, C., Hashimoto, M., Utsumi, H., Yamaguchi-Iwai, Y., Shinohara, A. and Takeda, S. (1998) Homologous recombination and non-homologous end-joining pathways of DNA double-strand break repair have overlapping roles in the maintenance of chromosomal integrity in vertebrate cells. *Embo J.*, **17**, 5497-5508.

Takata, M., Sasaki, M.S., Tachiiri, S., Fukushima, T., Sonoda, E., Schild, D., Thompson, L.H. and Takeda, S. (2001) Chromosome instability and defective recombinational repair in knockout mutants of the five Rad51 paralogs. *Mol. Cell Biol.*, **21**, 2858-2866.

Wood, R.D., Mitchell, M. and Lindahl, T. (2005) Human DNA repair genes, 2005. *Mutat. Res.*, **577**, 275-283.

Wyman, C., Ristic, D. and Kanaar, R. (2004) Homologous recombination-mediated double-strand break repair. *DNA Repair (Amst)*, **3**, 827-833.

Yamaguchi-Iwai, Y., Sonoda, E., Buerstedde, J.M., Bezzubova, O., Morrison, C., Takata, M., Shinohara, A. and Takeda, S. (1998) Homologous recombination, but not DNA repair, is reduced in vertebrate cells deficient in RAD52. *Mol. Cell Biol.*, **18**, 6430-6435.

Yamaguchi-Iwai, Y., Sonoda, E., Sasaki, M.S., Morrison, C., Haraguchi, T., Hiraoka, Y., Yamashita, Y.M., Yagi, T., Takata, M., Price, C., Kakazu, N. and Takeda, S. (1999) Mre11 is essential for the maintenance of chromosomal DNA in vertebrate cells. *Embo J.*, **18**, 6619-6629.

Chapter 19

ATM, A PARADIGM FOR A STRESS-RESPONSIVE SIGNAL TRANSDUCER IN HIGHER VERTEBRATE CELLS

Ken-ichi Yamamoto, Masahiko Kobayashi and Hiroko Shimizu
Department of Molecular Pathology and Center for the Development of Molecular Target Drugs, Cancer Research Institute, Kanazawa University, Kanazawa, Ishikawa 920-0934, Japan

Abstract: ATM, the gene mutated in ataxia telangiectasia, is related to a family of large phosphatidylinositol 3-kinase domain-containing protein kinases involved in cell cycle control and DNA repair. To define the physiological roles of ATM in higher vertebrate cells, we created an ATM-deficient DT40 cell line, which, despite of the lack of p53 expression, displays multiple p53-independent defects in cell cycle checkpoint control and in maintenance of chromosomal DNA. ATM-/- DT40 cells also show a mild impairment in homologous recombination repair, which is independent of its checkpoint control defects. These ATM deficient DT40 clones thus provide a useful model system for analyzing p53-independent ATM functions in cellular response to double-strand break. Furthermore, we observe various abnormalities in cellular response to noxious stress such as oxidative stress in ATM-/- DT40 cells, indicating that ATM plays important roles not only in cellular response to DNA damage but also in the maintenance of the cell homeostasis in response to oxidative damage.

Key words: Cell cycle checkpoint, apoptosis, homologous recombination repair, oxidative stress, double strand break.

J.-M. Buerstedde and S. Takeda (eds.), Reviews and Protocols in DT40 Research, 327–339.
© 2006 *Springer.*

1.   INTRODUCTION

Cell-cycle checkpoints are surveillance mechanisms that monitor the physical state of the genome and protect genome integrity by inducing cell-cycle arrest or programmed cell death (apoptosis) in response to genome damage or genome replication errors. Genetic studies in yeasts have identified six key checkpoint molecules: Rad3, Rad17, Rad1, Hus1, Rad9, and Rad26 in *S. pombe*; and Mec1, Rad24, Rad17, Mec3, Ddc1, and Ddc2/Lcd1/Pie1 in *S cerevisiae*. Recent studies have established that many features of cell-cycle checkpoints are conserved throughout evolution, and each of these yeast checkpoint proteins is represented in higher vertebrates by a closely related counterpart (Zhou and Elledge, 2000; Rouse and Jackson, 2002).

Table 19-1. Conserved cell cycle checkpoint proteins.

S. pombe	*S. cerevisiae*	mammal	protein functions
Rad1	Rad17	Rad1	PCNA-like proteins
Rad9	Ddc1	Rad9	PCNA-like proteins
Hus1	Mec3	Hus1	PCNA-like proteins
Rad17	Rad24	Rad17*	RFC-like proteins
Rad26	Ddc2/Lcd1	ATRIP	DNA binding proteins
Rad3	Mec1	ATR	PI3K-like kinases
Tel1	Tel1	ATM*	PI3K-like kinases
Chk1	Chk1	Chk1	Effector kinases
Cds1	Rad53	Chk2*	Effector kinases
Cut5	Dpb11	TopBP1	BRCT adaptor proteins
Crb2	Rad9	BRCA1	BRCT adaptor proteins
Mrc1	Mrc1	Claspin	Adaptor proteins

One example is the human gene mutated in ataxia telangiectasia (AT). AT is an autosomal recessive disorder with complex clinical features including progressive cerebellar ataxia, oculocutaneous telangiectasias, growth retardation, cellular and humoral immunodeficiency, increased predisposition to lymphoma and leukemias, and gonadal abnormalities. Cells derived from AT patients are characterized by a high level of chromosomal abnormalities, a defect that is greatly potentiated by ionizing radiation (IR), and by hypersensitivity to IR. In AT cells, DNA damage fails to induce an arrest in DNA synthesis (leading to the phenomenon of radioresistant DNA synthesis) or an appropriate arrest at the G_1/S or G_2/M cell cycle checkpoints, suggesting anomalous cell cycle regulation and/or DNA repair defects as a major underlying cause of the disease. The gene mutated in AT (designated ATM) was found to be closely related to Rad3 and Mec1. These proteins form a family of unconventionally large protein kinases that have a PI3 kinase domain at the C terminus, and play essential roles in DNA repair and cell-cycle checkpoint control by phosphorylating several key downstream effector molecules (Shiloh and Kastan, 2001). Another mammalian family member that is even more closely related to Rad3/Mec1 was later identified and designated as ATR (ATM-Rad3 related); the results of subsequent studies indicated that ATR also plays important roles in multiple cell-cycle checkpoint controls (Abraham, 2001). However, because ATR gene disruption is lethal in mice (Brown and Baltimore, 2000), the precise functions of ATR in cell-cycle checkpoint controls in vertebrates remain to be established.

1.1   p53-independent defects in ATM-/- DT40 cells

There have been some disagreements in the literature about various defects reported for ATM deficient cells derived from patients (Meyn, 1995; Lavin and Shiloh, 1997), probably as a consequence of the genetic heterogeneity inherent in human studies. While the results of recent ATM-knockout mouse studies provide a more clear picture of various defects in ATM deficient cells (Barlow et al., 1996; Xu and Baltimore, 1996), detailed biochemical and more quantitative analysis using an isogenic set of stable cell lines differing only in their ATM status is important to further define the physiological functions of ATM at the cellular level. For this purpose, we disrupted the *ATM* locus in a chicken B lymphocyte DT40 line which exhibits highly efficient targeted integration following the transfection of genomic DNA constructs (Buerstedde and Takeda, 1991). While p53 is one of most important ATM targets involved in cell cycle checkpoint control, apoptosis and DNA repair (Shiloh and Kastan, 2001), DT40 cells do not express p53 (Takao et al., 1999), due to enhanced expression of the Yin Yang 1 transcription factor, which is involved in p53 ubiquitination and degradation (Sui et al., 2004). However, the ATM deficient DT40 clones created display multiple functional defects in cell cycle checkpoint control, such as retarded cellular proliferation, defective G_2/M checkpoint control and radio-resistant DNA synthesis. Furthermore, ATM-/- DT40 cells were sensitive to IR and showed highly elevated frequencies of both spontaneous and radiation-induced chromosomal aberrations. In addition, a slight but significant reduction in targeted integration frequency was observed in ATM-/- DT40 cells. These results suggest that ATM has multiple p53-independent functions in cell cycle checkpoint control and in maintenance of chromosomal DNA (Takao et al., 1999). These ATM deficient DT40 clones therefore provide a useful model system for analyzing p53-independent ATM functions.

Table 19-2. p53-independent defects in ATM-/- DT40 cells.

Defect
Radiosensitivity
Retarded cell proliferation
Enhanced aopotosis
Increased chromosomal breakage
Defective G2/M checkpoint control
Defective S-phase checkpont control

1.2 Roles for ATM, c-Abl and Arg in DNA repair

Previous studies on nonisogenic cells from AT patients showed the presence of defects in processing of DSB, though there have been some disagreements in the nature and degree of these defects (Meyn, 1995; Lavin and Shiloh, 1997). We also found that ATM-/- DT40 cells display elevated frequencies of both spontaneous and radiation-induced chromosomal aberrations and that there is a slight but significant reduction in targeted integration frequencies in ATM-/- DT40 cell. These results therefore further support an idea that ATM plays some roles in DNA repair capacity, in addition to its established role in cell cycle checkpoint control (Jeggo et al., 1998). However, it is difficult to exclude completely the phenotypic effects of the cell cycle defects. In order to examine recombinational repair before any checkpoint deficiencies might manifest themselves, we monitored the radiation-induced formation of nuclear foci of Rad51 and Rad54, which associate closely in the eukaryotic response to DNA damage (Golub et al., 1997; Tan et al., 1999). We found that the initial kinetics of Rad54 focus formation were very similar between wild-type and ATM-null DT40 cells, while those of Rad51 were markedly different, there being a much slower activation phase in cells lacking ATM (Morrison et al., 2000; Takao et al., 2000b). We disrupted *ATM* along with other genes involved in the principal, complementary double-strand break (DSB) repair pathways of homologous recombination (HR) or non-homologous end-joining (NHEJ) in chicken DT40 cells. Ku70-deficient (NHEJ-deficient) ATM-/- chicken DT40 cells show radiosensitivity and high radiation-induced chromosomal aberration frequencies, while Rad54-deficient (HR-deficient) ATM-/- cells show only slightly-elevated aberration levels after irradiation, placing ATM and HR on the same pathway. These results reveal that ATM defects impair HR-mediated DSB repair and may link cell cycle checkpoints to HR activation (Morrison et al., 2000).

Among ATM target proteins involved in HR, c-Abl is of particular interest. c-Abl is an ubiquitously expressed non-receptor-type tyrosine kinase (Goff et al., 1980) and is activated by DNA damage in an ATM-dependent manner (Baskaran et al., 1997; Shafman et al., 1997). It plays important roles in growth arrest (Sawyers et al., 1994; Yuan et al., 1996) and apoptosis (Yuan et al., 1997; Agami et al., 1999; Gong et al., 1999; Yuan et al., 1999), and also function in DNA repair through the phosphorylation of Rad51 (Yuan et al., 1998; Chen et al., 1999), a key molecule in HR DNA repair (Shinohara and Ogawa, 1995; Sonoda et al., 1998). Previous studies showed that c-Abl interacts with and phosphorylates Rad51 on tyrosine in response to DNA damage. However, the biological significance of these findings in HR DNA repair is not yet clear, as c-Abl-mediated phosphorylation negatively affected Rad51's activity in one set of experiments, but enhanced its association with Rad52 in another (Yuan et al., 1998; Chen et al., 1999). More recently, the BCR/ABL oncogenic tyrosine kinase has been shown both to enhance Rad51 expression and to phosphorylate it, resulting in drug resistance (Slupianek et al., 2001). However, we found that Rad51 focus formation and DSB repair capacity in c-Abl-/- DT40 cells is not grossly impaired (Takao et al., 2000b), although various DSB repair defects have been well documented in ATM-/- DT40 cells (Takao et al., 1999; Morrison et al., 2000; Takao et al., 2000b). These results indicate that, while ATM is indispensable in DSB repair in eukaryotic cells, there must be redundant functions for c-Abl in DSB repair, at least in chicken DT40 cells and mouse fibroblasts (Liu et al., 1996). One possibility is that Arg (Abl-related gene), the only other known member of the c-Abl family, substitutes for c-Abl in the genome maintenance of the c-Abl-/- cells. Arg shares considerable structural and sequence homology with c-Abl in the N-terminal SH3, SH2, and tyrosine kinase domains (Kruh et al., 1990), and abnormal variants of Arg are implicated in some human lymphoid malignancies (Cazzaniga et al., 1999; Iijima et al., 2000). However, the roles played by Arg in the cellular response to DNA damage are unknown. We found that IR-induced Rad51 focus formation is reduced in Arg-deficient cells generated from a chicken B cell line by targeted disruption. This is consistent with the findings that Arg-deficient cells display hypersensitivity to IR, elevated frequencies of IR-induced chromosomal aberrations, and reduced targeted integration frequencies. All of these abnormalities in DNA damage repair are also observed in ATM-deficient cells but not in c-Abl-deficient cells. Finally, we show that Arg interacts with and phosphorylates Rad51 in 293T cells (Li et al., 2002). These results suggest that Arg plays a role in HR DNA repair by phosphorylating Rad51. Since both of Arg-/- and

ATM-/- DT40 cells display abnormalities in HR DNA repair, and since c-Abl is activated by DNA damage in an ATM-dependent manner, it is very possible that Arg is also activated by DNA damage in an ATM-dependent manner. The results of preliminary experiments in various DT40 gene-knockout cells using an antibody to human c-Abl which cross-reacts with chicken c-Abl and Arg, showed that Arg is activated by IR in an ATM-dependent manner (unpublished data). However, we have been unable so far to detect the direct phosphorylation of Arg by ATM (unpublished data). The dependency of Arg activation on ATM may therefore be more complex.

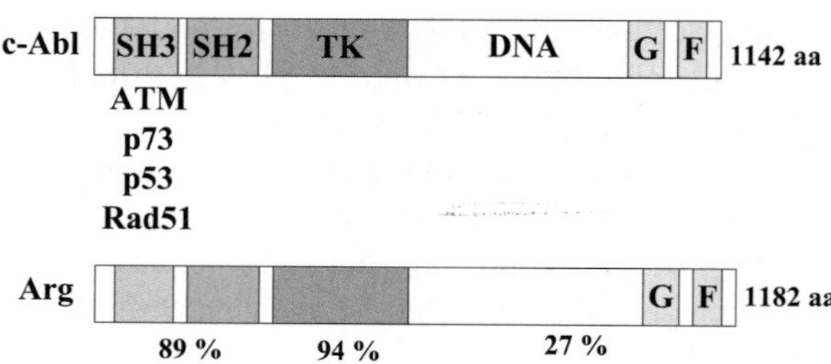

Figure 19-1. A schematic diagram for the c-Abl tyrosine kinase family, showing conserved, SH3, SH2, tyrosine kinase, DNA binding, and G/F-actin binding domains (see plate 46).

1.3 Roles for ATM in oxidative stress response

Cells derived from AT patients are characterized by a high level of chromosomal abnormalities, hypersensitivity to IR, and defective cell-cycle regulation, suggesting anomalous cell-cycle regulation and/or DNA repair defects as a major underlying cause of the disease (Shiloh and Kastan, 2001). However, the molecular mechanisms underlying the apoptosis of postmitotic cerebellar Purkinje cells in AT patients in the absence of DNA damage have remained elusive. Because DNA-damaging agents such as IR generate reactive oxygen intermediates (ROI), which can damage macro-molecules and induce cell death (Buttke and Sandstorm, 1994), ATM abnormalities may lead to apoptotic cell death due to oxidative damage. Consistent with this hypothesis, decreased levels of catalase and increased levels of lipid hydroperoxides were observed in several AT cell lines (Watters et al., 1999), and in increased levels of oxidant-modified proteins and lipids were detected in tissues, including brain, of ATM knockout mice (Barlow et al., 1999). To study the roles played by ATM in the cellular reaction to oxidative damage, we used an isogenic set of stable cell lines differing only in their ATM status, which we created previously from the chicken B cell line DT40. Although these stable DT40 cell lines, as with most transformed chicken cell lines, do not express p53, they display many of the characteristics of A-T cells, most notably a hypersensitivity to IR (Takao et al., 1999). We found that ATM-/- DT40 cells are more susceptible than wild-type cells to apoptosis induced not only by IR and bleomycin but also by C_2-ceramide. We further showed that, unlike IR and bleomycin, which induce DSB, C_2-ceramide does not induce significant DSB, indicating that ATM-/- DT40 cells are also more sensitive to the apoptosis induced by non-DNA-damaging apoptotic stimuli (Takao et al., 2000a). Although the precise mechanism underlying ceramide-induced apoptosis is still poorly defined (Hofmann and Dixit, 1998; Kolesnick and Hannun, 1999), ROI are candidate mediators (Garcia-Ruiz et al., 1997; Quillet-Mary et al., 1997). In support of this hypothesis, we found that ATM-/- DT40 cells also are more susceptible to H_2O_2-induced apoptosis, and that the apoptosis induced by C_2-ceramide and H_2O_2 is completely blocked by anti-oxidants such as catalase or N-acetyl cysteine (Takao et al., 2000a). Several groups of investigators recently reported evidence that ATM-deficient cells are more susceptible than wild-type cells to oxidative damage (Barlow et al., 1999; Reichenbach et al., 1999; Watters et al., 1999). These results raise the interesting possibility that ATM is involved in ROI-detoxification. In support of this hypothesis, we found that more ROI are generated in ATM-/- DT40 cells than in wild-type DT40 cells, following treatment with several apoptotic stimuli, and that a reduced level of catalase activity was observed in ATM-/- DT40 cells (Takao et al., 2000a). These results are also consistent with those of a recent study that showed increased levels of oxidant-modified products, decreased levels of catalase and enhanced heme oxygenase activity in ATM-deficient cells and tissues (Barlow et al., 1999; Watters et al., 1999). However, at the present time, we cannot exclude the possibility that ATM also functions in anti-apoptotic pathways.

1.4   Rad17, Rad9 and ATR/Chk1 in replication checkpoint control

While ATM-/- DT40 cells were highly sensitive to X-rays, they did not display sensitivity to UV, MMS, cis-platinum, and 4-NQO (Kobayashi et al., 2004). ATM, thus, seems to be more specifically involved in checkpoint responses to DSB in vertebrate cells, and ATR seems to play essential roles in checkpoint responses to DNA damage that do not induce DSB directly but rather cause DNA replication to stall. However, the lethality of ATR gene disruption both in mice and DT40 cells precludes detailed biochemical and more quantitative analysis for ATR at the cellular level. We therefore disrupted the *Rad17* and *Rad9* loci by gene targeting in the chicken B lymphocyte line DT40. The Rad17-replication factor C (Rad17-RFC) and Rad9-Rad1-Hus1 complexes are thought to function in the early phase of cell-cycle checkpoint control as sensors for ATR (Kondo et al., 1999; Caspari et al., 2000; Kondo et al., 2001; Melo et al., 2001). We found that Rad17-/- and Rad9-/- DT40 cells are highly sensitive to UV irradiation, MMS, *cis*-platinum, and 4-nitroquinoline oxide (4-NQO), which all stall DNA replication. Consistent with these results, we further found that Rad17-/- and Rad9-/- DT40 cells are defective in the slowing of the S-phase progression in response to these DNA-damaging agents (Kobayashi et al., 2004). Using Hus1-/-p21-/- MEFs, Hus1 has also been shown to be required for the repression of DNA replication that is induced in response to another genotoxic agent, benzo(a)pyrene dihydrodiol epoxide (BPDE), which causes bulky DNA adducts and stalls DNA replication (Weiss et al., 2003). However, Rad17-/- DT40 cells (unpublished data) and Rad9-/- or Hus1-/- MEFs (Roos-Mattjus et al., 2003; Weiss et al., 2003) did not display a clear defect in the IR-induced S-phase checkpoint control. These results therefore indicate that the vertebrate Rad17-RFC and 9-1-1 complexes are essential for the S-phase checkpoint response to UV irradiation, MMS, 4-NQO, *cis*-platinum, BPDE, and other genotoxic agents that do not induce DSB directly but rather cause DNA replication to stall. Although genetic studies are not available for vertebrate ATR, the results of studies with mammalian cells expressing kinase-dead ATR indicate that ATR also plays an important role in the checkpoint response to UV irradiation (Cliby et al., 1998; Heffernan et al., 2002). A more recent study with mammalian cells expressing kinase-dead ATR or Chk1 further have shown that ATR and Chk1 play critical roles in the S-phase checkpoint response to UV irradiation (Heffernan et al., 2002). These results therefore indicate a functional link between the Rad17-RFC/9-1-1 complexes and ATR/Chk1 in the S-phase checkpoint response to these types of DNA damage in vertebrate cells.

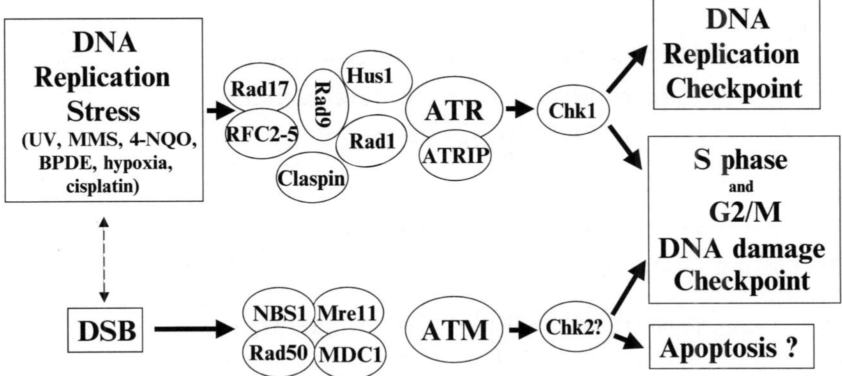

Figure 19-2. A schematic diagram for checkpoint responses to various DNA damage in higher vertebrate cells.

1.5   Future perspective

It is now clear that ATM plays important roles not only in cellular response to DSB DNA damage but also in the maintenance of the cell homeostasis in response to oxidative stress. Defective cellular response to oxidative stress may underlies the pathogenesis of cerebellar ataxia, premature aging and cancer predisposition in AT. However, it remains to be established how oxidative stress activates ATM. In particular, it is important to study whether or not ATM responds to oxidative damage to DNA or responds directly to a change in intracellular redox state independent of DNA damage.

c-Abl and Arg is activated by DNA damage in an ATM-dependent manner and regulates homologous recombination DNA repair by phosphorylating Rad51. However, we have been unable so far to detect the direct phosphorylation of Arg by ATM, though ATM can directly phosphorylate c-Abl. The dependency of Arg activation on ATM may therefore be more complex, involving some scaffold molecules such as BRCA1 and TOPBP1, or other kinases downstream of ATM, such as Chk2.

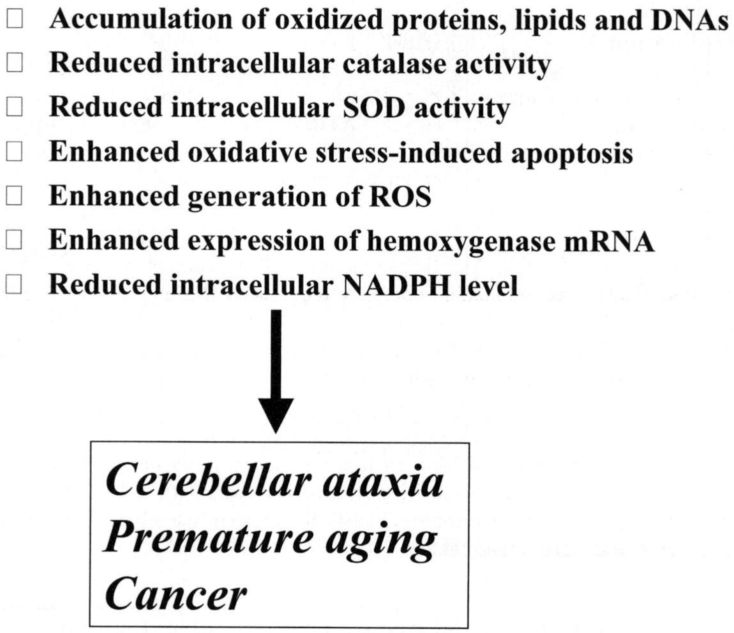

☐ **Accumulation of oxidized proteins, lipids and DNAs**
☐ **Reduced intracellular catalase activity**
☐ **Reduced intracellular SOD activity**
☐ **Enhanced oxidative stress-induced apoptosis**
☐ **Enhanced generation of ROS**
☐ **Enhanced expression of hemoxygenase mRNA**
☐ **Reduced intracellular NADPH level**

$$\downarrow$$

> *Cerebellar ataxia*
> *Premature aging*
> *Cancer*

Figure 19-3. Defective cellular response to oxidative stress may underlies the pathogenesis of cerebellar ataxia, premature aging and cancer in ataxia-telangiectasia.

REFERENCES

Abraham RT. 2001. Cell cycle checkpoint signaling through the ATM and ATR kinases. Genes & Dev. 15:2177-2196.

Agami R, Blandino G, Oren M, Shaul Y. 1999. Interaction of c-Abl and p73a and their collaboration to induce apoptosis. Nature 399:809-813.

Barlow C, Dennery PA, Shigenaga MK, Smith MA, Morrow JD, Roberts LJI, Wynshaw-Boris A, Levine RL. 1999. Loss of the ataxia-telangiectasia gene product causes oxidative damage in target organs. Proc. Natl. Acad. Sci. U.S.A. 96:9915-9919.

Barlow C, Hirotsune S, Paylor R, Liyanage M, Eckhaus M, Collins F, Shiloh Y, Crawley JN, Ried T, Tagle D, Wynshaw-Boris A. 1996. *Atm*-deficient mice: a paradigm of ataxia telangiectasia. Cell 86:159-171.

Baskaran R, Wood LD, Whitaker LL, Canman CE, Morgan SE, Xu Y, Barlow C, Baltimore D, Wynshaw-Boris A, Kastan MB, Wang JYJ. 1997. Ataxia telangiectasia mutant protein activates c-Abl tyrosine kinase in response to ionizing radiation. Nature 387:516-519.

Brown EJ, Baltimore D. 2000. ATR disruption leads to chromosomal fragmentation and early embryonic lethality. Genes & Dev. 14:397-402.

Buerstedde J-M, Takeda S. 1991. Increased ration of targeted to random integration after transfection of chicken B cell lines. Cell 67:179-188.

Buttke TM, Sandstorm PA. 1994. Oxidative stress as a mediator of apoptosis. Immunol. Today 15:7-10.

Caspari T, Dahlen M, Kanter-Smoler G, Lindsay HD, Hofmann K, Papadimitriou K, Sunnerhagen P, Carr AM. 2000. Characterzation of Schizosaccharomyces pombe Hus1: a PCNA-related protein that associates with Rad1 and Rad9. Mol. Cell. Biol. 74:1254-1262.

Cazzaniga G, Tosi S, Aloisi A, Giudici G, Daniotti M, Pioltelli P, Kearney L, Biondi A. 1999. The tyrosine kinase Abl-related gene ARG is fused to ETV6 in an AML-M4Eo patient with a t(1;12)(q25;p13): molecular cloning of both reciprocal transcripts. Blood 94: 4370-4373.

Chen G, Yuan SS, Liu W, Xu Y, Trujillo K, Song B, Cong F, Goff SP, Wu Y, Arlinghaus R, Baltimore D, Gasser PJ, Park MS, Sung P, Lee EY. 1999. Radiation-induced assembly of Rad51 and Rad52 recombination complex requires ATM and c-Abl. J Biol Chem 274:12748-12752.

Cliby WA, Roberts CJ, Cimprich KA, Stringer CM, Lamb JR, Schreiber SL, Friend SH. 1998. Overexpression of a kinase-inactive ATR protein causes sensitivity to DNA-damaging agents and defects in cell cycle checkpoints. Embo J 17:159-169.

Garcia-Ruiz C, Collel A, Mari M, Morales A, Fernandez-Checa JC. 1997. Direct effect of ceramide on the mitochondrial electron transport chain leads to generation of reactive oxygene species. Role of mitochondrial glutathion. J. Biol. Chem. 272:11369-11377.

Goff SP, Gilboa E, Witte ON, Baltimore D. 1980. Structure of the Abelson murine leukemia virus genome and the homologous cellular gene: studies with cloned viral DNA. Cell 22:777-785.

Golub EI, Kovalenko OV, Gupta RC, Ward DC, Radding CM. 1997. Interaction of human recombination proteins Rad51 and Rad54. Nucleic Acids Res. 25:4106-4110.

Gong J, Costanzo A, Yang H-Q, Melino G, Kaelin WGJ, Levrero M, Wang JYJ. 1999. The tyrosine kinase c-Abl regulates p73 in apoptic response to cisplatin-induced DNA damage. Nature 399:806-809.

Heffernan TP, Simpson DA, Frank AR, Heinloch AN, Paules RS, Cordeiro-Stone M, Kaufmann WK. 2002. An ATR- and Chk1-dependent S checkpoint inhibits replication initiation following UVC-induced DNA damage. Mol. Biol. Cell 22:8552-8561.

Hofmann K, Dixit VM. 1998. Ceramide in apoptosis-does it really matter? TIBS 23:374-377.

Iijima Y, Ito T, Oikawa T, Eguchi M, Eguchi-Ishimae M, Kamada N, Kishi K, Asano S, Sakai Y, Sato Y. 2000. A new ETV6/TEL partner gene, ARG (ABL-related gene or ABL2), identified in an AML-M3 cell line with a t(1;12)(q25;p13) translocation. Blood 95:2126-2131.

Jeggo PA, Carr AM, Lehmann AR. 1998. Splitting the ATM: distinct repair and checkpoint defects in ataxia-telangiectasia. Trends Genet. 14:312-316.

Kobayashi M, Hirano A, Kumano T, Xiang S, Mihara K, haseda Y, Matsui O, Shimizu H, Yamamoto K. 2004. Critical role for chicken Rad17 and Rad9 in the cellular response to DNA damage and stalled DNA replication. Genes to Cells 9:291-303.

Kolesnick R, Hannun YA. 1999. Ceramide and apoptosis. TIBS 24:224-225.

Kondo T, Matsumoto K, Sugimoto K. 1999. Role of a complex containing Rad17, Mec3, and Ddc1 in the yeast DNA damage checkpoint pathway. Mol. Cell. Biol. 19:1136-1143.

Kondo T, Wakayama T, Naiki T, Matsumoto K, Sugimoto K. 2001. Recruitment of Mec1 and Ddc1 checkpoint proteins to double-strand breaks through distinct mechanisms. Science 294:867-870.

Kruh GD, Perego R, Miki T, Aaronson SA. 1990. The complete coding sequence of *arg* defines the Abelson subfamily of cytoplasmic tyrosine kinases. Proc. Natl. Acad. Sci. USA 87:5802-5806.

Lavin MF, Shiloh Y. 1997. The genetic defect in ataxia telangiectasia. Annu. Rev. Immunol. 15:177-202.

Li Y, Shimizu H, Xiang S, Maru Y, Takao N, Yamamoto K. 2002. Arg tyrosine kinase is involved in homologous recombinational DNA repair. Biochem. Biophys. Res. Comm. 299:697-702.

Liu Z-G, Baskaran R, Lea-Chou ET, Wood LD, Chen Y, Karin M, Wang JYJ. 1996. Three distinct signalling responses by murine fibroblasts to genotoxic stress. Nature 384: 273-276.

Melo JA, Cohen J, Toczyski DP. 2001. Two checkpoint complexes are independently recruited to sites of DNA damage in vivo. Genes & Dev. 15:2809-2821.

Meyn MS. 1995. Ataxia telangietasia and cellular responses to DNA damage. Cancer Res. 55:5991-6001.

Morrison C, Sonoda E, Takao N, Shinohara A, Yamamoto K, Takeda S. 2000. The controlling role of ATM in recombinational repair of DNA damage. EMBO J. 19:463-471.

Quillet-Mary A, Jaffrezou J-P, Mansat V, Bordier C, Naval J, Laurent G. 1997. Implication of mitochondrial hydrogene peroxide generation in ceramide-induced apoptosis. J. Biol. Chem. 272:21388-21395.

Reichenbach J, Schubert R, Schwan C, Muller K, Bohles HJ, Zielen S. 1999. Anti-oxidative capacity in patient with ataxia telangiectasia. Clin. Exp. Immunol. 117:535-539.

Roos-Mattjus P, Hopkins KM, Oestreich AJ, Vroman BT, Johnson KL, Naylor S, Lieberman HB, Karnitz LM. 2003. Phosphorylation of human Rad9 is required for genotoxin-activated checkpoint signaling. J. Biol. Chem. 278:24428-24437.

Rouse J, Jackson SP. 2002. Interfaces between the detection, signaling, and repair of DNA damage. Science 297:547-551.

Sawyers CL, McLauphlin J, Goga A, Havlik M, Witte ON. 1994. The nuclear tyrosine kinase c-Abl negatively regulates cell growth. Cell 77:121-131.

Shafman T, Khanna KK, Kedar P, Spring K, Kozlov S, Yen T, Hobson K, Gatei M, Zhang N, Watters D, Egerton M, Shiloh Y, Kharbanda S, Kufe D, Lavin MF. 1997. Interaction between ATM protein and c-Abl in response to DNA damage. Nature 387:520-523.

Shiloh Y, Kastan MB. 2001. ATM: genome stability, neuronal development, and cancer cross paths. Adv. Cancer Res. 83:209-254.

Shinohara A, Ogawa T. 1995. Homologous recombination and the roles of double-strand breaks. Trends Biochem. Sci. 20:387-391.

Slupianek A, Schmutte C, Tombline G, Nieborowska-Skorska M, Hoser G, Nowicki MO, Pierce AJ, Fishel R, Skorski T. 2001. BCR/ABL regulates mammalian RecA homologs, resulting in drug resistance. Mol. Cell 8:795-806.

Sonoda E, Sasaki MS, Buerstedde J-M, Bezzubova O, Shinohara A, Ogawa H, Takata M, Yamaguchi-Iwai Y, Takeda S. 1998. Rad51-deficient vertebrate cells accumulate chromosomal breaks prior to cell death. EMBO J. 17:598-608.

Sui G, Affar EB, Shi Y, Brignone C, Wall NR, Yin P, Donohoe M, Luke MP, Calvo D, Grossman SR, Shi Y. 2004. Yin Yang 1 is a negative regulator of p53. Cell 117:859-872.

Takao N, Kato H, Mori R, Morrison C, Sonoda E, Sun X, Shimizu H, Yoshioka K, Takeda S, Yamamoto K. 1999. Disruption of ATM in p53-null cells causes multiple functional abnormalities in cellular response to ionizing radiation. Oncogene 18:7002-7009.

Takao N, Li Y, Yamamoto K. 2000a. Protective roles for ATM in cellular response to oxidative stress. FEBS Letters 472:133-136.

Takao N, Mori R, Kato H, Shinohara A, Yamamoto K. 2000b. c-Abl tyrosine kinase is not essential for ataxia telangiectasia mutated functions in chromosomal maintenance. J. Biol. Chem. 275:725-728.

Tan TLR, Essers J, Citterio E, Swagemakers SMA, De Wit J, Benson FE, Hoeijmakers JHJ, Kanaar R. 1999. Mouse Rad54 affects DNAcondormation and DNA-damage-induced Rad51 foci formation. Curr. Biol. 9:325-328.

Watters D, Kedar P, Spring K, Bjorkman J, Chen P, Gatei M, Birrel G, Garrone B, Srinivasa P, Crane DI, Lavin MF. 1999. Localization of a portion of extranuclear ATM to peroxisomes. J. Biol. Chem. 274:34277-34282.

Weiss RS, Leder P, Vaziri C. 2003. Critical role for mouse Hus1 in an S-phase DNA damage cell cycle checkpoint. Mol. Biol. Cell 23:791-803.

Xu Y, Baltimore D. 1996. Dual roles of ATM in the cellular response to radiation and in cell growth control. Genes Devel. 10:2401-2410.

Yuan Z, Shioya H, Ishiko T, Sun X, Gu J, Huang Y, Lu H, Kharbanda S, Weichsekbaum R, Kufe D. 1999. p73 is regulated by tyrosine kinase c-Abl in the apoptotic response to DNA damage. Nature 399:814-817.

Yuan Z-M, Huang Y, Ishiko T, Nakada S, Utsugisawa T, Kharbanda S, Wang R, Sung P, Shinohara A, Weichselbaum R, Kufe D. 1998. Regulation of Rad51 function by c-Abl in response to DNA damage. J. biol. Chem. 273:3799-3802.

Yuan ZM, Huang Y, Ishiko T, Kharbanda S, Weichselbaum R, Kufe D. 1997. Regulation of DNA damage-induced apoptosis by the c-Abl tyrosine kinase. Proc Natl Acad Sci U S A 94:1437-1440.

Yuan ZM, Huang Y, Whang Y, Sawyers C, Weichselbaum R, Kharbanda S, Kufe D. 1996. Role for c-Abl tyrosine kinase in growth arrest response to DNA damage. Nature 382: 272-274.

Zhou BS, Elledge SJ. 2000. The DNA damage response:putting checkpoints in perspective. Nature 408:433-439.

Method 1

STABLE NON-TARGETED TRANSFECTION OF DT40

Julian E. Sale
M.R.C. Laboratory of Molecular Biology, Hills Road, Cambridge, CB2 2QH, U.K.

Abstract: The introduction of stably integrated transgenes into cells is a central technique in any system for genetic analysis, in particular for complementation and for overexpression. This protocol makes use of relatively high electroporation energies that result in DNA breaks and which favour non-targeted integration of the construct into the genome.

Key words: Protocol, stable transfection.

Materials

Reagents
DT40 cell culture medium
2x antibiotic selection medium
Trypan blue
Phosphate buffered saline (PBS)

Equipment
96 well flat-bottom microtiter plates
15 ml tubes
Electroporator
4mm electroporation cuvettes
Laminar flow hood
Humidified CO2 incubator
multichannel pipette
Pipette aid

J.-M. Buerstedde and S. Takeda (eds.), Reviews and Protocols in DT40 Research, 341–344.
© 2006 *Springer.*

Method

1. Precipitate 10-30 µg of the plasmid to be transfected in a 1.5ml tube with 0.1vol 3M sodium acetate pH 5.2 and 2.5vol 100% ethanol, following normal molecular biological practice.
2. Pellet the DNA by centrifugation and wash with 70% ethanol. It is a good idea to remove this wash in the flow hood to keep the DNA sterile.
3. Allow the DNA pellet to dry in the flow hood and resuspend in 50µl sterile ddH$_2$O.
4. Pellet c. 2 x 10^7 cells at 250g for 5 minutes then wash once with cold PBS. From this stage on, keep the cells on ice.
5. After the wash, resuspend the cells in 600µl cold PBS.
6. Thoroughly mix the cells with the DNA solution and transfer to the electroporation cuvette.
7. Incubate the cells and DNA on ice for 20 minutes.
8. Vortex briefly then electroporate at 250V, 950µF. The time constant will be in the order of 20 - 25ms.
9. Immediately return the cells to ice for 5 minutes.
10. Using a wide bore Pasteur pippette, transfer the cells to 50ml DT40 medium. It is normal to find free DNA from lysed cells after the electroporation.
11. Incubate for 18 - 24 hours.
12. Add 50ml 2x selection medium and plate the cells into flat-bottomed 96-well plates, placing 200µl in each well.
13. Incubate in a humidified atmosphere until colonies are visible, usually after about 10 days. Generally the number of cells used in this protocol will yield plenty of wells with a single clone per well and avoid the need for subsequent subcloning. However, it may be necessary to vary the plating density depending on the construct being transfected.

Selection media

G418 (geneticin sulphate)
Selection cassette: neo^r (neomycin phosphotransferase)
Final concentration: 2mg/ml
Method for 2x media: Dissolve 2g G418 in c. 10ml water. Add to 500ml DT40 growth medium and adjust pH with 4N NaOH (need about 1ml).
Selection characteristics: G418 is rather slow to kill so cells may continue to divide and metabolise after selection for some time. This can result in depletion of nutrients in the medium and cell death. This is usually detectable by the media turning yellow. In this case it may be necessary to

feed the cells by removing c. 150µl medium with a multichannel pippete and then replacing it with fresh 1x selection medium. Having to do this does have the disadvantage of disrupting clone formation, making it difficult to tell if there is more than one clone in a well.

Puromycin
Selection cassette: puro^r or pac (puromycin acetyltransferase)
Final concentration: 0.5µg/ml
Method for 2x media: Puromycin stock solution is 2mg/ml in water, kept in aliquots at -20°C. Avoid freeze-thawing. For 2x selection medium, dilute puromycin stock solution 1:2000 in growth medium (i.e. 25µl per 50ml). Mix well and sterile filter.
Selection characteristics: Generally clean (i.e. drug resistant clones are clearly distinguishable from surrounding cells), but cells can take a number of days to stop metabolising completely.

Blasticidin S
Selection cassette: bs^r or bsd (blasticidin S deaminase)
Final concentration: 20µg/ml
*Method for 2x media:*Stock solution is 50mg/ml, kept at -20°C. The blasticidin has a tendency to precipitate out whilst frozen. The stock should be vortexed before use to ensure that the chemical is fully dissolved. For 2x selection medium, dilute blasticidin stock solution 1:1250 in growth medium (i.e. 40µl per 50ml). Mix well and sterile filter.
Selection characteristics: Generally very rapid and clean.

Histidinol
Selection cassette: his^r or hisD (histidinol dehydrogenase)
Final concentration: 20µg/ml
Method for 2x media: Stock is 200mg/ml in water. For 2x selection medium, dilute the stock 1:100 in growth medium. Adjust the pH with 4N NaOH. Sterile filter.
Selection characteristics: Killing is extremely rapid and selection is clean.

Hygromycin B
Selection cassette: hyg^r (hygromycin phosphotransferase)
Final concentration: 1.4mg/ml
Method for 2x media: Hygromycin usually comes as a brown liquid and concentrations are often presented as units/ml. In this case it is necessary to obtain the equivalence in mg/ml from the data sheet or manufacturer. For 2x selection medium, add the appropriate amount of hygromycin solution to give 2.8mg/ml. Adjust the pH with hydrochloric acid.

Selection characteristics: Generally very rapid and clean as long as dosing is accurate.

Mycophenolic acid
*Selection cassette:*gpt
Final concentration: 15µg/ml (plus xanthine 250µg/ml and hypoxanthine 20µg/ml)
Method for 2x media: Stock mycophenolic acid is 50mg/ml in methanol. In addition to the mycophenolic acid at 30µg/ml, the 2x selective medium is supplemented with xanthine 0.5mg/ml and hypoxanthine 40µg/ml. The stock solution of xanthine is 5mg/ml in 50mM NaOH and of hypoxanthine 2mg/ml in 10mM NaOH. It is therefore necessary to adjust the pH of the 2x selection medium with hydrochloric acid.
Selection characteristics: Generally rapid and clean.

Notes & troubleshooting

Unlike for targeted integration, it is not necessary to linearise the plasmid construct.

It is important that the volume of DNA in water does not exceed 10% of the total volume to avoid problems with hypotonicity.

For some of these antibiotics, the amount of active drug per unit mass or volume can vary between suppliers and batches. The final concentrations given above are for guidance and may need adjusting depending on supplier. If in doubt a titration should be carried out.

The electroporation conditions used in this protocol delivers roughly ten times the energy of the conditions used for transfection for targeted integration (550V, 25µF). This higher energy level results in more DNA breaks, which may help promote random integration. In some DNA damage sensitive mutants, it may be necessary to reduce the total energy delivered to prevent excessive cell death.

BASIC CELL CULTURE CONDITIONS

Huseyin Saribasak and Hiroshi Arakawa
GSF, Institute for Molecular Radiobiology, Ingolstaedter Landstr. 1, D-85764 Neuherberg-Munich, Germany

Abstract: The DT40 cell line is derived from ALV (avian leukosis virus)-transformed bursal B cell of chicken. This cell line is suspension cell, and can be cultured in cell culture bottles and in culture plates (6-, 24- and 96-well etc.). Optimal culture temperature for the DT40 is (39°C-) 41°C, because chicken has higher body temperature than human and mouse. Here we summarize protocols for culture medium and condition of the DT40 cell line.

Key words: Culture Conditions, Chicken Medium, Freezing Medium.

Materials

Reagents: Chicken Medium
RPMI
FBS
Penicillin (10000 units/ml)/Streptomycin (10mg/ml) solution (Gibco/BRL Cat No 15140-114)
200 mM L-Glutamin solution (Gibco/BRL Cat No 25030-024)
Chicken serum (Sigma Cat No C-5405
1 M β-mercaptoethanol solution

Reagents: Freezing Medium
RPMI
FBS
DMSO

Equipment
Laminar flow hood
Humidified CO_2 incubator

J.-M. Buerstedde and S. Takeda (eds.), Reviews and Protocols in DT40 Research, 345–346.

Method

1) DT40 cells can be cultured in cultures flasks, petri dishes, or in 24 well plates. Microtiter plates are suitable for transfection or subcloning. The optimum culture condition for the cells is $41°C$ with $5\%\,CO_2$.
2) For preparing chicken medium, add 50 ml of FBS, 10 ml of Penicillin (10000 units/ml)/Streptomycin (10mg/ml), 5 ml of 200 mM L-Glutamin, 5 ml of chicken serum and 0.05 ml of 1 M β-mercaptoethanol solution into 500 ml of RPM solution.
3) Freezing medium can be prepared by mixing 70 ml of RPMI, 20 ml of FBS and 10 ml of DMSO. Chicken medium can also be used instead of RPMI.

Troubleshooting

In order to control serum and other materials and to avoid fungi or yeast contamination each bottle of chicken serum needs to be tested. This can be done as follows:

1) Aliquate 1 ml of chicken medium into two wells of 24 well plate.
2) Add around 10000 cells of DT40 into 1 ml of chicken medium leaving another well as a blank control.
3) Track the efficiency of DT40 growth over the following days.

Method 3

EXCISION OF FLOXED-DNA SEQUENCES BY TRANSIENT INDUCTION OF MER-CRE-MER

Hiroshi Arakawa

GSF, Institute for Molecular Radiobiology, Ingolstaedter Landstr. 1, D-85764 Neuherberg-Munich, Germany

Abstract: Disruption of multiple genes and complementation of the phenotypes is restricted by the number of available drug-resistance genes. It is therefore highly desirable to recycle the drug-resistance genes using a site-specific recombination system like Cre/loxP. DT40^{Cre1} cell line, which is transgenic with an inducible Cre recombinase (MerCreMer) is useful for transient induction of Cre/loxP recombination. MerCreMer is active in the presence of the Mer ligand, 4-hydroxy tamoxifen, but sequestered in an inactive form by heat shock proteins in the absence of the ligand.

Key words: Protocol, Cre, loxP, drug-resistance marker recycling.

Materials

Reagents
4-hydroxy tamoxifen (SIGMA)
DT40 cell culture medium
selection drug

Equipment
Laminar flow hood
Humidified CO2 incubator

Sample
DT40^{Cre1} stably transfected with floxed vectors

J.-M. Buerstedde and S. Takeda (eds.), Reviews and Protocols in DT40 Research, 347–349.

Method

Protocol for complete excision out of floxed cassettes

1) You need around 10^5 cells of DT40^{Cre1} stably transfected with floxed vectors.

2) Culture the cells in 1 ml of chicken medium containing 0.01 mM 4-hydroxt tamoxifen for 2 days.

3) Subclone the cells with limiting dilution for final concentration of 0.3, 1, 3 and 10 cells/well in 96-well flat-bottom plates (see Protocol 'subcloning').

4) Around 7 – 8 days after subcloning, subclones can be observed as round colonies on flat-bottom 96-well plates.

5) Transfer 10 µl of stable transfectants into 1 ml of chicken medium. Then make duplicate (or triplicate) of the cells in 200 µl of chicken medium or selection drug-containing medium in flat-bottom 96-well plates in order to test the excision out of the drug-resistance gene cassettes.

Protocol for partial excision out of floxed cassettes

1) You need around 10^5 cells of DT40^{Cre1} stably transfected with floxed vectors.

2) Culture the cells in 1 ml of chicken medium containing 0.01 mM 4-hydroxt tamoxifen for overnight.

3) For partial excision of the floxed cassettes, you can culture the cells in selective-drug culture medium. For example, if you want to excise out bsr cassette but do not want to excise out puroR cassette, you can culture tamoxifen induced cells with puroR medium. Subclone the cells with limiting dilution for final concentration of 1, 3, 10, 30, 100 and 300 cells/well in 96-well flat-bottom plates (see Protocol 'subcloning').

4) Around 7 – 8 days after subcloning, subclones can be observed as round colonies on flat-bottom 96-well plates.

5) Transfer 10 µl of stable transfectants into 1 ml of chicken medium. Then make duplicate (or triplicate) of the cells in 200 µl of chicken medium or selection drug-containing medium in flat-bottom 96-well plates in order to test the excision out of the drug-resistance gene cassettes.

Troubleshooting

It is important to use a well humidified incubator to prevent evaporation of medium. Efficiency of Cre/loxP recombination depends on the distance of two loxP sites. Duration of tamoxifen induction needs to be optimized depending on the distance between two loxP sites.

Method 4

IMMUNOGLOBULIN GENE CONVERSION AND HYPERMUTATION ASSAY BY FACS

Hiroshi Arakawa

GSF, Institute for Molecular Radiobiology, Ingolstaedter Landstr. 1, D-85764 Neuherberg-Munich, Germany

Abstract: Immunoglobulin (Ig) gene conversion and hypermutation rates can be quantified by FACS. The spontaneous DT40 variant, Cl18, and Cl18-derived DT40 mutants can be used to analyze gene conversion phenotype. The Cl18 has a frameshift in its rearranged V segment. When this frameshift mutation is repaired by pseudogene-templated gene conversion, sIgM(–) DT40 reverts to sIgM (+) (Ig reversion assay). The sIgM (+) pseudogene knockout DT40 clone can be used to analyze Ig hypermutation phenotype. Pseudogene knockout DT40 accumulates Ig hypermutation, rapidly generating sIgM (–) population (sIg loss assay). These changes of sIgM expression can be measured by FACS. To minimize the fluctuation of gene conversion or hypermutation events, it is needed to analyze many subclones of the DT40 mutant for calculating average sIgM (–) or (+) percentage.

Key words: Protocol, gene conversion, hypermutation.

Materials

Reagents
anti-chicken Cμ monoclonal antibody M1 (Southern Biotechnology Associates)
R-PE-conjugated goat anti-mouse IgG polyclonal antibody (Southern Biotechnology Associates) [Another fluorescence-conjugated anti-mouse IgG polyclonal antibodies are also available.]
FACS staining buffer (PBS, 10% FCS, 0.05% sodium azide)
PBS

J.-M. Buerstedde and S. Takeda (eds.), Reviews and Protocols in DT40 Research, 351–352.
© 2006 *Springer.*

Equipment
multi-channel pippette
centrifuge for microtiter plate
flow cytometer

Sample
Cell lines for gene conversion or hypermutation assay

Method

Subcloning

1) Subclone DT40 mutant clones for gene conversion or hypermutation phenotype analysis by limiting dilution (see Protocol Subcloning').
2) Incubate the plates for 7-8 days without changing medium.
3) Pick-up 24 colonies for each DT40 clone. Transfer 10 µl of colonies into 1 ml of chicken medium in 24-well flat-bottom plates.
4) After 2 or 3 days, you need to change medium in order to avoid overgrowth of the cells.

Antibody staining

5) Two-weeks after subcloning, you can stain the subclones by anti-chicken Cµ antibody. Remove approximately 3/4 of supernatant by aspiration without mixing, and transfer the cells into round-bottom 96-well plates.
6) Once wash by PBS.
7) Incubate with 1/100 fold - 1/500 fold-diluted anti-chicken Cµ monoclonal antibody M1 on ice for 30 min. Twice wash with PBS.
8) Incubate with 1/100-diluted R-PE-conjugated goat anti-mouse IgG polyclonal antibody on ice for 30 min. Three times wash with PBS.
9) Analyze sIgM (+) or (–) percentage by flow cytomer.

Troubleshooting

Antibody badge and dilution needs to be tested and optimized before these FACS assay. Dying or dead cells can cause fluorescence pseudo-positive population because of auto-fluorescence or non-specific binding of antibodies. In order to avoid such background, the cells need to be carefully cultured for avoiding overgrow.

Predominantly sIgM(–) or (+) subclones needs to be excluded from the analysis, since they most likely originated from cells that were already sIgM(–) or (+) at the time of subcloning.

Method 5

TARGET SCREENING BY PCR

Hiroshi Arakawa
GSF, Institute for Molecular Radiobiology, Ingolstaedter Landstr. 1, D-85764 Neuherberg-Munich, Germany

Abstract: After the stable transfection of the knockout vector, the transfectants successful for targeted integration can be screened by PCR. A primer located upstream of the 5' targeting arm of the construct can be used for this PCR together with a primer from the resistance marker. PCR screening is more advantageous than Southern screening not only for speed but also for scaling up the screening.

Key words: Protocol, target screening.

Materials

Reagents
proteinase K
tween 20
PCR polymerase (Expand Long Template PCR System [Roche] recommended)
NTP
PCR primers designed for target screening
96 well PCR plates

Equipment
themal cycler for 96-well plate

Sample
DT40 clones stably transfected with knockout vector

J.-M. Buersteade and S. Takeda (eds.), Reviews and Protocols in DT40 Research, 353–354.
© 2006 *Springer.*

Method

Cell preparation

1) Transfect knockout vector into the DT40 cell line or its mutant (see Protocol 'electroporation'). Around 7 – 10 days after electroporation of knockout vector, stable transfectants can be observed as round colonies on flat-bottom 96-well plates.
2) Transfer 10 µl of stable transfectants into 300 µl of chicken medium in flat-bottom 96-well plates.
3) Incubate for 3 days.

Crude extract

4) Once wash the cells with PBS in 96-well PCR plates.
5) Resuspend the cells in 10 µl of K buffer [1 x PCR buffer, 0.1 mg/ml proteinase K and 0.5% Tween 20].
6) Incubate at 56°C for 45 min for proteinase K-mediated proteolysis.
7) Incubate at 95°C for 10 min to inactivate the proteinase K. Use 1µl of the crude extract for PCR.

PCR

8) Targeted integration can be screened by PCR using a primer located upstream of the 5' targeting arm of the construct together with a primer from the resistance marker. Expand Long Template PCR System (Roche) is recommended for PCR screening.

PCR can be performed with long-range PCR protocol: 2 min initial incubation at 93°C, 35 cycles consisting of 93°C for 10 sec, 65°C for 30 sec and 68°C for 5 min with cycle elongation of 20 sec per cycle, and a final 5 min elongation step at 68°C.

Troubleshooting

In order to achieve PCR products efficiently, primer sequence and annealing temperature needs to be optimized. The region, which is supposed to be deleted after targeted gene disruption, can serve as PCR positive control for heterozygous knockout screening and as PCR negative control for homozygous knockout screening.

Method 6

MITOTIC INDEX DETERMINATION
BY FLOW CYTOMETRY

David A.F. Gillespie and Mark Walker
CR-UK Beatson Laboratories, Beatson Institute for Cancer Research, Garscube Estate, Bearsden, Glasgow G61 1BD

Abstract: Chromosome condensation during mitosis is associated with phosphorylation of histone H3 at serine 10 ($pS^{10}H3$). Detection of $pS^{10}H3$ phosphorylation using phospho-specific antibodies in combination with DNA content flow cytometry provides a rapid and accurate means of determining the percentage of mitotic cells in a population. Using the spindle poison nocodazole to trap mitotic cells it is possible to determine the rate at which cells accumulate in mitosis with time. By comparing the rates of mitotic accumulation before and after DNA damage it is possible to gauge the efficiency with which the G2 checkpoint blocks entry to mitosis. Because DNA content is measured simultaneously with $pS^{10}H3$ fluorescence this method can also be used to identify cells which enter mitosis with incompletely replicated DNA as a result of S-M checkpoint failure during DNA synthesis inhibition.

Key words: Protocol, mitosis, cell cycle.

Materials

Reagents
Bovine Serum Albumin (BSA;Sigma A7888)
Phosphate Buffered Saline (PBS)
70% Ethanol
Nocodazole (Sigma M1404;stock 10mg/ ml in DMSO)
Propidium Iodide (PI;stock 1mg/ml in H2O)
Ribonuclease A (RNase A; Qiagen 19101; 100mg/ml)
Triton X-100 (Sigma T9284)
Rabbit polyclonal anti-phospho Serine 10 Histone H3 antibody (anti-$pS^{10}H3$;Upstate Ltd Cat No. 06-570)

J.-M. Buerstedde and S. Takeda (eds.), Reviews and Protocols in DT40 Research, 355–358.

Goat polyclonal anti-rabbit IgG FITC conjugate (anti-rabbit IgG FITC; Stratech Scientific Cat No. 111-096-047)

Equipment
15ml blue-capped tubes
polystyrene FACS tubes
Bench-top centrifuge
Vortex
Flow cytometer

Method *(Adapted from Xu et al., Mol. Cell. Biol. 22:1049-1059)*

1) Treat cells as required with nocodazole (1μg/ ml final concentration) plus or minus DNA damaging agents, DNA replication inhibitors etc according to experimental design.
2) Remove approximately 2×10^6 cells to a 15ml blue-capped tube and pellet at 1000rpm for 5min in bench-top centrifuge.
3) Resuspend cells in 200μl PBS then add 2mls of 70% EtOH at -20°C with vortexing. Leave at -20°C for at least 3h or ideally overnight.
4) Pellet cells at 1000rpm for 5min.
5) Resuspend cells in 1ml of PBS containing 0.25% Triton X-100 and incubate on ice for 15min.
6) Pellet cells at 1000rpm for 5min and resuspend in 100μl PBS/ 1%BSA containing anti-pS^{10}H3 antibody (1:30 dilution).
7) Incubate cells for 2h at room temperature.
8) Pellet cells at 1000rpm for 5min and wash with 1ml PBS/ 1%BSA - pellet cells again at 1000rpm for 5min.
9) Resuspend the cells in PBS/ 1%BSA containing goat-anti-rabbit IgG FITC conjugate (1:30dilution).
10) Incubate cells for 30min at room temperature in the dark.
11) Pellet cells 1000rpm for 5min, wash with 1ml PBS and pellet again - 1000rpm for 5min.
12) Resuspend cells in 1ml PBS containing PI at 25μg/ml and RNaseA at 250μg/ml.
13) Incubate cells for at least 30 min at room temperature in the dark.
14) Analyse by flow cytometry using two gates to pre-filter the data as shown in Fig. 1. Gate R1 on a FSC versus SSC dotplot followed by R2 on an FL2-H versus FL2-W dotplot. Acquisition of 20,000 gated events in R2 is generally sufficient for accurate determinations. The percentage of mitotic cells in the R2 population is then determined using R3 to gate the pS^{10}H3-positive population in an FL2-H versus FL1-H dotplot.

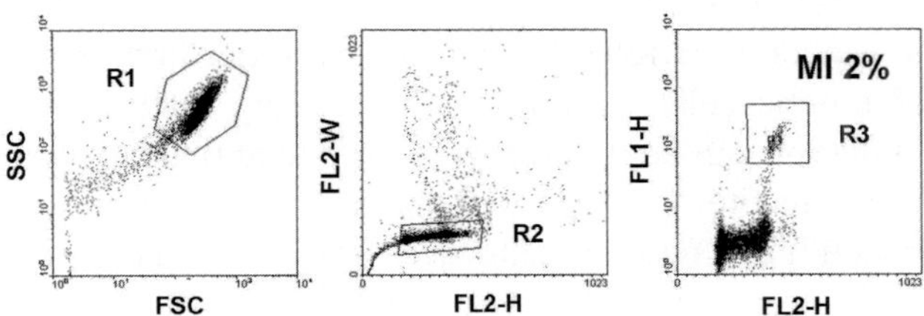

Figure 6-1. Plots and gating used for determination of mitotic indices using anti-pS10H3 staining and DNA content flow cytometry pS10H3-positive mitotic cells appear in gated region R3 and typically constitute between 2 and 4% of a DT40 culture during exponential growth phase.

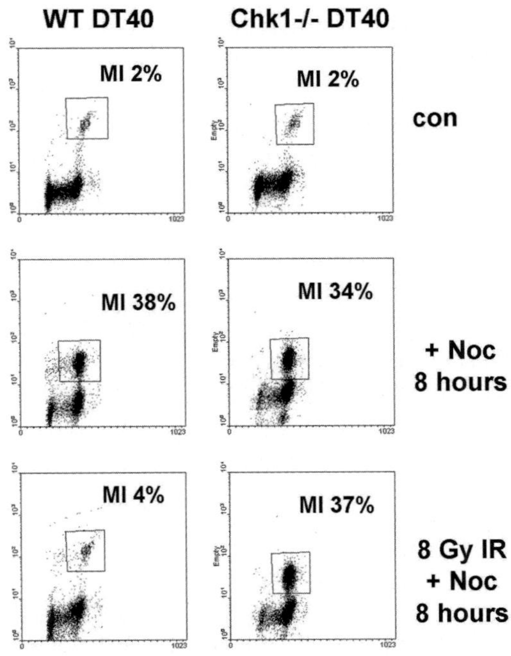

Figure 6-2. Effect of γ-irradation (8 Grays) on mitotic progression in WT and Chk1-/- DT40 cells. Cultures were incubated with nocodazole for 8 hours with or without irradiation immediately prior to addition of drug. Mitotic accumulation is effectively blocked by irradiation in WT DT40 cells owing to G2 checkpoint activation which is dependent on Chk1.

Notes and Troubleshooting

This is generally a reliable and robust procedure, however if large numbers of mitotic cells are present the staining intensity of the $pS^{10}H3$-positive population can be diminished owing to antibody starvation (evident to some degree in Fig. 6-1). In extreme cases this can make it difficult to distinguish and gate mitotic cells accurately – if necessary reduce the number of cells processed in affected samples.

Method 7

CENTRIFUGAL ELUTRIATION AS A MEANS OF CELL CYCLE PHASE SEPARATION AND SYNCHRONISATION

David A.F. Gillespie and Catarina Henriques
CR-UK Beatson Laboratories, Beatson Institute for Cancer Research, Garscube Estate, Bearsden, Glasgow G61 1BD

Abstract: The ability to purify cells in specific phases of the cell cycle in sufficient quantities for biochemical analysis or to obtain a "synchronized" population of living cells enriched in a particular phase is of great value for studying a wide range of cell cycle processes. Transformed tumour cell lines, such as DT40, can be arrested and released from specific points in the cell cycle using drugs which inhibit DNA or protein synthesis or interfere with mitotic spindle function, however such drugs are frequently toxic and may themselves perturb the cell cycle process or phenomenon under study. Centrifugal elutriation provides a means of separating an unperturbed culture of living cells into highly enriched G1-, S-, and G2/ M-phase fractions. The resulting cell fractions can be analysed directly for protein and mRNA expression or returned to culture and their subsequent progression through the cell cycle monitored by flow cytometry, microscopy, or other techniques.

Key words: Protocol, cell cycle, synchronization.

Materials

Reagents
2-4 litres of DT40 growth medium
Phosphate Buffered Saline (PBS)
70% ethanol

J.-M. Buerstedde and S. Takeda (eds.), Reviews and Protocols in DT40 Research, 359–361.
© 2006 *Springer.*

Equipment
Beckman JE-6B or JE-5.0 centrifugal elutriation rotor with standard (4ml) chamber plus centrifuge and peristaltic pump
250ml conical bottom plastic centrifuge bottles
15ml blue-capped tubes
Sterile 20ml plastic Universal bottles
Sterile 10ml plastic syringes

Method

Centrifugal elutriation separates a population of spherical objects of similar density according to size by subjecting them to opposing sedimentation (centrifugal) and frictional (flow) forces in a separation chamber contained within a specialised centrifuge rotor. The ratio of surface area to mass varies according to object size, therefore it is possible to separate small objects from large by varying the flow rate and/ or centrifugation speed. Because cells grow continuously as they progress from birth to mitosis, centrifugal elutriation effectively separates cells according to their position in the cell cycle.

We elutriate DT40 cells in growth medium at room temperature using a constant flow rate (40 ml/ min) and slowing the rotor to predetermined speeds to elute the desired fraction(s) of cells. The appropriate rotor speeds must be determined empirically and may vary according to the genotype of the cells. We suggest loading the cells at 4000rpm and collecting 150ml fractions as the rotor is slowed at intervals of 200 or 250rpm to 2000rpm as an initial separation test (see Fig. 7-1).

1) Assemble and prepare the rotor according to the manufacturer's instructions. Harvest the cells to be separated (10^9 maximum) by centrifugation and resuspend in 7ml of medium in a 20ml universal bottle. Begin to pump medium at room temperature through the rotor setting the flow rate to 40mls/ min (check using a measuring cylinder and timer) then start the rotor and adjust to the loading speed.

2) Disperse any clumps by pipetting the cell suspension gently a few times using a 10ml plastic syringe then introduce the cells into the rotor using the loading harness according to the instruction manual.

3) Allow approximately 450mls of medium to pass through the rotor, then slow the rotor to the desired elutriation speed(s) collecting 150ml fractions into a 250ml plastic centrifuge bottle at each interval.

4) Remove 10-15mls of each fraction to a 15ml blue-capped tube, collect the cells by centrifugation at 1000rpm for 5 min, resuspend in 200µl PBS and add 2mls of 70% ethanol at –20°C with vortexing. Process for flow cytometry as described in Method # 20.

5) Cells in the remainder of each fraction are recovered by centrifugation and can either be returned to culture or processed for protein or RNA extraction etc as appropriate.

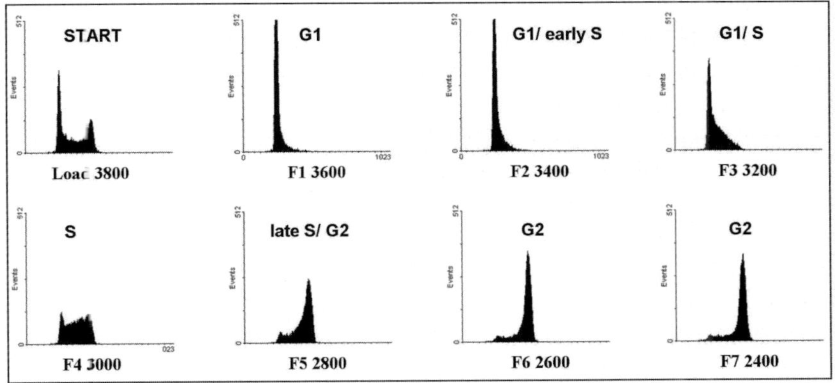

Figure 7-1. Example separation of 2 x 108 WT DT40 cells using a JE-5.0 rotor. Cells were loaded at 3,800 rpm and 150 ml fractions collected at the indicated speeds. 1/10 of each fraction was fixed and analysed by flow cytometry whilst the remainder was used to prepare cell extracts.

Notes and Troubleshooting

It is difficult to maintain complete sterility during elutriation, however by sterilising the rotor (follow manufacturer's instructions) and using sterile pipettes, tubes etc it is possible to culture elutriated cells for up to 24 hours without detectable contamination. If the pH of the elutriation medium becomes excessively alkaline through exposure to air it may be necessary to increase the buffering capacity (use 7.5% sodium bicarbonate at 50ml/ litre and HCl to neutralise). For biochemical analyses it may be desirable to elutriate at 4°C, however the rotor speeds at which specific cell fractions are eluted will differ from room temperature owing to changes in the viscosity of the medium.

Method 8

PREPARATION OF GENOMIC DNA FOR MICROARRAY-BASED COMPARATIVE GENOME HYBRIDIZATION

Robert Kimmel, Amalia Icreverzi and Paul Neiman
Fred Hutchinson Cancer Research Center, Seattle, Washington, U.S.A.

Abstract: Genome-wide assessment of DNA gains and losses can be accomplished by comparative hybridization using a variety of microarray platforms that employ oligo, cDNA, BAC and other sequences as probes. Here we describe the preparation of genomic DNA for hybridization to a set of chicken cDNA probes spotted on glass slide microarrays. *Method 1* can be used to assess DNA copy-number differences between two genomes, typically an experimental genome and a normal, diploid genome. We then present a specialized application of array CGH for detecting DNA amplifications containing long, inverted repeat structures (palindromes). *Method 2* describes the procedure for palindrome enrichment and the internal controls used to distinguish direct versus inverted, long repeat structures using the cDNA microarray platform.

Key words: Array CGH, genomic DNA, palindrome, protocol.

1. INTRODUCTION

Microarray-based comparative genome hybridization (array CGH) has been shown to yield DNA copy-number profiles having greater spatial and amplitude resolution than conventional, cytogenetic CGH (Pollack et al., 1999). In cDNA array CGH, two target genomes are compared by co-hybridizing their fluorescence-labeled representations to EST probe sequences spotted on polylysine-coated glass slides. When induced to fluoresce, the target genomes emit unique colors, one for the experimental

J.-M. Buerstedde and S. Takeda (eds.), Reviews and Protocols in DT40 Research, 363–371.
© 2006 *Springer.*

sample and the other for the control. The two signals are then combined electronically to produce a pseudocolor representation for visualization and a corresponding set of fluorescence ratios (FRs) that are linearly related to DNA copy-number ratios between the two samples. However, variations in array spot performance, and in the quality of input DNA, can result in noise and/or systematic bias in the output FRs. Noise can be dealt with in several ways, including careful study design, statistical analysis, high-quality microarray production, high-quality input DNA and enrichment for target sequences.

Array CGH has been used to detect genome-wide formation of palindromes (GAPF) (Tanaka et al., 2005). We have developed an array-based, structural assay for palindrome detection called palindrome-enriched array CGH (PEACH). This approach distinguishes a palindrome structure from a direct repeat configuration by use of internal control constructs that enable the monitoring of palindrome fold-back versus spontaneous reassociation of denatured DNA during the enrichment process. Enrichment is enhanced by digestion of non-duplex DNA and amplification of palindrome segments by ligation mediated-PCR. PEACH combines a DNA structural assay with the mapping power of the array CGH platform to detect less than one copy of a long palindrome per genome. PEACH results in a set of qualitative FRs that are very high for palindromes and very low for other sequences.

This chapter presents a basic protocol for the preparation of genomic DNA for input into the array CGH assay. We have obtained excellent array CGH results using genomic DNA extracted by various methods, including salt-based as well as classical, phenol-chloroform-based protocols. We then present a specialized application of array CGH that compares samples that are highly enriched for long, palindrome sequences that are thought to have a role in gene amplification (Ford and Fried, 1986; Tanaka et al., 2002). Long palindromes are best enriched from very high-quality, phenol-chloroform-extracted genomic DNA.

Materials for Array CGH

Equipment
Allcentrifugationsarecarriedoutinabench-topmicro-centrifugeattopspeed.
Microcon YM-30 Millipore Concentrators (Bedford, MA)
Ultra-free MC Amicon 0.45μm filters (Millipore #UFC30HV25, Bedford, MA)
QIAquick PCR Clean-Up Kit (Qiagen;, Valencia, CA)
37°C. and 65°C. water baths
96-100°C. heating block

Reagents
Nuclease-free distilled H_2O
*Dpn*II restriction enzyme 10,000U/mL and 10X buffer
Klenow 40U/μL (Bioprime DNA Labeling System, Invitrogen, Carlsbad, CA)
2.5x random primers solution (octamers) (Bioprime, Invitrogen, Carlsbad, CA)
TE, pH 8.0 & pH 7.4
10x BSA (from 100X purified)
Cy3-dUTP and Cy5-dUTP (Amersham, Piscataway, NJ)
dNTPs mix, Invitrogen 100mM each A, G, C, T (Carlsbad, CA)
yeast tRNA 10μg/μL
species-specific Cot-1 DNA 1μg/μL (optional)
Poly (dA:dT) at 5μg/μL (Amersham, Piscataway, NJ)

Samples for Array CGH
Genomic DNA, 100 – 2000 ng, from each of the two tissue samples to be compared.

Method 1: Array CGH

In brief, equal amounts of experimental and control gDNA are separately restricted, random primer-labeled and combined with blockers of repetitive sequences prior to hybridization on the cDNA microarray.

Prepare DNA and Random Primer Mix
1. Digest experimental and control genomic DNA samples to completion with *Dpn*II.
2. Clean up samples using QIAquick PCR Purification Kit (Qiagen), elute and check yield by spectrophotometry.
3. Ethanol precipitate, wash and resuspend each at 2.2 μg per 5μL 10 mM Tris, pH 8.0.
4. Add 20 μL 2.5X random primer mix to 5 μL DNA.

Prepare Labeling Master Mix
1. Make intermediate 10 x dNTP mix such that dATP, dGTP, dCTP are at 1.2 mM each and dTTP is at 0.6 mM:
   1 μL 100 mM dTTP
   2 μL 100 mM dATP
   2 μL 100 mM dGTP
   2 μL 100 mM dCTP
   160 μL water
2. 10 x Buffer (10 mM Tris, pH 7.5, 1 mM disodium EDTA)
   10 μL 1M Tris, pH 7.5

    2 μL 0.5M EDTA, pH 8.5
    988 μL water (qs 1000)
3.   Master Mix (22μL total per reaction)
    5 μL 10 x dNTP mix
    1 μL Klenow 40 U/μL
    5 μL 10 x Buffer
    11 μL water

Labeling Reaction
1. Denature DNA/primers at 100°C. for 5 min; quench on ice.
2. Combine the following (50μL total):
    25 μL denatured DNA/primer mix
    3 μL Cy5 or Cy3-dUTP
    22μL Master Mix
3. Incubate at 37°C. for 1.5 hours.

Clean Up
1. Increase volume and add blockers:
    440 μL TE pH 7.4
    25 μL Chicken Cot-1 DNA (1 μg/μL) (optional)
    4 μL Poly (dA-dT) 5 μg/μL
    10 μL yeast tRNA 10 μg/μL
2. Spin in Microcon YM-30 concentrator to a final volume of 10-15μL (typically 8-12 min.).
3. Invert and elute at 12K rpm for 2 min. Check volume and record.
4. Combine Cy5 and Cy3 pairs.
5. Add water qs 30.6 μL
6. Add 5.4 μL of 20xSSC, (3xSSC final).
7. Pre-wet a Microcon Ultrafree-MC 0.45μm filter unit with TE pH 7.4. Spin to remove TE.
8. Place each sample on side of cylinder and quickly place it in the rotor and spin for 2 min.
9. Deliver on ice to array lab or freeze at –20°C.
10. SDS (0.22% final) is added (after last thaw) prior to hybridization.

Materials for Palindrome-Enriched Array CGH

The following additional materials are needed for the application of array CGH to the detection of long, inverted repeats in palindrome-enriched genomic DNA.

Equipment
PCR thermocycler
30°C. water bath

Reagents
50 mM MgCl₂, PCR grade
10Mm Tris, 1mM EDTA (TE) , pH 7.4
Glycogen 20mg/mL Roche (Indianapolis, IN)
T4 Ligase HC 5U/μL and 5X buffer Invitrogen (Carlsbad, CA)
Taq DNA polymerase 5U/μL and 10x PCR buffer minus MgCl₂ Invitrogen
Cy3- and Cy5- 5'-labeled primers 0.2μmol MWG (High Point, NC)
Mung bean nuclease 10,000U/mL and 10X buffer, New England Biolabs
(Beverly,MA)
Linker adapter, annealed top and bottom strands (SA1T & SA2B) MWG
DMSO (Sigma, St. Louis, MO)

Samples for Palindrome-Enriched Array CGH
100 – 2000ng high-quality experimental and control genomic DNA.
Maximum yield of long, inverted repeat sequences is best obtained from
phenol- chloroform extracted genomic DNA (5).

Preparation for Palindrome-Enriched Array CGH

1. *Structural control constructs:* Potentially any DNA sequence can be
   used as the basis for the control constructs provided the sequence is
   represented among the microarray probes. We used the cDNA
   sequence from a human EST (Gene Bank Accession R98985) for
   oxidative stress response 1 (OSR1) to construct a direct repeat,
   pTOD, and an inverted repeat, or palindrome, pTOI. Briefly, a 1460
   bp sequence of OSR1 was ligated into pT7T3D-Pac1-OSR1 in an
   inverted or direct configuration such that the homologous segments
   of OSR1 were separated by 138 bp. When denatured and rapidly
   renatured, *Sca*I linearized pTOI forms a hairpin structure consisting
   of a 1460 bp duplex stem, a 138 bp loop, and two single-stranded
   tails (~1800 and ~1100 bp). Unlike pTOI, *Sca*I linearized pTOD
   cannot fold back on itself. The OSR1 EST was spotted on the
   chicken glass slide cDNA microarray to probe for the fold-back of
   linearized pTOI and the spontaneous reassociation of linearized
   pTOD. Linearized constructs were gel purified, phenol/chloroform
   extracted and tested for their ability to form fold-back or reassociated
   structures by agarose gel electrophoresis prior to use in this assay.

2. *Linker adapter:* Top strand SA1T 5'- TTCACTACACACC TCATCCTTCT-3' and bottom strand SA2B 5'-GATCAGAAGGA-3' linker component oligos were annealed slowly by gradually allowing equimolar quantities to cool from 100°C. to RT. The reconstituted linker adapter was frozen in single-use aliquots at 100pmol/μL.
3. *Labeled primers:* Cy3- and Cy5- 5'-labeled primers must match the top strand of the linker adapter to perform efficient PCR amplification. Re-suspend labeled primes at a concentration of 100pmol/μL in TE, pH 7.3, in single-use aliquots.

Method 2: Palindrome Enrichment and Labeling Procedure

In brief, typically 1 to 30 copies of pTOI/*Sca*I (pTOD/*Sca*I) are added to the experimental (control) gDNA, although it makes no difference which channel receives which control construct. Both samples are separately denatured and rapidly renatured on ice, thus favoring palindrome fold-back formation over the reassociation of untethered, single-stranded sequences. In rapid succession, the samples are digested with ssDNA-specific endonuclease, restricted with *Dpn*II, and ligated to linker adapters suitable for ligation-mediated-PCR (LM-PCR) using specific Cy-dye-labeled primers. PCR products are then combined with blockers of repetitive sequences prior to hybridization on the cDNA microarray.

Add internal controls to gDNA samples
1. Working dilutions of gDNA, pTOI and pTOD are in 100mM NaCl.
2. Combine 100ng (2000ng) gDNA with desired number of molar equivalent copies of control construct per mole of gDNA in a final volume of 10 (200) μL.

DNA denaturation and rapid renaturation
3. Denature gDNA/internal control at 100°C. for 10 minutes. Quench on ice for 90 seconds.

Nuclease digestion and clean-up
4. Add 8 (20) units of Mung bean nuclease to each tube.
5. Incubate at 30°C. for 10 minutes (or as empirically determined; see *Troubleshooting*).
6. Phenol-chloroform extraction, ethanol precipitation (with 1μL glycogen).

Rapid restriction digestion
7. Resuspend pellet with 18 units of *Dpn*II in 18μL.
8. Mix and incubate for 5 min at 37°C.
9. Inactivate *Dpn*II at 65°C. for 20 min.

Rapid Ligation
10. Cool tubes to RT, add 3.1 mcL linker (final 900 pmol/mcL), 10 units of ligase, and ligase buffer, in a volume of 10 mcL, to make a total ligase reaction volume of 28 mcL.

LM-PCR
11. Perform PCR in 70μL volume with 12.5 units Taq Polymerase, 10X PCR buffer, 3.5μL 50mM $MgCl_2$, 7μL BSA 10X, 3.5μL DMSO, 1.5μL 10mM dNTP mix, 2.8μL labeled primer.
12. Program 72°C. for 3 min. (to remove bottom linker strand and fill in), 94°C. for 3 min., *28 cycles:* 94°C. x 30 sec., 64°C. x 30 sec., 72°C. x 2 min., extend for 72°C. x 5 min.
13. *Clean-up and final preparation* as per Method 1.

Interpretation of OSR1 signals: Microarray hybridization and scanning are beyond the scope of this chapter. However, interpretation of the structural reporter signals is key to understanding and troubleshooting of this protocol. pTOI is the reporter for a palindrome and will produce a strong signal from the OSR1 microarray spot if its integrity is maintained throughout the assay. However, it will also produce a strong signal if it is not denatured and simply remains in its original linearized form, since it will be restricted, ligated, amplified and labeled. pTOD is the reporter for reassociation of a direct repeat and will produce only a background OSR1 signal if such reassociation is prevented. pTOD will produce a strong signal if it is not denatured or if significant reassociation occurs. By comparing OSR1 signals from both channels it is thus possible to monitor the behavior of direct and inverted repeat structures in this assay. The ideal array result is a set of fluorescence ratios that are low to slightly above background for most spots, high for OSR1 (due to high pTOI channel signal and very low pTOD channel signal) and reproducibly high for genomic palindromes that are present in the sample for one channel and not present in the other. In effect, palindromes result in bright spots in a sea of dim spots. Individual experiments are informative without the need for elaborate statistical analysis.

1.1 Troubleshooting

General considerations: Make all master mixes and dilute linkers, primers and gDNA to their working concentrations prior to starting the assay to facilitate accurate timing of each step of the assay. Nuclease digestion, restriction digestion and ligation reactions are intentionally not carried to completion for two reasons. Keeping the reactions short in duration minimizes spontaneous reassociation of non-palindrome DNA. Additionally, brief single-strand nuclease digestion minimizes the effects of degradation of duplex DNA. Although the latter is a much slower reaction, it can effectively destroy all the DNA.

Very low or no signals from most microarray spots: In this situation, most spots show little or no signal from either channel but some very strong, often reproducible, bright spots may be present. This indicates generalized failure of the assay. Contributing causes include expiration of the labeled primers, nuclease over-digestion or PCR failure. Primers should be ordered lyophilized and split into single-use aliquots after reconstitution in 7.3 pH TE. The shelf life is no more than one month after reconstitution. Mung bean nuclease activity exhibits considerable lot to lot variability despite unit definition. It may be necessary to test various digestion times from 5 to 20 minutes. Use only high-quality PCR reagents, with special attention to minimizing the freeze-thaw cycles of dNTPs. Finally, it may be necessary to increase input DNA.

No OSR1 microarray signal from pTOI: This situation is probably due to expired labeled primer (shelf life only about one month), excessive freeze-thaw cycles of primers, linker, or linearized control constructs (make single-use aliquots) or a PCR inhibitor (perform at least two chloroform extractions after phenol/chloroform extractions). Another common cause is the improper performance of serial dilutions of genomic samples. We recommend performing not more than 10-fold dilutions with the sample and diluent heated to 65°C.

High OSR1 microarray signal from pTOD: This indicates significant spontaneous reassociation of untethered, denatured DNA fragments or inadequate initial denaturation. Contributing causes include failure to accurately time each step of the assay or molecular crowding secondary to excessively high concentration of DNA in the working solutions or at the organic-aqueous interface of phenol/chloroform extractions. In difficult cases, it may be necessary to increase the volume in which the denaturation and rapid renaturation steps are performed, since the fold-back reaction is of zero-order kinetics, while reassociation of untethered DNA fragments is concentration-dependent.

REFERENCES

Pollack JR, Perou CM, Alizadeh AA, Eisen MB, Pergamenschikov A, Williams CF, Jeffrey SS, Botstein D, Brown PO. 1999. Genome-wide analysis of DNA copy-number changes using cDNA microarrays. Nat Genet 23:41-46.

Tanaka H, Bergstrom DA, Yao MC, Taps cott SJ 2005. Widespread and nonrandom distribution of DNA palindromes in cancer cells provides a structural platform for subsequent gene amplification. Nat Genet 37:320-327.

Ford M, Fried M. 1986. Large inverted duplications are associated with gene amplification. Cell 45:425-430.

Tanaka H, Tapscott SJ, Trask BJ, Yao MC. 2002. Short inverted repeats initiate gene amplification through the formation of a large DNA palindrome in mammalian cells. Proc Natl Acad Sci U S A 99:8772-8777.

Method 9

ANALYSIS OF CELLULAR Mg^{2+} IN DT40 CELLS

Anne-Laure Perraud[1], Carsten Schmitz[1] and Andrew M. Scharenberg[2]
1 Dept. of Immunology, University of Colorado and National Jewish Medical and Research Center, Denver, CO; 2 Depts. of Pediatrics and Immunology, University of Washington, Seattle, WA

Abstract: Study of Mg^{2+} homeostasis in continuously growing cells requires the capacity to measure total cellular Mg^{2+} and net Mg^{2+} fluxes per unit time. Our laboratory's protocols for measurement of total cellular Mg^{2+} by atomic absorption spectrophotometry and measurement of net Mg^{2+} fluxes using the stable Mg^{2+} isotope ^{26}Mg are described below.

Key words: Protocol, Mg^{2+}, magnesium homeostasis, magnesium uptake, ^{26}Mg.

1) Analysis of total cellular Mg^{2+} using atomic absorption spectrophotometry

Materials

Reagents
DT40 cell culture medium
Mg^{2+} free DT40 cell culture medium

Equipment
Laminar flow hood
Humidified CO$_2$ incubator
Standard laboratory pipetting equipment

Sample
Cell culture to be analyzed

J.-M. Buerstedde and S. Takeda (eds.), Reviews and Protocols in DT40 Research, 373–378.
© 2006 Springer.

Methods

a) Cell preparation and lysis

1) Cells are grown under the control and test conditions. In our laboratory, DT40 cells are grown in RPMI supplemented with either 10% FBS and 1% chicken serum or both types of sera. The test conditions will typically involve some type of manipulation of the extracellular $[Mg^{2+}]$ via the addition of varying amounts of MgSO4.

2) To initiate an experiment, each culture is started in their respective conditions with 2 x 10^6 cells in 10 mls of media and allowed to grow in the indicated media for the desired time.

3) After incubation, cells are spun down in a 15 ml falcon tube, and resuspended in 1 ml regular media in a microcentrifuge tube. They are then centrifuged for 30 seconds (approximately, minimum required to create a stable pellet), and then resuspended in 1 ml of PBS or Ringer's type buffer without Mg^{2+}, and repelleted in the same manner.

4) For lysis, cells are resuspended in 200 ul of 5% trichloroacetic acid (TCA), with a combination of vigorous pipetting and vortexing to fully extract the cell pellet.

5) The extractate is next centrifuged for 5 minutes at high speed to pellet the insoluble material, and the supernatant is carefully removed.

6) The supernatant is assayed for Mg^{2+} content using an atomic absorption spectrophotometer and appropriate standards to generate a standard curve encompassing the sample values. For our atomic absorption instrument, which utilized a carbon furnace as opposed to a flame, samples were diluted 1:100 prior to analysis.

b) atomic absorption spectrophotometry

1) The specifics of performing the atomic absorption measurements will necessarily depend on the instrument utilized, and are therefore will not be discussed here.

2) Standard curve preparation - For our samples, a standard curve was generated utilizing Mg^{2+} concentrations in μg per ml, as our goal was to quantitate μg of Mg^{2+} per cell. Typically, our max concentration standard was 1000 μg per ml, and our minimum was 25 μg per ml, with a no added Mg^{2+} sample serving as a "0" point. The standard curve was linear over this range in all experiments.

3) Data analysis

Once absorption measurements and an appropriate standard curve (e.g. linear over a range encompassing the sample values) are available, concentrations are derived from the standard curve by linear regression/interpolation.

For most experiments, we performed normalization of the resulting value of Mg^{2+} (in µg) on a per-cell basis by using a combination of visual counts of morphologically normal appearing cells and trypan blue staining. This was judged to be the most useful way to assess net cellular changes in total Mg^{2+}in our continuous growing cell system, as it would account for changes in Mg^{2+} uptake associated with changes in cell size/volume at different growth rates. However, for other systems or experimental conditions in which cell division rate or size changes are not likely to vary, it may be simpler to perform protein assays and normalize on a per mg protein basis.

Troubleshooting

We have observed that differential carry over of Mg^{2+} from control and test conditions in which the Mg^{2+} concentrations are significantly different is the most important variable affecting assay reproducibility. We performed several experiments to demonstrate that no significant carry over was occurring from our experimental growth media, and found the wash conditions above to be adequate. However, it is recommended that each investigator test this under their own laboratory conditions. If carryover is observed, then an additional PBS/0 Mg^{2+} Ringers wash may be added and should eliminate the problem.

2) Analysis of net Mg^{2+} flux using ^{26}Mg uptake measured with induction coupled plasma mass spectrometry (ICP-MS)

Materials

Reagents
Mg^{2+} free DT40 cell culture medium
^{26}MgO (available from Isotec, a division of Sigma-Aldrich, or Isoflex)

Equipment
Laminar flow hood
Humidified CO_2 incubator
Standard laboratory pipetting equipment

Sample
Cell culture to be analyzed

Methods

a) Dissolution of MgO
1) MgO is insoluble in water, and must be dissolved in HCL prior to use.
2) It is advisable to do this under a hood, and to take great care in the addition of the HCl solution, as the MgO/HCl reaction is vigorously exothermic and somewhat explosive, and a lack of care may result in particulization and loss of valuable ^{26}Mg material.

b) Caveats to media preparation and incubation conditions
1) Depending on the length of time of the desired measurement, one may utilize differing amounts of $^{26}Mg^{2+}$ enrichment in the incubation media. We have typically chosen to utilize smaller volumes of media highly enriched ($\approx$ 85%) in $^{26}Mg^{2+}$ in order to minimize our incubation times. However, our signals are of sufficient magnitudes that lesser degrees of enrichment could probably be tolerated.
2) Fetal bovine serum (FBS) contains Mg^{2+} in the natural isotope ratio. In order to minimize the influence of this on the degree of $^{26}Mg^{2+}$ enrichment of our media, we have typically reduced the serum concentration of our incubation media to 1%. However, as we have also noted that the magnitude of $^{26}Mg^{2+}$ uptake in DT40 cells is easily detectable even within a 20-40 minute time frame, it is probably reasonable to use higher amounts of FBS, accepting a slight decrease in signal due to the reduced incubation media isotope ratio.

c) Cell incubation lysis
1) 1×10^{7} cells are grown in their normal conditions, and spun down and washed once in 1 ml of test condition media to ensure that all incubation media are at a consistent isotope ratio.

2) The cells are pelleted and resuspended in 1 ml of the test condition media to initiate short incubations (0-2 hours). For longer incubations, we place the cells into 3 ml's of media.

   In our laboratory, the test media is created by adding a defined amount of ^{26}Mg^{2+} and 1% chicken serum to Mg^{2+}-free RPMI.

3) After the desired time of incubation, cells are spun down in a 15 ml falcon tube (for longer incubations) or a 1 ml microfuge tube, and resuspended in 1 ml regular media in a microcentrifuge tube. They are then centrifuged for 30 seconds (approximately, minimum required to create a stable pellet), resuspended in 1 ml of PBS or Ringer's type buffer without Mg^{2+}, and repelleted in the same manner.

4) For lysis, cells are resuspended in 200 ul of 0.1N HCl, with a combination of vigorous pipetting and vortexing to fully extract the cell pellet. This suspension is incubated overnight at room temperature to complete the extraction

5) The extractate is centrifuged for 5 minutes at high speed to pellet the insoluble material, and the supernatant is carefully removed.

6) The supernatant is then prepared for ICP-MS as below.

d) Preparation of cell lysates for ICP-MS

   1) The presence of organic substances may interfere with the ICP-MS measurements, or may cause problems with clogging of the ICP-MS vaporizer. Consequently, sample preparation for ICP-MS involves repeated treatment and lyophilization of the sample with distilled HNO3 to oxidize all sample organics.

   2) Our protocol has used a course of three repeated HNO3/lyophilization treatments, followed by resuspension of the sample in 200 ul of dI H$_2$O. However, different ICP-MS operators may have different protocols for oxidization of organic material, and it is likely that you will need to work with them to generate a protocol they are comfortable with.

e) ICP-MS

   1) The ICP-MS instrument we have used to assay our samples is a seven-collector VG Sector, and the performance characteristics of this instrument allows very high precision assays on small samples. It may be that other instruments will require larger sample sizes than specified in this protocol.

2) Other specifics of performing the atomic absorption measurements will necessarily depend on the instrument utilized, and therefore will not be discussed here.

Troubleshooting

As with the total Mg^{2+} measurements above, differential carry over of Mg^{2+} from control and test conditions in which the Mg^{2+} concentrations are significantly different is the most important variable affecting assay reproducibility, and can be handled in a similar manner by additional washings.

Method 10

TRANSIENT TRANSFECTION OF DT40

Roger Franklin and Julian E. Sale
M.R.C. Laboratory of Molecular Biology, Hills Road, Cambridge, CB2 2QH, U.K.

Abstract: Transient transfection of cell lines can be extremely useful in a number of applications. However, until recently, obtaining useful and reproducible percentages of transiently expressing DT40 was almost impossible. Standard techniques such as electroporation usually give low transfection efficiencies (can be up to 40%, but ususally less than 5%) while many liposome-based protocols result in the death of all cells. Recently Amaxa Biosystems GmbH (Cologne, Germany) have developed the 'nucleofector' system which has been shown to be effective in transfecting previously resistant cell lines. This system has, for the first time, allowed efficient and non-toxic transient transfection of DT40, routinely achieving transient transfection efficiencies of over 50%.

Key words: Protocol, transient transfection.

Materials

Amaxa Nucleofector II (Cat No. AAD-1001)
Amaxa Nucleofector Kit T (Cat No. VCA-1002)
benchtop centrifuge
humidified incubator
DT40 culture medium
6 well plates
15ml tubes

J.-M. Buerstedde and S. Takeda (eds.), Reviews and Protocols in DT40 Research, 379–382.
©2006 *Springer.*

Method

1. Grow up the required number of DT40 cells. This protocol is optimised for three million (3×10^6) per nucleofection.
2. Prepare the DNA. You will need 30µg DNA per nucleofection in a volume of not more than 15µl. This is normally obtained by ethanol precipitation and resuspension to the appropriate concentration.
3. Pre-warm standard DT40 media in 6 well plates ready for resuspension of cells following nucleofection. Use 5mls media per well.
4. Pre-warm the nucleofector solution T to room temperature.
5. Spin down three million cells per nucleofection in a 15ml Falcon tube at 250g for 5 minutes in an benchtop centrifuge. Use a separate tube for each nucleofection to be undertaken.
6. Suck off **all** the media and resuspend cells in 100µl of pre-warmed nucleofector solution T. **Do not** wash the cells in PBS or add cold nucleofector solution to the cells. Pipette the cells up and down gently to make sure they are fully resuspended.
7.   (a) Add 30µg of the DNA to be nucleofected in a volume of not more than 15µl. Mix the DNA with the cell suspension by gentle pipetting up and down.
   (b) Transfer the sample into an Amaxa cuvette (supplied in the Nucleofector kit T)
   (c) Gently tap the cuvette on the bench to ensure that the sample covers the bottom and that there are as few bubbles as possible. A few bubbles on the surface of the sample is ok.
   (d) Insert the cuvette into the Nucleofector II set to programme B-23. Press 'X' to activate the nucleofection.
   (e) Remove the cuvette from the nucleofector and add approximately 500µl pre-warmed media. Then transfer the whole sample to a well of pre-warmed media in a 6 well plate. To do this use the Pasteur pipettes provided in the kit to avoid damage to the cells. Incubate the cells under standard conditions for DT40.
8. Expression of the transfected gene should be detectable within 24hrs.

Notes and troubleshooting

The whole of step 7 should be performed separately for each sample. i.e. do not add DNA to all the samples and then perform all the nucleofections.

The number of samples should be small enough such that you can perform all the nucleofections within 15-20 minutes, after which the solution starts to kill DT40 cells. If you have a large number of samples (more than 16) you should perform the experiment in batches.

The most important parameter in this protocol is the ratio of DNA to number of cells. The higher this ratio, the higher the transfection efficiency but also the higher the level of cell death. 30µg of DNA is far more than the 1-5µg suggested in the protocol from Amaxa but we have found this modification gives a much higher transfection efficiency, normally around 60-70% as measured by FACS following transfection with a CMV-driven GFP plasmid.

For most applications, particularly where the detection method is FACS analysis, transfection efficiency and not survival is the most important parameter hence the use of 30µg of DNA with 3 million cells. However if cell or nuclear extracts must be prepared from cells following nucleofection survival may be more important. In this case it may be advisable to try using more cells and less DNA, for example 10 million cells with 15µg DNA. Alternatively multiple parallel nucleofections as described may be performed and the cells then pooled giving enough material for a cell/nuclear extract.

Another important feature of this protocol is the volume in which the DNA is dissolved. This should be as small as possible and must not exceed 15µl otherwise the nucleofector solution becomes diluted such that the carefully determined conditions are altered. It is also important to use high quality DNA with A260:A280 ratio of 1.8 or higher.

Although a massive improvement on liposome based approaches, this protocol is still toxic to DT40 and so cells should spend as little times as possible in the nucleofector solution. Poor transfection efficiencies are often caused by too lengthy exposure to the solution prior to nucleofection. Additionally it is crucial that the cells are very healthy before nucleofection: cells which have been overgrown or only recently defrosted are unsuitable for nucleofection.

Finally, two tips on the use of the Amaxa Nucleofectors. Both the Nucleofector II and the older Nucleofector I have a rotating wheel in which the cuvette is mounted prior to nucleofection. However, in the Nucleofector II this wheel automatically rotates on depression of the 'X' button which activates nucleofection. In the Nucleofector I the wheel must be rotated manually. It is essential that you **do not** attempt to rotate the wheel of the Nucleofector II manually.

When nucleofection is activated there is pause as the machine electroporates the cells. Following the pause the screen displays 'Ok' to indicate that the nucleofection has been successful. The length of the pause for electroporation often varies being long for the first sample and shorter thereafter. This is not a cause for concern: transfection efficiencies are independent of the length of this pause.

Method 11

RETROVIRAL TRANSDUCTION OF DT40

Felix Randow and Julian E. Sale
M.R.C. Laboratory of Molecular Biology, Hills Road, Cambridge, CB2 2QH, U.K.

Abstract: Retroviral transduction of DT40 provides an easy way to obtain population of cells stably expressing a transgene without causing cellular stress or the levels of cell death seen in transfection protocols. By employing Moloney Murine Leukemia Virus based constructs and pseudotyping the viral particles with VSV envelope glycoprotein, it is a highly efficient procedure, routinely resulting in transgene expression in the majority of cells. It is also rapid. From production of the transgene-containing virus to expression takes only four days.

Key words: Protocol, retroviral transduction, transgene expression.

Materials

Reagents & Solutions
293 cells
Growth medium for 293 cells
   [IMDM plus 10% fetal calf serum, penillin/streptomycin & 50μM 2-mercaptoethanol]
293 transfection vehicle e.g. Lipofectamine 2000® (Invitrogen)
Retroviral helper plasmids, e.g. pMD.GAGPOL and pMD.VSVG [1]
Retroviral expression plasmid e.g. M5P [2]
Polybrene [hexadimethrine bromide] 8mg/ml in water [0.2 μm filtered]
Liquid nitrogen
−70°C freezer
DT40 growth medium

J.-M. Buerstedde and S. Takeda (eds.), Reviews and Protocols in DT40 Research, 383–386.

Apparatus
Suitable containment facility
Benchtop centrifuge capable of carrying tissue culture plates
24 well plates
freezing vials

Method

Note this procedure is likely to require Category 2 containment facilities, depending on local regulations.

This protocol describes virus production in 24 well plates using Lipofectamine 2000 transfection of 293 cells, but the procedure can be scaled up easily if required.

Day 1

1. Split 293 cells for transfection on Day 2, with reference to the Lipofectamine 2000 transfection protocol i.e. aiming for 80% confluence at transfection.

Day 2

2. Transfect 293 cells following the Lipofectamine 2000 protocol with 0.5µg M5P, 0.35µg pMD.GAGPOL and 0.15µg pMD.VSVG.

Day 3

3. Change medium on 293 cells, replacing with 0.5ml medium per well of a 24 well plate.

Day 4

4. Beginning on Day 4 the supernatant containing virus can be harvested daily as long as a healthy culture can be maintained (2 - 4 days). Supernatants need to be freed from contaminating cells by either centrifugation or 0.2µm filtration. They can be snap-frozen in liquid nitrogen and stored at −70°C. Frozen supernatants can remain active for months, but repeated freeze-thaw cycles will dramatically decrease viral titres.

5. Place 2×10^5 DT40 cells in a 24 well plate in 0.5ml of standard DT40 growth medium containing 16µg/ml Polybrene.

6. Add 0.5ml viral supernatant. Since transduction is dependent on target cell proliferation and growth of 293 cells will have depleted nutrients, this is the maximum amount of viral supernatant it is prudent to add.

7. Wrap plates in Saran wrap and spin for 1 hour at room temperature and 700g.

8. Move plates into 37°C, 5-10% CO_2 humidified incubator.

Day 5

9. Change medium, replacing with standard DT40 medium. Split cells if necessary.

Day 6

10. Cells are ready for use. 48 hours after transduction expression levels will have plateaued.

Notes & Troubleshooting

This method relies on co-transfection of three plasmids into 293 cells. The retroviral expression plasmid resembles a provirus in which the viral genes gag, pol, and env have been replaced with the gene of interest (Figure 1). Only RNA expressed from this plasmid is packaged into viral particles due to the presence of a specific packaging signal, Ψ. This RNA is also translated and it is therefore possible to check expression of newly cloned genes by simply transfecting 293 cells. The viral genes deleted from the retroviral expression plasmid are provided by helper plasmids, which provide a source of protein but do not contribute their genetic information towards the viral particles. The resulting viruses are therefore infectious but replication incompetent. In order to infect chicken cells which are not a natural target of MLV, the envelope glycoprotein of Vesicular Stomatitis Virus, VSV-G, is used for pseudotyping. Note that human cells can also be transduced by VSV-G pseudotyped viruses.

It is important that the transfection of 293 is efficient, optimally above 90%. If this step does not work well, the resulting supernatant will be of low titre. This can be easily checked by either using a GFP-containing virus (for example M5P-GFP[2]), or indeed any GFP expression plasmid.

There are cell lines, such as NIH 3T3, which are easier to transduce than DT40 and these can provide a worthwhile control if transduction of DT40 does not work.

Production of a GFP virus control is also useful in measuring the titre of virus produced by the 293 cells by transduction into DT40.

If it is necessary to scale up virus production, 293 transfections can be done using HBS/calcium chloride instead of Lipofectamine 2000 as this will significantly reduce costs.

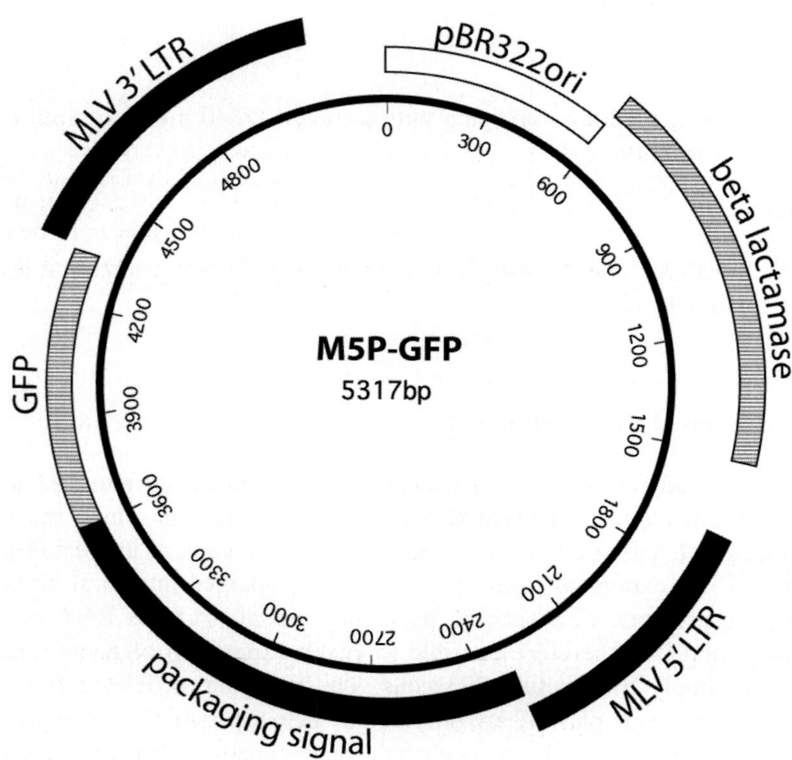

Figure 11-1. The retroviral expression plasmid M5P, containing the GFP cDNA cloned into the unique NcoI/NotI sites of the plasmid.

REFERENCES

1. Ory DS et al. (1996). A stable human-derived packaging cell line for production of high titer retrovirus/vesicular stomatitis virus G pseudotypes. Proc Natl Acad Sci USA. 93:11400-6.

2. Felix Randow, unpublished, but available on request.

Method 12

COLONY SURVIVAL ASSAY

Laura. J. Simpson and Julian E. Sale
M.R.C. Laboratory of Molecular Biology, Hills Road, Cambridge, CB2 2QH, U.K

Abstract: For studies of DNA repair networks the sensitivity of mutants and combinations of mutants to varying forms of DNA damaging agents has formed a mainstay of genetic analysis in bacteria and yeast. Likewise, this form of epistasis analysis has proved immensely informative in DT40. Because DT40 is non-adherent, it is necessary to restrict the movement of cells by growing them in a viscous medium containing methylcellulose. Here we present methods for carrying out DNA damage survival assays in DT40 with chemical mutagens, ionising radiation and ultraviolet irradiation.

Key words: Protocol, DNA damage, colony survival.

Materials

Reagents
D-MEM:F-12 powder with L-Glutamine (Gibco cat. 32500-043)
NaHCO$_3$
4000 centipoise methylcellulose
Fetal calf serum
Chicken serum
Penicillin/streptomycin
50µM 2-mercaptoethanol

J.-M. Buerstedde and S. Takeda (eds.), Reviews and Protocols in DT40 Research, 387–391.

Equipment
1 litre glass bottle
Magnetic stir bar
0.2µm vacuum filters
Laminar flow hood
Magnetic stir plate
Cold room
Humidified CO2 incubator
254 nm UV lamp and enclosure
X-ray or γ-ray set

Method (General)

Preparation of methylcellulose medium

1. Dissolve 11.9g D-MEM powder and 2.44g sodium bicarbonate in 500ml water. Sterilize by 0.2µm filtration.
2. Autoclave 10g methylcellulose powder and a stir bar in a 1L bottle.
3. Add approximately 220mls warm (50 – 65°C) sterile water to the methylcellulose powder. Shake well to mix, ensuring an even suspension. This is VERY important.
4. When temperature of methylcellulose solution is ≤50°C, add the DMEM/HAM solution from (1) little by little while mixing well. Shake the bottle vigorously.
5. Mix this solution overnight on a magnetic stirrer in a cold room.
6. Add the 0.2µm-filtered serum/antibiotic/2-mercaptoethanol cocktail.
7. Check pH is 7.0 to 7.4 and adjust with NaOH (4N solution) as necessary. This is not usually necessary. Top up to 1L with sterilised water.
8. Stir for at least 3 hours (preferably overnight) in a cold room and keep at 4°C for up to 1 month

Preparing the methylcellulose plates

1. Label the appropriate number of 6 well plates. One 6 well plate will be required for each dose for every cell line e.g. testing 4 cell lines at 5 different UV doses will require 20 6 well plates.

2. Put 5 ml of methylcellulose in each well. Stir the methylcellulose for 30 minutes at room temperature before use.
3. Half fill the central gaps between the wells with sterile water.
4. Place the plates in a 37°C incubator until ready for use.

Plating out samples

1. Pipette 100µl of sample onto the appropriate well in a figure of 8 pattern.
2. When the all the samples have been pipetted on the plate swirl the plate to ensure that all the cells are evenly distributed across the methylcellulose.
3. The plates are then incubated in a humid box in a CO_2 incubator until the colonies are clearly visible and countable. This usually takes 10 days to 2 weeks depending on the growth characteristics of the cell line. An example plate is shown in Figure 2. For consistency it is generally best to take the count from the pair of wells with between 0 and 100 colonies. Counting above 100 is unlikely to be accurate and colony formation is likely to be compromised at high densities. The whole experiment should be repeated three times. Survival is plotted as a percentage of the cells surviving on the untreated control.

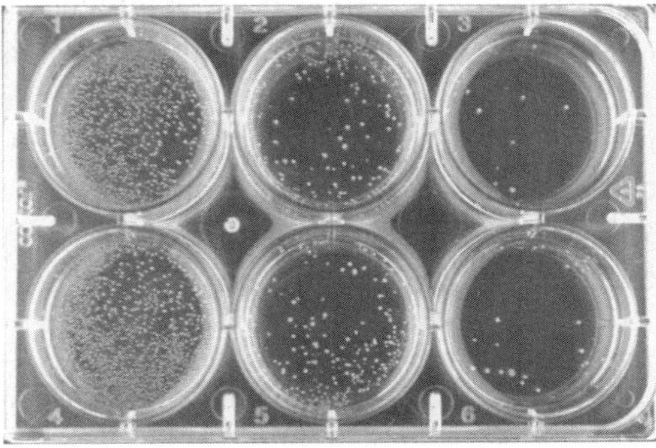

Figure 12-1. Example plate from a colony survival assay (see plate 47).

Method (UV)

1. Spin down (5 minutes, 250g) 1 million cells for each UV dose plus a spare million
2. Resuspend the cells in PBS so the cells are at a density of 1million/ml.

3. Take one 6 well plate for each UV dose and plate 1 million cells (1ml) of each cell type in a separate well.

4. The cells are now ready to be UV irradiated. The lamp needs to warm up/stabilise before use (allow 10 minutes). Precautions must be taken to avoid exposure of the skin or eyes to the lamp when it is on. Ideally, the lamp will be mounted in an enclosure with a shutter that allows precise exposure times, while protecting the operator.

5. The UV output should be measured and the appropriate exposure time calculated. A good starting point is 0 to 10 J/m^2.

6. The cells are now ready to be irradiated. Place them under the UV lamp for the required number of seconds remembering to take the lid off the plate.

7. When the irradiation is complete the plates are transferred to the laminar flow hood and 4ml of media is added to each well.

8. The UV will make the cells adherer to the bottom of the plate. They should be pipetted up and down vigorously at this stage to ensure that no cells become lost.

9. The plates are then stored in the incubator until they are ready for use.

10. For each dose two further dilutions need to be carried out. This can be carried out in 24 well plates or FACs tubes. In either case 0.9ml of media needs to be added to tube/well.

11. Transfer 100µl of the cells to the first tube/well.

12. Well mix (vortex if using tubes) the cells before transferring 100µl to these cells to the second tube/well. Make sure these are well mixed.

13. These 3 dilutions of cells can now be plated on to the appropriate methylcellulose wells. 100µl of dilution per well.

14. Repeat this with all UV doses and cell types.

Method (Chemical mutagens)

1. Count the cells. For each cell line dilute the cells to 2×10^5 in media.

2. Put 1ml of cells in a FACS tube for each dose to be tested.

3. For each dose of each cell line take a further 2 tubes with 0.9ml of media in.

4. Add the appropriate dose of chemical to the tubes containing the cells. Make sure the chemical is well mixed into the cells.

5. Incubate all the tubes for 1 hour at 37°C
6. Wash the cells; top up the tubes with media and then spin them to pellet the cells (7 minutes, 250g)
7. Resuspend the cells in 1ml media. This is the highest dilution.
8. For each dose two further dilutions need to be carried out. Transfer 100μl of the cells to one of the tube containing 0.9ml.
9. Well mix (vortex) the cells before transferring 100μl to these cells to the second tube. Mix well
10. These 3 dilutions of cells can now be plated on to the appropriate methylcellulose wells. 100μl of dilution per well.
11. Repeat this with all chemical doses and cell lines.

Method (Irradiation)

1. Dilute the cells to 2×10^5/ml.
2. Carry out two further 1/10 dilutions.
3. Plate 100μl of each cell dilution (as Figure 1) into the appropriate wells. This will result in 20,000 2000 or 200 cells per well.
4. Incubate the plates for 1 hour in the humidified CO_2 incubator at 37°C.
5. The plates can now be exposed to the appropriate dose of either X- or γ-irradiation.

Troubleshooting

Once the samples have been plated out make sure the plates are not moved/shaken as this will result in the colonies being visible as streaks rather than tight round colonies. It is essential that the plates are kept humidified. As well as including sterile water between the wells of the plate, we also keep the plates inside pizza boxes humidified with a beaker of water when they are in the incubator. The inclusion of copper sulphate in the humidifying water helps reduce fungal contamination.

Inadequate mixing or the formation of precipitates in the methylcellulose medium will reduce plating efficiency and may render the assay useless.

Adequate swirling of the plates following application of the cells is important to ensure even distribution of surviving clones. Failure to do so will result in uneven patches of cells making colony counting very difficult. The cells are initially on the surface of the medium, but with 1% methylcellulose, sink to the bottom over the course of the next couple of days.

Method 13

SUBCLONING DT40 BY LIMITING DILUTION

Jean-Marie Buerstedde
GSF, Institute for Molecular Radiobiology, Ingolstaedter Landstr. 1, D-85764 Neuherberg-Munich, Germany

Abstract: Subcloning by limited dilution can be used to derive clonally related cell populations from a heterogeneous DT40 cell culture. For example, if one suspects that a drug resistant population may represent the progeny of more than one transfectant, the protocol can be used to isolate genetically homogeneous mutant clones. Other uses are the excision of floxed DNA sequences after Cre recombinase expression or fluctuation analysis to determine mutation rates (see Protocols 'Excision of floxed-DNA sequences by transient induction of Mer-Cre-Mer' and 'Analysis of sIgM expression by FACS').

Key words: Protocol, subcloning.

Materials

Reagents
96 well flat-bottom microtiter plates
15 ml tubes
DT40 cell culture medium
Trypan blue

Equipment
Laminar flow hood
Humidified CO2 incubator
Pipette aid

Sample
Cell culture to be subcloned

J.-M. Buerstedde and S. Takeda (eds.), Reviews and Protocols in DT40 Research, 393–394.

Method

1) Using Trypan blue count the viable cell density of the culture to be subcloned.
2) Prepare three tubes each containing 10 ml DT40 cell culture medium and respectively 30, 100 and 300 viable cells.
3) Use one 96 well flat bottom microtiter plate for each of the three cell dilutions. Add 100 microliter of the cell suspension to each well of the plate.
4) Incubate the plates for 7-8 days without changing medium. Subclones should be visible by then as round colonies. Pick clones from a plate showing colonies in a low percentage of wells. The precision of subcloning may be increased by transferring only cells from the center of a colony using a 10 microliter pipette.

Troubleshooting

Subcloning of wild-type DT40 is relatively straightforward, since the cells grow without feeder layer and the need of conditioned medium. No change of medium is required, until the subclones are picked from the wells of the microtiter plate.

It is important to use a well humidified incubator to prevent evaporation of medium. The quality of the sample cell culture is also critical and one should use exponential growing cells of high viability. Certain knock-out mutants of DT40 with slowed proliferation capacity may have decreased cloning efficiencies and visible colonies may appear later than in the case of wild-type DT40.

Method 14

SUBNUCLEAR IMMUNOFLUORESCENCE

Dávid Szüts and Julian E. Sale
M.R.C. Laboratory of Molecular Biology, Hills Road, Cambridge, CB2 2QH

Abstract: Despite their relatively small size and the fact that they are naturally non-adherent, it is possible to obtain good immunofluorescence staining of subnuclear structures in DT40. This, combined with their genetic tractability, provides a powerful combination for the study of DNA replication and repair. Here we provide a general protocol for immunofluorescence of molecules such as RAD51 and phosphorylated histone H2Ax. We also present a modification to this protocol that allows visualisation of chromatin bound PCNA and hence sites of DNA synthesis.

Key words: Protocol, immunofluorescence, DNA replication, DNA repair.

Materials

Reagents
poly-L-lysine (1mg/ml in water)
phosphate buffered saline (PBS)
PBSTS (PBS + 0.1% Tween 20 + 0.02% sodium dodecyl sulphate)
4% paraformaldehyde (see below for method of preparation)
methanol chilled to -20°C
relevant primary and secondary antibodies
mounting medium (e.g. Vectormount)
nail varnish

Apparatus
13mm round coverslips
24 well plates
37°C incubator
humid box
confocal microscope

J.-M. Buerstedde and S. Takeda (eds.), Reviews and Protocols in DT40 Research, 395–398.
© 2006 *Springer.*

Method

1. Place poly-L-lysine-coated 13 mm round coverslips on the bottom of individual wells of a 24-well tissue culture dish.
2. Add up to one million cells in medium to each well.
3. Spin the plate in a suitable centrifuge at 400g for 5 minutes
4. Remove the medium, and wash the cells with PBS
5. Fix with 4% paraformaldehyde for 5 minutes
6. Wash with PBS
7. Block with PBSTS with or without additional blocking agents (see below) for 10 minutes
8. Carefully remove the coverslips from the wells, blot the edge on absorbent paper, and place across the top of two neighbouring wells on a 24-well plate.
9. Add 35 µl primary antibody mix in PBSTS on top of the coverslip, so that it is held by surface tension.
10. Incubate for 1-3 hours at 37°C in a humidified box in the dark.
11. Remove the coverslips and blot the edge on absorbent paper.
12. Place the coverslips into wells containing PBSTS for 10 minutes
13. Remove and blot the coverslips as in step 8.
14. Add 35 µl secondary antibody mix.
15. Incubate for 1 hour at 37°C in a humidified box in the dark.
16. Remove the coverslips and blot the edge on absorbent paper.
17. Place the coverslips into wells containing PBSTS for 10 minutes
18. Label microscope slides and add small drops of mounting medium
19. Carefully blot the coverslips as dry as possible, and place, upside down, on the drop of mounting medium.
20. Seal the edges of the coverslips with nail varnish if necessary.
21. Once dry the slides are ready for viewing under the confocal microscope

Additional steps for the visualisation of chromatin-bound PCNA

In place of step 5 above, carry out the following

5a. Incubate in 0.2% Triton in PBS for 5 minutes.
5b. Wash with PBS.
5c. Fix with 4% paraformaldehyde for 5 minutes.
5d. wash with PBS.

    5e.   wash with dH$_2$O.

    5f.   add –20°C methanol to fix for 5 minutes.

    5g.   wash with dH$_2$O

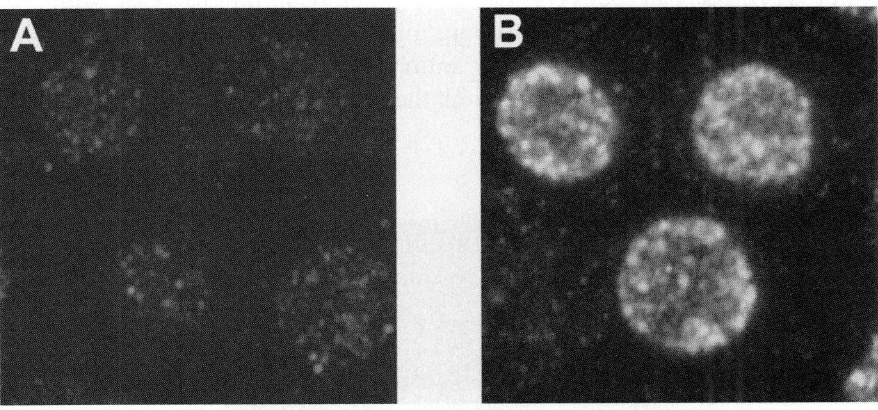

Figure 14-1. Example of subnuclear immunofluorescence in DT40. RAD51 in red, γH2Ax in green. Areas of co-localisation appear yellow. A. Unirradiated B. 2 hours after 254nm UV irradiation (see plate 48).

Notes and troubleshooting

Poly-L-lysine-coated coverslips can be prepared by incubating coverslips in 1mg/ml poly-L-lysine in water for 20 minutes, rinsing them in dH$_2$O, and drying them individually in air.

4% paraformaldehyde solution is prepared by heating, in a fume hood, paraformaldehyde powder suspended in PBS to about 50°C followed by the dropwise addition of 4M NaOH (or addition of NaOH pellets) until the paraformaldehyde is dissolved. The pH is adjusted back to 7.4 with HCl. The paraformaldehyde solution can be aliquoted and frozen at -20°C and remains good for several months.

Additional blocking agents in addition to PBSTS are frequently unnecessary. However, some antibodies do require additional blocking. It is worth trying bovine serum albumin, fetal calf serum or non-fat dried milk. This needs to be optimised for each antibody.

When solutions are added to or removed from the coverslips, try to be as gentle as possible to avoid dislodging the cells.

The antibody incubations can also be performed by leaving the coverslips in the wells, though a much higher volume of antibody solution is needed.

It is much easier to remove the coverslips from the wells if the 24-well plate is specially modified earlier by adding a drop of superglue to the middle of each well and letting it dry. These plates can be reused.

Nuclear DNA can be visualised by, amongst other dyes, propidium iodide (PI). PI can be very useful for confocal microscopes with no ultraviolet laser line. To use PI, 100 µg/ml RNase A needs to be added to the coverslip together with the primary antibody, to remove cytoplasmic RNA. Then, 1µg/ml PI is added together with the secondary antibody.

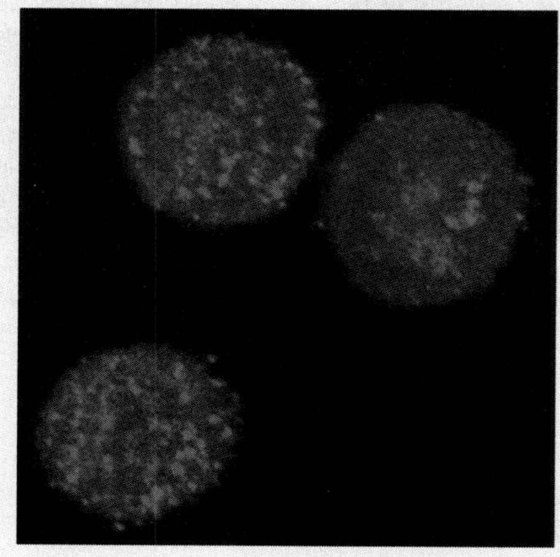

Figure 14-2. PCNA foci in an asynchronous culture of DT40. PCNA red, DNA blue (see plate 49).

Method 15

SISTER CHROMATID EXCHANGE ASSAY

Laura J. Simpson and Julian E. Sale
M.R.C. Laboratory of Molecular Biology, Hills Road, Cambridge, CB2 2QH, UK

Abstract: This method results in the differential staining of sister chromatids during replication and permits the direct visualisation of genetic exchanges between sister chromatids. It therefore provides a direct visual readout for crossover recombination. The generation of SCE is dependent on homologous recombination and is elevated in a number of circumstances including following exposure to DNA damaging agents and in some genetic backgrounds that result in increased dependence on recombination based pathways. It is important to remember that SCEs are probably only generated by a fraction of recombination events and their elevation may reflect increased use of recombination-based pathways or an increase in the resolution of recombination intermediates with cross-over.

Key words: Sister chromatid exchange, homologous recombination.

Materials

Solutions & Reagents
Bromodeoxyuridine [1mM in water]
Colcemid solution10μg/ml
75mM KCl
2x SSC pH 7.0 [300mM NaCl, 30mM Na citrate]
Freshly prepared ice cold methanol:acetic acid 3:1
0.1M Phosphate buffer pH 6.8
MacIlvaine's solution pH 7.0 [164mM Na2HPO4,15mM citric acid]
Hoechst 33258 (bisbenzimide H) 1mg/ml in water
Leishman's Stain 3g/litre in methanol [stirred for 24 hours prior to use]
10% Tween 20 in water
Mounting medium e.g.Vectormount®

J.-M. Buerstedde and S. Takeda (eds.), Reviews and Protocols in DT40 Research, 399–403.

Equipment
Laminar flow hood
Humidified CO2 incubator
benchtop centrifuge
Oven
365nm UV (UVA) source
Pasteur pipettes
Microscope slides
22 x 64mm glass coverslips
Microscope capable of 1000x magnification

Method

1) Set up a flask containing $10 - 20$ million cells.
2) Add BrDU to a final concentration of $10\mu M$ to each flask. Incubate the cells for 2 cell cycles in a $37°C$ incubator. For wild-type cells, this is approximately 20 hours.
3) If damage induction of SCE is to be studied, add mutagen of choice to cells 8 hours before harvest.
4) 2 hours before harvest add colcemid to a final concentration of $0.1\mu g/ml$.
5) Harvest the cells by centrifugation in a benchtop centrifuge (5 minutes at 250g) in a 50ml tube. Remove the supernatant and resuspend in 20ml 75mM KCl. Leave the cells in this solution (at room temperature) for $15 - 30$ minutes.
6) Spin cells down (5 minutes, 250g) and resuspend in the ice cold methanol: acetic acid (3:1). At this stage the cells are VERY fragile; pipetting up and down should be avoided.
7) Spin cells down (7 minutes, 250g) and resuspend in ice cold methanol: acetic acid (3:1). Transfer cells to a 15ml falcon tube.
8) Repeat step 7 a further 2 times each time reducing the volume of methanol: acetic acid until the cells are resuspended in approximately one ml. The cells are now ready to be 'splatted' onto the slides although can be stored at this stage for weeks at $4°C$ until use.
9) Using a wide bore Pasteur Pippette to gently take up the fixed cell suspension, drop a single drop of cells onto a clean microscope slide. The slides should be at approx $30 - 45°$ angle and the cells need be dropped from no more than approximately $20 - 30cm$.
10) Allow slides to dry then bake them in a rack over a heated block set to $50°C$ for at least 20 minutes.

11) Stain the slides by immersion in bisbenzimide diluted to 10µg/ml in 0.1M phosphate buffer (pH 6.8) for 20 minutes. Wrap the container in foil to reduce light exposure.

12) Wash the slides in MacIlvaine's solution.

13) Expose the slides for 1 hour on the UV box on UVA. Put cover slips on prior to baking.

14) Incubate the slides at 62°C in 2xSSC for an hour. Make sure all the cover slips are removed prior to incubation

15) Dilute the stock Leishman's stain 1 in 4 with phosphate buffer (pH 6.8). Filter before adding Tween to a final concentration of 0.05%. Mix well. Flood the slides with stain and incubate for 1 and a half minutes.

16) Pour off the stain, allow the slides to dry, and then bake them over the heated block at 50°C.

17) Put on the cover slips using Vectormount. Allow slides to dry overnight before viewing under oil immersion at 1000x magnification.

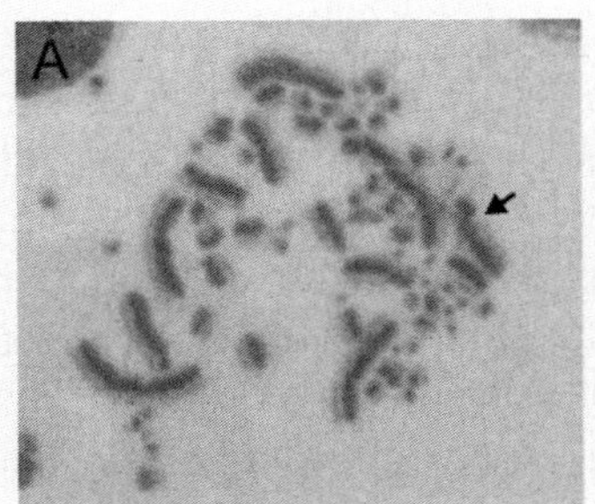
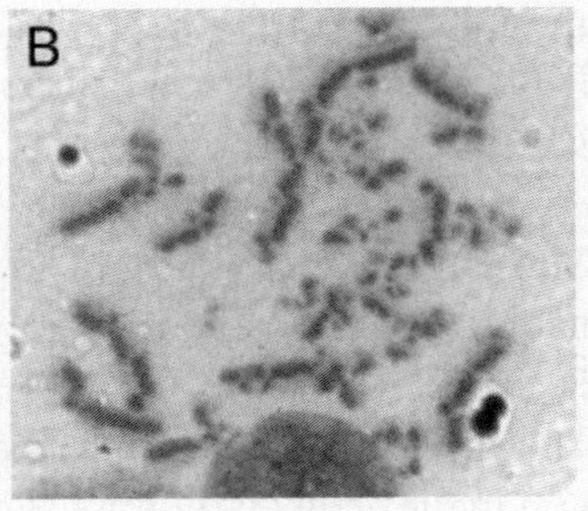

Figure 15-1. Example of sister chromatid exchanges in DT40. A) Metaphase from wild type cell. An example of an exchange is indicated with the arrow. B) Metaphase from *blm* DT40 showing elevated spontaneous SCEs (see plate 50).

18) An exchange is scored each time the dark and light stained chromatids change sides (Figure 1). Scoring should be undertaken blind and in an ideal world is done from randomised coded photographs of individual metaphases. However, if scoring from slides directly, it is important that there is a random mix of control and experiment slides. Scoring from photographs is probably more

accurate in situations of very high levels of exchange. Generally, exchanges are only scored in the macrochomosomes as they are almost impossible to see in the minichromosomes.

Notes & Troubleshooting

The principle of the method is summarised in Figure 2. It relies on the differential binding of bisbenzimide to DNA singly or doubly substituted with bromodeoxyuridine. Following the first round of replication in the presence of BrDU, the newly synthesised strand in both daughter DNA molecules will contain BrDU. After the second round, one grand-daughter has both strands containing BrDU, the other only one. The doubly substituted strand does not bind bisbenzimide as effectively as the singly substituted strand. While bisbenzimide staining alone can be visualised under UV light, it fades rapidly and using a mordant like Leishman's makes for much more convenient viewing.

This can be a troublesome technique and can go wrong at several steps. Here we provide some pointers that should maximise the chances of success.

Do not leave the cells for longer than 30 minutes in the hypotonic KCl solution or they are liable to lyse.

It is important to make sure that the cells are treated with care once they have swollen up as they are very fragile and will easily burst. This will result in few metaphases.

Make sure only a single drop hits each slide. This will minimise the chances of chromosomes from one cell getting muddled with those from other cells.

The time in colcemid may have to be varied depending on the length of the cell cycle, especially important for some mutants.

Sometimes the SCEs can look rather "fuzzy". We believe this is often due to inadequate dehydration and fixing. Make sure the methanol:acetic acid is freshly prepared as on cooling it will take up water from condensation on the bottle. There is some evidence that leaving the cells for several days in fixative before spreading can improve the quality of metaphases.

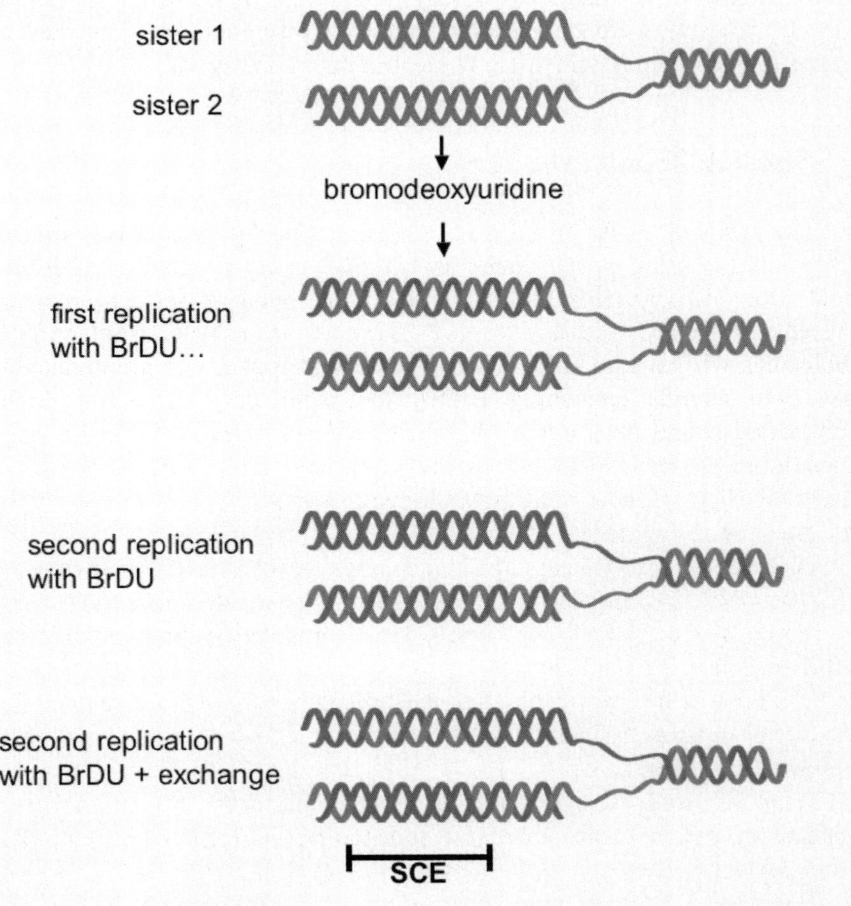

Figure 15-2. Principle of differential staining of sister chromatids (see plate 52).

Method 16

2D CELL CYCLE ANALYSIS

Roger Franklin and Julian E. Sale
M.R.C. Laboratory of Molecular Biology, Hills Road, Cambridge, CB2 2QH, U.K.

Abstract: This method provides a snapshot of the cell cycle stage of cells within a population. It relies on simultaneously monitoring DNA synthesis, through the ability to take up bromodeoxyuridine, and the total DNA content of the cells. DT40 cells cycle rapidly and 50 - 60% of cells in an asynchronous population will be in S phase. Cell cycle analysis is useful for assessing the success of cell synchronisation and for examining the response of cells to DNA damage in terms of cell cycle arrest and apoptosis. However, some caution needs to be exercised since DT40 are deficient in p53, which plays a key role in the G1/S checkpoint.

Key words: Protocol, cell cycle.

Materials

Reagents
Bromodeoxyuridine (BrDU) 1mM in water
DT40 culture medium
Phosphate buffered saline (PBS)
70% ethanol
Hydrochloric acid
Triton X-100
Sodium tetraborate
Tween 20
bovine serum albumin (BSA)
propidium iodide
anti-BrDU antibody (e.g. BD Biosciences at no. 555627)
anti-mouse Ig, FITC conjugated (e.g. Pharmingen cat no. 12064B)

J.-M. Buerstedde and S. Takeda (eds.), Reviews and Protocols in DT40 Research, 405–408.
© 2006 *Springer.*

Equipment
15ml tubes
polystyrene FACS tubes
Vortex
Flow cytometer

Method

1. Add 200µl 1mM BrDU to 5-6 x 10^6 cells in 10ml medium (final concentration 20µM) and incubate at 37°C for 20 minutes.
2. Pellet cells for 5 minutes at 250g in a benchtop centrifuge. Remove medium and then wash twice in cold PBS.
3. Resuspend cells in 250 µl cold PBS then add 2ml 70% ethanol (pre-chilled to –20°C) while vortexing to fix.
4. Leave at –20°C for at least 5 hours and preferably overnight
5. Wash once in 2ml PBS then resuspend in 1ml 2M HCl/0.5% Triton X-100 with vortexing
6. Incubate at room temperature for 30 minutes
7. Pellet cells then resuspend in 1ml 0.1M sodium tetraborate pH8.5 to neutralise.
8. Incubate at room temperature for 1 minute
9. Pellet cells and wash in 1ml PBS/0.1% Tween 20/1% BSA
10. Pellet cells and resuspend in 50 µl PBS/0.1% Tween 20/1% BSA + anti-BrDU antibody at the appropriate dilution (determined by titration).
11. Incubate at room temperature for 20 minutes
12. If the primary anti-BrDU antibody is not directly conjugated, wash cells again in PBS/0.1% Tween 20/1% BSA and incubate cells for a further 30 minutes in the secondary antibody.
13. Add 1ml PBS, pellet cells then wash again in PBS
14. Resuspend in 500µl PBS containing 20µg/ml propidium iodide
15. The cells are now ready to be analysed by FACS. Gate cells on FL2-A and FL2-W as shown in Figure 1 to ensure analysis of single cells
16. The percentage of cells in different stages of the cell cycle can then be assessed as shown in Figure 2 and mutants or treatments compared (Figure 2).

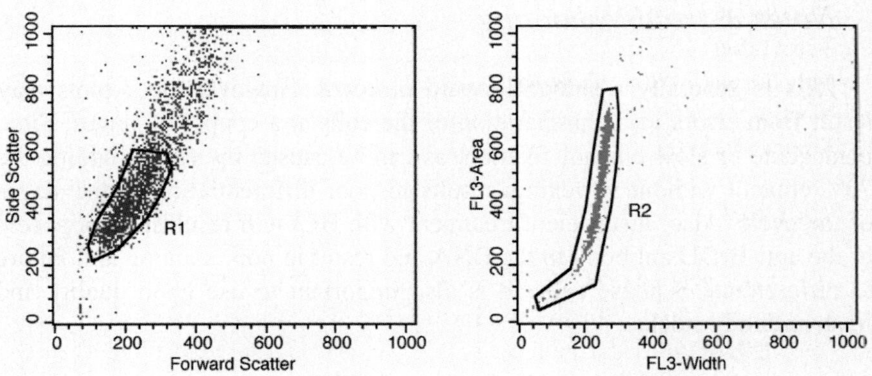

Figure 16-1. Gating of cells for cell cycle analysis. R1 is the gate for forward and side scatter. R2 is the gate for FL3-A/FL-3W, ensuring only single cells are analysed in the final 2D plot (see plate 51).

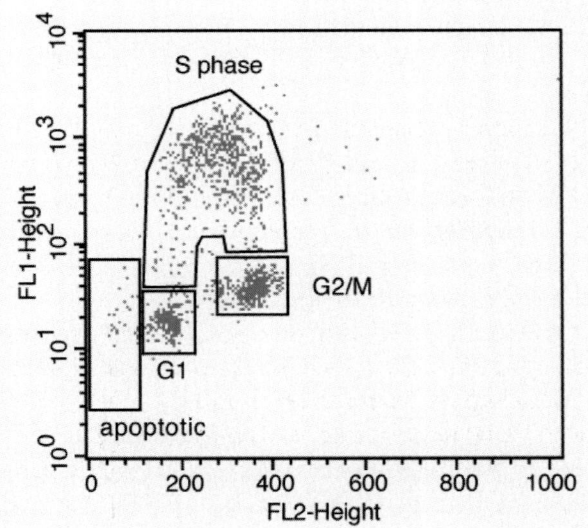

Figure 16-2. Typical cell cycle profile for wild type DT40. FL1 is the green (FITC) channel measuring BrDU uptake, FL2 is the red (PI) channel measuring DNA content (see plate 53).

Notes & troubleshooting

This is generally a straightforward protocol. However, poor plots may result from errors in the preparation of the cells at a couple of stages. First, inadequate or slow ethanol fixation, as can be caused by simply adding the 70% ethanol without vortexing, results in poor differentiation of the stages of the cycle. Also, insufficient treatment with HCl will result in poor access of the anti-BrDU antibody to the DNA and result in poor staining and failure to differentiate S phase cells. It is also important to use good quality and fresh antibody stocks.

Method 17

PURIFICATION OF TAP-TAGGED PROTEINS BY TWO-STEP PULL DOWN FROM DT40 CELLS

Hiroyuki Kitao and Minoru Takata
Kawasaki Medical School, Department of Immunology and Molecular Genetics, 577 Matsushima, Kurashiki, Okayama 701-0192, Japan

Abstract: For proteomic analysis, protein purification from cell extracts is an important step. Since production of high quality antibody is time consuming and not guaranteed to be successful, expression of epitope-tag conjugated protein of interest followed by immunoprecipitation using anti-epitope-tag antibody is a common method for protein purification. Here we describe use of an epitope-tag called TAP (tandem affinity purification) in DT40, which consists of Protein A IgG-binding motif and calmodulin binding motif separated by TEV cleavage site. Tandem purification using two different epitopes should eliminate non-specific binding and help identifying physiological protein-protein associations.

Key words: Protocol, TAP tag, IgG-agarose, calmodulin affinity resin, TEV protease.

Materials

Reagents
T75 culture flask
T225 culture flask
DT40 cell culture medium
50 ml tube
15 ml tube
Large centrifuge tube (for large scale culture)

J.-M. Buerstedde and S. Takeda (eds.), Reviews and Protocols in DT40 Research, 409–413.
© 2006 *Springer.*

Poly-Prep Chromatography column (BIO-RAD, Cat #731-1550)
Rabbit IgG-agarose (Sigma, Cat # A2909)
Calmodulin affinity resin (Stratagene, Cat # 214303-52)
Trichloroacetic Acid (TCA)
AcTEV enzyme (Invitrogen, Cat # 12575-015)
Anti-Calmodulin binding protein (CBP) antibody (Upstate, Cat # 07-482)
PAG Mini 4/20(13W) (Daiichi Pure Chemicals co., ltd., Cat # 301506)

Buffers
Lysis buffer
   10 mM Tris-HCl (pH 8.0)
   150 mM NaCl
   0.5% NP40
   + Proteinase inhibitors
      Complete, EDTA-free (Roche, Cat # 11 873 580 001)
         1 tablet/50 mL solution
      1 mM PMSF (Sigma, Cat # P-7626)
IPP150 buffer
   10 mM Tris-HCl (pH 8.0)
   150 mM NaCl
   0.1% NP40
   + Proteinase inhibitors
TEV cleavage buffer
   10 mM Tris-HCl (pH 8.0)
   150 mM NaCl
   0.1% NP40
   0.5 mM EDTA
   + 1 mM DTT (should be added just before use)
CaM binding buffer
   10 mM Tris-HCl (pH 8.0)
   150 mM NaCl
   0.1% NP40
   1 mM Mg Acetate
   1 mM Imidazol
   2mM $CaCl_2$
   + 1 mM 2-Mercaptoethanol (should be added just before use)
CaM elution buffer
   10 mM Tris-HCl (pH 8.0)
   150 mM NaCl
   0.1% NP40
   1 mM Mg Acetate

1 mM Imidazole
2 mM EGTA
+ 1 mM 2-Mercaptoethanol (should be added just before use)
5 x sample buffer
  0.312 M Tris-HCl (pH 6.8)
  10% SDS
  0.05% Bromophenol Blue
  25% 2-Mercaptoethanol
  5% Glycerol
2 x sample buffer
  0.125 mM Tris-HCl (pH 6.8)
  4% SDS
  0.02% Bromophenol Blue
  10% 2-Mercaptoethanol
  2% Glycerol

Equipment
Laminar flow hood
Humidified CO_2 incubator
Pipette aid
Centrifuge machine
Ultracentrifuge machine

Sample
DT40 cells expressing TAP-tagged protein.

Method

1) Prepare 500 mL of ~1×10^6 cells/mL DT40 cells expressing TAP-tagged protein. Total cell number will be 5×10^8.
   Cell culture volume should be determined based on the expression level of TAP-tagged protein.
2) Centrifuge 5 min at 1,200 rpm. Discard the supernatant by decantation and save the pellet.
3) Resuspend the pellet with PBS.
4) Centrifuge 5 min at 1,200 rpm. Aspirate the supernatant with a gel-loading pipet tip and save the pellet.
5) Lyse cells by incubating on ice for 1 hr in 10 mL lysis buffer.
6) Clarify the lysed cells by centrifugation 1 hr at 100,000 x g at 4°C. Save 100 µL of supernatant in an Eppendorf tube as input. Transfer the rest of supernatant into a new 15 mL tube.

7) Take 0.25 mL (50% slurry) per sample of rabbit IgG-agarose and equilibrate it with IPP150 buffer (wash 3 times with 0.5 mL IPP150 buffer).

8) Add the equilibrated rabbit IgG-agarose into the clarified lysates in 15 mL tube of step 6.

9) Incubate the mixture by rotation for 2 hr at 4°C.

10) Centrifuge 5 min at 1,000 rpm. Save 100 microL of supernatant in an Eppendorf tube as IgG unbound fraction. Aspirate the rest of supernatant and save the agarose.

11) Stand Poly-Prep Chromatography column in cold room and transfer the agarose into the column by resuspending in IPP150 buffer.

12) Wash the column with 15 mL IPP150 buffer.

13) Wash the column with 5 mL TEV cleavage buffer.

14) Plug the bottom of the column. Using 0.5 mL TEV cleavage buffer, transfer the agarose into a new Eppendorf tube. Add 1.25 microL AcTEV enzyme to the agarose.

15) Incubate overnight at 4°C with gentle agitation.

16) Centrifuge 5 min at 1,000 rpm. Transfer supernatant into a new 15 mL tube and save the agarose.

17) Add 0.5 mL TEV cleavage buffer to the agarose of step 16. Mix well with gentle agitation for 5 min. Centrifuge 5 min at 1,000 rpm. Transfer the supernatant into the 15 mL tube of step 16 (Total 1 mL eluate). Save the agarose.

18) Add 3 mL CaM binding buffer and 3 microL of 1 M $CaCl_2$ to the 15 mL tube of step 17. Save 100 microL of supernatant in an Eppendorf tube as TEV cleaved eluate.

19) Take 0.125 mL (50% slurry) Calmodulin affinity resin per sample and equilibrate it with CaM binding buffer (wash 3 times with 0.5 mL CaM binding buffer).

20) Add the equilibrated resin to the 15 mL tubes of step 18. Incubate 1 hr at 4°C by rotation.

21) Centrifuge 1,000 rpm 5 min. Save 100 microL of supernatant in an Eppendorf tube as Calmodulin affnity resin unbound fraction. Aspirate the rest of supernatant and save the resin. Stand Poly-Prep Chromatography column in cold room and transfer the resin into it using CaM binding buffer.

22) Wash the resin with 15 mL CaM binding buffer.

23) Elute with 0.5 mL of CaM elution buffer to 1.5 mL Eppendorf tube. Save the agarose in an Eppendorf tube.

24) Add 55 µL of 100% TCA to the eluate of step 23.

25) Incubate 30 min on ice. Vortex occasionally.

26) Centrifuge 13,000 rpm 30 min at 4°C. Aspirate the supernatant and save the pellet.

27) Add 1 mL Acetone. Vortex briefly and centrifuge 13,000 rpm 10 min at 4°C. Aspirate the supernatant and save the pellet.

28) Repeat step 27 once more.

29) Briefly air-dry the pellet. This pellet can be stored at −80°C.

30) Add 25 μL each of 5 x SDS sample buffer to input (step 6), IgG-unbound fraction (step 10), TEV cleaved eluate (step 18) and calmodulin affinity resin-unbound fraction (step 21). Add the equal volume of 2 x SDS sample buffer to the agarose (step 17) and the calmodulin resin (step 23). Boil 5 min at 100°C. Load samples on SDS-PAGE and perform Western blotting using anti-calmodulin binding protein (CBP) antibody to check each step of the experiment.

31) Resuspend the pellet of step 29 into 1 x SDS sample buffer and load it on SDS-PAGE.

We usually use 4-20 % gradient gel to separate proteins.

32) Stain the gel with silver or with Coomassie blue.

33) Excise stained bands of your interest with a clean blade. The gel slices are treated with trypsin. The eluted peptides are subject to mass spectrometry.

Troubleshooting

If the protein of your interest cannot be solubilized by the lysis method shown above, you have to modify the protocol to solubilize the protein. We often solubilize the protein in chromatin fraction by use of RNaseA-free DNaseI (available from Roche).

You likely need to optimize the condition of TEV protease digestion so as to increase the yield of your protein of interest (different incubation time etc). After the TAP-tagged protein is trapped with rabbit IgG-agarose, the buffer composition should be gradually changed to IPP150 buffer, which is used in the following steps.

Method 18

SYNCHRONIZATION OF CELLS

Eiichiro Sonoda
Radiation Genetics, Graduate School of Medicine, Kyoto University, Yoshida Konoe, Kyoto 606-8501, Japan

Abstract: Synchronization of cells is essential to study cell cycle specific events. If, for example, one suspects that a given DNA repair pathway is used in a particular cell cycle phase, the protocol can be used to enrich cells in each phase of the cell cycle and analyze the cellular response to DNA damage. Synchronization is also useful, when a gene is essential for a particular phase of the cell cycle. If a gene is, for example, essential for mitosis, synchronization of the cells in G1 phase with concomitant inactivation of the gene enables us to study the function of the gene in interphase, and to follow synchronous cell cycle progression to M phase. Two synchronization methods: centrifugal elutriation to enrich G1, S or G2 phase cells and nocodazole-mimocine sequential treatment to enrich cells at the G1/S boundary are described. Centrifugal elutriation can be achieved in less time (0.5–2 h) and with very little physiological stress to the cells whereas synchronization by drugs, such as nocodazole and mimocine, may result in unfavorable side effects.

Key words: Protocol, synchronization.

Materials

1. SAMPLE

DT40 cells ($1 \sim 10 \times 10^7$) in culture.

J.-M. Buerstedde and S. Takeda (eds.), Reviews and Protocols in DT40 Research, 415–418.
© 2006 *Springer.*

2.1   FOR ELUTRIATION

Reagents
DT40 cell culture medium
Elutriation buffer (0.01%EDTA/1%FBS/PBS, filtrated with 0.22um filter); 2~3L
70% ethanol; 0.5L
Sterilized water; 1L
Propidium iodide
Ribonuclease (RNase)

Equipment
Elutriator (Hitachi Koki R5E system: 4.7ml Separation chamber; equivalent to Beckman JE-5.0 System)
Flowcytometry

2.2   FOR NOCODAZOLE-MIMOCINE BLOCK

Reagents
DT40 cell culture medium
Nocodazole (10 mg/ml in DMSO, stock at –20°C)
L-Mimocine (100mM in acidic water, stock at –20°C)

3.1   Method for Elutriation

1) Assemble the elutriator rotor, elutriator chamber and the elutriator centrifuge according to manufacturers' directions.
2) Sterilize the apparatus by running 70% ethanol at 20mL/min for 30min, and then sterilized water for 30min without turning on the centrifuge.
3) Replace the water with the elutriation buffer and run the centrifuge at 2,000 r.p.m. at 4°C.
4) Calibrate the speed of pump and the actual flow rate.
5) Harvest 5~10 x 10^7 cells and resuspend in 3~5ml of ice-cold elutriation buffer. Pass through nylon mesh to remove clumped cells. Maintain the cells on ice.
6) Inject the cells into a sample reservoir and then load them into the separation chamber. Start with a flow rate of 10mL/min. Increase the flow late incrementally and collect the cell fraction serially as follows;

Fraction 1:10mL/min, 200ml
Fraction 2:11mL/min, 200ml
Fraction 3:12mL/min, 200ml
Fraction 4:14mL/min, 200ml
Fraction 5:16mL/min, 200ml
Fraction 6:18mL/min, 200ml
Fraction 7:20mL/min, 200ml

7) Concentrate the cells by low-speed centrifugation (1.200 r.p.m. for 5min) at 4°C. Resuspend each fraction in 10ml of ice-cold elutriation buffer and count the numbers of cells. Keep the cells on ice to prevent cell cycle progression.

8) To determine the purity of cells, take a small aliquot from each fraction ($\sim$1 x 10^5) and spin down. Fix the cells with 70% ethanol for 5~10 min and resupsend the cells in PBS containing 5μg/ml of propidium iodide and 100μg/ml of RNAse. Analyze the content of DNA by flowcytometry. In most cases, cells in G1, S and G2/M are enriched in the fraction 1~2, 3~5 and 6~7, respectively.

3.2 METHOD FOR NOCODAZOLE-MIMOCINE BLOCK

1) Add nocodazole to the cell culture at a final concentration of 0.5 μg/ml and incubate for 4 hr.
2) Wash the cells three times with warm medium, and incubated with medium containing 0.8 mM mimosine for 12 hours.
3) Wash the cells three times with warm medium and release them from the G1/S block.

4. Troubleshooting

1) Poor enrichment of synchronized population; Before cell separation, it is essential to calibrate the actual flow rate and the speed of pump. Measure the actual flow rate with measuring cylinder when the speed of rotation becomes constant.

2) Low recovery of cell; In DT40 cells, the percentages of cells in G1, S and G2 phase is 10~15, 60~70 and 15~20%, respectively, in an asynchronous population. Accordingly, the number of the G1-enriched cells is small, i.e. around 1~5 x 10^6 from 5 x 10^7 asynchronous cells. In addition, the quality of the sample cell culture is critical and one

should use exponential growing cells of high viability. Before elutriation, cells should be resuspended completely in the elutriation buffer to avoid clogging.

3) Low viability of cells in mimosine-nocodazole method; Since mimosine is toxic to cell culture, longer exposure of cells to mimocine causes massive cell death. Do not treat cell with mimosine longer than 12 hours.

4) Poor synchronization of the cell in mimosine-nocodazole method; Residual drugs often cause aberrant cell cycle progression. Do not use cold medium or PBS for washing.

Method 19

TARGETED TRANSFECTION
OF DT40 CELLS

Huseyin Saribasak and Hiroshi Arakawa
*GSF, Institute for Molecular Radiobiology, Ingolstaedter Landstr. 1, D-85764
Neuherberg-Munich, Germany*

Abstract: In order to understand the function of genes, transfection can be used as a
method in which artificially prepared knockout and expression constructs
are being introduced into cell lines. Since many genes are essential for
embryonic development and a homozygous deletion results in non-viable
embryos, gene disruption in a cell line by using transfected constructs
might be an alternative choice. Electroporation is often used to stably
transfect knockout and knockin vectors into DT40 cell line.

Key words: Transfection, DT40.

Materials

Reagents
Chicken medium
Trypan blue
Selective drug

Equipment
Electroporator (Gene Pulser I, II or Xcell [BioRad] is recommended.)
Electroporation cuvettes (Gene Pulser Cuvettes, 0.4 cm, BioRad)
Falcon tubes
96 well flat-bottom microtiter plates
Laminar flow hood

J.-M. Buerstedde and S. Takeda (eds.), Reviews and Protocols in DT40 Research, 419–421.
© 2006 *Springer.*

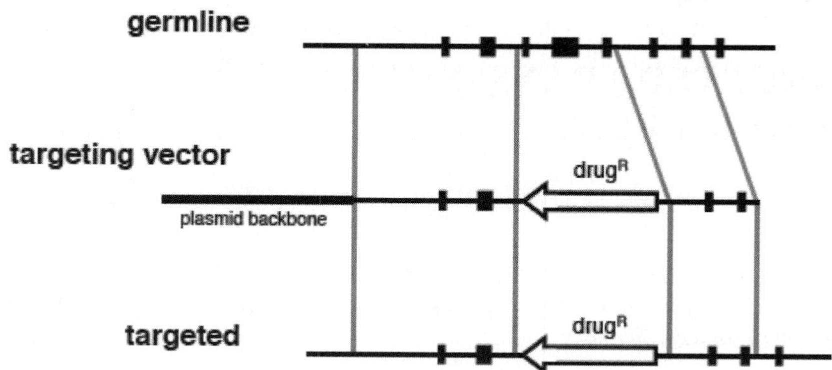

Figure 19-1. After the stable transfection of the knockout vector into the DT40 cell line, the vector is integrated into the target locus by homologous recombination at high efficiency. The transfectants successful for targeted integration can be screened by PCR (see Method 4).

Humidified CO2 incubator
Stepper
Stepper tips
Sample cell culture to be transfected

Method

Plasmid Preparation:
1) Linearize plasmid DNA (40 µg per each electroporation) outside of the genes of interest by an appropriate restriction enzyme for overnight. On the following day, check the quality of digested DNA by electrophoresis. You can run 0.5 µg of digested and undigested DNA side-by-side on an agarose gel.
2) Purify DNA once with phenol/chloroform extraction, once with chloroform extraction and precipitate with propanol. Rince the DNA with 70% ethanol. Dry up the pellet naturally for 10 min inside the laminar.
3) Resuspend the pellet in distilled water for final 1 µg/µl concentration.

Electroporation:
1) Determine the concentration of cells which you want to use for transfection. Transfer 10 million cells to a 50 ml tube and spin down for 5 min at 1500 rpm, 4°C.
2) Remove supernatant and resuspend the cell pellet into 800 µl of chicken medium.
3) Transfer cell suspension and 40 µg of linearized DNA into an electroporation cuvette.
4) Electroporate using 25 µF and 700 V.
5) Add the cell/DNA solution to the tube containing 9.5 ml of chicken medium and distribute into a 96-well flat-bottom microtiter plate adding 100 µl into each well.
6) The following day (12-24 h after electroporation), add 100 µl of selective medium (containing twice the final concentration of the drug) to each well.
7) Leave the plates for about seven to ten days in the incubator without changing the medium. Drug resistant colonies should be visible by then.

Troubleshooting

It is important to use exponentially growing cells for transfection. It is recommended to dilute cells one day before transfection. It is important to use a well humidified incubator to prevent evaporation of medium.

In the case you get too many transfectants per plate (>100), less DNA can be used or alternatively more plates can be used for distribution of the cells after electroporation for further trials.

Certain knock-out mutants of DT40 with growth defect may have decreased electroporation efficiencies and visible colonies may appear later than in the case of wild-type DT40.

If the electroporation parameters mentioned above does not work in your facilities, you can titrate voltage within the range 500V up to 1000V by keeping capacitance value at 25 µF using a reporter plasmid like GFP-expression vector.

Method 20

LUCIFERASE REPORTER ASSAY

Frank R. Wettey and Antony P. Jackson
Department of Biochemistry, University of Cambridge, Tennis Court Road, Cambridge, CB2 1QW, UK.

Abstract: This luciferase reporter assay provides a simple and highly sensitive method to determine the responsiveness of the Tet-system. It is quantitative, reproducible and easy to use in DT40 screens with both transient and stable transfections. The reaction catalyzed by firefly luciferase is the oxidation of luciferin in the presence of coenzyme A with concomitant production of a photon that can be measured by a luminometer as relative light units (RLU) or with a less sensitive scintillation counter.

Key words: Luciferase, Tet-system.

Materials

Reagents
DT40 culture medium
Phosphate buffered saline (PBS): 137mM NaCl, 2.7mM KCl, 4.3mM Na_2HPO_4, 1.4mM KH_2PO_4, pH7.3
Promega Luciferase Cell Culture Lysis Reagent: 25mM Tris-phosphate pH7.8, 2mM DTT, 2mM 1,2-diaminocyclohexane-N,N,N',N'-tetraacetic acid, 10% (vol) glycerol, 1% (vol) Triton X-100
Promega Luciferase Assay Reagent: 20mM tricine, 1.07mM $(MgCO_3)_4Mg(OH)_2 \cdot 5H_2O$, 2.67mM MgSO4, 0.1mM EDTA, 33.3mM DTT, 270µM coenzyme A, 470µM luciferin, 530µM ATP pH7.8
Regulatory plasmid encoding the luciferase gene (e.g. pUHC13-3 or pBI-L from Clontech)

423

J.-M. Buerstedde and S. Takeda (eds.), Reviews and Protocols in DT40 Research, 423–424.
© 2006 *Springer.*

Equipment
BioRad Gene Pulser
Bradford Assay Reagent (e.g BioRad)
Luminometer (e.g. LB9501Berthold)
Polycarbonate tubes 75x12mm, 5ml (e.g. Sarsted)

Method

1) For transient transfections centrifuge 1×10^7 cells for 5min, 300g, 4°C
2) Resuspend the pellet in 800µl PBS
3) Incubate the cells with 10 µg of reporter plasmid for 5min on ice and then transfect according to the standard DT40 protocols (eg Method 10)
4) Transfer cell suspension to 3-5ml warm culture medium and incubate in the presence or absence of 1µg/ml doxycycline
5) Dilute the cell culture as required after 24hrs
6) Harvest cells 48hrs post transfection for the luciferase assay as described below
7) For stably transfected cells incubate the cells during the exponential phase in culture medium either in the presence or absence of 1µg/ml doxycycline for 48hrs
8) Harvest the cells for the luciferase assay by centrifugation as above
9) Wash cells once in 5ml PBS as above
10) Lyse the cell pellet in 100µl lysis reagent for 20min on ice
11) After a brief spin in a benchtop centrifuge, samples can be either frozen at -80°C, or assayed immediately with the luminometer according to manufacturer's instructions
12) Using the LB9501(Berthold), set the measuring intervals to 10 seconds and dispense 25 µl of sample and 100 µl of Luciferase Assay Reagent into 5ml tubes
13) To ensure that measurements are taken in the linear range of the light detection, a standard curve with serial dilutions of luciferase in 1x Luciferase Cell Culture Lysis Reagent supplemented with 1mg/ml BSA is recommended
14) Use 1x Luciferase Cell Culture Lysis Reagent for the blank reading
15) Normalise the data by measuring the protein content of each sample (eg by Bradford assay)

Troubleshooting

Readings with the LB9501 luminometer can range between 500 relative light units (RLU) for the blank up to 20,000,000 RLU for a highly active sample. However, excessive enzyme activity, i.e. close to or above the detection limit of around 20,000,000 RLU, can cause instrumental artefacts and reduce the detected signal, thus mimicking a decrease in enzyme activity. Therefore the sample volume of concentrated samples can be reduced to 5µl or the samples diluted appropriately 10- to 100-fold.

Method 21

INDIRECT IMMUNOFLUORESCENCE MICROSCOPY

Frank R. Wettey and Antony P. Jackson
*Department of Biochemistry, University of Cambridge, Tennis Court Road, Cambridge,
CB2 1QW, UK.*

Abstract: This is a standard protocol for the detection of intracellular proteins by indirect immunofluorescence microscopy in DT40. It has been used extensively to investigate the intracellular distribution of various proteins of the endocytic machinery.

Key words: Indirect immunofluorescence microscopy, intracellular.

Materials

Reagents
Appropriate secondary dye-conjugated antibody (e.g. Alexa488- from Molecular Probes or Cy3- from Jackson Immuno Research)
Fixation buffer: Phosphate buffered saline (PBS), 4% (wt/vol) paraformaldehyde pH 7.0 (e.g. Sigma-Aldrich)
Hoechst 33342 (e.g. Sigma-Aldrich) in Wash Buffer at 2.5μg/ml
Incubation buffer: PBS, 5% (vol) foetal calf serum (FCS), 0.02% (wt/vol) Na-Azide, 0.01% (wt/vol) saponin, 0.1% (vol) Triton X-100
Mounting medium: 0.2M Tris pH8.3, 80% (vol) glycerol, 3% (wt/vol) N-propylgallate
0.01% (wt/vol) Poly-L-lysine solution MW150.000-300.000 (e.g. Sigma-Aldrich)
Primary antibody against desired protein
Wash buffer: PBS, 0.01% (wt/vol) saponin, 0.1% Triton X-100

J.-M. Buerstedde and S. Takeda (eds.), Reviews and Protocols in DT40 Research, 427–429.
© 2006 *Springer.*

Equipment
Centrifuge with rotor compatible to spin culture plates
Fluorescence light microscope
Cover slips
Lab Tek Chamber Slides™ (Nunc™, Life Technologies)
Nail varnish (clear)

Method

1) All handling and incubations are done at room temperature
2) Coat the glass chamber slides with poly-l-lysine by incubating them for 10-15min with the 0.01% solution at room temperature, aspirate the solution and dry the slides at 65°C
3) Centrifuge about 2.5×10^5 cells per sample for 5min at 300g and wash 2x in PBS to remove media proteins and amino acids
4) Adjust the cell concentration to 1×10^6 ml
5) Transfer 250µl of cell suspension per chamber for an 8-chamber slide and spin the slide for 1min at 200g.
6) Carefully aspirate the supernatant and fix the cells in fixation buffer for 10min at room temperature
7) Wash cells 6 x 2min in 500µl wash buffer
8) Block the chambers for 1 hr at room temperature with Incubation Buffer
9) Incubate the cells for 45min with incubation buffer containing the diluted primary antibody (usually at around 1-2 µg/ml antibody)
10) Wash cells 4x 5min with 500µl wash buffer
11) Incubate the cells for 45min with incubation buffer containing the diluted dye-conjugated secondary antibody (usually diluted 1:500-1:1000)
12) Wash the cells 4x 5min with 500µl wash buffer containing 2.5µg/ml Hoechst 33342 in the last wash for nucleic staining
13) Aspirate carefully all liquid, use a soft tissue if required and air dry for a couple of minutes
14) Apply a small drop of mounting medium, cover with a cover slip and seal edges with nail varnish
15) Analyse optical sections with a fluorescence microscope

Notes & troubleshooting

Poly-l-lysine at 0.01% is toxic for DT40, if left for extended periods on the coated glass chamber slides. Immediate fixation after attachment of the cells is recommended.

Paraformaldehyde needs to be dissolved in PBS at 65°C under a fume hood. Adjustment of the pH to 7-7.5 is usually required with NaOH and helps to dissolve the powder.

For some antibodies 0.3% Tween-20 plus 0.02% SDS works better than Triton X-100 in the incubation and wash buffers. Once the dye-conjugated antibody is applied, one should avoid direct exposure to bright light and keep the samples in the dark during incubation times.

The mounting medium needs stirring over night for the N-propylgallete to dissolve completely. Commercial mounting mediums can be even better to preserve the signals (e.g. Vecta Shield from Vector Laboratories).

Method 22

QUANTIFICATION OF RECEPTOR-MEDIATED ENDOCYTOSIS

Frank R. Wettey and Antony P. Jackson
Department of Biochemistry, University of Cambridge, Tennis Court Road, Cambridge, CB2 1QW, UK.

Abstract: This method allows measuring of the receptor-mediated internalization of ^{125}I-labeled conalbumin, the chicken egg white isoform of transferrin. Kinetic data, i.e. the rate constant k_i for the initial internalization process, can be extracted from the data by linear curve fitting using an In/Sur plot (intracellular label/cell surface label over time).

Key words: Quantification, receptor-mediated endocytosis, transferrin.

Materials

Reagents

^{125}I-Iodine (e.g. Amersham)
Buffer T1: RPMI1640 medium pH 7.4 without phenol red, 10mM HEPES, 0.1% (wt/vol) BSA (bovine serum albumin)
2mg/ml Chloramine-T in Na-phosphate buffer
Incubation Buffer: RPMI1640 medium pH7.4 without phenol red, 20mM HEPES, 5mM glucose, 0.1% (wt/vol) BSA
Iron-complexed conalbumin (e.g. Sigma-Aldrich)
0.3M Na-phosphate Buffer pH7.4
Neutralisation Buffer: RPMI1640 medium pH7.4 w/o phenol red, 20mM HEPES, 10% (vol) chicken serum
Phosphate buffered saline (PBS), 1% BSA
PBS, 0.1% (vol) Triton X-100
PD-10 Columns (e.g. Pharmacia-Biotech)
Thiosulfate solution
Saturated sodium thiosulphate solution

J.-M. Buerstedde and S. Takeda (eds.), Reviews and Protocols in DT40 Research, 431–434.
© 2006 *Springer.*

Equipment
Gamma counter
2ml conical screw cap tubes
Tube rack
Tube stand
50ml tube

Method

Conalbumin Iodination

1) Prior to commencing the experiment prepare the working area, in particular the fume hood and workbench, and wear all the necessary safety garments including lab coat, safety spectacles, two layers of gloves, over-shoes and also dose-monitors (finger and body badges) according to your institutional safety instructions

2) Conalbumin should be saturated with iron using the method described by Karin and Mintz (Karin and Mintz, 1981), if desired

3) Iron-complexed conalbumin is labeled by the chloramine-T method (Greenwood et al., 1963) as follows:

4) Mix 32μg of conalbumin stock solution with Na-phosphate buffer to give 50μl final volume in an Eppendorf tube

5) Remove a 10μl aliquot (=1mCi) from the $Na^{125}I$ lead pig following the manufacturer's guidelines and mix it with the buffered conalbumin by carefully pipetting up and down

6) Add 10μl of the chloramine-T solution and mix by carefully pipetting up and down and incubate the solution for 2.5min

7) Add 25μl of the saturated tyrosine solution, mix by pipetting carefully up and down and incubate for a further 2.5min

8) Add 250μl PBS to the tube and layer the entire content onto a PD-10 column, which has been equilibrated with PBS/1% BSA

9) After the radioactive solution has run into the gel bed, layer 250μl PBS/BSA onto the column

10) Collect approximately 0.5ml fractions in screw cap tubes and continue eluting with 2X 5mls PBS/1% BSA

11) Pool the peak fractions by mixing in the 50ml tube and then aliquot into 2ml screw cap tubes

12) Store the iodinated conalbumin in a lead container at 4°C

13) Reduce any ^{125}I remaining on the column with thiosulfate

14) Discard the waste as recommended in your local safety instructions

15) The amount of labeled versus free radioactivity can be determined by trichloroacetic acid precipitation (90µl PBS:10µl pooled conalbumin:11µl trichloroacetic acid). Incubate 10min on ice and centrifuge for 10min at 4°C at 13,000 rpm in a benchtop centrifuge. Take off the supernatant containing any free ^{125}I and add an equal volume of PBS/BSA to the pellet.

16) Determine the amount of radioactivity in all samples using a gamma counter and calculate the specific activity

Internalisation Assay

17) Take 10% more than the required amount of cells (see 21) and centrifuge 5min at 300g

18) Wash the cells 2x in 10-20ml PBS

19) Resuspend the cells in 20ml Buffer T1 and incubated for 20 min at 37°C to release all internal transferrin

20) Centrifuge as above and wash the cells 2x in ice-cold Buffer T1

21) Resuspend in ice-cold Incubation Buffer to give a concentration of 1×10^7 cells/ml

22) The following steps are best done in the cold room at 4°C and the cell suspension as well as any other solution are ideally kept on ice, when not in use

23) Add ^{125}I-conalbumin to give a final concentration of ≈ 20 nM ($=4 \times 10^7$ cpm/ml)

24) Aliquot the cell suspension into 100µl per sample/time point in triplicates, each in a 2ml screw cap tube, and incubate for 1hr on ice to allow for binding of the ligand to its receptor

25) Transfer the tubes at once in a rack to a 37°C water bath to induce internalization

26) Stop the uptake at the indicated times by the addition of 900µl ice-cold PBS/0.1% BSA and transfer each tube to iced water

27) Centrifuge the cells for 3min at 400g and aspirate the supernatant without touching the pellet, thus leaving a residual amount of ≤50µl

28) Wash the cells 5x with 1ml Buffer T1 (resuspend the cells just by adding the buffer with a little pressure, pipetting up and down is not required)

29) After the final wash solubilise the pellet in 1 ml PBS/TritonX-100 0.1%
30) Determine the cell associated radioactivity with a gamma counter
31) Determine non-specific binding by adding radioactive conalbumin and 100-fold excess of unlabelled conalbumin to cells at 4°C
32) The kinetic constant k_i for the initial rate of I-conalbumin internalisation is derived by linear curve fitting of the shown In/Sur plot in figure #9.6 as described in detail elsewhere (Wiley and Cunningham, 1982).

Notes & troubleshooting

The preparation in a cold room is necessary to inhibit any intracellular trafficking during the pipetting and handling steps.

REFERENCES

Karin M, Mintz B. (1981) Receptor-mediated endocytosis of transferrin in developmentally totipotent mouse teratocarcinoma stem cells. J Biol Chem 256: 3245-52

Greenwood FC, Hunter WM, Glovers JS (1963) The preparation of[131] I-labelled human growth hormone of high specefic radioactivity. Biochem J 89: 114-23

Wiley HS, Cunningham DD (1982) The endocytotic rate constant. A cellular parameter for quantitating receptor-mediated endocytosis. J Biol Chem 257: 4222-9

Method 23

MEASUREMENT OF DNA SYNTHESIS AND STRAND BREAKS USING ALKALINE SUCROSE DENSITY GRADIENT CENTRIFUGATION

Kouichi Yamada and Jun Takezawa
The National Institute of Health and Nutrition, Toyama, Shinjuku-ku, 162-8636 Tokyo, Japan

Abstract: Alkaline sucrose density gradient (ASDG) centrifugation is probably an only method to detect elongation of "pulse-labeled" replication products in cells. If the cells are pulse-labeled after being exposed to some DNA-damaging agents, their "post-replication repair" can be measured by ASDG technique. With non-damaged cells, normal replication in replicon size can be observed, too. In addition, the method is also applicable to measure single strand breaks. We have modified this classical method to reproducibly detect very long single-stranded DNA at the megabase level. Here, the protocols are optimized to DT40 cells.

Key words: Protocol, alkaline sucrose density gradient centrifugation.

Materials

Reagents

DT40 cell culture medium [RPMI1640, 10% heat-inactivated fetal calf serum, 1% chicken serum, 25 mM Hepes, pH7.4]
Ca^{2+},Mg^{2+}-free phosphate-buffered saline (PBS)
[methyl-^{14}C]thymidine (Amersham CFA532, 50 µCi/ml)
[U-^{14}C]thymidine (Moravek MC267, ca.470 mCi/mmol)
100 mM NaN$_3$ in PBS (stock solution)
"5% sucrose solution"[5% sucrose, 0.3 M KOH, 2.0 M KCl, 1 mM EDTA, 0.1% N-lauroylsarcosine]
"20% sucrose solution"[20% sucrose, 0.3 M KOH, 2.0 M KCl, 1 mM EDTA, 0.1% N-lauroylsarcosine]

J.-M. Buerstedde and S. Takeda (eds.), Reviews and Protocols in DT40 Research, 435–438.
© 2006 *Springer.*

"80% sucrose cushion"[80% sucrose, 0.3 M KOH, 2.0 M KCl, 1 mM EDTA, 0.1% N-lauroylsarcosine]
1% sucrose in PBS
"Cell lysis solution"[0.6 M KOH, 2.0 M KCl, 10 mM EDTA, 1.0% N-lauroylsarcosine]
Whatmann No.17 paper circles
5% trichloroacetic acid, ethanol, acetone
Toluene-based scintillator

Equipment
Laminar flow hood
Humidified CO_2 incubator
UV lamp or X-ray generator… etc.
Polyallomer centrifuge tubes (Beckman 326819)
Gradient maker
Ultracentrifuge and the rotor (Beckman SW50.1 or Hitachi RPS50-2)
Peristaltic pump
Liquid scintillation counter

Sample
DT40 Cells culture to be analyzed

Method

Preparation of cell samples for detection of replication products
1) Sediment the cells and resuspend in small volume of PBS-1% FCS solution.
2) Place the cell suspension center of culture dish, expose to UV or X-ray and add DT40 medium into the dish.
3) After appropriate incubation, sediment the cells, resuspend in culture medium containing 10 μCi/ml [U-^{14}C]thymidine and incubate for 20 min. (pulse-label) (Non-damaged cells are incubated for 10 min.)
4) Sediment the cells, and recover the labeling medium. Resuspend the cells in culture medium containing 10 μM thymidine and incubate further (chase). Do sampling at appropriate intervals.
5) Rinse the sampled cells once with PBS-1 mM NaN^3 and count the cells.
6) Resuspend the cells in PBS-1 mM NaN^3 (1.2×10^7 cells/ml).
7) Stand the cell samples on ice until loading onto gradient.

Preparation of cell samples for detection of DNA single strand breaks
1) Add [methyl-^{14}C]thymidine to cell culture to final 0.025 µCi/ml, and incubate the cells for 1 day. (prelabel)
2) Sediment the cells, resuspend in medium containing 10 µM thymidine and incubate further 6 hr (chase).
3) Sediment the cells and resuspend in small volume of PBS-1% FCS solution. Place the cell suspension center of culture dish, expose to UV or X-ray and add DT40 medium into the dish as above.
4) After appropriate incubation, do sampling.
5) Rinse the sampled cells once with PBS-1 mM NaN$_3$ and count the cells.
6) Resuspend the cells in PBS-1 mM NaN$_3$ (1.2 x 10^7 cells/ml).
7) Stand the cell samples on ice until loading onto gradient.

Preparation of alkaline sucrose density gradient
1) Place 0.4 ml of "80% sucrose cushion" at the bottom of polyallomer centrifuge tube.
2) Make up 4.35 ml of 5-20% alkaline sucrose gradient in the tube using a gradient maker.
3) Balance the tubes.

Centrifugation and assay
1) Gently place 100 µl of "cell lysis solution" onto the 5-20% alkaline sucrose gradient.
2) Gently layer 50 µl of "1% sucrose in PBS" onto the "cell lysis solution".
3) Gently apply 50 µl of cell sample (6.0 x 10^5 cells) on the top of the gradient. (Step 1-2-3 are done in sequence for each sample.)
4) Set the tubes in buckets of rotor, and set the buckets to the rotor.
5) Centrifuge the gradient at 6 krpm (4,320 x g) for 15.6 hr at 15°C.
6) Fractionate the gradient onto Whatmann No.17 paper circles using a peristaltic pump.
7) Dry the paper circles, immerse in cold 5% trichloroacetic acid for 10 min, wash 3times with ethanol and once with acetone and dry.
8) Measure the radioactivity in a toluene-based scintillator.

Troubleshooting

[U-^{14}C]thymidine is too expensive, but reusable for 10 times or more from our experience. You had better request Moravek to "dry up" the isotope, and it is easily resolved in medium without increasing the volume. The labeling medium is frozen-stocked until next experiment. We do not recommend the use of [^{3}H]thymidine, because it has cytotoxity to prevent the growth of DT40 cells. We think it also important not to increase the centrifugal velocity, because it causes, with higher probability, artificial aggregation of sticky DNA and proteins in the gradient [please refer Mutat.Res.<u>364</u>, 125-131, 1996].

Method 24

ISOLATION OF NUCLEAR AND CYTOPLASMIC PROTEINS FROM DT40 CELLS

Yan-Dong Wang and Randolph B. Caldwell
GSF, Institute for Molecular Radiobiology, Ingolstaedter Landstr. 1, D-85764 Neuherberg-Munich, Germany

Abstract: Nuclear and cytoplasmic proteins from DT40 B cells are extracted by serial fractionation in this protocol. The protein extracts are suitable for the detection of DNA-protein interaction, protein-protein interaction, DNase I footprinting analysis, and related techniques. This is a general protocol that was adapted as shown here to our use due to the fragile nature of DT40 cells and their nuclei.

Key words: Protocol, nuclear proteins, cytoplasmic proteins.

1.   MATERIALS AND REAGENTS

- Cell viability counter (ViCell)
- Microscope
- Centrifuge tubes and centrifuges (Heraeus Multifuge 3 S-R and Biofuge *Pico*)
- Vortex mixer/agitator
- Buffers:
  1. 1x PBS: Dissolve the following in 800ml distilled H_2O: 8g of NaCl, 0.2g of KCl, 1.44g of Na_2HPO_4, 0.24g of KH_2PO_4, adjust pH to 7.4 with HCl and adjust final volume to 1L with additional H_2O. Sterilize by autoclaving.
  2. Isotonic lysis buffer: 10 mM Tris HCl, pH 7.5, 2 mM $MgCl_2$, 3 mM $CaCl_2$, and 0.32 M Sucrose supplemented with a 1/100 protease inhibitor cocktail (Roche; Complete Mini, EDTA-free) and 1mM DTT.
  3. Extraction buffer: 20 mM HEPES, pH 7.7, 1.5 mM $MgCl_2$, 0.42 M NaCl, 0.2 mM EDTA, and 25% (v/v) Glycerol supplemented with a 1/100 protease inhibitor cocktail and 1mM DTT.

J.-M. Buerstedde and S. Takeda (eds.), Reviews and Protocols in DT40 Research, 439–440.

4. Dilution & Equilibration Buffer: 20 mM HEPES, pH 7.7, with 1.5 mM $MgCl_2$, 0.2 mM EDTA, 10mM KCl and 25% (v/v) Glycerol

2. METHOD

1. Determine cell viability; cell viability should be 80% or better.
2. Centrifuge desired culture volume at 1500rpm for 5min to harvest.
3. Discard supernatant and wash cells with equal volume cold PBS as in step 2 and repeat wash 3 times.
4. Add 5× PCV (packed cell volume) equivalent of isotonic lysis buffer to the cell pellet and resuspend gently to avoid foaming. Note: for DT40 cells, 1ul PCV= ~1×10^6 cells.
5. Incubate the resuspended cells in lysis buffer on ice for 15 minutes to allow the cells to swell. This may be checked by microscopy.
6. To the swollen cells, add 10% IGEPAL CA-630 to a final concentration of 0.3% and mix thoroughly, but gently.
7. Centrifuge immediately. (For small scale, using 1.5-ml eppendorf tube, centrifuge 30sec at 5500rpm; for large scale, using 50-ml tube, centrifuge 2min at 5500rpm.)
8. Transfer the supernatant (cytoplasmic protein fraction) to a fresh chilled tube.
9. Resuspend the pellet (containing the nuclei) with 5 × PCV equivalent of isotonic lysis buffer and mix gently.
10. Centrifuge the pellet at 1500rpm for 5min and remove the supernatant.
11. Resuspend the pellet (containing the crude nuclei) in 2/3 × PCV equivalent extraction buffer.
12. Place tube on a mixer and agitate for 15min at 700rpm followed by 15min at 1400rpm all the while being careful to avoid foam formation.
13. Centrifuge for 15min at 9400rpm and then carefully transfer the supernatant to a fresh chilled tube.
14. Snap-freeze the supernatant in aliquots with liquid nitrogen and store at −80°C for long term and −20°C for short term.

2.1 Troubleshooting

DT40 cells and nuclei seem to be quite fragile and so it is advisable that this protocol be monitored via microscopy to ensure a quality specimen. If the protein(s) of interest has pH or salt concentration issues, the extraction buffer can be diluted with 10mM $MgCl_2$ and 50mM Tris-HCl (pH7.5) or with dilution & equilibration buffer and then adjust the pH accordingly.

COLOR PLATES

Plate 1

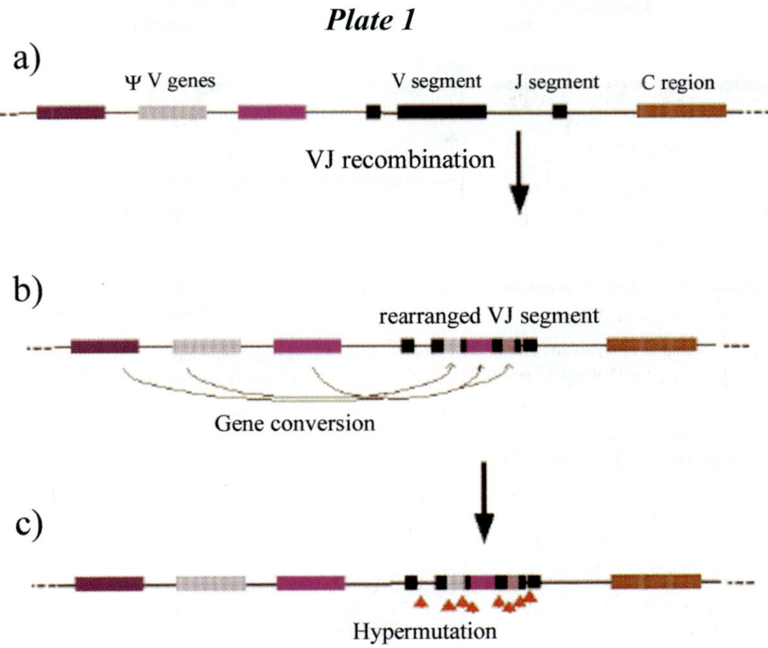

Figure 2-1. Development of a diverse Ig light chain gene repertoire by gene conversion and hypermutation.

Plate 2

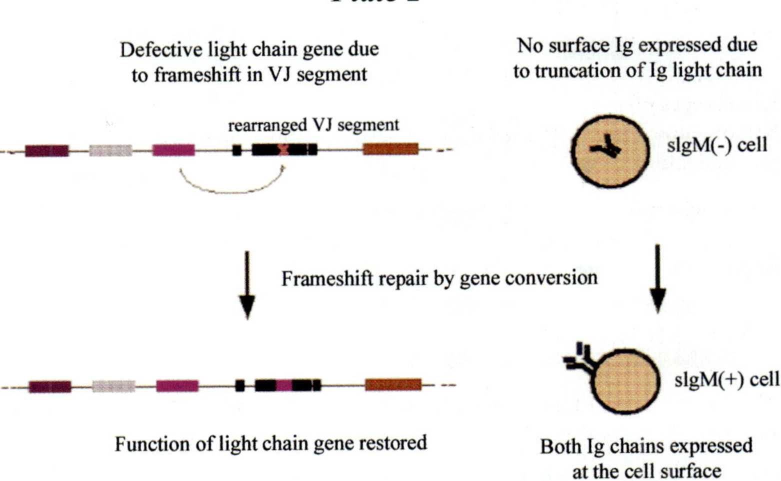

Figure 2-2. Surface Ig reversion assay to quantify Ig gene conversion activity.

Plate 3

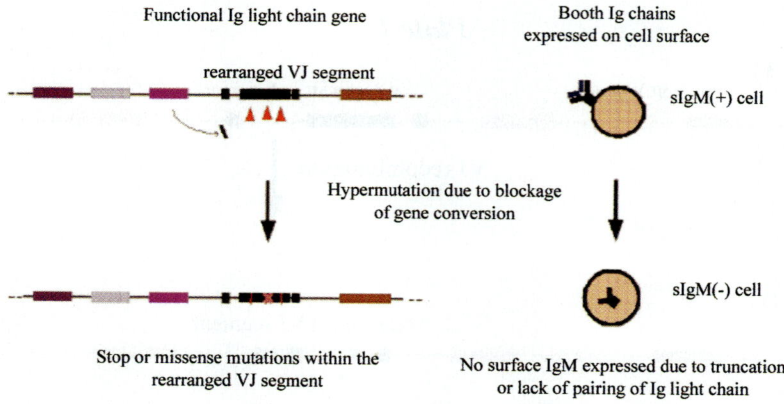

Figure 2-3. Surface Ig loss assay to quantify Ig hypermutation activity.

Plate 4

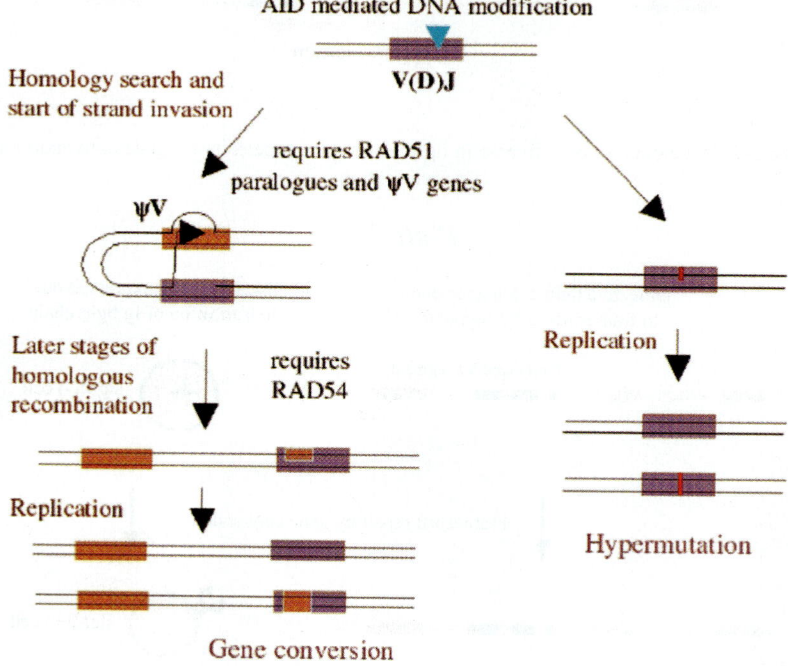

Figure 2-4. Model for Ig gene conversion and hypermutation.

Plate 5

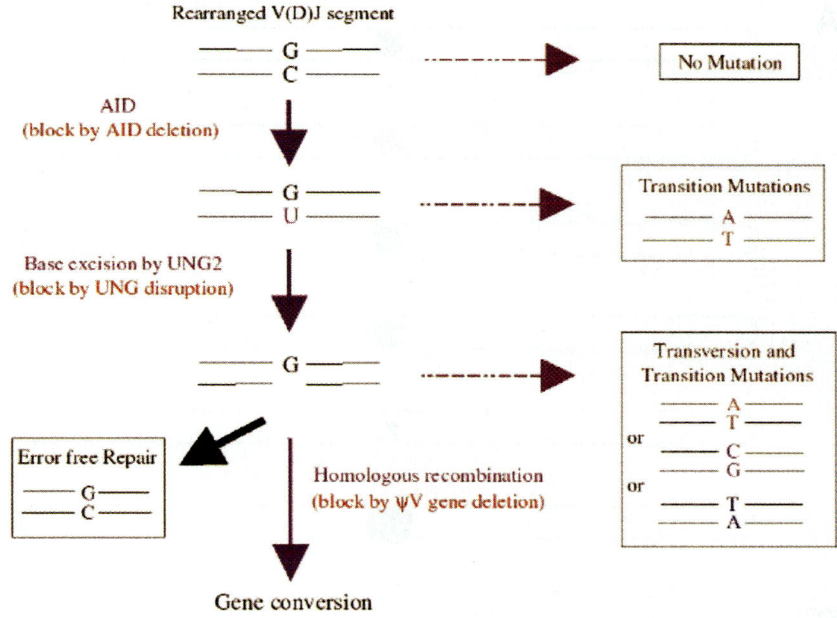

Figure 2-5. Phenotypes of DT40 mutants affecting Ig gene conversion and hypermutation.

Plate 6

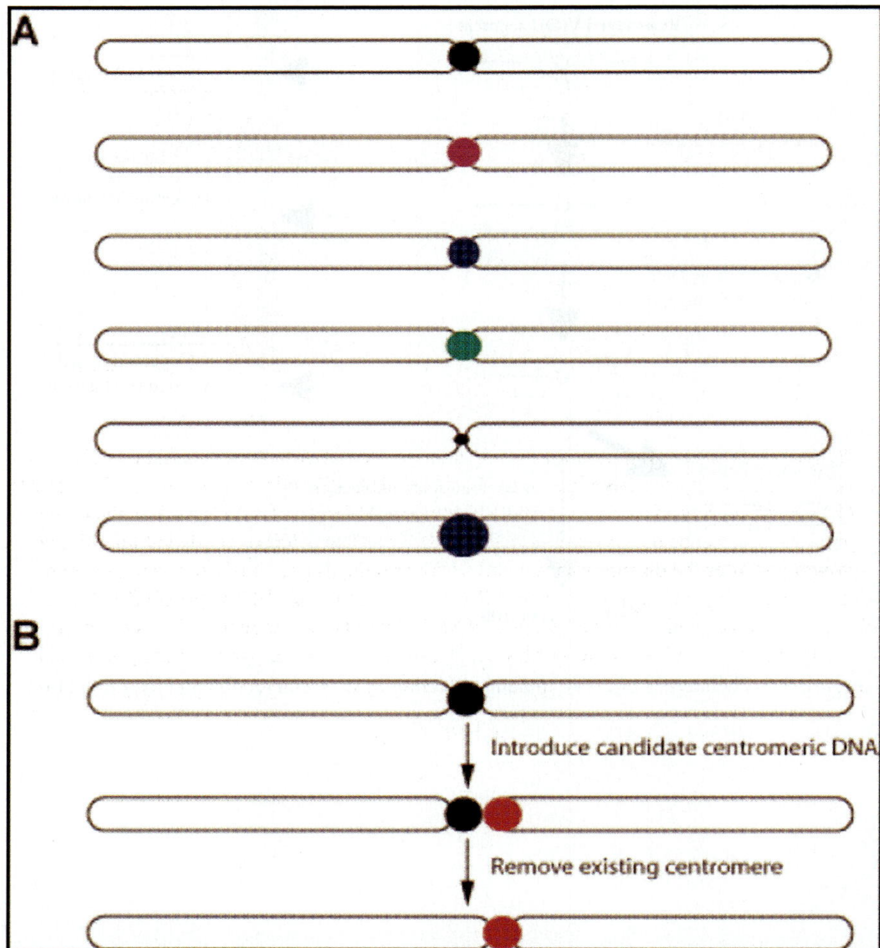

Figure 4-1. Exchanging sequences at the vertebrate centromere.

A. The ideal type of manipulation needed for the study of vertebrate centromeres is one where the pre-existing centromeric DNA, represented by a solid disc is exchanged for a variety of different sequences of varying sequence identity and size.

B. The methodological approach that we are attempting to achieve the goals set out in A. A candidate centromeric sequence (red disc) is introduced near the pre-existing centromere (black disc) and then the pre-existing centromere is removed. Both the construction of the candidate centromeric DNA and removal of the pre-existing centromere depend upon the use of unidirectional serine recombinases.

Plate 7

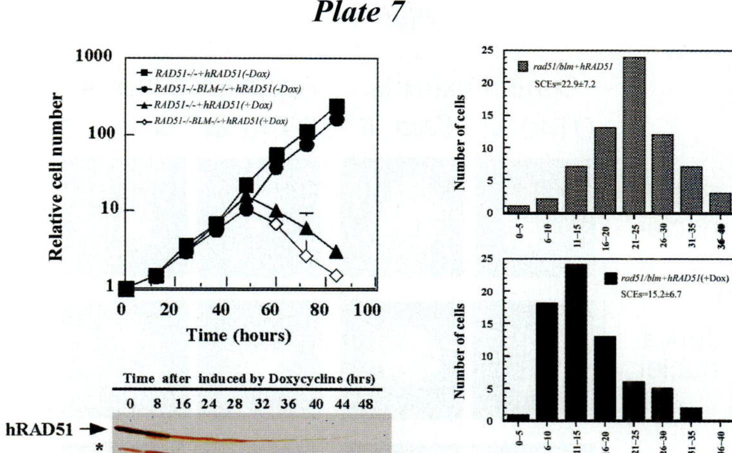

Figure 5-3. RAD51 is required for the elevated levels of SCE in blm cells. The rad51/blm tet-off hRAD51 cells are derivatives of rad51 tet-off hRAD51 cells (Sonoda et al., 1998, 1999). (A) Growth curves of rad51 tet-off hRAD51 and rad51/blm tet-off hRAD51 cells in the presence or absence of doxycycline (Dox). (B) After Dox addition, both cell lines cease growth just after the disappearance of hRAD51 protein, as previously described (Sonoda et al., 1998, 1999). The arrow indicates hRAD51 protein in rad51/blm tet-off hRAD51 cells after Dox addition, and an asterisk shows hRAD51 degradation products. (C) Sister chromatid exchange (SCE) in rad51/blm tet-off hRAD51 cells in the absence (upper panel) or presence (lower panel) of Dox.

Plate 8

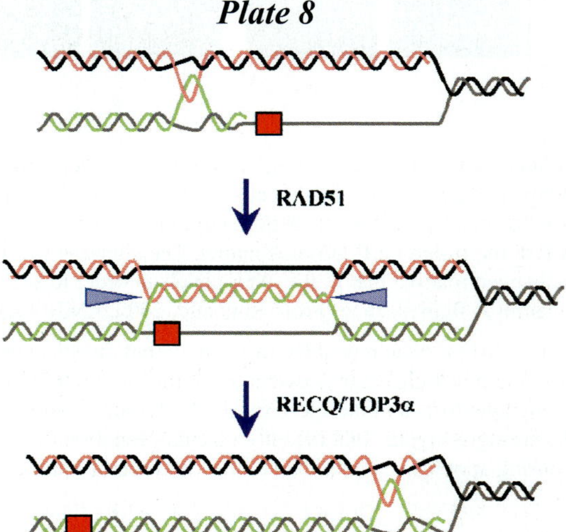

Figure 5-4. Template switching model for replication fork restart on damaged templates. The model is adapted from a published report (Liberi et al., 2005). Replication forks on a damaged template bypass DNA lesions through template switching mediated by hemicatenanes (structure A). A Rad51 homologous recombination-dependent pathway converts/stabilizes the intermediates as a pseudo double Holliday junction by extending pairing between the newly synthesized strands (structure B). The RECQ/Top3α complex (arrow heads) may mediate the conversion of this structure to a replication fork containing hemicatenanes (structure C).

Plate 9

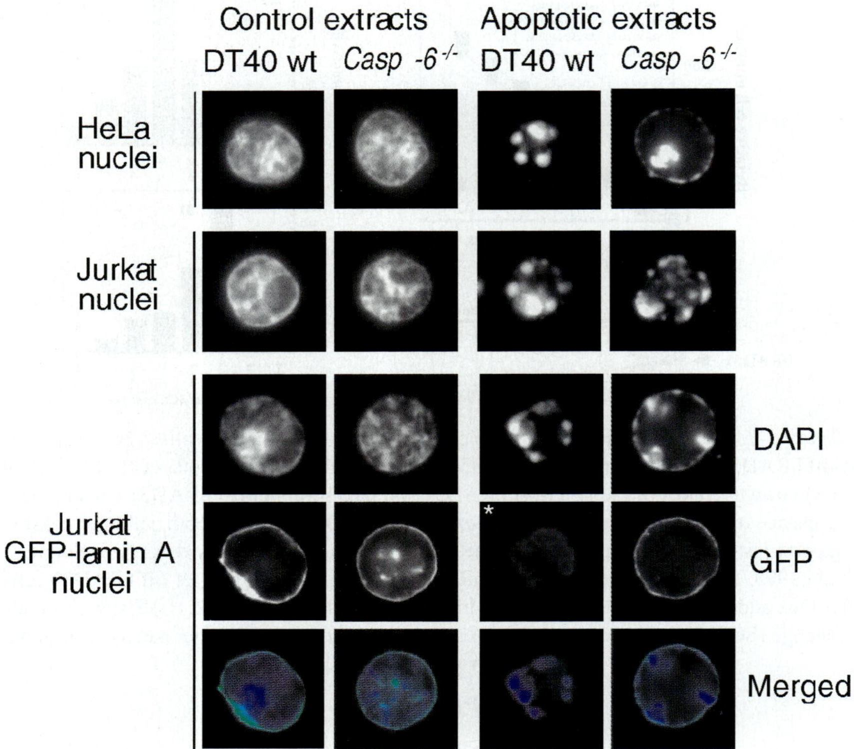

Figure 6-2. Isolated HeLa, Jurkat or Jurkat:GFP-lamin A nuclei were incubated for 2 h in extracts from wild type and *caspase-6*[-/-] DT40 cells treated with 1 µM staurosporine or diluent for 8 h. A nucleus representative of the major population is shown for each condition (DAPI) along with the GFP in the Jurkat:GFP-lamin A nuclei. The signal in the panel indicated by a white star was enhanced relative that in the other panels in order to see the residual GFP-lamin A fluorescence (From Ruchaud, 2002, EMBO J.).

The use of the cell-free system enabled us to show that, in cells expressing lamin A, caspase-6-mediated lamin A cleavage is essential for the chromatin to undergo complete condensation during apoptosis and therefore for the formation of apoptotic bodies. Taken together our results strongly suggest that the only essential function of this particular caspase during apoptosis is restricted to the cleavage of Lamin-A.

Plate 10

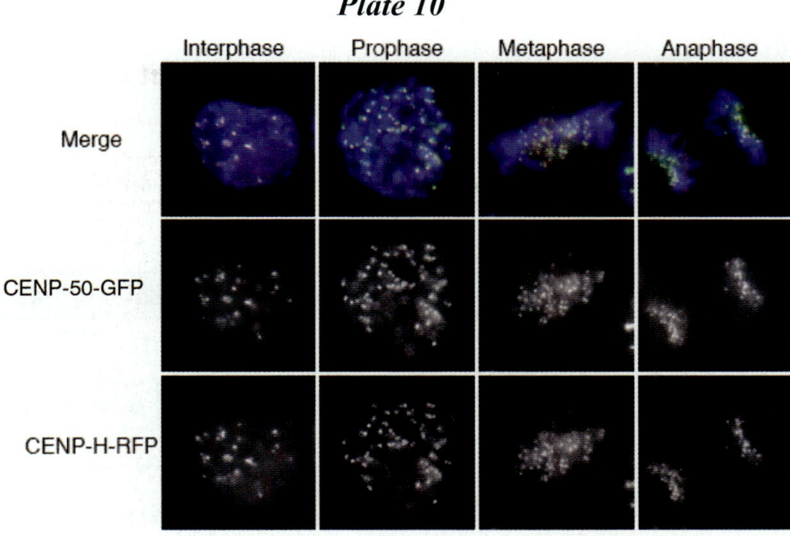

Figure 7-2. Localization of CENP-50, a member of the CENP-H/I complex proteins. GFP tagged CENP-50 protein (green) was expressed in DT40 cells that express chicken CENP-H-RFP (red). CENP-50 protein co-localizes to CENP-H throughout the cell cycle (Minoshima et al., 2005).

Plate 11

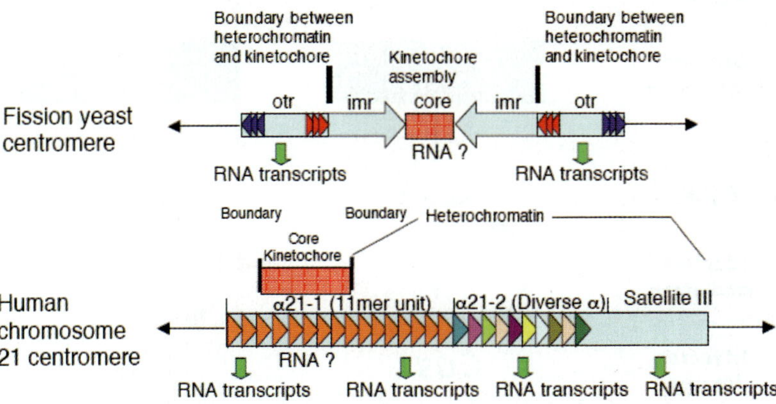

Figure 7-3. Schematic representations of the human chromosome 21 centromere region in chicken-human hybrid DT40 cells and of the fission yeast centromere. α-satellite arrays are composed of the α21-I array that comprises repeating 11-mer units and the α21-II array that contains diverged α satellite repeats. RNA transcripts were detected from α21-I, α21-II and satellite III arrays. We are not sure that RNA molecules are trasscribed from core kinetochore region. We also suppose that the boundary between the core kinetochore and heterochromatin structure is not determined by the primary DNA sequence in vertebrate cells (Fukagawa et al., 2004).

Plate 12

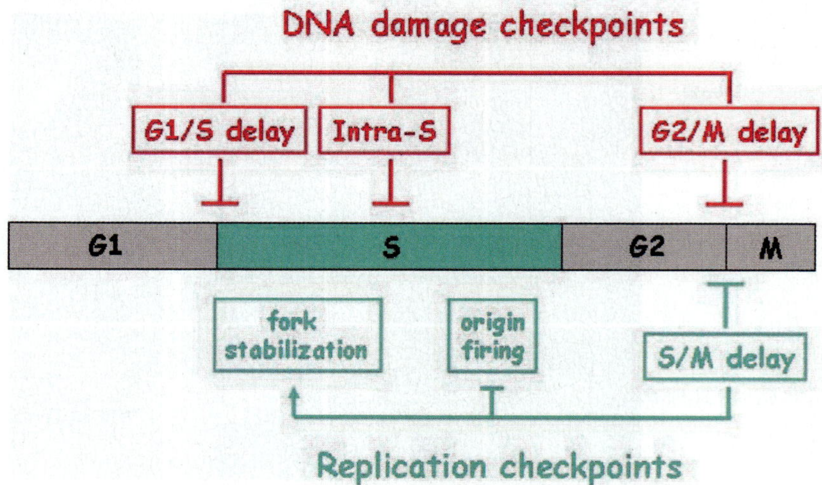

Figure 8-1. Overview of the various branches of the DNA damage (red) and Replication (green) checkpoints which exist in vertebrate cells. See text for a detailed explanation.

Plate 13

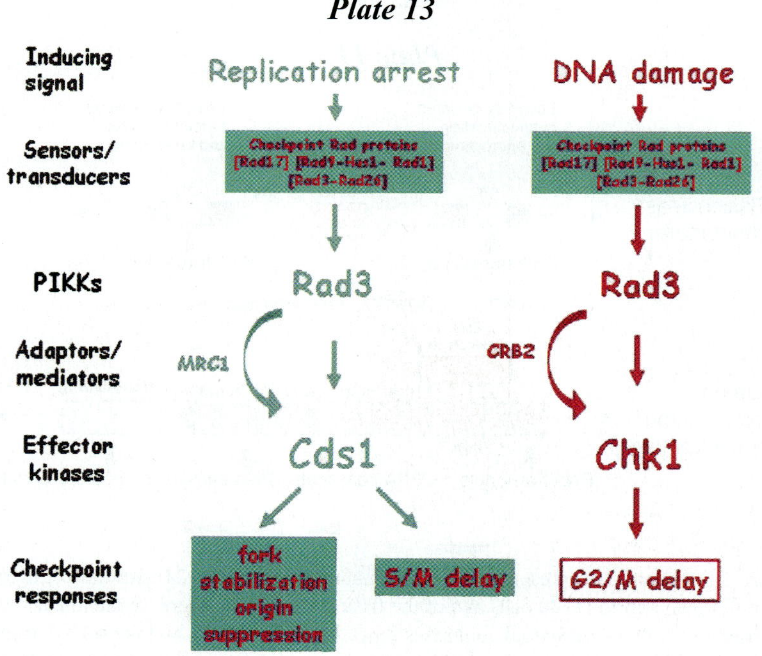

Figure 8-2. Overview of the principal components and functional organization of checkpoint pathways in the fission yeast, *S. pombe*. See text for a detailed description.

Plate 14

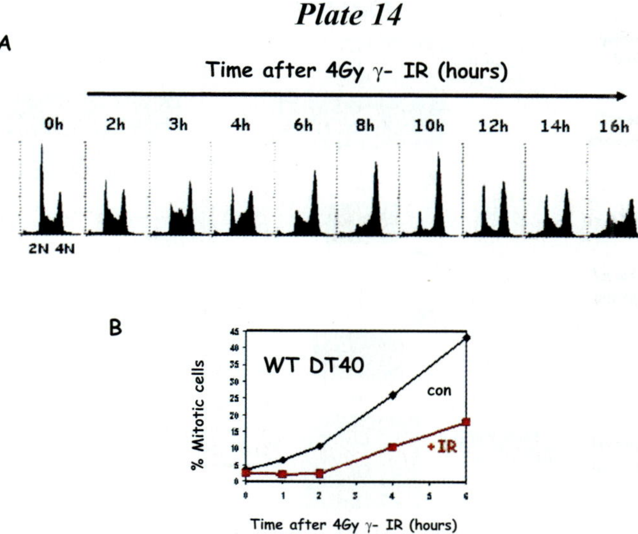

Figure 8-3. Visualizing ionizing radiation (IR)-induced G2/ M checkpoint arrest in WT DT40 cells by A) DNA content flow cytometry (Method 20), or B) quantifying the rate at which mitotic cells accumulate in the presence of nocodazole with or without prior irradiation using anti-pH3 antibodies (Method 6). Cultures were exposed to 4 Grays of IR in each case.

Plate 15

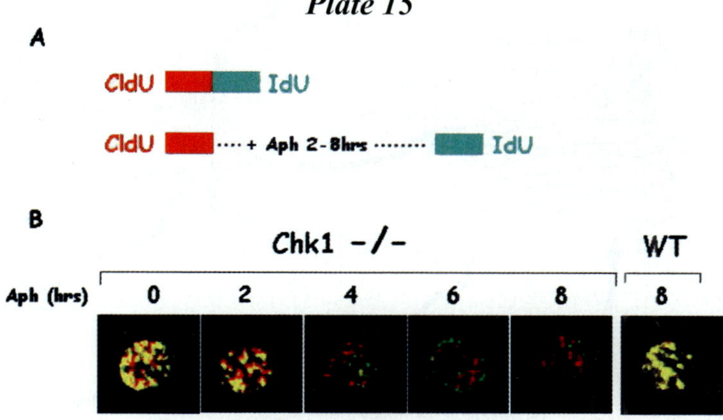

Figure 8-4. Visualizing replication fork stability and origin firing during aphidicolin-induced replication arrest in WT and Chk1-/- DT40 cells using dual labelling with CldU and IdU. A) Principle of the method; cells are exposed to a short pulse of CldU, replication is then blocked with aphidicolin for various periods of time. The inhibitor is then removed and cells are exposed to a short pulse of IdU. Incorporation of CldU and IdU into discrete replication foci in individual nuclei is then detected using monoclonal antibodies specific for each halogenated nucleotide and visualized by confocal microscopy. Foci which appear as yellow are interpreted to represent replication forks which were able to restart after aphidicolin was removed (ie viable), red alone indicates previously active forks which became incapable of resuming replication (collapsed), whilst green signifies new forks generated by origin firing during the period when elongation was blocked (futile origin firing). A more detailed description of the protocol employed can be found in (Zachos et al., 2003).

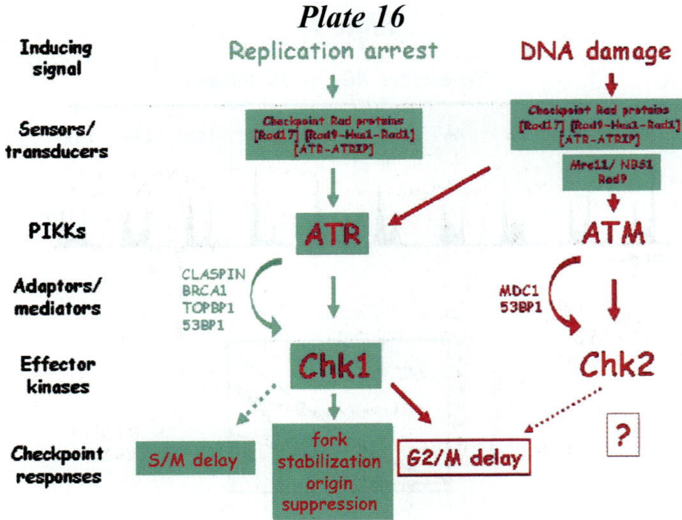

Figure 8-5. Overview of the principal components and proposed functional organization of checkpoint pathways in vertebrate cells, focusing on the functions controlled by the effector kinases, Chk1 and Chk2. See text for a detailed description.

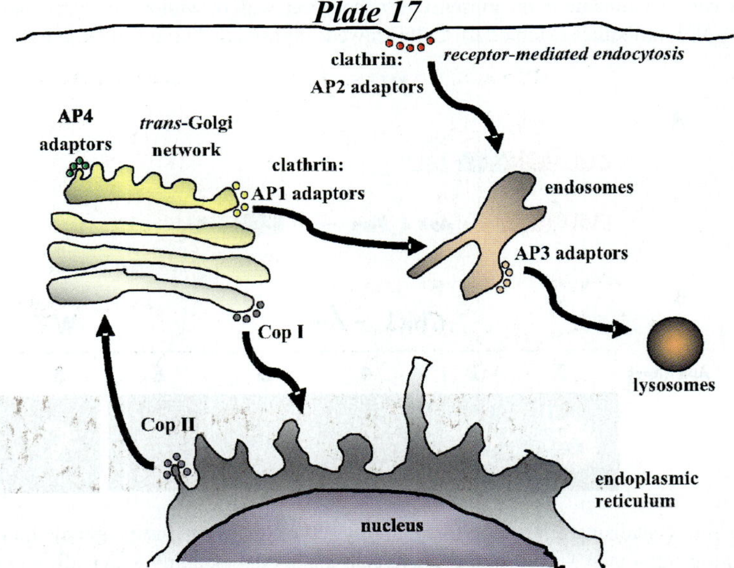

Figure 9-1. Major trafficking pathways in a typical eukaryotic cell. Proteins move from the endoplasmic reticulum (ER) through the secretory pathway via distinct classes of coated transport vesicles. For example, transport between the ER and the Golgi is largely carried out by Cop I and Cop II vesicles. Clathrin-coated vesicles are required for receptor-mediated endocytosis from the plasma membrane and transport of lysosomal proteins from the *trans*-Golgi network to endosomes/lysosomes. Clathrin-coated vesicles at different cellular locations are associated with different classes of adaptor proteins (APs) (see text). The diagram is highly simplified. For example, only one endosome compartment is shown and multiple pathways to and from each compartment are omitted for clarity.

Plate 18

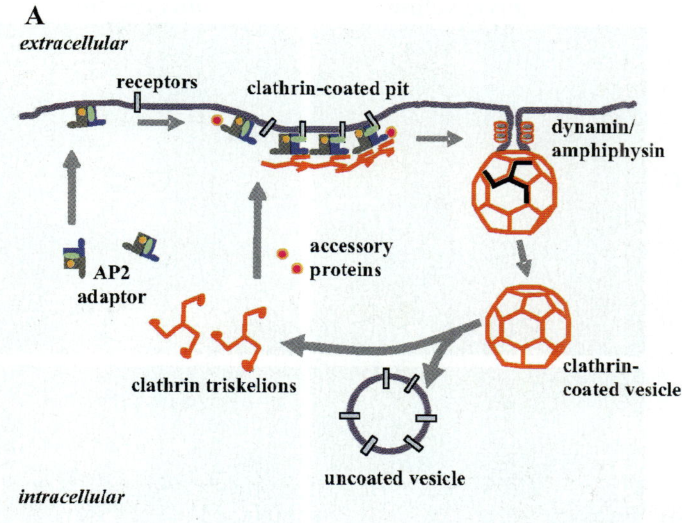

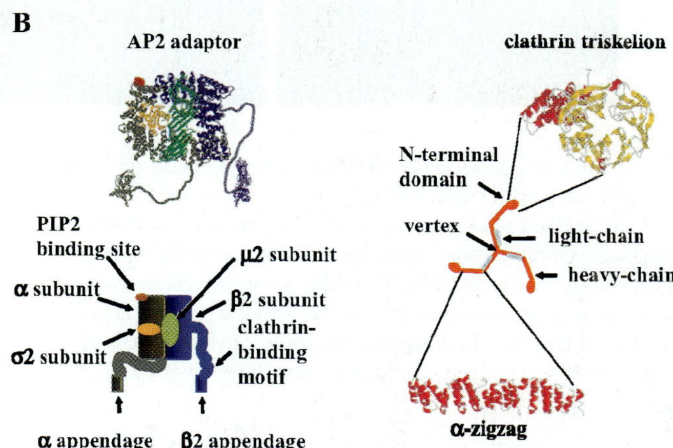

Figure 9-2. A. Outline of the main events in receptor-mediated endocytosis. The AP2 adaptor attaches to the inner-surface of the plasma membrane, initially via its PIP2 binding site. A conformational change allows the μ2 subunit to bind the cytoplasmic domain of the receptors (see section 3.5). Clathrin, together with other accessory proteins, binds the adaptors and drives membrane curvature to form the coated pit. The final conversion of the coated pit into the coated vesicle is catalysed by dynamin, which assembles with amphiphysin on the membrane neck of the highly invaginated pit. The black outline on the coated vesicle shows the location of an individual triskelion on the assembled coat.B. Structures of the AP2 adaptor and clathrin triskelion. Structures were downloaded from http://pdb.ccdc.cam.ac.uk/pdb/.

Plate 19

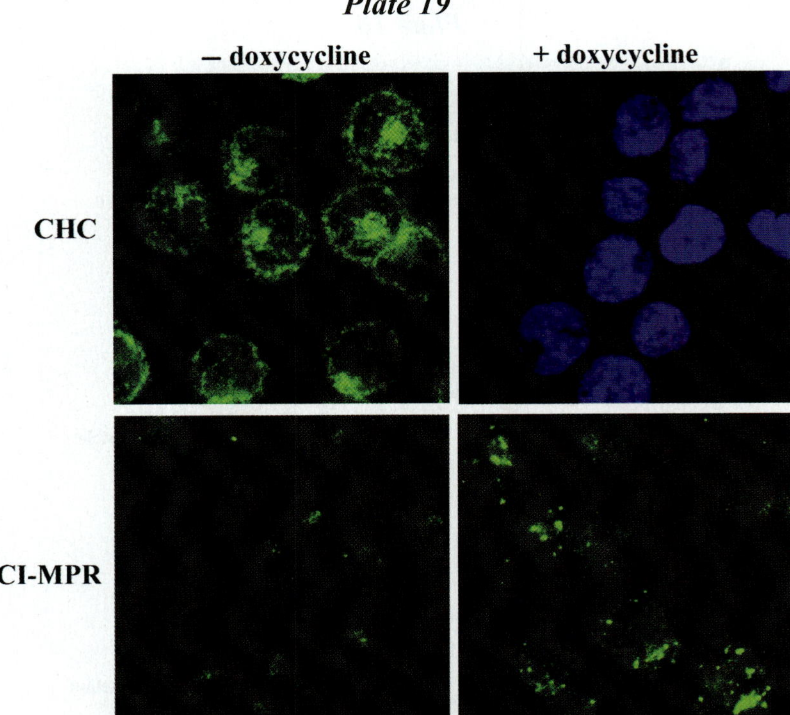

Figure 9-8. Effect of CHC depletion on the distribution of the cation-independent mannose 6-phosphate receptor (CI-MPR). DKO-R cells were incubated for 96 hrs either with or without 50 ng/ml doxycycline and separately assayed by indirect immunofluorescence microscopy as described (see methods section) using specific antibodies against clathrin heavy-chain (monoclonal antibody TD1) and CI-MPR (rabbit polyclonal antibody PL603-CI) followed by Alexa 488-conjugated secondary antibodies. No specific staining is visible in doxycycline-treated cells stained for CHC. In this panel, the nuclei have been stained with Hoechst 33342 dye to reveal the cells. Bar = 5 μm.

Plate 20

a

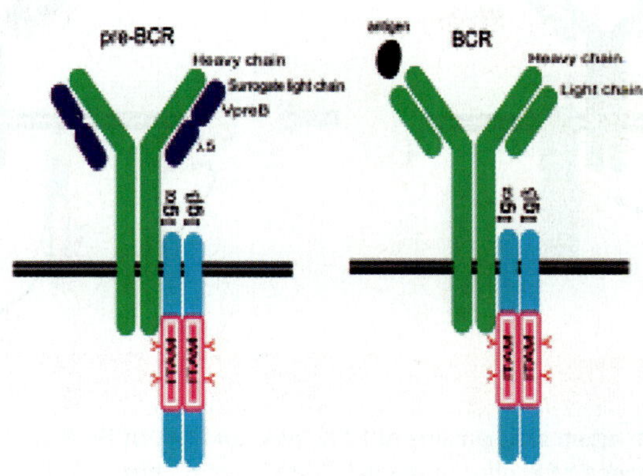

b

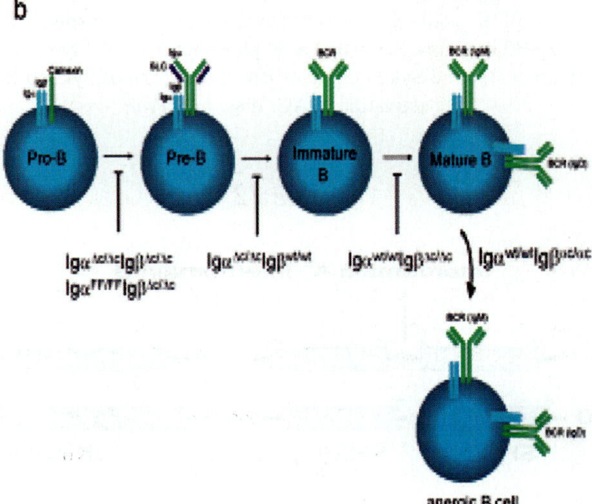

Figure 10-1. a) Simplified scheme of pre-BCR and BCR. Immunoreceptor tyrosine-based activation motif (ITAM) contains two functional tyrosines. b) Selection steps during B cell development and the various mutated IgαIgβ involved in the developmental steps, as revealed by genetic studies. Vertical 'T-bars' indicate blocks in development. The $Ig\alpha^{wt/wt} Ig\beta^{\alpha c/\alpha c}$ mice were exchanged the cytoplasmic domain of Igα for the cytoplasmic domain of Igβ by gene targeting.

Plate 21

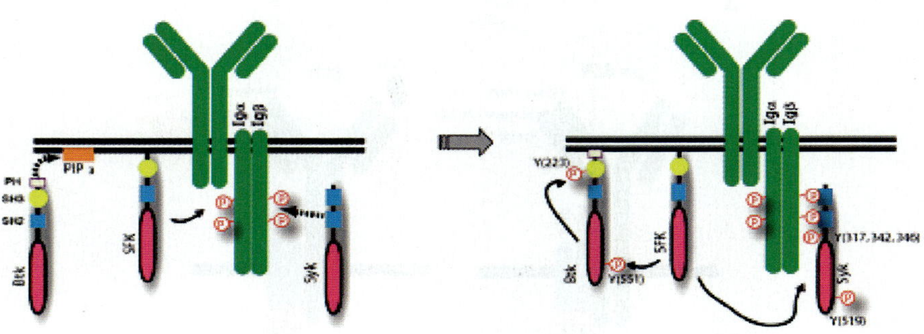

Figure 10-2. Activation mechanisms of SFK, Syk, and Btk in BCR signaling. After BCR aggregation, activated SFK phosphorylates ITAMs within Igα-Igβ. The activity of SFK at resting condition is determined by the phosphorylation status of its carboxyterminal tyrosine, which is regulated by Csk and CD45. Carboxyterminal phosphorylation renders SFK inactive. Btk is recruited to the plasma membrane by interaction of its PH domain with PI-3,4,5-P3 (PIP3), a product of PI3K, while Syk is recruited to the doubly-phosphorylated ITAMs within Igα-Igβ by its SH2 domains. Activated SFK phosphorylates the tyrosine residue in the activation loop of Btk and Syk (Tyr551 of Btk and Tyr520 of Syk) so leading to their activation. DAG: diacylglycerol.

Plate 22

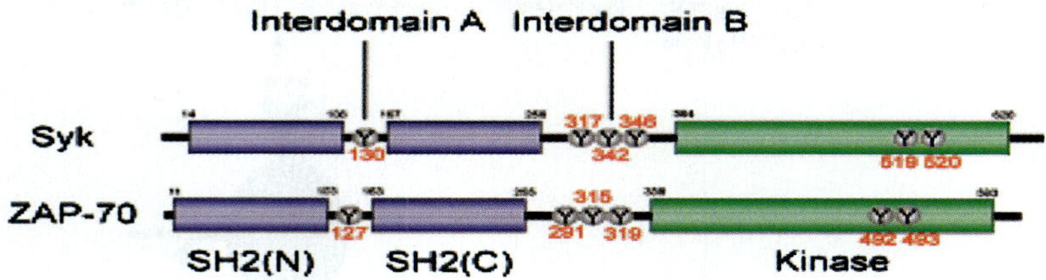

Figure 10-3. Schematic structure of Syk/ ZAP-70 family protein tyrosine kinases. The tandem SH2 domains are shown as blue boxes and the kinase domains as green boxes. Black bars depict the interdomains connecting the SH2-SH2 (A) and SH2-kinase domains (B). The tyrosines in Syk which have been shown to undergo phosphorylation are indicated. These sites may be important in regulating enzymatic activity or recruiting other signalling proteins. Phosphorylation of Y342 and Y346 is dependent on the SFK; phosphorylation of Y130, Y317, Y519 and Y520 is dependent on Syk itself and thus these may be sites of auto-phosphorylation. Homologous tyrosines.

Plate 23

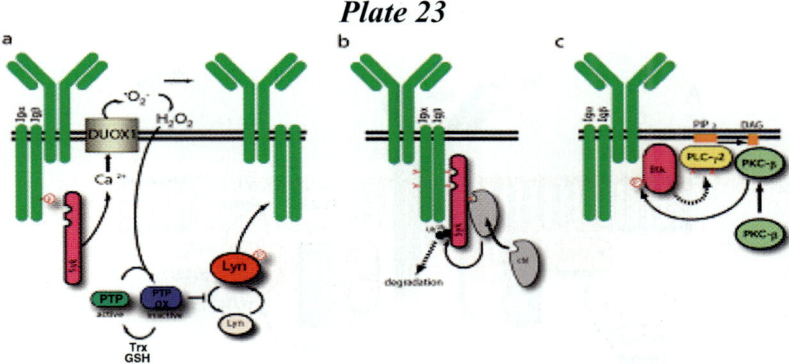

Figure 10-4. Positive and negative feed-back loops in BCR-signaling. a). Redox regulation of Protein tyrosine phosphatases (PTPs) determines the activation status of the protein tyrosine kinase, Lyn, which in turn controls signal output emanating from BCR. Thioredoxin (Trx) and glutathione (GSH) systems control PTP oxidation status. Recently, an annotated murine cDNA sequence identified as DUOX1 was discovered that displayed 87% sequence identity with that of its human DUOX (dual oxidases) counterpart. DUOX represents the second family of nonphagocytic NADPH oxidases that possesses two EF hand motifs. And DUOX1 is thought to be a connector between calcium and supreoxide (H2O2) generation.
b). Activated Syk autophosphorylates Tyr317, in addition to its substrates such as BLNK. This Tyr317 phosphorylation provides a binding site for Cbl, which acts as E3 ubiquitin ligase. Ubiquitin-mediated degradation of Syk results in the attenuation of BCR signaling. c). In this model, PLC-γ2 is activated by virtue of Btk action, leading to generation of DAG. Then, the generated DAG recruits PKCβ to the plasma membrane, wherein PKCβ is activated. Phosphorylation of Btk by PKCβ down-regulates Btk kinase activity and decreases its membrane localization, both of which in turn inhibit PLC-γ 2 activity.

Plate 24

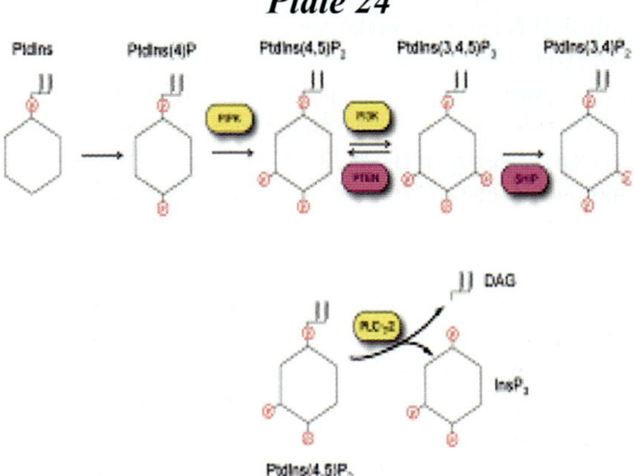

Figure 10-5. Phosphoinositide metabolism mediated by PIPK, PI3K, SHIP, PTEN and PLC-γ2. Both Phosphoinositide 3-kinase (PI3K) and phospholipase Cγ2 (PLC-γ2) share the common substrate, phosphoinositide 4,5 bis-phosphate (PtdIns(4,5)P2), the product of phosphatidylinositol-4 phosphate 5-kinase (PIPK). PI3K phosphorylates PtdIns(4,5)P2 to give rise to PtdIns(3,4,5)P3; (PLC-γ2) hydrolyses PtdIns(4,5)P2 to produce InsP3 and diacylglycerol (DAG); SRC-homology-2-domain-containing inositol 5-phosphatase (SHIP) and phosphatase and tensin homologue (PTEN) hydrolyse PtdIns(3,4,5)P3 to PtdIns(3,4)P2 and PtdIns(4,5)P2, respectively.

Plate 25

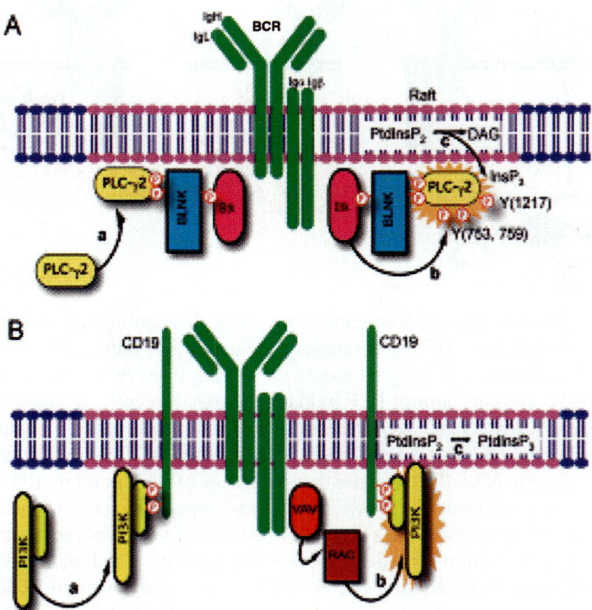

Figure 10-6. Two-step models for PLC-γ2 and PI3K activation. The first step of activation of (A) phospholipase Cγ2 (PLC-γ2) and (B) phosphoinositide 3-kinase (PI3K) requires their recruitment to the rafts (step a), presumably because their substrate, phosphatidylinositol-4, 5-bisphosphate (PtdInsP2) is enriched in the rafts. Tyrosine phosphorylation of (A) B-cell linker (BLNK) and (B) CD19 participates in this process. Then, in the case of PLC-γ2 activation, this enzyme undergoes tyrosine phosphorylation by Btk (step b in A), which is important for PLC-γ2 activation (step c, in A). In the case of PI3K, the recruited PI3K is then activated by activated RAC (step b, in B). BCR, B-cell receptor; DAG, diacylglycerol; IgH, immunoglobulin heavy chain; IgL, immunoglobulin light chain; P, phosphate.

Plate 26

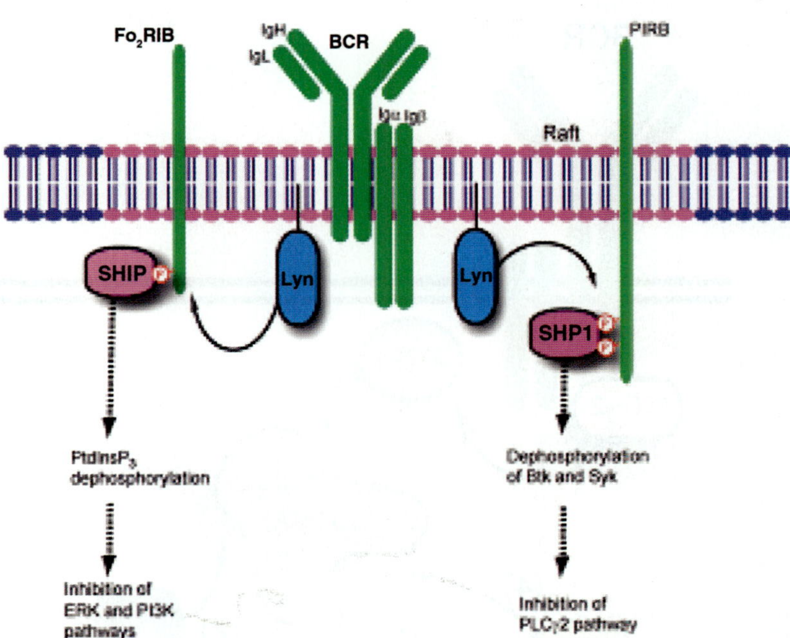

Figure 10-7. Negative-regulatory loops mediated by inhibitory receptors on B cells. Once the immunoreceptor tyrosine-based inhibitory motif (ITIM) of FcγRIIB is phosphorylated, SRC-homology-2 (SH2)-domain-containing inositol 5-phosphatase (SHIP) is recruited, which, in turn, have negative influences on phosphoinositide 3-kinase (PI3K) and extracellular signal-regulated kinase (ERK) pathways. By contrast, tyrosine phosphorylation of paired immunoglobulin-like receptor B (PIRB) ITIMs recruites SH2-domain-containing protein tyrosine phosphatase 1 (SHP1), which, in turn, dephosphorylates various protein-tyrosine kinases, including Syk and Bruton's tyrosine kinase (Btk). BCR, B-cell receptor; IgH, immunoglobulin heavy chain; IgL, immunoglobulin light chain; PLC-γ2, phospholipase Cγ2; PtdInsP3, phosphatidylinositol-3,4,5-triphosphate.

Plate 27

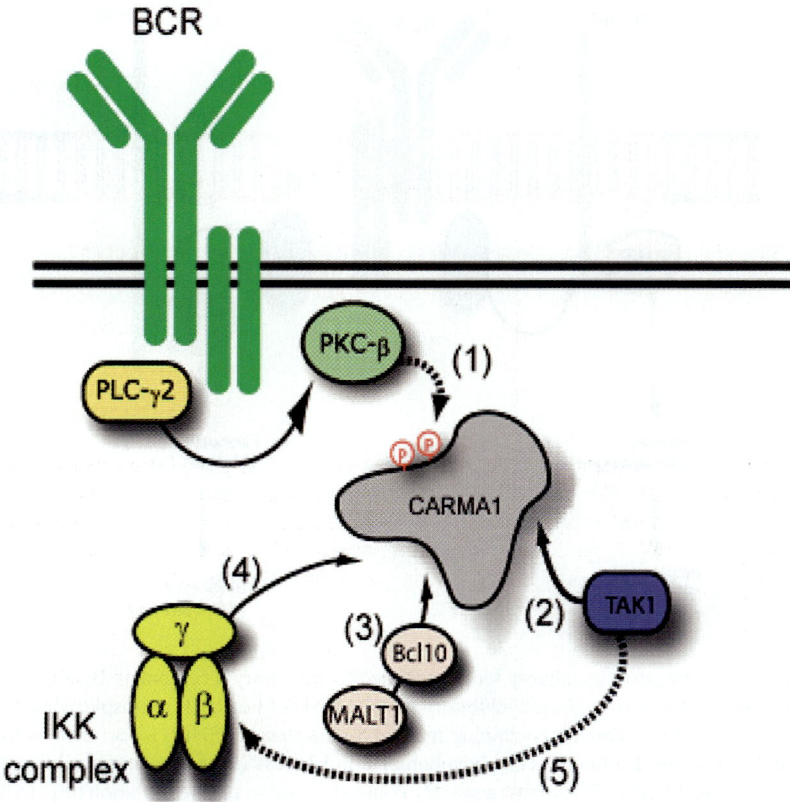

Figure 10-8. A model for BCR-mediated NF-κB activation. Stimulation of BCR leads to activation of proximal protein tyrosine kinases including Syk and Btk. Btk phosphorylate several tyrosine residues on phospholipase Cγ2 (PLC-γ2), and subsequent activation of protein kinase Cβ (PKCβ). Activated PKCβ phosphorylates CARMA1 (1), directly or indirectly, which is able to recruit TAK1 to the phosphorylated CARMA1 (2). Meanwhile, the IKK complex probably through the Bcl10/ MALT1 complex is recruited to the phosphorylated CARMA1 (3 and 4). These interactions (CARMA1-IKK and CARMA1-TAK1) contribute to access of two key protein kinases, TAK1 and IKK, leading to activation of the IKK complex (5).

Plate 28

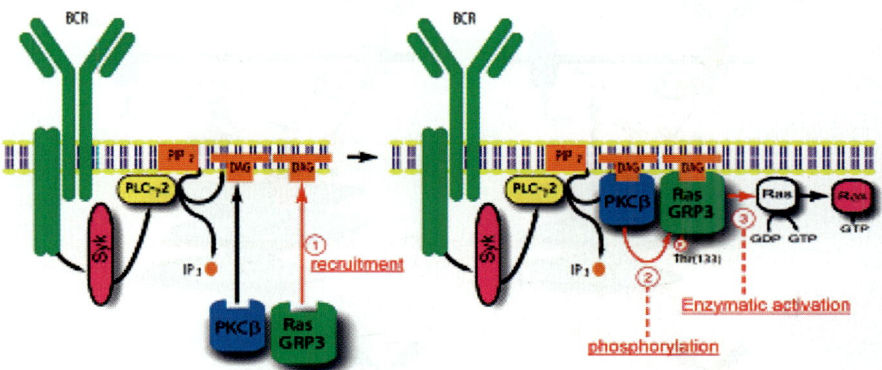

Figure 10-9. A model of Ras regulation in B cells by a Ras-guanyl nucleotide exchange factor, RasGRP3. After BCR stimulation, RasGRP3 is recruited to plasma membrane by binding to DAG, a PLC-γ2 product (1). PKC is also recruited to membrane and phosphorylate RasGRP3 at Thr-133, thereby resulting in its activation (2). Then, fully activated RasGRP3 turns RasGDP into its active RasGTP form (3).

Plate 29

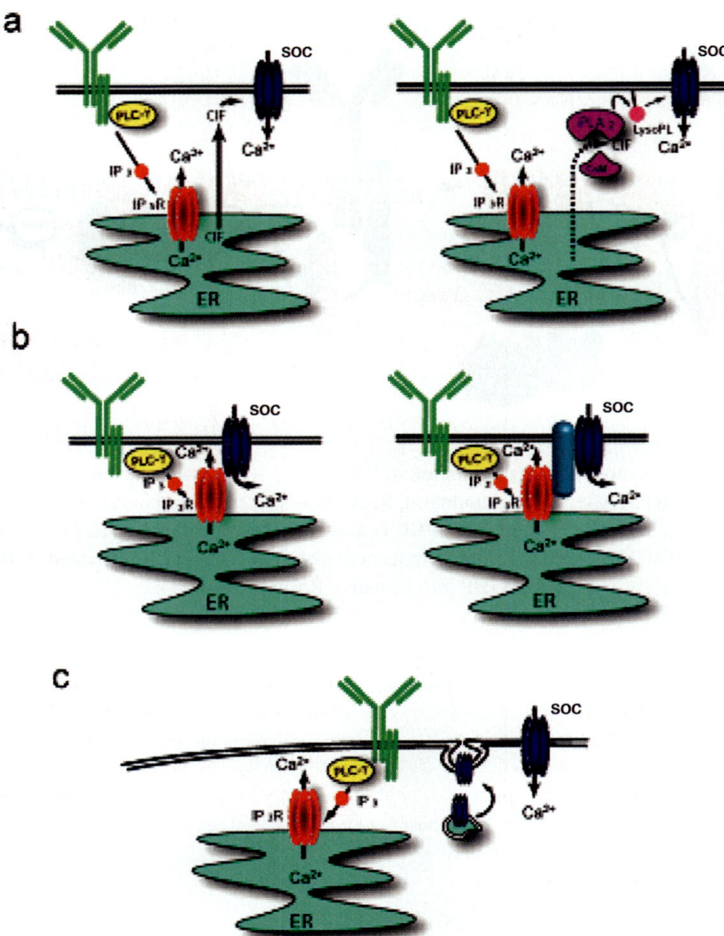

Figure 10-10. Models of SOCs activation. (A) Left, The direct activation model by the diffusible-messenger. In this model, it is postulated that depletion of calcium from intracellular store induces production of calcium influx factor (CIF). CIF diffuses to the plasma membrane and activates SOCs directly. Right, The indirect activation model by the diffusible-messenger. CIF produced by depletion of the internal calcium pool diffuses and activates iPLA2 by displacement of inhibitory calmodulin (CaM). Once iPLA2 gets activated, it generates lysophospholipids, which in turn activate SOCs. (B) Left, The direct conformational-coupling model. In this model, it is postulated that conformational change of the IP3 receptor takes place by it opening and this change is transmitted to SOCs by direct interaction and activation of SOCs. Right, The indirect conformational-coupling model. Instead of the IP3 receptor by itself, other ER-resident molecules sense the conformational change of the IP3 receptor and directly interact to the SOC, thereby activating them. (C) The secretion model. This model suggests a mechanism by which depletion of calcium from intracellular store initiates the vesicular translocation and insertion of calcium channels to the plasma membrane. Then, delivery of these channels to the membrane might be a trigger for cation influx.

Plate 30

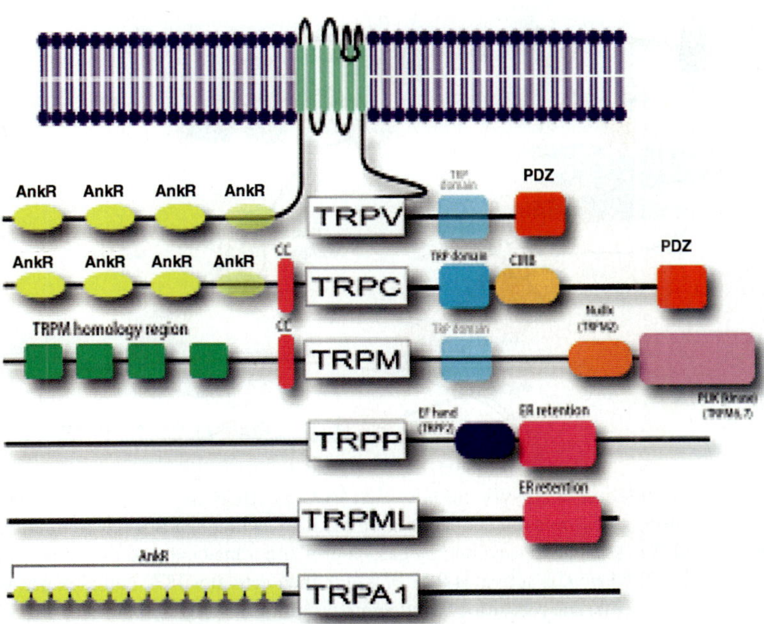

Figure 10-11. Schematic structure of TRP-family ion channels. Schematic structure of TRP-ion channels. All the members of the family consist of six transmembrane domain-containing ion channels flanked with two cytoplasmic tails that are characteristic to each member. The TRP box is EWKFAR in TRPC, but is less conserved in TRPV and TRPM. CC indicates coiled-coil domain. Ankyrin repeats (AnkR) range from 0 to 14 in number. CIRB stands for putative calmodulin- and IP3 receptor-binding domain. EF hand, canonical helix-loop-helix Ca2þ-binding domain; PDZ, amino acids-binding PDZ domains; PLIK, phospholipase C-interacting kinase; Nudix, NUDT9 hydrolase protein homologue-binding ADP ribose.

Plate 31

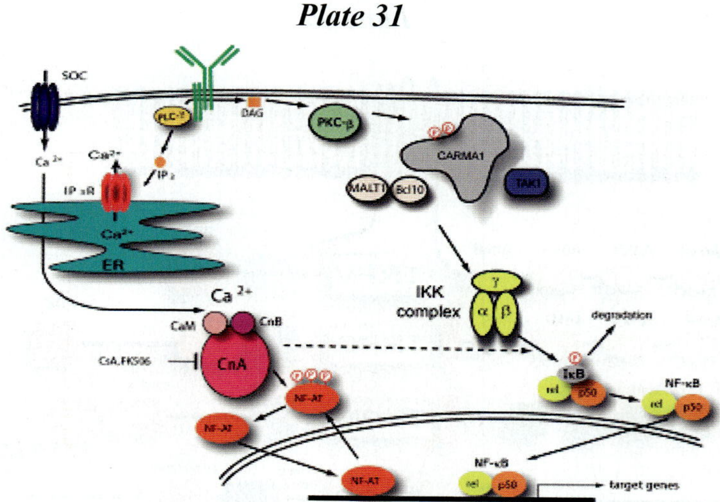

Figure 10-12. Regulation of NF-κB and NFAT by PLC-γ2. BCR-mediated PLC-γ2 activation causes elevation of cytoplasmic concentration of Ca2+, which leads to the activation of calcineurin. NFAT is dephosphorylated by calcineurin and translocates to the nucleus. DAG, generated by PIP2hydrolysis, activates PKCβ, which then activates IKK complex presumably by promoting membrane recruitment and aggregation of the CARMA1/Bcl10/MALT1 complex. This complex activates IKKs, resulting in the phosphorylation and ubiquitin-mediated degradation of IκB. Released rel/p50 complex translocates to the nucleus. It has been reported that inhibition of calcineurin blocks the activation of NF-κB suggesting that calcineurin is involved in NF-κB activation pathway (Biswas et al., 2003).

Plate 32

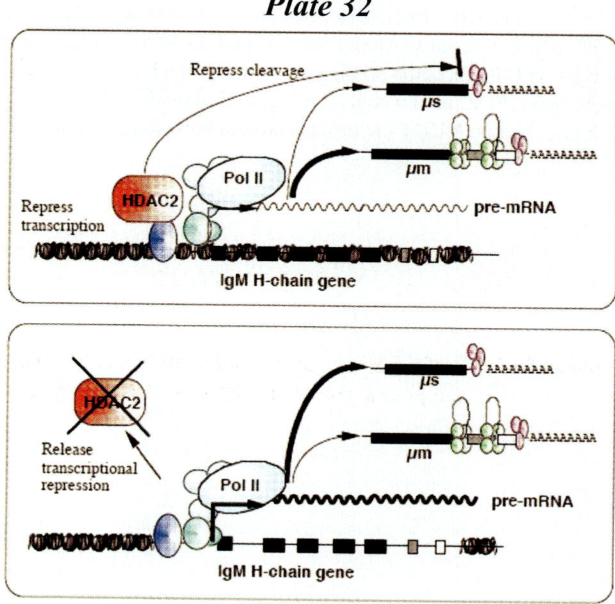

Figure 13-1. A model for a role of HDAC-2 in the control of the amount of the IgM H-chain.

Plate 33

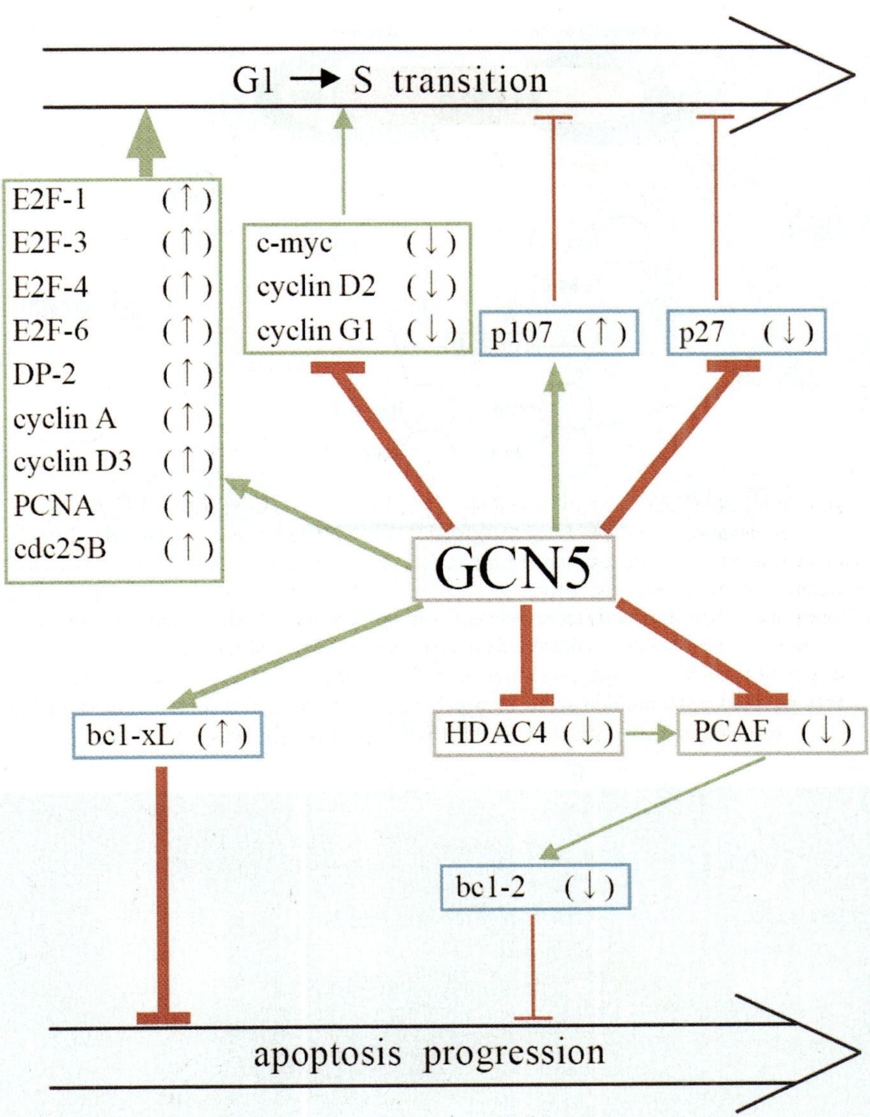

Figure 13-2. A model for a role of GCN5 as a supervisor in all-inclusive control of cell cycle progression of vertebrate cells.

Plate 34

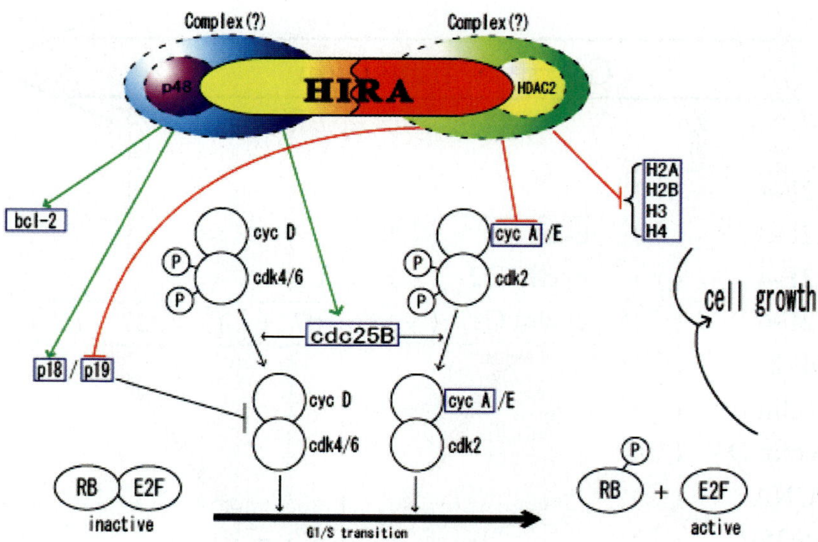

Figure 13-3. A model for different roles of N-terminal and C-terminal helves of HIRA in control of vertebrate cell growth.

Plate 35

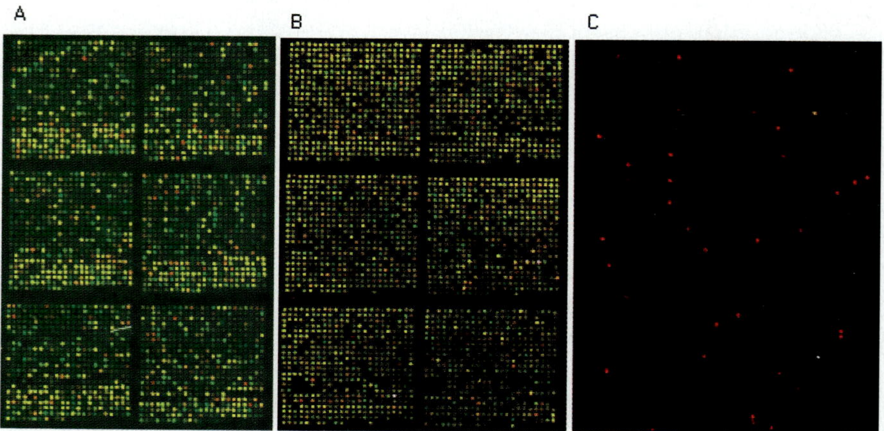

Figure 14-1. Microarray fluorescence scans images of DT40 or bursal lymphoma (Cy-5 dye label, red) vs. normal chicken bursa (Cy-3 dye label, green) experiments. Six of 32 feature blocks are shown (A): DT40 expression array, original DT40 and immune system ESTs are in the lower seven rows of each block, (B): DT40 arrayCGH, (C): DT40 GAPF.

Plate 36

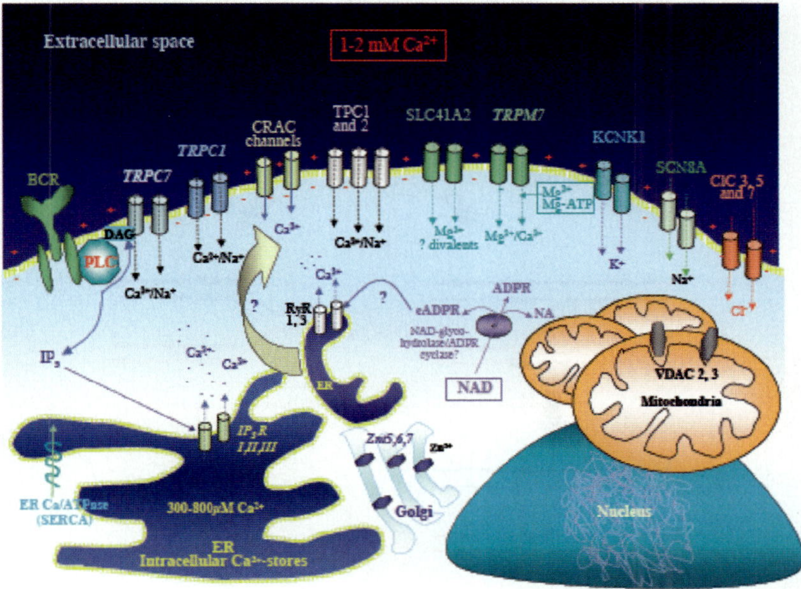

Figure 15-1. The known DT40 channelome: The genes encoding channels and transporters in bold italic have been deleted in the DT40 system, and the phenotype published.

Plate 37

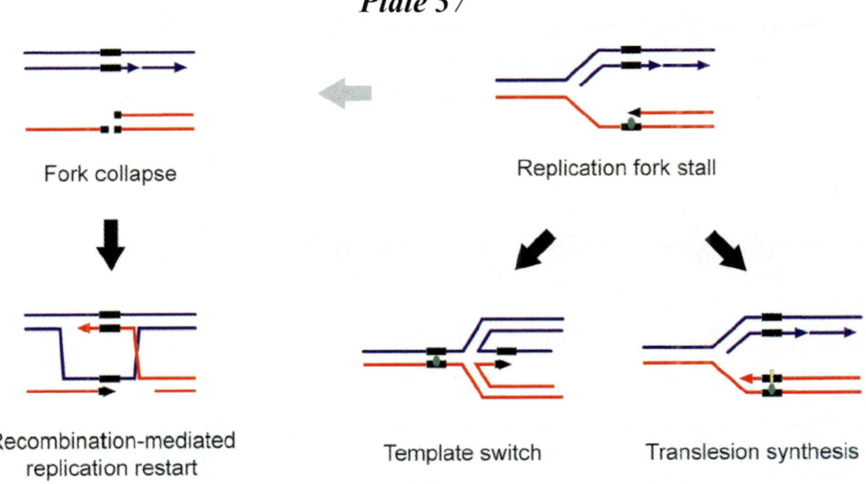

Figure 16-1. A simplified scheme of DNA damage bypass during replication.

Plate 38

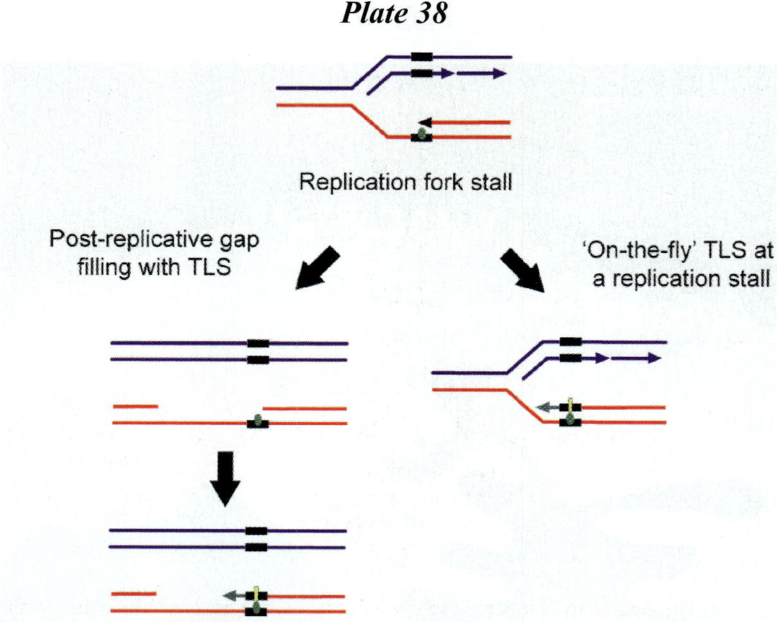

Figure 16-2. The timing of TLS use - rescue of stalled replication vs. post-replicative gap filling.

Plate 39

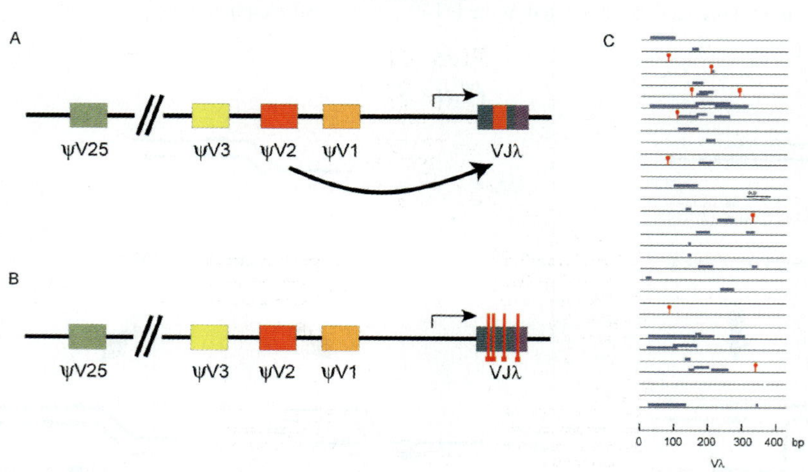

Figure 16-3. Gene conversion and point mutation in the Ig light chain locus of DT40. A. Gene conversion results in the introduction of blocks of sequence change into the rearranged V gene, templated from one of 25 upstream pseudogene donors in the case of the light chain illustrated here. B. Gene conversion is accompanied by point mutations which cannot be explained by templation from the pseudogenes. C. Light chain sequences from DT40 showing a typical patchwork of gene conversion and point mutation. Each grey line represents a single Vλ sequence. Gene conversion tracts are shown as blue lines and point mutations as red lollipops.

Plate 40

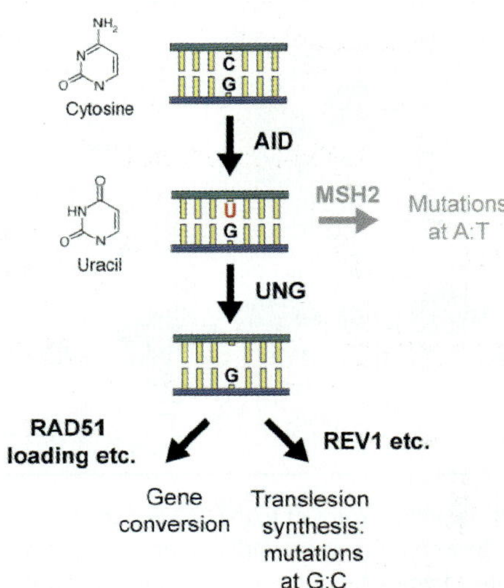

Figure 16-4. The deamination model of Ig diversification applied to DT40. While in mouse and human mutations at A:T, arising in consequence of MSH2 recognition of U:G mismatches, make a substantial contribution, mutation in DT40 is largely the result of TLS across abasic sites formed at C. Gene conversion is also predominantly initiated from abasic sites. (Adapted from Petersen-Mahrt et al., 2002).

Plate 41

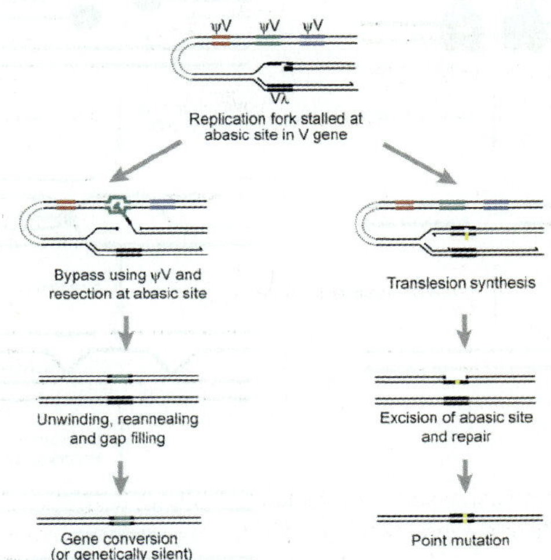

Figure 16-5. A model for gene conversion and point mutation initiated by abasic sites.

Plate 42

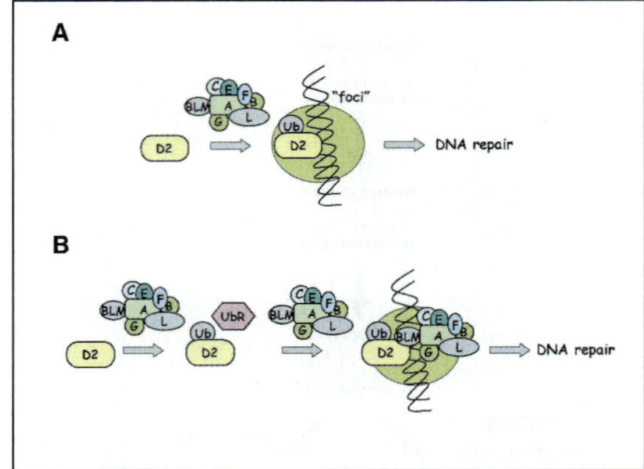

Figure 17-2. Models depicting roles of FANCD2 and the FA core complex. (A) A conventional model. The core complex only works in monoubiquitination of FancD2. (B) Our proposed model. The core complex functions in monoubiquitination, chromatin targeting of FancD2, and DNA repair in DNA-damage induced foci. UbR, putative ubiquitin receptor.

Plate 43

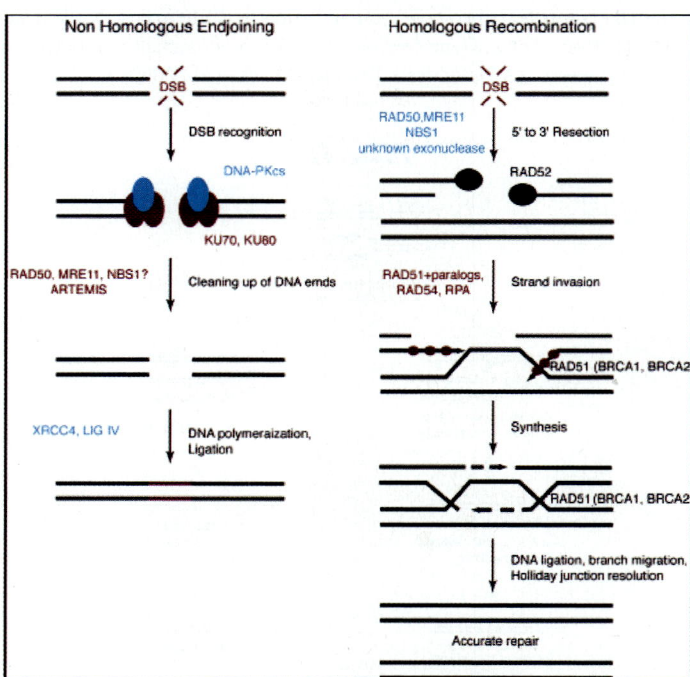

Figure 18-1. NHEJ and HR are the two major pathways of DSB repair. During NHEJ the two broken ends are ligated back together, often resulting in loss of information. HR uses a homologous sequence such as the sister chromatid to copy the sequence lost at the break and repair the lesion with accuracy.

Plate 44

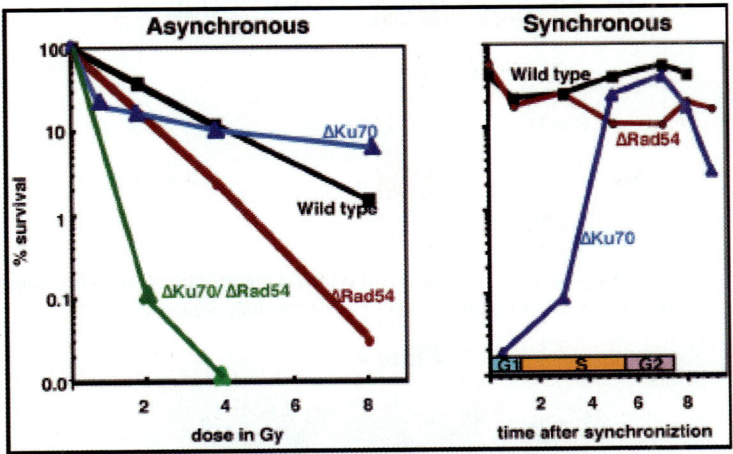

Figure 18-2. Cell cycle specific sensitivity of HR and NHEJ mutants. The left panel shows the sensitivity of *ku70* and *rad54* cells in asynchronous culture. The left panel represents IR-sensitivities of cell taken at different times after synchronization by elutriation. *Ku70* cells are highly sensitive in G1 but more resistant than WT in G2. *Rad54* cells, show the opposite pattern, being IR sensitive only in late S and G2-phase.

Plate 45

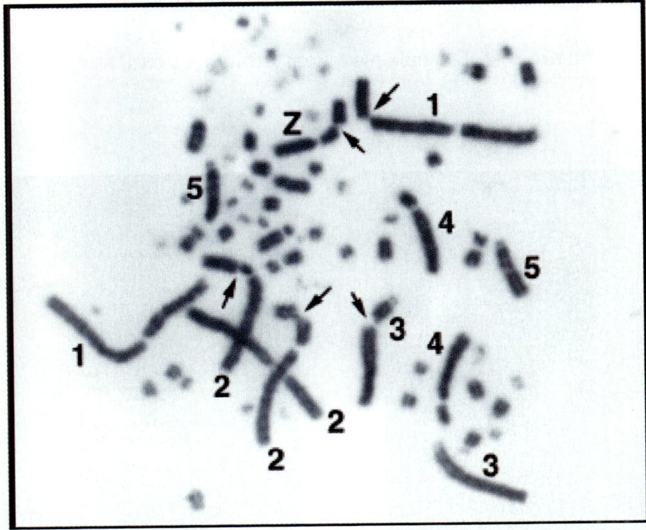

Figure 18-3. Chromosome aberrations in *rad51* cells after depletion of the RAD51 transgene. The cells die rapidly displaying a more then 10-fold increase in breaks and gaps that span all of the chromosome. This Picture was published by (Sonoda et al., 1998).

Plate 46

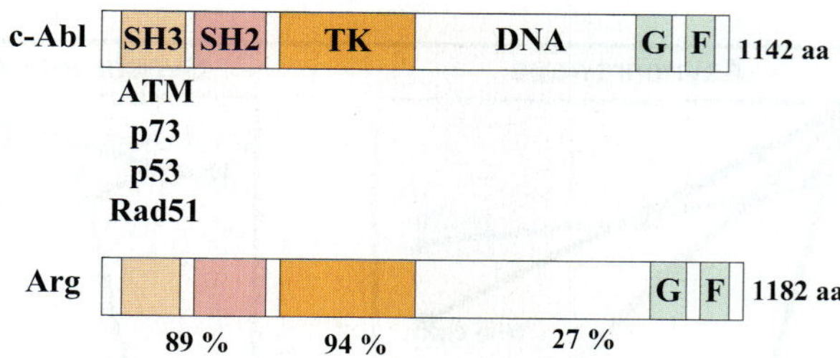

Figure 19-1. A schematic diagram for the c-Abl tyrosine kinase family, showing conserved, SH3, SH2, tyrosine kinase, DNA binding, and G/F-actin binding domains.

Plate 47

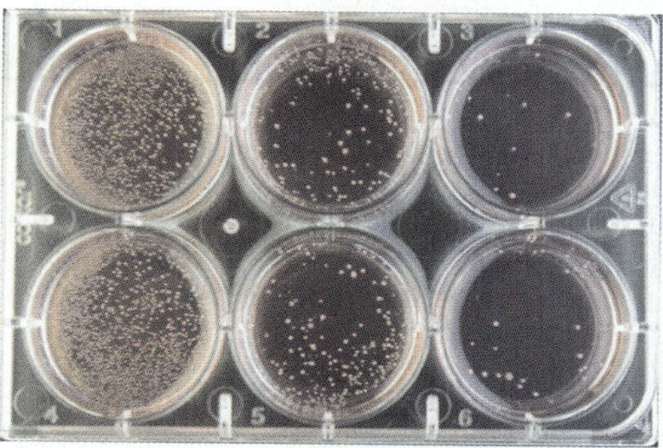

Figure 12-1. Example plate from a colony survival assay.

Plate 48

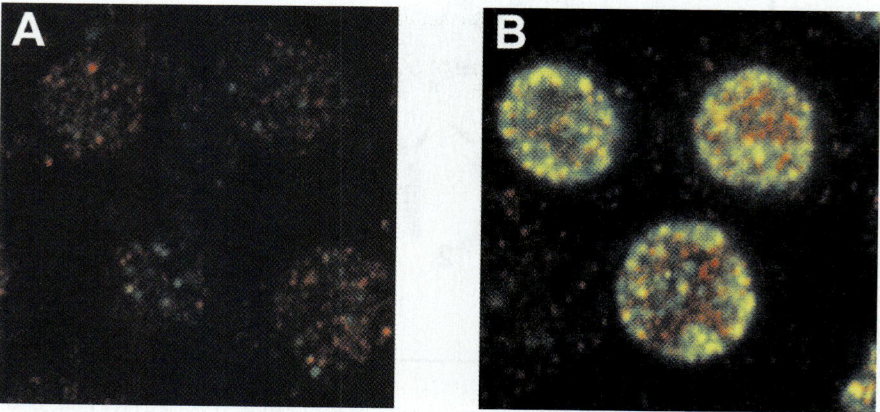

Figure 14-1. Example of subnuclear immunofluorescence in DT40. RAD51 in red, γH2Ax in green. Areas of co-localisation appear yellow. A. Unirradiated B. 2 hours after 254nm UV irradiation.

Plate 49

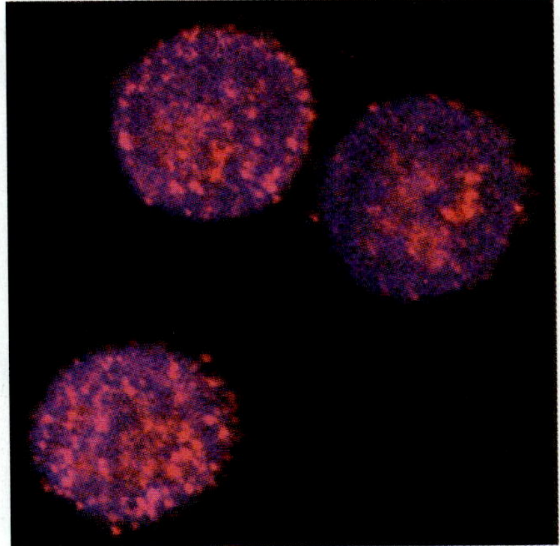

Figure 14-2. PCNA foci in an asynchronous culture of DT40. PCNA red, DNA blue.

Plate 50

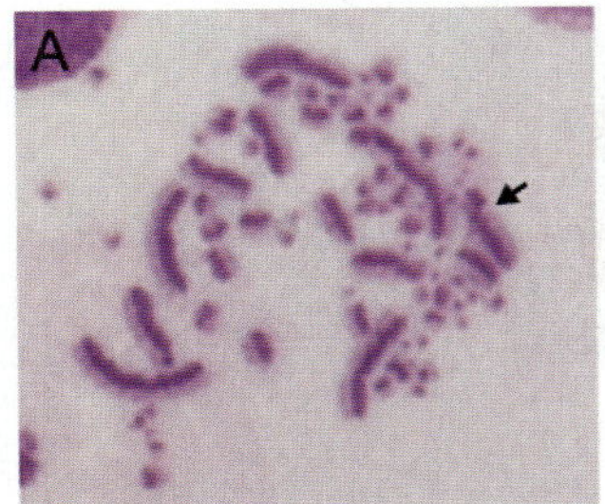

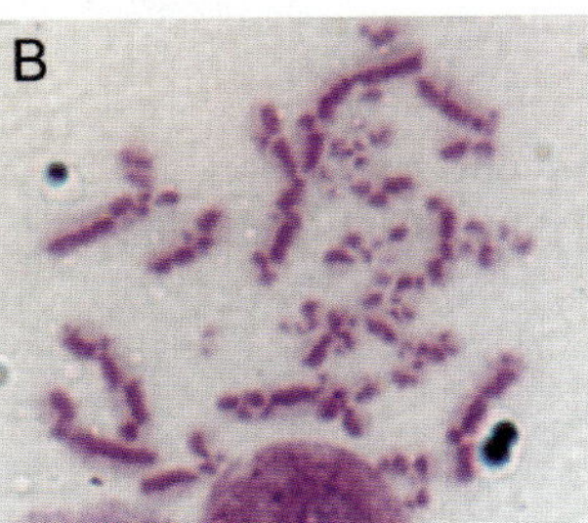

Figure 15-1. Example of sister chromatid exchanges in DT40. A) Metaphase from wild type cell. An example of an exchange is indicated with the arrow. B) Metaphase from *blm* DT40 showing elevated spontaneous SCEs.

Plate 51

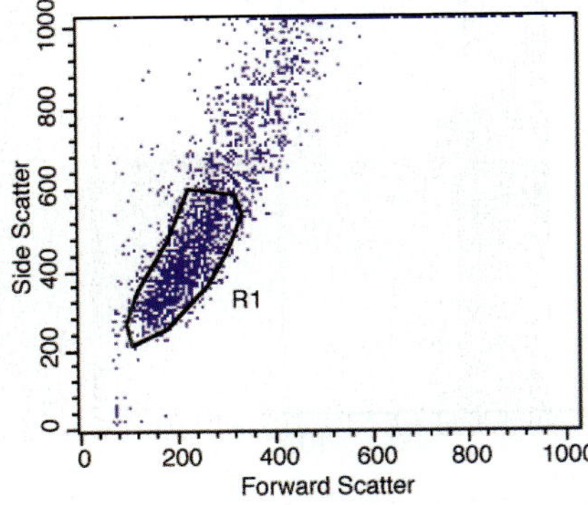

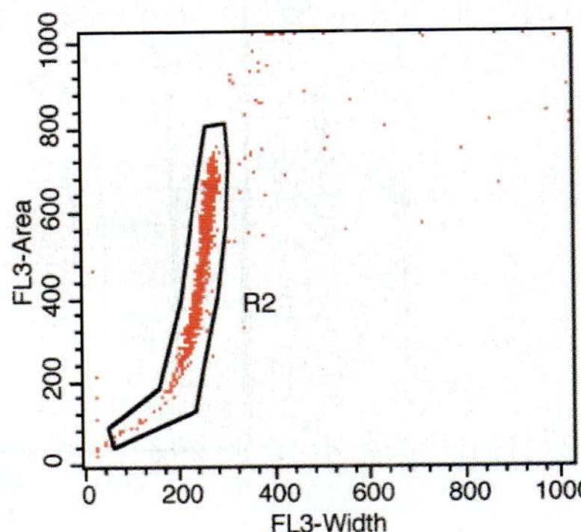

Figure 16-1. Gating of cells for cell cycle analysis. R1 is the gate for forward and side scatter. R2 is the gate for FL3-A/FL-3W, ensuring only single cells are analysed in the final 2D plot.

Plate 52

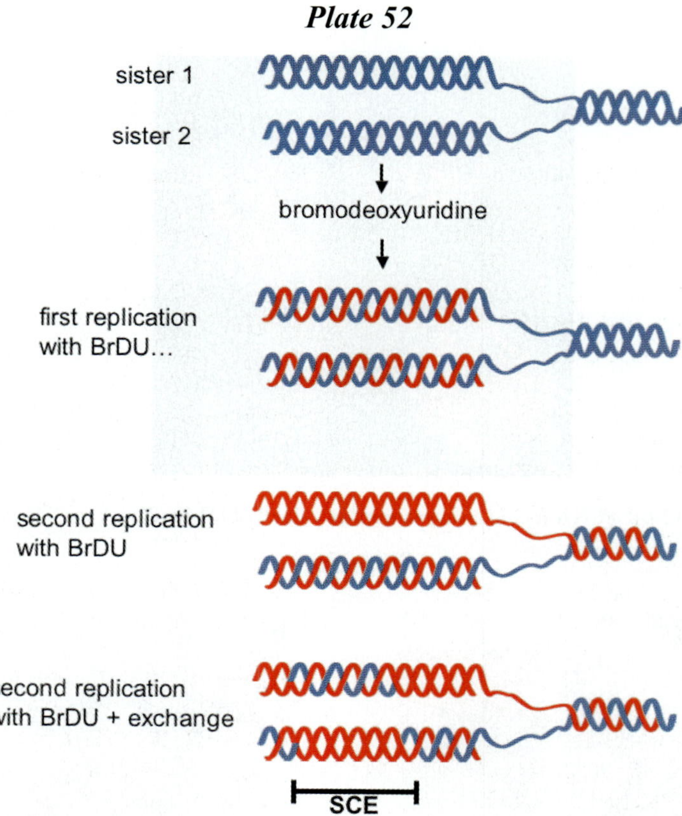

Figure 15-2. Principle of differential staining of sister chromatids.

Plate 53

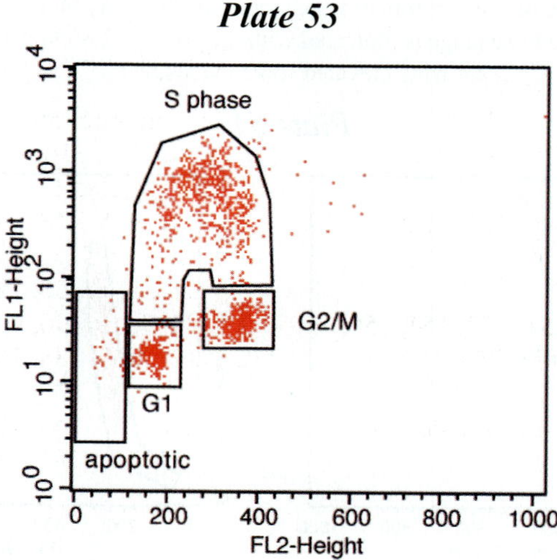

Figure 16-2. Typical cell cycle profile for wild type DT40. FL1 is the green (FITC) channel measuring BrDU uptake, FL2 is the red (PI) channel measuring DNA content.

^{26}Mg 373, 375-377
2D cell cycle analysis 405-408
4-hydroxy tamoxifen 347
activation induced deaminase (AID)
11, 13-15, 17, 18, 20-22, 194, 200,
283, 284, 287, 301
adaptor molecule 145, 157, 162
Ailos 189, 190, 192-195, 199, 200
alignment 25, 29, 32, 33, 278
alkaline 268, 276, 361, 435, 437
annotation 25, 27, 29, 30, 33, 34, 198
apoptosis 75-79, 81, 83-86, 95, 116,
133, 138, 193-195, 197, 210,
215-218, 234, 239, 241, 252, 271,
314, 327-329, 331-333, 335, 336,
405, 446
apoptotic execution 75-78, 81, 84, 85, 87
array CGH 245, 246, 249-253, 363-367
ATM 56, 61, 63, 114, 314, 327-33,
335, 336, 470

B cell 1, 11, 13-15, 25, 77, 84, 135,
145-149, 156, 158, 160, 162, 171,
189-201, 259, 264, 331, 332, 345,
453, 456, 457
B cell signaling 145
BCR signal 145, 146, 151, 155, 191,
200
bic 245-249
BLAST 27-29, 32, 34, 94
BLAT 2, 32
Bloom Syndrome (BS) 49, 50, 52, 56,
60, 303
bursa of Fabricius 11, 31, 189, 191,
199, 200, 282, 301

calcium 145, 146, 151, 152, 154, 156,
162, 166-175, 193, 257, 260, 261,
386, 455, 460
calcium mobilization 145, 146, 151,
166, 167, 175, 193

calcium signaling 167, 173-175, 257,
261
calmodulin affinity resin 409, 410,
412, 413
cDNA 2, 6-8, 25-31, 33, 34, 40, 82, 99,
127, 129, 154, 209, 214, 230-232,
234, 236, 238, 246, 248, 298, 363,
367, 386, 455
cDNA microarray 245, 246, 249, 253,
363, 365, 367, 368
cell cycle 7, 55, 79, 93, 94, 97, 98,
107, 109, 114, 138, 195, 196, 207,
209-218, 225, 229, 232, 234, 235,
237-239, 241, 279, 314, 315, 137,
318, 320, 327-332, 334, 336, 355,
359, 360, 402, 405-407, 415, 417,
418, 447, 463, 469, 471, 472
cell cycle checkpoint 107, 211, 327-331,
334
cell cycle synchronization 415
cell physiology 198-200, 257, 263
CENPs 91-99, 101, 447
centrifugal elutriation 111, 114, 359,
360, 415
centrifugation 276, 342, 360, 361, 364,
384, 400, 411, 417, 424, 435, 437
centromere 39, 41-47, 91-97, 99-102,
444, 445
centromere function 39, 41, 43, 47, 93,
102
checkpoints 55, 56, 59, 65, 91, 96, 97,
102, 107-116, 145, 195, 210, 211,
314, 321, 327-331, 333-336, 355,
357, 405, 448, 449, 450
chicken 1, 2, 5, 6, 11-14, 17, 19, 25-28,
31, 33, 49, 50, 52, 63, 75, 80, 82,
84, 92, 94-96, 98-102, 124, 126,
127, 133, 134, 139, 145, 149, 189,
198-200, 227, 229, 230, 232, 234,
236, 238, 245-247, 249, 253, 257,
265, 278, 281-283, 285, 289,